TETRAHEDRON ORGANIC CHEMISTRY SERIES
Series Editors: J E Baldwin, FRS & P D Magnus, FRS

VOLUME 12

Enzymes in Synthetic Organic Chemistry

Related Pergamon Titles of Interest

BOOKS

Tetrahedron Organic Chemistry Series:
CARRUTHERS: Cycloaddition Reactions in Organic Synthesis
DEROME: Modern NMR Techniques for Chemistry Research
DESLONGCHAMPS: Stereoelectronic Effects in Organic Chemistry
GAWLEY: Asymmetric Synthesis*
HASSNER: Organic Syntheses based on Name Reactions and Unnamed Reactions
PAULMIER: Selenium Reagents & Intermediates in Organic Synthesis
PERLMUTTER: Conjugate Addition Reactions in Organic Synthesis
SIMPKINS: Sulphones in Organic Synthesis
WILLIAMS: Synthesis of Optically Active Alpha-Amino Acids

JOURNALS

BIOORGANIC & MEDICINAL CHEMISTRY
BIOORGANIC & MEDICINAL CHEMISTRY LETTERS
JOURNAL OF PHARMACEUTICAL AND BIOMEDICAL ANALYSIS
TETRAHEDRON
TETRAHEDRON: ASYMMETRY
TETRAHEDRON LETTERS

Full details of all Elsevier Science publications/free specimen copy of any Elsevier Science journal are available on request from your nearest Elsevier Science office

* In Preparation

Enzymes in Synthetic Organic Chemistry

C. H. WONG

The Scripps Research Institute

and

G. M. WHITESIDES

Harvard University

PERGAMON

U.K.	Elsevier Science Ltd, The Boulevard, Langford Lane, Kidlington, Oxford OX5 1GB, U.K.
U.S.A.	Elsevier Science Inc., 660 White Plains Road, Tarrytown, New York 10591-5153, U.S.A.
JAPAN	Elsevier Science Japan, Tsunashima Building Annex, 3-20-12 Yushima, Bunkyo-ku, Tokyo 113, Japan

First Edition 1994

Library of Congress Cataloging in Publication Data

Wong, C.-H.
Enzymes in synthetic organic chemistry/C. H. Wong and
G. M. Whitesides. -- 1st ed.
p. cm. -- (Tetrahedron organic chemistry series ; v. 12)
1. Organic compounds--Synthesis. 2. Enzymes. I. Whitesides, G. M.
II. Title. III. Series.
QD262. W65 1994 547. 7'0459--dc20   94-2329

British Library Cataloguing in Publication Data

A catalogue record for this book is available from the
British Library

ISBN 0 08 035942 6 Hardcover
ISBN 0 08 035941 8 Flexicover

Printed and Bound in Great Britain by Redwood Books, Trowbridge

CONTENTS

Preface

This book is about using enzymes as catalysts in organic synthesis.

Why should synthetic chemists make the effort to learn the unfamiliar techniques required to use this class of catalysts? Organic synthesis has, after all, been one of the most successful of scientific disciplines, and has also been of enormous practical utility. New synthetic reagents, catalysts and strategies now make possible the synthesis of molecules of a degree of structural complexity that would have been unthinkable only 10 years ago. The types of problems at which non-biological organic synthesis has excelled—the synthesis of natural products, drugs, polymers, functional molecules—will continue to be important. Catalysis—especially non-biological catalysis with acids, bases and metals—has always been one of the foundations of the success of organic synthesis. Why bother now with biological catalysts, and with a new and quite different set of associated reagents and techniques?

There are three answers to the question "Why use enzymes?": necessity, convenience and opportunity. New synthetic and catalytic methods are necessary to deal with the new classes of compounds that are becoming the key targets of molecular research. Compounds relevant to biology—especially carbohydrates and nucleic acids—pose particular (and sometimes insurmountable) challenges to non-biological synthetic methods, but are natural targets for biological methods. For some types of compounds (for example, high molecular weight RNA), it may only be possible to synthesize these molecules by biological methods; for others, both biological and non-biological methods may offer synthetic routes, but it may simply be much more convenient to use enzymes. The ability to carry out synthetic transformations that are otherwise impossible or impractical, especially in key areas of biochemistry, is clearly one of the best opportunities now available to chemistry.

Now synthetic methods incorporating new catalysts are also necessary to deal with the increasing constraints imposed by environmental concerns. Many of the new reagents and catalysts that have benefited organic synthesis in the last years have contained transition metals or heavy elements. When these materials are used with great efficacy, they may still be environmentally acceptable, but their handling and disposal poses problems, and their replacement with environmentally acceptable catalysts would almost always be an advantage. The additional constraints on the design of synthetic processes that come from environmentally based restrictions on the use of organic solvents have made water enormously attractive as a solvent for reactions. Enzymes are intrinsically environmentally benign materials that operate best in water.

The high interest in enantioselective synthesis provides another reason for considering enzymes as catalysts. The active sites of enzymes are chiral, and enzymes are now well accepted as catalysts for reactions generating the enantiomerically pure intermediates and products demanded by the pharmaceutical industry (and being found increasingly useful in other areas). There are, of course, excellent non-biological catalysts for many chiral reactions, but if an enzyme is the best catalyst available for the synthesis of a chiral compound, why not use it? Synthetic chemists have never avoided using other naturally occurring materials with valuable catalytic activities (e.g. platinum black); they should not avoid enzymes.

In broader and more strategic terms, enzymes fill an important part of the spectrum of catalysts available to synthetic chemists. Catalysis is one of the most important activities in chemistry: it permeates all branches of chemistry and chemical engineering. Enzymes are among the most active and selective of catalysts. From that vantage alone, they must be a part of synthesis in the future. In addition, however, they offer other interesting characteristics. As one example, because most enzymes operate at room temperature in aqueous solution at pH 7, they are, as a group, intrinsically compatible with one another. Numbers of enzymes can therefore be used together, in sequence or cooperatively, to accomplish multistep reaction sequences in a single reaction vessel. In contrast, many useful non-biological catalysts are intrinsically incompatible with one another, or operate under incompatible conditions, and opportunities for using multiple non-biological catalysts at the same time are relatively limited.

In the long term, enzymes provide the basis for one approach (although certainly not the only approach) to one of the Holy Grails of chemistry: that is, to catalysis by design. The idea that one could design catalysts that would act specifically in any reaction of interest is one that would, if it were realized generally, change the face of synthesis. The generation of new classes of biological catalysts—catalytic monoclonal antibodies produced by immunization using a transition state analog, tailored enzymes produced by site-specific mutagenesis, catalytic RNA s selected by taking advantage of the enormous power of the polymerase chain reaction—suggest entirely new approaches to the production of new catalysts with specific activities. Powerful methods of screening microorganisms for enzymatic activities also provide new approaches to the discovery of useful catalysts.

Finally, there is important instructional value in using enzymes in synthesis. Some of the most exciting problems available to chemistry now come from biology, and enzymes are often the object or the solution to these problems. It is difficult to see how one can be an organic chemist in the future without a keen interest in molecules important in biology. Using enzymes in organic synthetic schemes provides, of course, an approach to the solution of certain specific problems in synthetic biochemistry; perhaps as importantly, however, it provides a method of learning biochemistry. Molecular recognition and selective catalysis are the key chemical processes in life; these processes both are embodied in enzymes. Organic chemists must learn about molecular biology, and using enzymes in the familiar activity of synthesis provides an excellent method of beginning to do so.

The book is organized into one introductory chapter dealing with the characteristics of enzymes as catalysts, and five chapters dealing with different types of chemical transformations. The first chapter is not intended to be a general introduction to enzymology–this function is much better served by the many excellent textbooks in enzymology. Instead, it is a summary of some of the types of information that are necessary or useful in applying enzymes in organic synthesis. Enzymes are unlike many catalysts routinely used in organic chemistry, as they are often well-defined structurally and thoroughly analyzed kinetically. It usually does not pay to try to analyze the kinetic behavior of most of the non-biological catalysts that are used in synthesis. In contrast, considering the kinetic behavior of enzymes may make it possible to optimize their use, and to proceed in a quite rational way to design reaction conditions that avoid catalyst poisoning (called in enzymology "enzyme inhibition") and that optimize catalytic performance.

The subsequent chapters are organized to group together related, useful information concerning the application of enzymes in important types of reactions. One of the difficulties that synthetic chemists have encountered in trying to use enzymatic catalysts has been that of trying to identify the right enzymes to accomplish a particular transformation. The literature of enzymology is organized along lines based in biochemistry, and is remarkably obscure to someone interested in synthetic applications. By grouping together enzymes that carry out related types of synthetic transformations, it should be easier to search for synthetically useful catalytic activities.

Chi-Huey Wong and George M. Whitesides

Acknowledgements

We thank the following coworkers who helped assemble and proofread the original manuscript: Chris Fotsch, Randy Halcomb, Ella Bray, Yi-Fong Wang, S.-T. Chen, Curt Bradshaw, Ziyang Zhong, Jeff Bibbs and Jim-Min Fang. Without their help, this book would not have been completed.

Chapter 1. General Aspects

The development of synthetic organic chemistry have made possible the stereocontrolled synthesis of a very large number of complex molecules. As the field has developed, its targets and constraints have changed. Two problems now facing organic synthesis are the development of techniques for preparing complex, water-soluble biochemicals, and the development of environmentally acceptable synthetic processes that are also economically acceptable. Enzymes are able to contribute to the resolution of both of these issues, and they should be considered as one useful class of catalysts to be used, when appropriate, for organic synthesis.

Enzymes are proteins; they catalyze most biological reactions *in vivo*[1,2] They also catalyze reactions involving both natural and unnatural substrates *in vitro*.[3-25] As catalysts, enzymes have the following characteristics:

1. They accelerate the rate of reactions, and operate under mild conditions.
2. They can be highly selective for substrates and stereoselective in reactions they catalyze, selectivity can range from very narrow to very broad.
3. They may be subject to regulation; that is, the catalytic activity may be strongly influenced by the concentrations of substrates, products or other species present in solution.
4. They normally catalyze reactions under the same or similar conditions.
5. They are generally unstable (relative to man-made catalysts).
6. They are chiral, and can show high enantiodifferiation.

The characteristics of instability, high cost, and narrow substrate specificity have been considered to be the most serious drawbacks of enzymes for use as synthetic catalysts. As a result, application of enzymes has been focused primarily on small-scale procedures yielding research biochemicals. The perception, however, that they are intrinsically limited as catalysts has changed dramatically in the past fifteen years due to new developments in chemistry and biology and new requirements in industry.

1. Large numbers of enzymatic reactions have been demonstrated to transform natural or unnatural substrates stereoselectively to synthetically useful intermediates or final products.[3-25] Table 1 is a list of enzymes commonly used in synthesis.
2. To scale up enzymatic reactions, new techniques have been developed to improve the stability of enzymes and to facilitate their recovery for reuse.[26]
3. Advances in molecular and cell biology, computation, and analytical chemistry have also created new tools for the manipulation of genetic materials to construct genes for expression of desired proteins.[27,28]

1

4.   New enzymes have been discovered that are key elements of molecular genetics and recombinant DNA technology. These enzymes and associated techniques have made it possible to construct genes for expression of the desired proteins.

5.   Recombinant DNA technology has made possible, in principle, the low-cost production of proteins and enzymes and the rational alteration of their properties.

6.   The area of enzymatic catalysis is further stimulated by the new discovery of catalytically active antibodies.[29]

Table 1. Enzymes commonly used for organic synthesis.

Not Requiring Cofactors	Not Requiring Added Cofactors	Cofactor Requiring
1) Hydrolytic Enzymes:	1) Flavoenzymes:	1) Kinases - ATP
Esterases	Glucose Oxidase	2) Oxidoreductases - NAD(P)(H)
Lipases	Amino Acid Oxidases	
Amidases	Diaphorase	3) Methyl Transferases - SAM
Phospholipases		
Epoxide Hydrases		4) CoA-Requiring Enzymes
Nucleoside Phosphorylase	2) Pyridoxal Phosphate Enzymes:	
SAM Synthetase		5) Sulfurylyases - PAPS
	Transaminases	
2) Isomerases and Lyases:	Tyrosinase	
	δ-Aminolevulinate Dehydratase	
Glucose Isomerase	Cystathionine Synthetase	
Aspartase		
Phenylalanine Ammonia Lyase	3) Metalloenzymes:	
Fumarase		
Cyanohydrin Synthetase	Galactose Oxidase	
	Monooxygenases	
3) Aldolases	Dioxygenases	
	Peroxidases	
4) Glycosyl Transferases	Hydrogenases	
	Enoate Reductases	
5) Glycosidases	Aldolases	
	Carboxylyases	
6) Oxynitilase	Nitrile Hydrase	
	4) Thiamin Pyrophosphate dependent enzymes:	
	Transketolases	
	Decarboxylases	
	5) Others:	
	SAH Hydrolase	
	B_{12}-Dependent Enzymes	
	PQQ (Methoxatin) Enzymes	

Among important challenges now facing synthetic organic chemists is that of understanding important biological processes in full molecular detail, and using this understanding to design and produce chemically well-defined molecules that are useful in medicine, agriculture, and biology. Since the biological activity of most drugs is due to their ability to interfere with receptor-ligand or enzyme-substrate interactions, a rational approach to the design and synthesis of drugs will require studies involving a range of substrates, inhibitors, ligands, and derivatives, some of which will be difficult to manipulate using classical synthetic methodology. Catalysis by enzymes may offer practical routes to these classes of molecules, and enzyme-based organic synthesis has become an attractive alternative to classical synthetic methods. It offers, when it is applicable, regio- and enantioselectiviey, low cost, and environmentally compatible reaction conditions.

1. Rate Acceleration in Enzyme-Catalyzed Reactions.

The fundamental concept, proposed by Eyring[30] in 1935, that for a reaction to proceed the reactant molecules must overcome a free energy barrier has provided the basis for quantitative approaches to enzyme kinetics. Once the reactants have reached this state of highest free energy-- the transition state--they proceed on to products at a fixed rate. Free energy contains both enthalpic and entropic terms. In general, the lower the activation energy, the faster the overall reaction will proceed. If a reaction proceeds through two or more steps, the one that has the highest free energy will often, but not always, be the rate-limiting step: in consecutive bi- and unimolecular reactions, for example, changes in concentration can shift the rate-limiting step from one to the other.

The assumptions in transition-state theory that the reactant ground state is in equilibrium with the transition state, and that the transition state proceeds to products at a fixed rate, have led to the development of the Eyring equation (eq 1).

$$k \cong (kT/h) \exp(-\Delta G^{\ddagger}/RT) \qquad (1)$$

In this equation, k, **k**, R and h are the rate, Boltzmann, gas, and Planck constants, respectively, where T is the temperature and $\Delta G^{\ddagger}$ represents the activation energy for the reaction. Since $\Delta G^{\ddagger}$ is related to $\Delta H^{\ddagger}$ and $\Delta S^{\ddagger}$, the enthalpy and enthropy of activation, by equation 2, equation 1 can be rearranged to equation 3.

$$\Delta G^{\ddagger} = \Delta H^{\ddagger} - T\Delta S^{\ddagger} \qquad (2)$$
$$k \cong (kT/h)\cdot\exp(-\Delta H^{\ddagger}/RT)\cdot\exp(\Delta S^{\ddagger}/R) \qquad (3)$$

The enthalpy of activation usually is dominated by changes in the energies of bonds, although non-bonding interactions can also be important. The entropy of activation is the non-enthalpic contribution to free energy and includes the costs of orienting the reactants, losses in conformational flexibility, and various effects of concentrations and solvent.[31]

Transition-state theory has proven to be an excellent, durable model with which to analyze basic principles of enzyme action.[32-34] One role of enzymes can be considered to be the reduction in the free energy of activation by stabilizing the rate-limiting transition state. This reduction in $\Delta G^{\ddagger}$ results in an acceleration in reaction rate. Enzymes accomplish this reduction by either reducing the enthalpy of activation ($\Delta H^{\ddagger}$), setting up more favorable interactions between substrates (an entropy effect, $\Delta S^{\ddagger}$), or by modifying interactions with solvent, or all of these.

2.   Michaelis-Menten Kinetics[1]

The multistage reaction process in enzyme catalysis requires that the substrate(s) initially bind noncovalently to the enzyme at a special site on its surface called a specificity pocket. The collection of specificity pockets for all the reactants is called the active site of the enzyme. The complex of substrates and enzyme is called the Michaelis complex and provides the proper alignment of reactants and catalytic groups in the active site. It is this active site where, after formation of the Michaelis complex, the chemical steps take place. Because each molecule of enzyme has only a limited number of active sites (usually one), the number of substrate molecules that can be processed per unit of time is limited.

After an enzyme is mixed with a large excess of substrate(s) and before equilibrium is reached, the reactive intermediates have different concentrations than they do at equilibrium. This short time interval is called the pre-steady state. Once the concentrations of the intermediates have reached equilibrium, the system is considered to be in the steady state. The steady state is the period in which the concentration of the reactive intermediates change slowly, and these conditions are known as steady-state conditions. Since there is a slow depletion of substrate, the steady-state assumption--that the rate of change of intermediates is small--is of course not always valid; however, restriction of rate measurements to this time interval is a good approximation to conditions used in synthesis. Steady-state rates are measured because these data are easier to collect (as compared to most pre-steady state rates) and generate the most reliable and relevant enzymatic rate constants.

Many reactions of enzymes follow a pattern of kinetic behavior known as Michaelis-Menten kinetics. By applying Michaelis-Menten kinetics, the measured reaction rates or velocities (v) can be transformed into rate constants that describe the enzymatic mode of action. Useful constants such as k_{cat}, K_m, and k_{cat}/K_m (below) can be determined. In most systems, the rate of reaction at low concentration of substrates is directly proportional to the concentration of enzyme $[E]_0$ and substrate $[S]$. As the concentration of substrate increases, a point will be reached where further increase in substrate concentration does not further increase v (as shown in Figure 1). This phenomena is called substrate saturation. The reaction velocity that is obtained under saturating concentrations of substrate is called V_{max}. Equation 4 is the Michaelis-Menten equation; it expresses quantitatively these characteristics of enzyme kinetics.

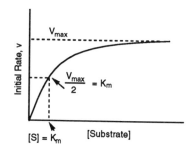

Figure 1. Relationship between the initial rate anad substrate concentration.

$$v = [E]_0 k_{cat}[S]/(K_m + [S]) \tag{4}$$

In this equation $k_{cat}[E]_0 = V_{max}$, [S] is the substrate concentration, K_m represents the concentration of substrate at which $v = V_{max}/2$, and k_{cat} is the apparent first-order enzyme rate constant for conversion of the enzyme-substrate complex to product; k_{cat} is also called the turnover number. At high concentrations of substrate, equation 4 simplifies to equation 5. Correspondingly, at low concentrations of substrate, equation 4 simplifies to equation 6. In equation 6, k_{cat}/K_m represents the apparent second-order rate constant for enzyme action.

$$V_{max} = k_{cat}[E]_0 \tag{5}$$
$$v = (k_{cat}/K_m)[E]_0[S] \tag{6}$$

Although not all enzyme systems follow the same mechanistic pathway, most systems can be reduced at least approximately to the above relationships, and they are widely used in considering applications of enzymes in synthesis.

Figure 2 illustrates the relationship of k_{cat} to k_{cat}/K_m. The value k_{cat}/K_m relates the reaction rate to the free enzyme and substrate rather than to the ES complex, and is the second-order rate constant. For the above system, k_{cat}/K_m is equal to $k_1 k_2/(k_{-1}+k_2)$. This rate constant includes kinetic constants associated with substrate binding: The ratio k_{cat}/K_m is sometimes referred to as the specificity constant, and is often used to assess the overall efficiency and specificity of enzyme action, especially when substrates are being compared.

The upper limit of k_{cat}/K_m is k_1, the diffusive rate of substrate binding ($10^8 - 10^9$ $M^{-1}s^{-1}$). At this upper limit in rate, k_{cat} is no longer rate limiting and the Michaelis-Menten kinetics changes to Briggs-Haldane kinetics.[1]

A)

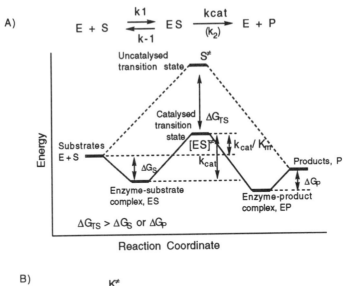

B)

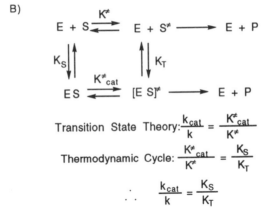

Figure 2. (A) Relationship of the apparent first-order (k_{cat}) and the second-order (k_{cat}/K_m) rate constants in enzyme-catalyzed reactions and comparison with nonenzymatic reactions. (B) Thermodynamic cycle relating the enzyme-catalyzed reaction of S to P. $K^{\neq}$, equilibrium constant for S and the transition state $S^{\neq}$ in a noncatalyzed reaction; $K^{\neq}_{cat}$, equilibrium constant for the Michaelis complex ES and its transition state $[ES]^{\neq}$ in the catalyzed reaction; K_S and K_T are the respective dissociation constants for the ground state and the transition state complexes.

One can apply transition-state theory to relate the first-order rate constants for the enzymatic (k_{cat}) and nonenzymatic (k) reactions to the corresponding equilibrium constants ($K^{\ddagger}_{cat}$ and $K^{\ddagger}$) for the formation of the transition-state complex, that is $k_{cat}/k \sim K^{\ddagger}_{cat}/K^{\ddagger}$. According to the thermodynamic cycle, these equilibrium constants are related to the dissociation constants for the transition state (K_T) and for the substrate (K_S), so that $K_S K^{\ddagger} = K_T K_{cat}^{\ddagger}$.[35] This simple analysis concludes that the enzyme binds to the transition state $S^{\ddagger}$ more strongly than to the ground state S by a factor approximately equal to the rate acceleration; that is, $k_{cat}/k \sim K_S/K_T$.[36]

The concept of transition-state binding has led to the development of transition-state analogs[33,36-37] for use as enzyme inhibitors and for the identification of possible groups involved in transition-state binding. X-ray crystal structures of enzyme-inhibitor complexes have played a vital role in these developments.[38] The enzymatic functional groups interacting with the transition-state analog are postulated to be those involved in transition-state binding. The active-site geometries obtained in these studies also provide information essential for enzyme engineering using the techniques of site-directed mutagenesis. The concept of transition-state binding has also led to an experimental approach to the design and synthesis of immunogenic transition-state analogs used in eliciting monoclonal antibodies that catalyze the reaction.[29]

Understanding the significance of kinetic constants allows the synthetic chemist to analyze an enzyme reaction so that the proper adjustments in concentration can be made to optimize the synthetic potential of the system. It is possible to adjust reaction conditions to increase the productivity and/or to alter the selectivity of the enzymatic system.

Since the catalytic activities of enzymes are sensitive to reaction conditions, it is very often necessary to determine the kinetic parameters under the synthetic conditions being used (or as close to these conditions as possible) to obtain the best performance. There are many ways to determine kinetic parameters, and most begin by measuring initial velocities at various concentrations of substrates while maintaining pH, enzyme concentration, volume of cosolvent, etc. constant. Probably the most straightforward procedure for generating the kinetic parameters is to use a rearranged Michaelis-Menten equation (7) and to plot $1/v$ versus $1/[S]$.

$$1/v = (K_m/V_{max})(1/[S]) + 1/V_{max} \qquad (7)$$

This treatment of the data is often referred to as the Lineweaver-Burk procedure and the plot called the Lineweaver-Burk plot. From this plot, V_{max} (1/y-intercept), K_m (-1/x-intercept), and V_{max}/K_m (1/gradient) can be obtained. Figure 3A shows a typical Lineweaver-Burk plot. This plot has the disadvantage of compressing the data points at high substrate concentrations into a small region. The Eadie-Hofstee plot, based on a different method of plotting the same data, (Figure 3B) will not have this problem. This type of plot is generally considered more accurate, but is historically the less commonly used in enzymology.

The initial velocities can be determined in a number of ways and the experimental procedure used depends upon the system under investigation. Standard textbooks in enzymology outline these procedures fully.

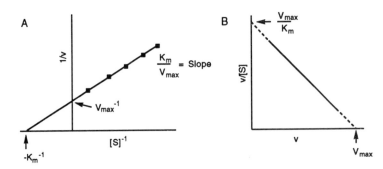

Figure 3. Typical Lineweaver-Burk (A) and Eadie-Hofstee plot (B) for determination of kinetic constants.

3. Enzyme Inhibition

Enzyme inhibition is decrease in catalytic activity of an enzyme as a result of a change of reaction conditions (i.e., pH, temperature, concentration of substrate or product, etc.). These conditions can cause conformational changes, blocking of active sites, or unfolding of the enzyme. Inhibition can also be caused by the substrates and/or products. It may be reversible or irreversible.

There are three general modes of inhibition[1]: Competitive (C), noncompetitive (NC), uncompetitive (UC), and mixed types of inhibition. These types of inhibition can be distinguished experimentally and are usually characterized using the Lineweaver-Burk plots (Figure 4).

Competitive inhibition reflects the binding of an inhibitor to the enzyme near or at the active site; this binding prevents the substrate(s) from binding properly or at all. The inhibitor and the substrate are thus competing for the active site of the enzyme. With this type of inhibition, the values of K_m increase with increasing concentration of the inhibitor in the Lineweaver-Burk plot; V_{max} (or k_{cat}) does not change (a common intersection on the y-axis). By increasing the concentration of substrate, eventually V_{max} can be reached.

For the other two types of inhibition, the Lineweaver-Burk patterns are different. With noncompetitive inhibition, there is a common intersection on the x-axis as opposed to the y-axis, indicating an effect on V_{max} rather than an effect on K_m. This type of inhibition can be observed if the inhibitor and the substrate are not at the same site. This pattern can be observed, for example, if the inhibitor binds at a site removed from the active site, but causes a change in the shape of the active site when it binds. An interpretation of non-competitive inhibition is that the inhibitor binds equally well to the enzyme and the enzyme-substrate (Michaelis) complex. The inhibitor will bind to the enzyme with or without the substrate present. The uncompetitive inhibition pattern is a collection of parallel lines indicating an influence of the inhibitor on *both*

K_m and V_{max}. One interpretation of uncompetitive inhibition is that the inhibitor binds only to the Michaelis complex and not to the enzyme.

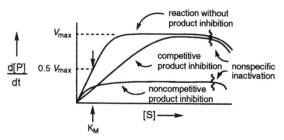

P is the kinetic product; S is the substrate

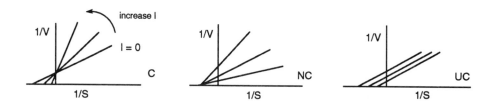

Figure 4. Schematic representation of the relation between the reaction rate d[P]/dt and the substrate concentration [S] for a simple reaction following Michaelis-Menten kinetics, and the Lineweaver-Burk plots for the three common inhibition patterns.

Of the different types of production inhibitions described, noncompetitive and mixed types of inhibition are the most serious problems in synthetic applications, since they cannot be overcome simply by increasing the concentrations of substrates. In a theoretical analysis of the relative reaction rate as a function of the extent of reaction in the presence of an inhibitor, it is difficult to achieve high rates and high conversions simultaneously in the reaction when $K_m/K_i>1$, whereas when $K_m/K_i<1$ the reaction can proceed rapidly to completion (K_i = inhibition constant).[39]

4. Specificity

Many important types of organic reactions have equivalent enzyme-catalyzed reactions. The major synthetic value of enzymes as catalysts is their selectivity. Because enzymes are large chiral molecules with unique stereo-structures in the active site; they can be highly selective for certain types of substrate structures and reactions. Useful types of enzyme-catalyzed reactions include the chemoselective reaction of one of several different functional groups in a molecule, the regioselective reaction of one of the same or similar groups in a molecule, the enantioselective reaction of one enantiomer of a racemic pair or one of the enantiotopic faces or groups, and the diastereoselective reaction of one or a mixture of diastereomers or one of the diastereomeric faces or groups. All such selective reactions occur because during a reaction, the prochiral or chiral reactants form diastereomeric enzyme-transition-state complexes that differ in transition-state ($\Delta G^{\ddagger}$) energy.

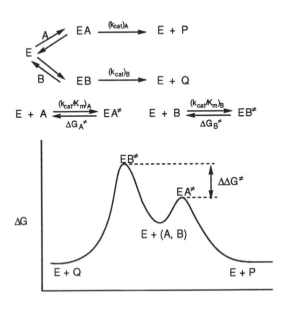

Figure 5. Enzyme-catalyzed enantioselective reactions with a racemic mixture (A + B). E, enzyme; P, product from A; Q, product from B.

For example, in an enantioselective transformation, the two enantiomeric substrates or two enantiotopic faces or groups compete for the active site of the enzyme (Figure 5). Using the steady-state or Michaelis-Menten assumptions, the two competing reaction rates are:

$$\upsilon_A = (k_{cat}/K_m)_A[E][A] \qquad\qquad (8)$$
$$\upsilon_B = (k_{cat}/K_m)_B[E][B] \qquad\qquad (9)$$

The ratio of these two reaction rates is therefore:

$$\upsilon_A/\upsilon_B = (k_{cat}/K_m)_A[A]/(k_{cat}/K_m)_B[B] \qquad\qquad (10)$$

This analysis shows that the ratio of specificity constants $[(k_{cat}/K_m)_A/(k_{cat}/K_m)_B]$ determines the enantioselectivity of the reaction. Since these specificity constants are related to free-energy terms (that is, $\Delta G^{\ddagger}_A = -RT \ln (k_{cat}/K_m)_A$ and $\Delta G^{\ddagger}_B = -RT \ln (k_{cat}/K_m)_B$), the enantioselectivity of the reaction is related to the difference in energy of the diastereomeric transition states by eq 11:

$$\Delta\Delta G^{\ddagger} = (\Delta G_A^{\ddagger} - \Delta G_B^{\ddagger}) = -RT \ln (k_{cat}/K_m)_A/(k_{cat}/K_m)_B \qquad (11)$$

In an enzyme-catalyzed kinetic resolution which proceeds irreversibly, the ratio of specificity constants (or the enantioselectivity value, E) can be further related to the extent of conversion $(c)^{40}$ and the enantiomeric excess (ee) as shown in equation 12. The parameter E is commonly used in characterizing the enantioselectivity of a reaction.

$$\frac{\ln [(1-c) (1 - ee_A)]}{\ln [(1-c) (1 + ee_A)]} = \frac{\ln [1-c (1 + ee_P)]}{\ln [1-c (1 - ee_P)]} = E \qquad (12)$$

Experimentally, one can use equation 12 to determine the E value, which in turn can be used to predict the ee of the product or remaining substrate at a certain degree of conversion. Quantitative expressions that describe the kinetic and thermodynamic parameters that govern the selectivity of enzyme catalyzed reversible esterification of enantiomers in organic solvents have also been developed (Figure 6).[41a] The E values determined on the basis of ee at high degrees of conversion using these expressions may not be accurate, and a new method based on the initial rates of reaction for mixtures of enantiomers has been reported.[41b]

$$Enz + A \underset{k_2}{\overset{k_1}{\rightleftharpoons}} Enz + P \qquad K = \frac{k_2}{k_1} = \frac{k_4}{k_3} = \frac{A}{P} = \frac{B}{Q}$$

$$Enz + B \underset{k_4}{\overset{k_3}{\rightleftharpoons}} Enz + Q$$

$$\frac{\ln [1 - (1 + K)(c + ee_S\{1 - c\})]}{\ln [1 - (1 + K)(c - ee_S\{1 - c\})]} = \frac{\ln [1 - (1 + K) c (1 + ee_P)]}{\ln [1 - (1 + K) c (1 - ee_P)]} = E$$

Where k_1, k_2, k_3, k_4 are second-order rate constants. A and B are enantiomeric substrates, ee_S and ee_P are the ee of remaining substrate and product respectively, and E is the enantioselectivity value.

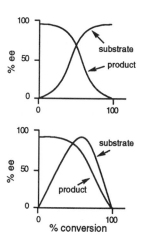

Figure 6. Reversible kinetic resolution of enantiomers A and B.
Insert: top, irreversible case (equation 12); bottom: reversible case.

For the enantioselective hydrolysis of meso diesters, the enantiomeric monoesters obtained are often not further hydrolyzed (Figure 7). In some cases, however, further hydrolysis of the enantiomeric monoesters, catalyzed by the same enzyme, occurs. The combination of enantioselective hydrolysis and kinetic resolution can result in the enhancement of the ee of the monoesters. Quantitative analysis of this case allows the optimization of optical and chemical yields of these enantioselective transformations.

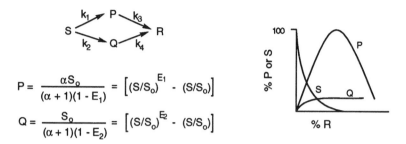

$$P = \frac{\alpha S_o}{(\alpha + 1)(1 - E_1)} = \left[(S/S_o)^{E_1} - (S/S_o)\right]$$

$$Q = \frac{S_o}{(\alpha + 1)(1 - E_2)} = \left[(S/S_o)^{E_2} - (S/S_o)\right]$$

S is the mesodiester substrate; S_o is the initial S concentration; R is the diol or diacid product; P and Q are enantiomeric monoesters; k_1, k_2, k_3, and k_4 are second-order rate constants;

$E_1 = k_3/(k_1 + k_2)$; $E_2 = k_4/(k_1 + k_2)$, $\alpha = k_1/k_2$,
$k_1 + k_2 = (k_{cat}/K_m)_S$; $k_3 = (k_{cat}/K_m)_P$; $k_4 = (k_{cat}/K_m)_Q$

Figure 7. Enantioselective conversion of meso diesters.

Sequential irreversible kinetic resolutions of racemic substrates using enzymatic catalysis have also been utilized in obtaining enantiomerically enriched products.[42,43] (2R,4R) and (2S,4S)-2,4-pentanediols, for example, have been prepared by sequential enantioselective esterification in anhydrous isooctane. Quantitative expressions describing this model system have been developed for the calculation of the relative kinetic constants that allow optimization of the chemical and enantiomeric yields (Figure 8).[42] Sequential kinetic resolution has also been applied to hydrolysis[43] and it has been shown that improvement of overall enantioselectivity can be achieved with a proper choice of solvents so that the rates for the two steps are close.[43]

Other sequential enzymatic resolutions involve hydrolysis-esterification[42b] or alcoholysis-esterification[42c] sequences. In each case, the enzyme displays the same enantioselectivity for the two sequential reactions. The desired product can be obtained with higher enantiomeric yield as a result of the double resolution process. For the hydrolysis-esterification sequence, the reaction is often carried out in an organic medium containing a minimum amount of water. The alcohol generated in this reaction reacts with the acid and forms the ester product (Figure 8b).

Although the stereoselectivity in most enzymatic reactions is dictated by the particular tertiary structure of the catalyst, it is difficult to predict the stereochemistry of a reaction and a change over in the sense of the stereoselectivity from one substrate to another is not uncommon.[18] The only approach to the prediction of stereoselectivity at present is to develop a reliable, empirical active-site model for the enzyme. Based on studies of enzyme selectivity reported in the literature, one can sometimes use computer graphics analysis to develop such a model. Horse liver alcohol dehydrogenase[44] and pig liver esterase[45] models are now available that are simple and reasonably reliable both for prediction of new reactions and rationalization of literature results. Empirical models are particularly useful for enzymes for which X-ray crystal structures are not available.

(a)

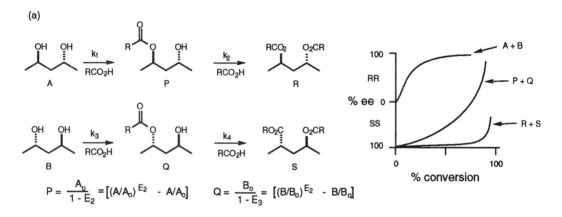

$$P = \frac{A_0}{1 - E_2} = [(A/A_0)^{E_2} - A/A_0] \qquad Q = \frac{B_0}{1 - E_3} = [(B/B_0)^{E_2} - B/B_0]$$

$E_1 = k_1/k_3;\ E_2 = k_2/k_1;\ E_3 = k_4/k_3$. A, B, P and Q are the corresponding concentrations at certain degree of conversion. A_0 and B_0 are the initial concentration of A and B.

(b)

Enzyme: lipase from *Mucor miehei*

Figure 8. Sequential irreversible kinetic resolution.

5. Improvement or Alteration of Enzyme Specificity

As mentioned previously, the enantioselectivity chacterizing an enzyme catalyzed reaction is due to the formation of diastereomeric transition states that differ in free energy $\Delta\Delta G^{\ddagger}$. The ratio of two enantiomeric products is equal to the ratio of the two corresponding second-order rate constants. To achieve an enantioselective reaction with 99.9% ee for the products requires $\Delta\Delta G^{\ddagger}$ = 4.5 kcal/mol; to achieve a 90% ee of product, $\Delta\Delta G^{\ddagger}$ = 1.74 kcal/mol (Table 2). This magnitude of free energy is equivalent to one or two hydrogen bonds and reaction conditions can often be altered to improve the enantioselectivity of a given reaction. One can also sometimes modify the substrate to improve the enantioselectivity by introducing different substituents or protecting groups. The types of alterations of reaction conditions that have proven useful in synthesis range from increasing the amount of organic solvent[46] to an adjustment in pH[47] and, occasionally, even a change in reaction temperature.[48] Examples of these kinds of alterations will be discussed in the following sections.

Table 2. Energy Requirements for Calculated Enantiomeric Excess.

ee	P_1/P_2 (k_1/k_2)	$\Delta\Delta G^{\ddagger}$, cal/mol
10	1.22	118
50	3	651
90	19	1740
95	39	2170
99	199	3140
99.9	1999	4500

5.1. The Effect of pH

By varying the pH of an enzymatic reaction one can change the conformation and/or the ionization status of the enzyme and reactants. The new conformation and charge distribution may or may not correspond to an active enzyme or it may or may not alter substrate selectivity, depending upon the particular protein. Another consideration involving pH is the effect it will have on possible side reactions. It has been shown[47] with many reactions of pig liver esterase (PLE) that, although the catalytic rate of hydrolysis is much faster at pH 8.0, it was beneficial to lower the pH so that the contribution of the non-enzymatic hydroxide-induced hydrolysis was small. By reducing the uncatalyzed hydrolysis, the ee could be increased. Optimization of pH can be extremely helpful in optimizing enzyme-assisted synthetic reactions.

5.2. Effect of Solvent

When the reaction medium is changed from water to an organic solvent, the overall efficiency of the enzyme can change dramatically. This change in medium can also affect stereoselectivity. The results of a study of esterase-catalyzed hydrolyses in water, and transesterification in butyl ether, are shown in Table 3. The large decrease in preference of the enzyme for the L-substrate when the solvent is changed from H_2O to butyl ether is particularly relevant to synthetic applications. In these reactions, the change in solvent increases the free energy of activation more for the L-substrates than for the R-substrates: this change has the effect of making $\Delta\Delta G^{\ddagger}$ smaller.

Table 3. Comparison of enzyme catalyzed hydrolysis in water and transesterification in butyl ether

SYSTEM	Water k_1/k_2	Butyl ether k_1/k_2
Subtilisin NAACE[a]	1800	4.4
Subtilisin NAPCE[b]	15000	5.4
Elastase NAACE	>1000	4.5
α-lytic protease NAACE	10000	8.3
Subtilisin (BPN') NAPCE	16000	7.3
α-chymotrypsin NAACE	710	3.2
Trypsin NAPCE	>4000	3.2

Each rate constant represents k_{cat}/K_m where k_1 is for the L-substrate and k_2 is for the D-substrate. [a]N-Ac-Ala-OEtCl. [b]N-Ac-Phe-OEtCl.

This and another study[49] have indicated that the enantioselectivity of subtilisin- and chymotrypsin-catalyzed hydrolyses of L and D esters in aqueous solution is higher for hydrophobic substrates than for hydrophilic substrates. In organic solvents, the enantioselectivity, however, drops substantially and hydrophilic substrates become more reactive than hydrophobic substrates. This phenomenon--solvent-induced change of substrate selectivity--can often be rationalized in terms of differences in partitioning of the substrate between the active site and medium; this change is reflected in the K_m values. In this instance, the productive binding of the L-ester to the active site of subtilisin was interpreted to release more water molecules from the hydrophobic binding pocket of the enzyme than did that of the D-isomer. This release of water is less favorable in hydrophobic media than in water. Thus, the reactivity of the L-ester in hydrophobic media decreases substantially, and the discrimination between the D- and L-esters is diminished.

Other studies on the effect of organic solvents on enzyme selectivity have been reported;[49b] relationships between solvent properties and selectivity can usually not be generalized. A change of solvent polarity, for example, may or may not affect the selectivity of the enzyme.[49b] Interestingly, inversion of enzyme enantioselectivity by organic solvents has also been reported,[49c,49d] although the effect was not large, whether this inversion is directly caused by the solvent or by the change in the structure of the enzyme caused by the solvent is not clear.

5.3. Temperature Effect

Changing reaction temperature is a less obvious approach for optimization of stereoselectivity than changing pH or solvent, since enzymes are temperature-labile. The enzyme *Thermoanaerobium brockii* alcohol dehydrogenase catalyzes the reduction of 2-pentanone to (*R*)-2-pentanol at 37 °C, while at 15 °C the product is (*S*)-2-pentanol.[50] Similarly, in the oxidation of 2-butanol catalyzed by the enzyme *Thermoanaerobacter ethanolicus* alcohol dehydrogenase, the (*S*)-enantiomer is preferred at <26 °C while the (*R*)-enantiomer is preferred at >26 °C.[48] The diastereoselectivity of the horse liver alcohol dehydrogenase-catalyzed reduction of 3-cyano-4,4-dimethylcyclohexanone is decreased at 45 °C relative to that observed at 4 °C.[51] A study of the temperature-dependent enantioselectivity of the alcohol dehydrogenase from *Thermoanaerobacter ethanolicus* revealed a linear relation between temperature (°K) and the difference in transition-state energies of the two enantiomers ($\Delta\Delta G^{\ddagger}$) examined. Since $\Delta\Delta G^{\ddagger}$ is related to the ratio of specificity constants as described previously [($\Delta\Delta G^{\ddagger}$ = -RT ln (k_{cat}/K_m)R/(k_{cat}/K_m)S], $\Delta\Delta G^{\ddagger}$ could be determined from the values of k_{cat} and K_m of each enantiomer at different temperatures. Establishing this linear relationship determined $\Delta\Delta G^{\ddagger} = \Delta\Delta H^{\ddagger} - T\Delta\Delta S^{\ddagger}$ and allowed prediction of (*R*) or (*S*)-enantioselectivity at different temperatures. It also indicated the temperature at which there would be no discrimination between (*R*) and (*S*)-enantiomers (the so-called the "racemic temperature").[48]

5.4. Site-Directed Mutagenesis and Natural Selection

For enzymes with known X-ray structure, the use of site-directed mutagenesis and computer-assisted molecular modeling has allowed an approach to the rational alteration of enzyme specificity. This field was in its infancy and progress has been difficult. There have been interesting successes, nonetheless. For example, aspartate aminotransferase, a pyridoxal phosphate-dependent enzyme that catalyzes the transamination of Asp or Glu, was converted to lysine-arginine transaminase by the replacement of the active-site Arg with Asp.[52] L-Lactate dehydrogenase was converted (by mutation of Gln-102 to Arg) to L-malate dehydrogenase; this conversion doubled the enzymatic activity of the natural malate dehydrogenase.[53] The lactate dehydrogenase from *Bacillus stearothermophilus* has been altered to accomodate a broader spectrum of substrates.[54] The coenzyme specificity of glutathione reductase for NADP was altered to make it selective for NAD,[55] and that of NAD-dependent glyceraldehyde 3-phosphate dehydrogenase was altered (*via* Leu-187→Ala, Pro-188→Ser) to that it accomodated both NAD and NADP.[56] Perhaps the most extensively engineered enzyme is the serine protease subtilisin BPN'.[57-61] Almost every catalytic property of this enzyme--substrate specificity, pH-rate profile, stability--has been altered. Even here, however, a radical change in substrate specificity (e.g., from L-specific to D-specific) has not been accomplished using site-directed mutagenesis. The major problems in this area are the difficulty of predicting protein tertiary structure from primary sequence and of predicting selectivity and catalytic activities from tertiary structure.

Traditional screening based on natural selection can lead to the discovery of new enzymes with interesting specificity. As examples, a thermostable NADP-dependent secondary alcohol dehydrogenase from *Thermoanaerobium brockii* was found at a hot spring site in Yellowstone Park;[50] a nitrile hydrolyzing enzyme was found at an acrylonitrile plant;[62] interesting monooxygenases were discovered in toxic waste sites;[63] the antimicrobial agent β-chloroalanine was used to screen for resistant organisms that contained pyridoxal phosphate-dependent enzymes using β-chloroalanine as a substrate for β-replacement;[16] new NAD-dependent secondary alcohol dehydrogenases with pro-R specificity for NADH and (R)-selectivity for alcohol substrates were discovered from microorganisms using selected alcohols as carbon sources;[64] a D-amino acid esterase was discovered for use in the synthesis of D-amino acid containing peptides;[65] an L-specific N-acyl proline acylase[66] and an enzyme for selective deamidation of peptide amides[67] were discovered for use in amino acid and peptide synthesis; an enzyme for asymmetric decarboxylation of disubstituted malonic acids was discovered by screening for microorganisms that utilized phenylmalonic acid.[68]

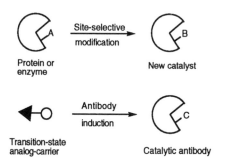

Figure 9.

6. Enzyme Stabilization and Reactor Configuration

Enzymes are often unstable in solution. They can be inactivated by denaturation (caused by increased or decreased temperature, by an unfavorable pH or dielectric environment, or by organic solvents), dissociation of cofactors such as metals, and covalent changes such as oxidation, disulfide interchange and proteolysis.[69] It is generally believed that the three-dimensional structure of a protein in a given environment is determined by its primary sequence[70] and is the thermodynamically most stable structure.

Thermal denaturation is the most studied mode of enzyme inactivation. Enzymes from thermophilic organisms (heat tolerant microorganisms) usually differ from those of mesophilic species (organisms existing in the usual range of temperatures) by only small changes in primary structures, and the three-dimensional structures of such enzymes are essentially the same.[71] Mesophilic enzymes usually retain their native structures in aqueous solution only at temperatures below 40 °C, while the thermophilic enzymes may not denature until 60 to 70 °C. This difference corresponds to an increase in stability of 5-7 kcal/mol. Free-energy changes of this order can be derived from a few additional salt bridges, hydrogen bonds, or hydrophobic interactions.

Mesophilic enzymes, in principle, can be made more thermally stable by introducing additional binding forces. Site-directed mutagenesis, chemical cross-linking, and immobilization have been explored as techniques to increase the stability of enzymes. A subtilisin variant incorporating multiple site-specific mutations, for example, is several thousand times more stable than is the wild type in both aqueous solution and in high concentrations of dimethylformamide.[61]

Of the different techniques available for enzyme stabilization,[26] immobilization is currently the most commonly used.[26a] The procedures generally involve the covalent or noncovalent attachment of enzymes to a support. Cross-linking of enzymes[26a] or enzyme crystals,[26b] and entrapment or encapsulation of enzymes have also been used. There is, however, no general procedure available for immobilization of enzymes, and substantial trial and error is usually required to find the best method. Functional ceramics, such as glass beads treated with 3-aminopropyltriethoxysilane,[26a] a cross-linked copolymer of acrylamide and acryloxysuccinimide (PAN),[72] epoxide-containing acrylamide beads (Eupergit C),[73] and carbohydrate-based supports[74] are commonly used for covalent immmobilization. In many cases, a spacer is often employed (and may be required) to link the enzyme to the support. Glutaraldehyde is often used to link amino groups of the support to those of the enzyme; it will also form crosslinks within the enzyme. Other bifunctional linkers containing reactive groups such as epoxide (specific for NH_2, SH, OH), succinimide (specific for NH_2) and maleamide (specific for SH) are also often used.[26] Ion-exchange resins, glass beads, and XAD-8[75] are often used for adsorption of enzymes to be used in organic solvent or biphasic systems. Enclosure of enzymes in a dialysis bag[76] is another particularly convenient method of enzyme immobilization in the laboratory.

These techniques for immobilization have been applied to large-scale processes. Continuous flow systems based on column, membrane, and hollow fiber reactors are often used in large-scale enzymatic reactions.[26a] Batch reactions in mono- or biphasic systems are also used. An additional advantage of enzyme immobilization is the ease of recovery of the enzyme for reuse.

7. Cofactor Regeneration

A number of synthetically useful enzymatic reactions require cofactors such as adenosine triphosphate (ATP), nicotinamide adenine dinucleotide (NAD) and its 2'-phosphate (NADP), acyl coenzyme A, S-adenosylmethionine, sugar nucleotides and 3'-phosphoadenosine-5'-phosphosulfate (PAPS).[77] These cofactors are too expensive to be used as stoichiometric reagents. Regeneration of the cofactors from their reaction products is thus required to make processes using them economical. Cofactor regeneration can also reduce the cost of synthesis by:[77] (i) influencing the position of equilibrium; that is, a thermodynamically unfavorable reaction can be driven toward product by coupling it with a favorable cofactor regeneration reaction; (ii) preventing the accumulation of cofactor by-products that inhibit the forward process; (iii) eliminating the need for stoichiometric quantities of cofactors and thus simplifying the workup of the reaction; and (iv) increasing enantioselectivity relative to stoichiometric reactions.[23]

The cofactor ATP and other nucleoside triphosphates have been used in selective enzymatic phosphorylations catalyzed by phosphoryl transfer enzymes; the products derived from the cofactors in such reactions are nucleoside di- or monophosphates. In a recycling scheme, these must be converted to the corresponding triphosphates, using an appropriate phosphorylating reagent, in a reaction catalyzed by another enzyme. Acetylphosphate coupled with acetate kinase or phosphoenol pyruvate coupled with pyruvate kinase has been used in the regeneration of nucleoside triphosphates from their diphosphates.[77] Both acetate kinase and pyruvate kinase accept virtually all nucleoside diphosphates, including ADP, GDP, UDP, and CDP. Phosphoenol pyruvate is much more stable in solution than acetyl phosphate and is also a more favorable phosphate donor. Procedures are available for large-scale syntheses of acetyl phosphate and phosphoenol pyruvate[78]. These two phosphorylating reagents have been used in large-scale syntheses of nucleoside di- and triphosphates, sugars and their phosphates, oligo-

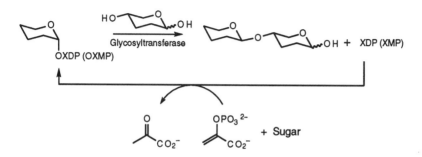

Figure 10. Formation of glycoside bonds catalyzed by glycosyltransferases with regeneration of sugar nucleotide.

saccharides, and related substances;[5,18,79,80] these are discussed in greater detail in subsequent chapters. Figure 10 illustrates general procedures for the synthesis of oligosaccharides that involve the regeneration of nucleoside monophosphate and diphosphate sugars and nucleoside triphosphate.

The nicotinamide cofactors are involved in enzymatic oxidoreductions. Several practical enzymatic methods are available for regeneration of NADH from NAD; these include reduction with formate/formate dehydrogenase, glucose/glucose dehydrogenase and isopropanol/*Pseudomonas* alcohol dehydrogenase. For regeneration of NADPH from NADP, glucose/glucose dehydrogenase and isopropanol/*Thermoanaerobium brockii* alcohol dehydrogenase are considered the most useful systems. For regeneration of NAD(P) from the corresponding reduced forms, the systems based on α-ketoglutarate/glutamate dehydrogenase (for NAD and NADP) and pyruvate/lactate dehydrogenase (for NAD) are the most useful. The system based on flavin mononucleotide (FMN)/FMN reductase is also very useful.[18,77] When a nicotinamide cofactor-dependent enzyme is used for synthesis, regeneration of the cofactor can be catalyzed by a second enzyme or by the same enzyme as that used in the synthesis (provided that the overall equilibrium is favorable). The isopropanol/alcohol dehydrogenase systems are typical examples of the one-enzyme system. When a second enzyme is used for cofactor regeneration,

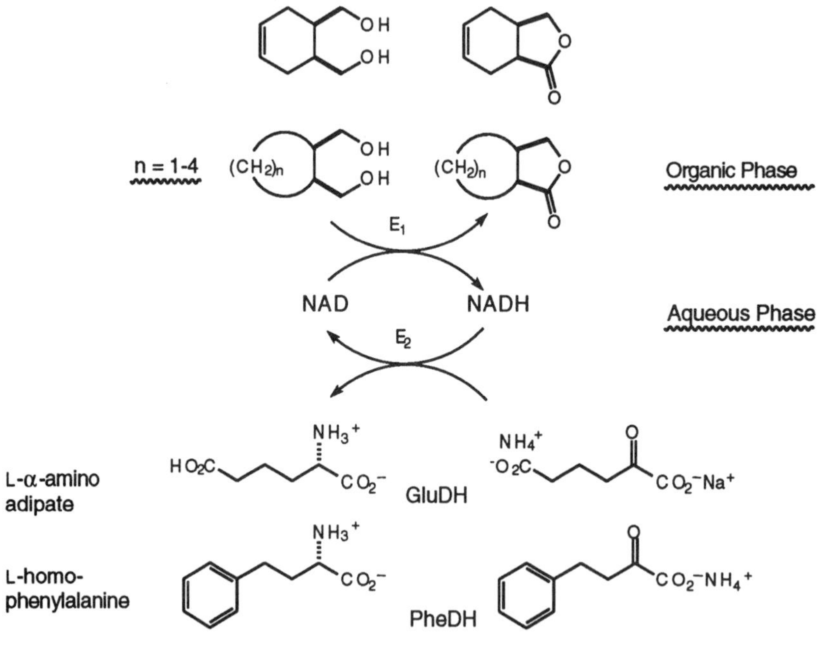

Figure 11. Biphasic system. E_1: horse liver alcohol dehydrogenase, E_2: amino acid dehydrogenase.

the regeneration system, in some instances, can also be synthetically useful if the overall reaction is favorable and the products are easily separated. One example is the one-pot synthesis of lactones and amino acids in a biphasic system in which the NAD-dependent oxidation of meso diols is coupled with NADH-dependent reductive amination of α-ketoacids (Figure 11).[81] In these reactions, the chiral lactones were extracted into the organic phase (hexane) while the amino acids were retained in the aqueous phase; separation of products was thus straightforward in this instance. Furthermore, inhibition of the enzyme caused by the product lactone was minimized, and the overall yield was increased.[82]

To date, the most highly developed large-scale process for NADH regeneration is that based on formate dehydrogenase in a membrane reactor.[83] In this system, the cofactor NAD was modified by covalent attachment to polyethylene glycol to prevent its leakage from the reactor.

The reduced forms of nicotinamide cofactors are also involved indirectly in many oxygenase-catalyzed reactions. These oxygenases are either metalloenzymes or flavoenzymes that are able to activate molecular oxygen and insert an oxygen atom stereoselectively into inactive molecules such as alkanes, aromatics, and olefins. For synthetic transformations, whole cells instead of free enzymes are used because of the instability of the enzymes. Figure 12 illustrates the regeneration of NADPH in a cyclohexanone monooxygenase catalyzed Baeyer-Villiger oxidation[84] and in a ω-monooxygenase catalyzed oxidative deprotection of a methyl ether.[85]

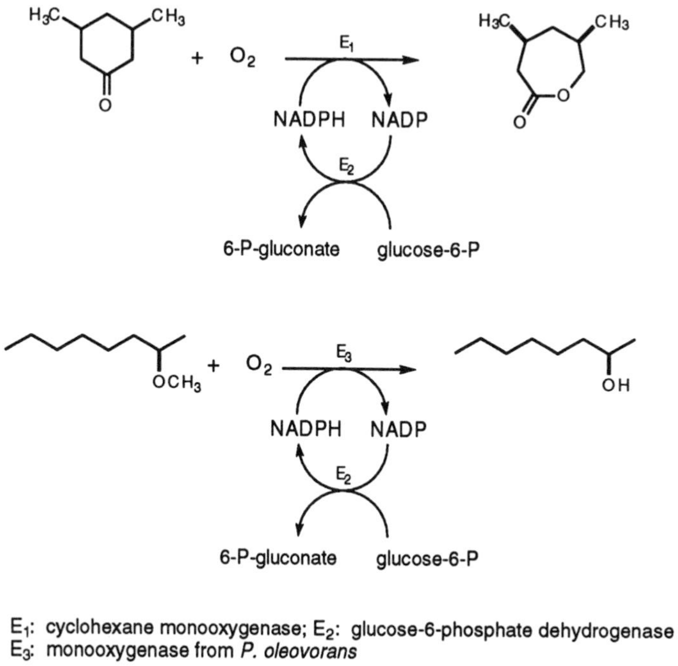

E_1: cyclohexane monooxygenase; E_2: glucose-6-phosphate dehydrogenase
E_3: monooxygenase from *P. oleovorans*

Figure 12. Regeneration of NADPH in monooxygenase reactions.

Although acyl coenzyme A-dependent enzymes may have value in organic synthesis, regeneration of acyl coenzyme A has only recently been developed, and there are few examples of synthetic applications.[86]

The most important problems remaining to be solved in cofactor regeneration are the regeneration of S-adenosylmethionine for enzymatic methylation and the regeneration of PAPS for enzymatic sulfation.

8. Enzyme Catalysis in Organic Solvents

The ester forming properties of lipases[87a] and amide forming properties of proteases[87b] were confirmed experimentally for many years. Crystalline chymotrypsin and xanthine oxidase were found catalytically active in anhydrous dioxane.[87c] Formation of N-acetyl-L-tryptophan ethyl ester from the corresponding acid and ethanol in water-immiscible biphasic organic solvent systems was observed where the volume of the aqueous phase could be as low as 1% of that of the organic phase.[87d] Immobilized thermolysin has been used in water-saturated ethyl acetate for the synthesis of Z-Asp-Phe-OMe from Z-Asp and L-Phe-OMe.[87e] A regioselective interesterification of a triglyceride in hexane with a lipase immobilized on photo-cross linkable resins, silica beads or Celite was used for the preparation of cocoa butter-like fat from olive oil and steric acid.[87f] The oleic acid at 1- and 3-positions of the lipid was replaced with steric acid, in a reaction catalyzed by the lipase from *Rhizopus delemar*. Kinetic resolution by esterification and transesterification reactions was carried out with an immobilized *Candida* lipase and hog liver esterase in a biphasic system containing a small amount of water.[87g] Enhancement of thermostability and change of substrate specificity has been observed in hydrophobic organic solvents containing <1% water.[87h] *Mucor miehei* lipase-catalyzed lactonization of 15-hydroxypentadecanoic acid and 4-hydroxybutyric acid has also been reported in organic solvents.[87i] Pancreatic lipase and yeast lipase were suspended in organic solvents for enantioselective esterifications and transesterifications,[87j] and enantioselective esterification of racemic menthol with a lipase and 5-phenylvaleric acid was also reported.[87k] A slow formation of amide bond in hexane between octylamine and a carboxylic acid of C_2-C_{16} units was reported.[87l]

These early developments of enzymatic reactions in media containing high volume fractions of organic solvents have stimulated further work in the field and major improvements have been reported subsequently (see Chapter 2). The enzymes used in water-miscible organic solvents, water-immiscible organic solvents, and reverse micelles,[88a] are in most cases *not* in organic media: they are still functioning within a pool of water as with a shell of associated water molecules.[88b] It was not clear how water affected enzyme stability and catalytic activity until a systematic study on the hydration of dry lysozyme was conducted.[89] In that study, it was inferred that when water was added to the enzyme it initially interacted with the charged functional groups of the protein, and subsequently with the uncharged polar and nonpolar groups. In the presence of 505 molecules of water per molecule of protein, the protein structure was nearly identical with that found in solution in bulk water. This study indicates that less than a monolayer of water may be all that is required for an enzyme to be catalytically active.

The stability of enzymes in organic solvents depends on the hydrophobicity of the solvent.[90] In general, enzymes are more stable when suspended in nonpolar solvents that have low solubility for water than in polar solvents. Many enzymatic transformations indeed have been performed in organic solvents containing minimum amounts of water.[91] Further studies that examined the role of water in enzymatic reactions in a number of anhydrous polar and nonpolar organic media concluded that in general, enzymes only need a thin layer of water on the surface of the protein to retain their catalytically active conformation.[92] The most useful nonaqueous media are hydrophobic solvents that do not displace these essential molecules of water from enzymes. Water-immiscible solvents containing water below the solubility limit (about 0.02 to 10% by weight depending on the solvent used) permit certain dry enzymes (crystalline or lyophilized powder) to be catalytically active. Lyophilization of enzymes dissolved in optimal pH solution provides the most active forms to be used in organic solvents.[92] Within this range of water content, the enzymatic activity in an appropriate organic solvent can be optimized and, in some cases, is comparable to that in aqueous solution, and the catalysis follows Michaelis-Menten kinetics. Mechanistic investigations of serine protease-catalyzed reactions in organic solvents by solid-state NMR,[93] linear free energy correlation,[94] and by kinetic isotope effect[95a] studies suggest that the transition-state structure in nonpolar organic solvents is nearly the same as that in aqueous solution, indicating the microenvironment of the enzyme active site in nonpolar organic solvents is the same as that in water. When the crystals of the serine protease subtilisin Carlsberg were cross-linked and washed with acetonitrile, the structure was found to be essentially the same as that in water.[95b] Higher thermostability of some enzymes in organic solvents than in water has also been reported, presumably because enzymes are conformationally less flexible in nonaqueous media.[92] Changes of stereoselectivity in going from water to organic solvent was also observed.[96-98] The change in stereoselectivity comes mainly from the different importance of the release of water during the binding of isomeric substrates to the enzyme. Many reactions that are sensitive to water, or are thermodynamically impossible to perform in water, become possible in organic media. Enzyme-catalyzed dehydrations, transesterifications, aminolyses, and oxidoreductions in organic solvents are now common. Novel enzymatic reactions in gases and supercritical fluids have also been exploited,[99a] and a change of enantioselectivity in such environments was also observed.[99b] Product or substrate inhibition can be lessened in certain enzymatic reactions in which products or substrates partition preferentially into the nonaqueous phase. In most cases, enzymes are insoluble in organic media; they can therefore be recovered by centrifugation or filtration and used repeatedly. Precautions should be taken, however, in interpreting the results of these studies, as in most cases enzymes are insoluble and may be protected by water and salts (the counterions) from contact with the organic solvent. The results of these studies regarding the behavior of enzymes in organic solvents therefore may not be very informative. An enzyme associated with different salts suspended in an organic solvent may have different properties (e.g. solubility, activity, and stability) due to different dissociation constants and other physical properties of the counterions. Subtilisin BPN' lyophilized from a sodium phosphate solution (50 mM, pH 8.0) and suspended in DMF, for example, is completely insoluble and much more stable (by a factor of ~1000) than that prepared from a Tris-HCl solution under the same conditions.

The combination of enzyme catalysis in an organic solvent with that in an aqueous solution provides a new route to both enantiomers in enantioselective transformation of meso- or racemic substrates. For example, both the (R) and (S)-chiral glycerol derivatives can be easily prepared from 2-O-benzylglycerol and the diacetate through lipase-catalyzed transesterification and hydrolysis, respectively (Figure 13).[100] In each case, the enzyme possesses the same enantiotopic group selectivity. The use of an enol ester as the transesterification reagent makes the process irreversible (thus preventing the loss of enantioselectivity due to the reverse reaction) and eliminates product inhibition.

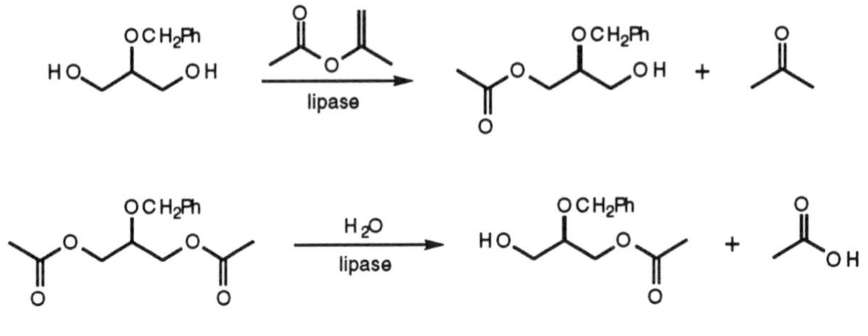

Figure 13. Transesterification using enol esters to obtain one enantiomer. Hydrolysis of the *meso*-diester provides the other enantiomer.

Despite the advantages of enzymatic transformations in organic solvents, there are some disadvantages: (i) organic solvents may not dissolve charged or polyfunctional species; (ii) adjusting the pH is difficult in large-scale processes; (iii) enzymes are generally unstable and less active in organic solvents, particularly in hydrophilic solvents; (iv) loss of stereoselectivity may occur or because of the reversible nature of the reaction; (v) severe substrate or product inhibition may occur for those enzymes that accept hydrophilic compounds (e.g., sugars) as natural substrates.[101]

Although some of these problems can be overcome (for example, immobilization to improve stability, and use of enol esters for transesterification to facilitate the reaction and to avoid reverse reaction and product inhibition) some still remain to be solved. Recent approaches to improving enzyme stability in polar organic solvents based on site-directed mutagenesis seem to be promising.[61,102] Modification of subtilisin BPN' by increasing the internal binding forces favorable in organic solvents (e.g. increasing H-bond interactions, metal binding, and conformational restriction) and minimizing the surface charge (to reduce the solvation energy) significantly improves the stability of the enzyme in dimethylformamide.[61] Improvement of subtilisin activities in organic solvents can also be accomplished by random[102b] or directed mutagenesis.[61,102c] Further study along this line may provide principles of design for engineering enzymes to be used in organic media.

9. Multienzyme Systems and Metabolic Engineering

Since all enzymes generally function under the same or similar conditions (aqueous solution, pH ~ 7, rt), two or more enzymatic reactions can be carried out in one pot. The multienzyme systems can be used not only to facilitate and simplify reaction processes but also to shift an unfavorable equilibrium to produce the desired product. Many multienzyme systems have been developed,[5,18,103,104] and the ones that require cofactors (such as nicotinamide cofactors and sugar nucleotides) have been used in large scale synthesis. In oligosaccharide synthesis, it has been demonstrated that more than six enzymes can be used in one pot to produce oligosaccharides effectively with concurrent regeneration of sugar nucleotides.[103] The systems both reduce the cost of the sugar nucleotides and eliminate the problem of product inhibition caused by the nucleoside phosphates generated during the reaction. All enzymes used in the one-pot syntheses can be co-immobilized to a solid support to improve their stability and to facilitate their recovery for reuse. The efficiency of the multienzyme systems can be further improved by cross-linking or by gene fusion.[105]

Another approach to multienzyme reactions is based on the whole cell or fermentation processes. Cells containing the desired multienzyme systems can be reconstructed through metabolic engineering via recombinant DNA methods[106-112] or via selection pressure.[113] In principle, genes coding for the enzymes responsible for the synthesis of a target molecule can be cloned and localized in one species or in a single plasmid. As one example, the gene coding for 2,5-diketo-D-gluconic acid reductase from *Corynebacterium sp.* has been cloned and expressed in *Erwinia herbicola*. Although the wild type of this organism lacks this reductase, the recombinant *E. herbicola* is able to produce α-keto-L-gulonate, a precursor to L-ascorbic acid, from glucose.[107] By localizing the genes encoding transketolase and 3-deoxy-D-arabino-heptulosonate phosphate (DAHP) synthetase on a single plasmid, a new *E. coli* strain is able to produce high levels of DAHP.[108] Further manipulation of the cells has led to the high-level microbial production of intermediates used in the shikimate pathway. "Inter-species" cloning of antibiotic biosynthesis genes has also been used to express in the same cell two biosynthetic pathways to

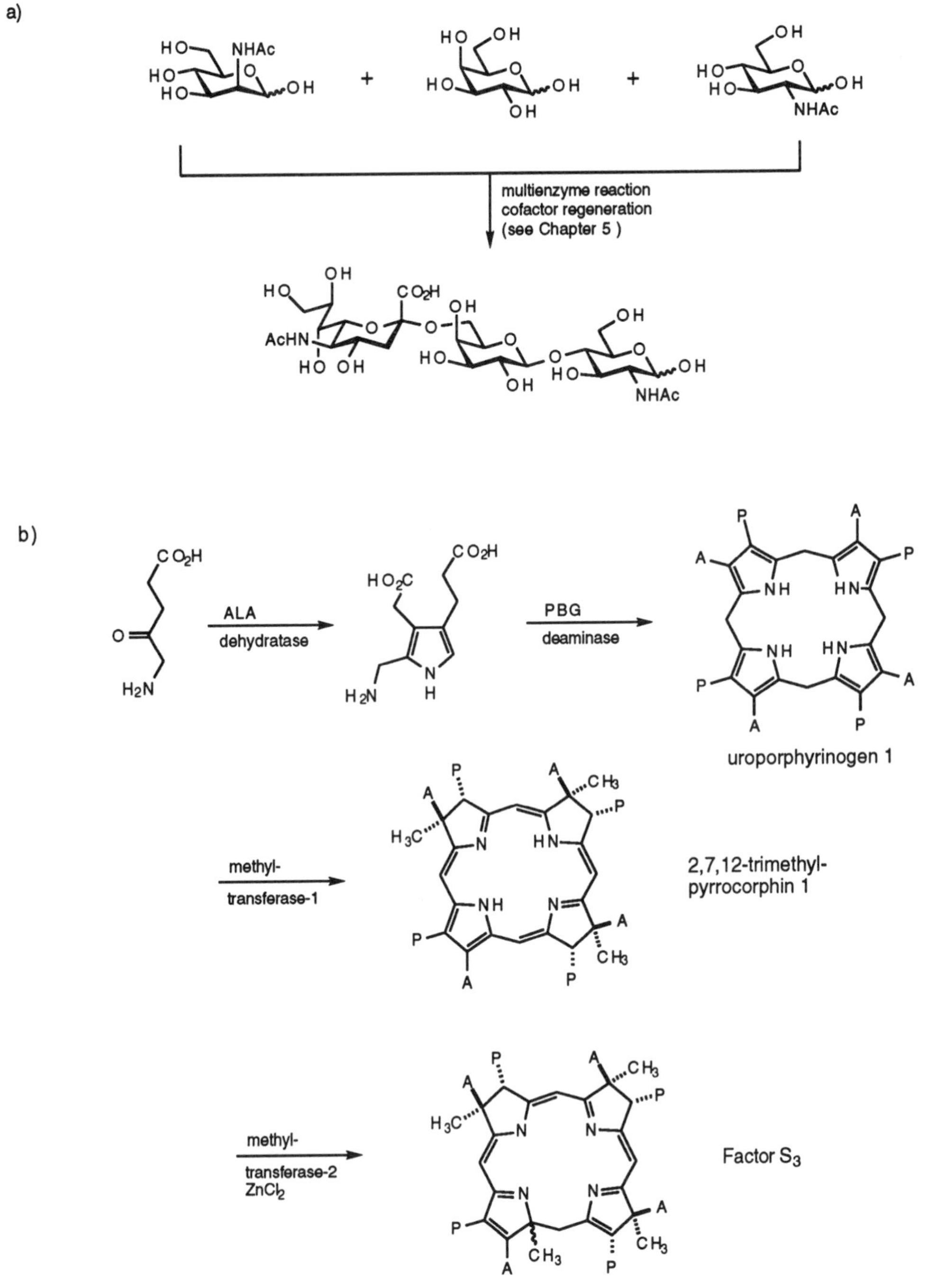

a)

b)

Figure 14. Examples of the use of multienzyme systems (a[103], b[104]) and reconstructed cells (c[107], d[108]) for synthesis.

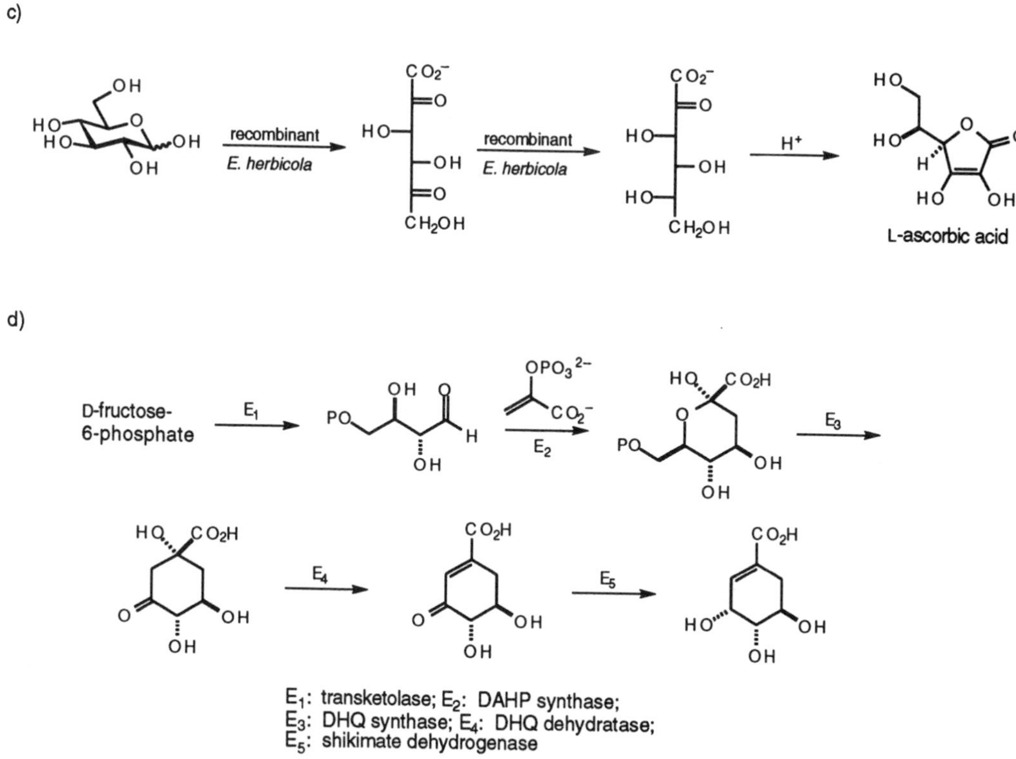

c)

d)

E_1: transketolase; E_2: DAHP synthase;
E_3: DHQ synthase; E_4: DHQ dehydratase;
E_5: shikimate dehydrogenase

Figure 14. Continued.

make hybrid antibiotics structurally different from those produced by the parent organisms.[109] Figure 14 illustrates some examples of reactions with multienzyme systems.

10. Rational Design of New Enzymatic Catalysts

A goal in enzymatic organic synthesis is to develop protein catalysts with tailored activities and selectivities. Despite a significant amount of work directed toward the understanding of protein folding and structure-function relationships, there is still little predictive understanding of how the protein primary sequence translates into its catalytically active tertiary structure. To construct a protein with a designed catalytic activity and selectivity from a designed sequence of amino acids is still very far from reality.[114a] Design of peptide enzymes based on surface-simulation synthetic peptides that mimic the chymotrypsin and trypsin active sites has recently been reported;[114b] this approach is, however, still very speculative. The alternative methods based on catalytic antibodies and site-selective modification of existing proteins, however, represent useful approaches toward the rational design of new enzymatic catalyts.

10.1. Catalytic Antibodies

One mechanism by which enzymes act as structure-selective catalysts is to provide steric and electronic complementarity to a rate-determining transition state of a given reaction.[32] This concept led to the suggestion that an antibody elicited against a stable transition-state analog of a reaction should catalyze that reaction.[33] This concept has been realized experimentally, and monoclonal antibodies elicited against a number of molecules designed to resemble the transition states of specific reactions are capable of catalyzing those reactions with rate accelerations of several orders of magnitude (some even approaching that of enzyme catalysis).[29] With appropriate design of the antigens, specific functional groups can be induced in the binding site of an antibody to perform general acid/base or nucleophilic/electrophilic catalysis. With this new technique, new protein catalysts can be designed and prepared for reactions that may be disfavorable or not attainable otherwise, or have different reaction mechanisms or specificities compared to the corresponding enzyme-catalyzed reactions.

The most successful reactions using catalytic antibodies are selective ester hydrolysis (and transesterification) and pericyclic reactions. The former take advantage of the availability of good transition-state analogs for ester hydrolysis (i.e. phosphonate) and the latter are due to the capability of a created antibody binding site to overcome the entropic barrier characterizing the highly ordered transition state of pericyclic reactions. Figure 15 illustrates representative reactions catalyzed by antibodies that are synthetically useful.

From the point of view of synthetic chemistry, approaches based on catalytic antibodies may become a powerful strategy for the generation of a new protein catalyst for a desired reaction, if an immunogenic transition-state analog of that reaction can be synthesized. A bottleneck in this rapidly evolving field, however, is the inefficiency and inaccessibility of the hybridoma technology for the production of desired monoclonal antibodies in large quantities for chemical transformations. Furthermore, the number of antibodies induced to a synthetic antigen based on the hybridoma technology is quite limited (only a few hundred antibodies are generated), and the methods used to detect the antigen binding are not efficient. The antibodies produced may therefore not represent the whole group of antibodies that might be induced by a given antigen. The probability of finding antibodies that are highly effective in catalysis is thus relatively low. Additionally, if the synthetic antigen does not closely resemble the transition-state of the reaction, the antibodies induced may not have the appropriate functional groups to participate in catalysis.

A fraction of these problems have been solved by the creation of a highly diverse library of heavy- and light-chain antigen binding fragments (Fab's) for use in screening for catalytically active Fab's.[116] The methodology depends on the use of the polymerase chain reaction (PCR) in the presence of designed DNA primers to amplify the mRNA's from the spleens of immunized mice, followed by cloning the resulting DNA into λ phage and construction of a plasmid-expression library in *E. coli.* Since Fab fragments are only one tenth of the entire antibody in size, but behave as whole antibodies in terms of antigen recognition, and since site-directed mutagenesis of Fabs can be easily conducted to try to improve the binding and catalysis, the field of catalytic antibodies based on Fab fragments may experience substantial development in the near future. The use of "single-chain antibodies",[117] and the optimization of the catalytic activity of antibodies by inserting their antibody genes into microorganisms lacking the corresponding enzyme may also become useful approaches to the development of efficient antibody enzymes.[118]

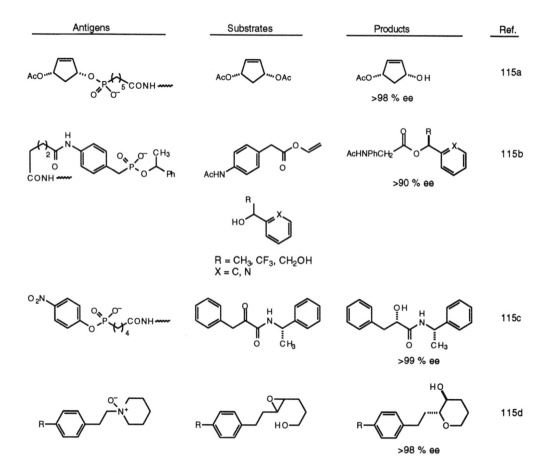

Figure 15. Representative antibody-catalyzed reactions.

10.2. Site-Selective Modification of Proteins or Enzymes using Chemical or Biological Strategies

Based on understanding the mechanisms of protein binding and catalysis, it is possible to try to alter the active site rationally to accomodate new substrates or to catalyze new reactions.

Although enzymes integrate both binding or recognition pockets and a catalysis pocket into one active site, these two functions--binding and catalysis--can be altered individually to create new catalysts with altered selectivity and/or catalytic activity. Site-directed mutagenesis has been used in such alterations. A limitation of this methodology is that the replacement of residues are restricted to the twenty naturally occurring amino acids. The recently developed technique of in vitro site-directed mutagenesis allows replacement of residues with unnatural amino acids.[119]

In a different strategy, a nonselective enzyme can be altered to show higher selectivity by covalent attachment of a substrate recognition component to the enzyme, either within[120] or outside the active site.[121] The cleavage of RNA by such a hybrid enzyme was performed by the attachment of an oligonucleotide binding site to a nonspecific staphylococcal nuclease.[122] Alternatively, selective chemical modification of functional groups in the active site of proteins may provide a route to new catalysts with useful properties.[120] Attachment of an organometallic Rh catalyst to avidin, the biotin-binding protein, converted avidin to a selective hydrogenation catalyst.[123] Papain has been converted into an oxidoreductase by the covalent modification of the active site with a molecule of flavin.[120] Subtilisin and chymotrypsin, which normally catalyze hydrolysis, have been modified to catalyze acyltransferase activity[124] by the selective methylation of the active site His at ε–2N or by converting the active-site Ser to Cys[125] or selenocysteine.[126]

11. Conclusion

Many kinds of protein catalysts can be constructed from the 20 common amino acids. Although the relation of the structure and catalytic activity of a protein to its amino acid sequence is still impossible to predict, Nature does provides an unlimited number of protein catalysts that are either known or remain to be explored. New protein catalysts specific for a predetermined reaction are now becoming available through the catalytic antibody approach. Existing enzymes or proteins can be rationally altered through site-directed mutagenesis or chemical or enzymatic modifications to provide additional new protein catalysts with novel catalytic activities and selectivities.

It is clear that enzymes represent a valuable class of catalysts for organic transformations, and a number of organic reactions can be performed with the use of enzymes: synthesis of chiral intermediates; transformations of sugars, nucleotides, and related species; synthesis of compounds important in metabolism and analogs of these metabolites (amino acids, sugars and their phosphates, etc.); transformations of peptides and proteins; and other transformations in which classical chemical methodology is constrained. The question for organic synthesis is whether an enzymatic approach to a particular synthetic problem is more practical than a non-biological approach. In many instances, enzymatic transformations represent only an alternative or improved process compared to an existing chemical methodology, and the value replacing the existing process with a new enzymatic method has to be evaluated in terms of the merits of each.

In some cases, however, enzymatic processes are more cost-effective than non-biological methods, and have clear advantages. For example, the use of DNA ligase and restriction enzymes to close and open nicks in DNA has no effective competition from other techniques.

Several large-scale enzymatic processes used in the industry (resolution of amino acids with acylase, production of high-fructose corn syrup with glucoamylase and glucoisomerase, transformation of pig insulin to human insulin catalyzed by trypsin, preparation of penicillin analogs with penicillin acylase, synthesis of aspartame with thermolysin, and preparation of acrylamide from acrylonitrile catalyzed by nitrilehydrolase) have demonstrated that enzymatic catalysis can be a route to fine chemicals, medicinals and commodity chemicals. Where else will enzymes be used? Synthesis of optically active fine chemicals, medicinals and intermediates; synthesis and modification of sugars, oligosaccharides, polysaccharides and their conjugates and analogs; regioselective transformation of complex molecules; modification of recombinant DNA products such as glycoproteins; and other transformations effected in food, agriculture, and materials chemistry may be best conducted enzymatically. Enzymatic synthesis, like other catalysis technology, is fundamentally a process technology. It is frequently less convenient ` to develop an enzymatic route to a substance required in milligram quantities than to carry through a classical chemical synthesis. When large quantities of that material are required, however, it will be worth the effort to develop appropriate enzymes for that synthesis.

After deciding to evaluate enzymatic catalysis for a given transformation, one can select one or more enzymes capable of performing that type of transformation and optimize the reaction conditions through proper choices of temperature, pH, solvent, regulators, and protecting groups. An astonishing number of enzymes are already commercially available, and many more can be easily isolated. The next step is to evaluate if the available enzymes possess adequate catalytic activities for the desired transformation. Selectivity and specific activity are important factors at this stage in an evaluation. The specific activity of an enzyme is often (but unfortunately not always) presented in a standard system of units called international units: one unit of enzyme is that amount of catalytic activity that will produce 1 μmol of product per minute at its maximum velocity, and the number of units per milligram of enzyme represents the specific activity of that enzyme. If the specific activity of an enzyme is low, the quantity of that enzyme required to accomplish a particular synthesis may be unrealistically large, and the time of reaction is so long that side reactions dominate. In that event, it is necessary to use a more active enzyme, or to develop another enzyme by screening or by site-directed mutagenesis, or to abandon the project to chemical methodology. Similar considerations are also given to the evaluation of selectivity.

If the specific activity and selectivity are high enough to be practical, the remaining steps in developing an enzymatic process are relatively straightforward, at least conceptually. One can usually find procedures that will stabilize an enzyme adequately for bench-scale synthesis.

Immobilization is usually possible, and when it is unsuccessful it may be perfectly practical to use the free enzyme in a homogeneous solution (perhaps contained behind a membrane). Site-directed or random mutagenesis can also be used to improve enzyme stability. Large-scale production of the enzyme can then be executed using recombinant DNA technology. With all kinetic parameters (e.g., k_{cat}, K_m, rate of enzyme inactivation) available for a desired transformation, one can, in principle, construct a large-scale synthesis.

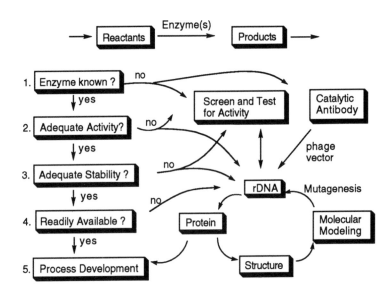

Figure 16. Strategy for the development of enzymatic catalysts.

In the case where there is no known enzyme for the desired transformation, it is more cumbersome to find an appropriate enzyme, and chemical methodology may be the method of choice. Approaches based on screening for new enzymes and catalytic antibodies may, however, be worth the effort to explore if the desired transformation is sufficiently important. Once a new protein catalyst is found or developed, its activity and stability can be improved as described previously. Proteins can also be displayed on the surface of phage and screened for binding and catalysis[127a-d] and as substrates.[127e] The catalytically active RNA's recently discovered have created a new dimension in biocatalysis and non-protein biocatalysts based on RNA are now, in principle, available to be exploited as synthetic catalysts.[128]

Table 4 summarizes the current status of application of enzymatic catalysis in organic synthesis, and Figure 16 provides a strategic approach to the development of enzymatic catalysts for reactions of interest. In Table 4, the heading "solved" does not imply that further and substantial improvements may not be possible, but only that these processes normally do not limit applications of enzymatic catalysis in the synthesis of fine chemicals.

Table 4. Status of certain generic problems in enzymatic synthesis.

Solved	Current Research	
• Immobilization	• Development of new laboratory scale uses of enzymes	• Mutagenesis for changes in properties
• Cofactor Regeneration NAD(P), ATP, CoA, Sugar Nucleotides	• Design of unnatural substrates	• Exploration of enzymes from new species for synthetic utility
• Enzyme Deactivation Thiol Oxidation Protease Action	• Use with organic cosolvents; one- and two-phase systems	• Scale-up in pharmaceutical synthesis
	• Enzyme stabilization	• Multi-enzyme systems in synthesis
• Use of recombinant DNA for large-scale production of enzymes	• Overcoming product inhibition	
	• Cofactor regeneration: PAPS, SAM	• Applications in synthesis of fine and specialty chemicals
	• Modification of enzyme activiey and semi-synthetic enzymes	• Antibody catalysts

Immediate Future

• Design and development of new enzymatic catalysts

• Chemoenzymatic synthesis of complex molecules

• Multienzyme systems via metabolic pathway engineering

• New catalysts from monoclonal Fab fragments

In summary, biocatalysts represent a new class of chiral catalytic activities potentially useful for a broad range of organic transformations. Synthetic chemists capable of using this class of catalysts will have a clear advantage over those limited to non-biological methods in their ability to tackle the new generation of synthetic problems appearing at the interface between chemistry and biology, both from academic and practical points of view.

References

1.   (a) Walsh, C. *Enzymatic Reaction Mechanisms*; W.H. Freeman and Co.: San Francisco, 1977. (b) Fersht, A. *Enzyme Structure and Mechanism, 2nd ed.*; W.H. Freeman and Co.: New York, 1985.

2.   Dixon, M.; Webb, E. *Enzymes*; Academic Press: New York, 1979.

3.   (a) Suckling, C. J.; Suckling, K. E. *Chem. Soc. Rev.* **1974**, 387. (b) Suckling, C. J.; Wood, H. C. S. *Chem. Br.* **1979**, *15*, 243.

4.   Jones, J. B.; Sih, C. J.; Perlman, D. *Application of Biochemical Systems in Organic Chemistry*; Wiley: New York, 1976.

5.   (a) Wong, C.-H.; Whitesides, G. M. *Angew. Chem. Int. Ed. Engl.* **1985**, *24*, 617. (b) Akiyama, A.; Bednarski, M.; Kim, J. J.; Simon, E. S.; Waldmann, H.; Whitesides, G. M. *Chem. Br.* **1987**, *23*, 645. (c) Toone, E. J.; Simon, E. S.; Bednarski, M. B.; Whitesides, G. M. *Tetrahedron* **1989**, *45*, 5365.

6.   Simon, H.; Bader, J.; Gunther, H.; Neumann, S.; Thanos, J. *Angew. Chem. Int. Ed. Engl.* **1985**, *24*, 539.

7.   Jakubke, H.-D.; Kuhl, P.; Konneckle, A. *Angew. Chem. Int. Ed. Engl.* **1985**, *24*, 85.

8.   Jones, J. B. *Tetrahedron* **1986**, *42*, 3341.

9.   Klibanov, A. M. *CHEMTECH* **1986**, 354; Klibanov, A. M. *Acc. Chem. Res.* **1990**, *23*, 114.

10. Morihara, K. *TIBTECH* **1987**, *5*, 164.

11. Kullmann, W. *Enzymatic Peptide Synthesis*; CRC Press: 1987.

12. *Enzymes in Organic Synthesis* ; Porter, R.; Clark, S., Eds., Ciba Foundation Symposium 111; Pitman: London 1985.

13. *Biocatalysis in Organic Synthesis*; Tramper, J.; von der Plas, H.C.; Linko, P., Eds.; Elsevier: Amsterdam, 1985.

14. *Enzymes as Catalysts in Organic Synthesis*; Schneider, M., Ed.; Reide, Dordrecht: The Netherlands, 1986.

15. Butt, S.; Robert, S. M. *Chem. Br.* **1987**, 127.

16. Yamada, H.; Shimizu, S. *Angew. Chem. Int. Ed. Engl.* **1988**, *27*, 622.

17. Eberle, M.; Egli, M. *Helv. Chim. Acta* **1988**, *71*, 1.

18. (a) Wong, C.-H. *Science* **1989**, *244*, 1145. (b) Wong, C.-H. *Chemtracts - Organic Chemistry* **1990**, *3*, 91.

19. Pratt, A. J. *Chem. Br.* **1989**, 282.

20. Dordick, J. S. *Enz. Microb. Technol.* **1989**, *11*, 194.

21. Hummel, W.; Kula, M.-R. *Eur. J. Biochem.* **1989**, *184*, 1.

22. Ohno, M.; Otsuka, M. *Org. React.* **1989**, *37*, 1.

23. Chen, C.-S.; Sih, C. J. *Angew. Chem. Int. Ed. Engl.* **1989**, *28*, 695.

24. Ward, O. P.; Young, C. S. *Enz. Microb. Tech.* **1990**, *12*, 482.

25. Hutchinson, C. W. *TIBTECH* **1990**, *8*, 348.

26. (a) For various immobilization techniques, see Mosbach, K. *Method. Enzymol.* Vol. 44, 1976; Vol 135-137, 1987; Chibata, I. *Immobilized Enzymes*; Wiley: New York, 1978. (b) Clair, N. L. S.; Navia, M. A. *J. Am. Chem. Soc.* **1992**, *114*, 7314.

27. Gerhartz, W. Enzymes in Industry, VCH: Berlin, 1990.

28.   (a) Wu, R.; Grossman, L. *Method. Enzymol.*, Vol. 153-155, **1987**. (b) Goeddel, D.V. *Method. Enzymol.*, Vol. 185, **1990**, 185.

29.   (a) Lerner, R. A.; Benkovic, S. J.; Schultz, P. G. *Science*, **1991**, *252*, 659.

30.   Eyring, H. *J. Chem. Phys.* **1935**, *3*, 107. (b) Schultz, P. G.; Lerner, R. A. *Acc. Chem. Res.* **1993**, *26*, 391. (c) Hilvert, D. *Acc. Chem. Res.* **1993**, *26*, 552.

31.   Schowen, R. L. In *Transition States of Biochemical Processes*; Gandour, R. D.; Schowen, R. L., Eds.; Plenum: New York, 1978.

32.   Pauling, L. *Am. Sci.* **1948**, *36*, 51.

33.   Jencks, W. P. In *Current Aspects of Biochemical Energetics*; E.P., Kennedy, E. P., Ed.; Academic Press: New York, 1966, p. 273.

34.   Kraut, J. *Science* **1988**, *242*, 533.

35.   Kurz, J. L. *J. Am. Chem. Soc.* **1963**, *85*, 987.

36.   (a) Wolfenden, R. *Nature* **1969**, *223*, 704. (b) Leinhard, G. E. *Ann. Rep. Med. Chem.* **1972**, *7*, 249.

37.   Bartlett, P. A.; Marlowe, C.K. *Science* **1987**, *235*, 569.

38.   Tronrud, D. E.; Holden, H. M.; Matthews, B. W. *Science* **1987**, *235*, 571.

39.   Lee, L. G.; Whitesides, G. M. *J. Org. Chem.* **1986**, *51*, 25.

40.   Chen, C.-S.; Fujimoto, Y.; Girdaukas, G.; Sih, C. J. *J. Am. Chem. Soc.* **1982**, *104*, 7294.

41.   a) Chen, C. S.; Wu, S.-H.; Girdaukas, G.; Sih, C. J. *J. Am. Chem. Soc.* **1987**, *109*, 2812. b) Jongejan, J. A.; van Tol, J. B. A.; Geerlof, A.; Duine, J. A. *Recl. Trav. Chim. Pays-Bas* **1991**, *110*, 247 and 255.

42.   a) Guo, Z.-W.; Wu. S.-H.; Chen, C.-S.; Girdaukas, G.; Sih, C. J. *J. Am. Chem. Soc.* **1990**, *112*, 4942. b) Macfarlane, E. L. A.; Roberts, S. M.; Turner, N. J. *J. Chem. Soc., Chem. Commun. 1990*, 569. Fowler, P. W.; Macfarlane, E. L. A.; Roberts, S. M. *J. Chem. Soc., Chem. Commun.* **1991**, 453. c) Chen, C.-S.; Liu, Y.-C. *J. Org. Chem.* **1991**, *56*, 1966.

43.   Caron, G.; Kazlauskas, R. J. *J. Org. Chem.* **1991**, *56*, 7251.

44.   (a) Jones, J. B.; Jakovac, I. J. *Can. J. Chem.* **1982**, *60*, 19. (b) Jakovac, I. J.; Goodbrand, H. B.; Lok, K. P.; Jones, J. B. *J. Am. Chem. Soc.* **1982**, *104*, 4659.

45.   Toone, E. J.; Werth, M. J.; Jones, J. B. *J. Am. Chem. Soc.* **1990**, *112*, 4946.

46.   Zaks, A.; Klibanov, A. M. *J. Biol. Chem.* **1988**, *263*, 3194; 8017.

47.   Bjorkling, F.; Boutelje, J.; Gatenbeck, S.; Hult, K.; Norin, T. *App. Microbiol. Biotech.* **1985**, *21*, 16.

48.   Pham, V. T.; Phillips, R. S.; Ljungdahl, L. G. *J. Am. Chem. Soc.* **1989**, *111*, 1935.

49.   a) Sakurai, T.; Margolin, A. L.; Russel, A. J.; Klibanov, A. M. *J. Am. Chem. Soc.* **1988**, *110*, 7236. b) Terradas, F.; Teston-Henry, M.; Fitzpatrick, P. A.; Klibanov, A.

M. *J. Am. Chem. Soc.* **1993**, *115*, 390. c) Wu, S.-H.; Wang, K.-T. *Bioorg. Med. Chem. Lett.* **1991**, *1*, 399. d) Tawaki, S.; Klibanov, A. M. *J. Am. Chem. Soc.* **1992**, *114*, 1882.

50.   Lamed, R.; Keinan, E.; Zeikus, J. E. *Enzyme Microb. Technol.* **1981**, *3*, 144.

51.   Willaert, J. J.; Lemiere, G. L.; Joris, L. A.; Lepoiure, J. A.; Alderweireldt, F. C. *Bioorg. Chem.* **1988**, *16*, 223.

52.   Cronin, C. N.; Malcolm, B. A.; Kirsch, J. F. *J. Am. Chem. Soc.* **1987**, *109*, 2222.

53.   Wilks, H. M.; Hart, K. W.; Freeney, R.; Dunn, C. R.; Muirhead, H.; Chia, W. N.; Barstow, D. A.; Atkinson, T.; Clarkee, A. R.; Holbrook, J.J . *Science* **1988**, *242*, 1541.

54.   Luyten, M. A.; Bur, D.; Wynn, H.; Parris, W.; Gold, M.; Friesen, J. D.; Jones, J. B. *J. Am. Chem. Soc.* **1989**, *111*, 6800.

55.   Scrutton, N. S.; Berry, A.; Perham, R. N. *Nature* **1990**, *343*, 38.

56.   Corbier, C.; Clermont, S.; Billard, C. P.; Skarzynski, T.; Branlant, C.; Wonacott, A.; Branlant, G. *Biochemistry* **1990**, *29*, 7101.

57.   Wells, J. A.; Estell, D. A. *Trends Biochem. Sci.* **1988**, *13*, 291.

58.   Russell, A. J.; Fersht, A. R. Nature 1987, 328, 496; Fersht, A.; Winter, G. *TIBS* **1992**, 92.

59.   Pantoliano, M. W.; Whitlow, M.; Wood, J. F.: Dodd, S. W.; Hardman, K. D.; Rollence, M. L.; Bryan, P. N. *Biochemistry* **1989**, *28*, 7205.

60.   Bonneau, P. R.; Graycar, T. P.; Estell, D. A.; Jones, J. B. *J. Am. Chem. Soc.* **1991**, *113*, 1026.

61.   Zhong, Z.; Liu, J. L.-C.; Dinterman, L. M.; Finkelman, M. A. J.; Mueller, W. T.; Rollence, M. L.; Whitlow, M.; Wong, C.-H. *J. Am. Chem. Soc.* **1991**, *113*, 683.

62.   Nagasawa, T.; Yamada, H. *Pure Appl. Chem.* **1990**, *62*, 1441.

63.   Gibson, D. T.; Hensley, M.; Yoshioka, H.; Mabry, T. J. *Biochemistry*, **1970**, *9*, 1626.

64.   (a) Bradshaw, C.; Shen, G.-J.; Wong, C.-H. *Bioorg. Chem.* **1991**, *19*, 398. (b) Bradshaw, C. W.; Fu, H.; Shen, G.-J.; Wong, C.-H. *J. Org. Chem.* **1992**, *57*, 1526. (c) Bradshaw, C. W.; Hummel, W.; Wong, C.-H. *J. Org. Chem.* **1992**, *57*, 1532.

65.   Asano, Y.; Nakazawa, A.; Kato, Y.; Kondo, K. *Angew. Chem. Int. Ed. Engl.* **1989**, *28*, 450.

66.   Groeger, U.; Drauz, K.; Klenk, H. *Angew Chem. Int. Ed. Engl.* **1990**, *29*, 417.

67.   Steinke, D.; Kula, M-R. *Angew. Chem. Int. Ed. Engl.* **1990**, *29*, 1139.

68.   Miyamoto, K.; Ohta, H. *J. Am. Chem. Soc.* **1990**, *112*, 4077.

69.   (a) Schmid, R. D. *Adv. Biochem. Eng.* **1979**, *12*, 41. (b) Martinek, K.; Berezin, I. B. *Russ.Chem. Rev.* **1980**, *49*, 385. (c) Ahern, T. J.; Klibanov, A. M. *Science* **1985**, *228*, 1280.

70. Anfinsen, C. B. *Science* **1973**, *181*, 223.

71. Mozhaev, V. V.; Martinek, K. *Enzyme Microb. Tech.* **1984**, *6*, 50.

72. Pollak, A.; Blumenfeld, H.; Wax, M.; Baughn, R. L.; Whitesides, G. M. *J. Am. Chem. Soc.* **1980**, *102*, 6324.

73. A product of Rohm GmbH. For example, see Hannibal-Friedrich, O.; Chun, M.; Sernetz, M. *Biotech. Bioeng.* **1980**, *22*, 157.

74. a) Wang, P.; Hill, T. G.; Wartchow, C. A.; Huston, M. E.; Oehler, L. M.; Smith, M. B.; Bednarski, M. D.; Callstrom, H. R. *J. Am. Chem. Soc.* **1992**, *114*, 378. b) Clair, N. L. S.; Navia, M. A. *J. Am. Chem. Soc.* **1992**, *114*, 7314.

75. For use in biphasic systems: Oyama, K.; Nishimura, S.; Nonaki, Y.; Kihara, K.; Hasimoto, T. *J. Org. Chem.* **1981**, *46*, 5241; in organic solvent: Barbas, C. F., III; Wong, C.-H. *J. Chem. Soc., Chem. Commun.* **1987**, 533.

76. Bednarski, M. D.; Chenault, H. K.; Simon, E. S.; Whitesides, G. M. *J. Am. Chem. Soc.* **1987**, *109*, 1283.

77. Chenault, H. K.; Simon, E. S.; Whitesides, G. M. *Biotech. Genetic Engineering Rev.* **1988**, *6*, 221.

78. Crans, D. C.; Kazlauskas, R. L.; Hirschbein, B. L.; Wong, C.-H.; Abril, O.; Whitesides, G. M. *Method. Enzymol.* **1987**, *136*, 263.

79. Drueckhammer, D. G.; Hennen, W. J.; Pederson, R. L.; Barbas, C. F., III; Gautheron, C. M.; Krach, T.; Wong, C.-H. *Synthesis* **1991**, *7*, 499.

80. Heidlas, J. E.; Williams, K. W.; Whitesides, G. M. *Acc. Chem. Res.* **1992**, *25*, 307.

81. Matso, J. R.; Wong, C.-H. *J. Org. Chem.* **1986**, *51*, 2388.

82. Lee, L. G.; Whitesides, G. M. *J. Am. Chem. Soc.* **1985**, *107*, 6999.

83. Wichmann, R.; Wandrey, C.; Buckmann, A. F.; Kula, M. R. *Biotech. Bioeng.* **1981**, *23*, 2789.

84. Taschner, M. J.; Black, D. J. *J. Am. Chem. Soc.* **1988**, *110*, 6892.

85. Colbert, J. E.; Katopodis, A. G.; May, S. W. *J. Am. Chem. Soc.* **1990**, *112*, 3993.

86. (a) Patel, S. S.; Conlon, H. D.; Walt, D. R. *J. Org. Chem.* **1986**, *51*, 2842. (b) Billhardt, U.-M.; Stein, P.; Whitesides, G. M. *Bioorg. Chem* **1989**, *17*, 1.

87. a) Kastle, J. H.; Loevenhart, A. S. *Am. Chem. J.* **1900**, *24*, 491; Iwai, M.; Tsujisaka, Y.; Fukumoto, J. *J. Gen. App. Microbiol.* **1964**, *10*, 13. b) Bergmann, M.; Fraenkel-Conrat, H. *J. Biol. Chem.* **1937**, *119*, 707; Fruton, J. S. *Adv. Enzymol.* **1982**, *53*, 239 and references cited. c) Dastol, F. R.; Musto, N. A.; Price, S. *Arch. Biochem. Biophys.* **1966**, *115*, 44; Dastol, F. R.; Price, S. *Arch. Biochem. Biophys.* **1967**, *118*, 163. d) Klibanov, A. M.; Samokhin, G. P.; Martinek, K.; Berezin, I. V. *Biotechnol. Bioeng.* **1977**, *19*, 1351; Martinek, K.; Semenov, A. N.; Berezin, I. V. *Biochim. Biophys. Acta* **1981**, *658*, 76. e) Oyama, K.; Nishimura, S.; Nonaka, Y.; Kihara, K.-

I.; Hashimoto, T. *J. Org. Chem.* **1981**, *46*, 5242. f) Yokozeki, K.; Yamanaka, S.; Takinami, K.; Hirose, Y.; Tanaka, A.; Sonomoto, K.; Fukui, S. *Eur. J. Appl. Microbiol. Biotechnol.* **1982**, *14*, 1. g) Cambou, B.; Klibanov, A. M. *J. Am. Chem. Soc.* **1984**, *106*, 2687. j) Kirchner, G.; Scollar, M. P.; Klibanov, A. M. *J. Am. Chem. Soc.* **1985**, *107*, 7072; Langrand, G.; Secci, M.; Buono, G.; Baratti, J.; Triantaphylides, C. *Tetrahedron Lett.* **1985**, *26*, 1857. h) Zaks, A.; Klibanov, A. M. *Science* **1984**, *224*, 1249. i) Gatfield, I. L. *Ann. N.Y. Acad. Sci.* **1984**, *434*, 569. k) Koshiro, S.; Sanomoto, K.; Taneka, A.; Fukui, S. *J. Biotechnol.* **1985**, *2*, 47; Fukui, S.; Tanaka, A. *Endeavor* **1985**, *9*, 10. l) Zaks, A.; Klibanov, A. M. *Proc. Natl. Acad. Sci.* **1985**, *82*, 3192.

88.   a) Torchilin, V. P.; Martinek, K. *Enzyme Microbiol. Technol.* **1979**, *1*, 74. (b) Luis, P. L.; Laane, C. *TIBTECH* **1986**, 153. (c) Luisi, P. L.; Luthi, P. *J. Am. Chem. Soc.* **1984**, *106*, 7285. (d) For a review, see Waks, M. *Proteins* **1986**, *1*, 4.

89.   (a) Careri, G.; Gratton, E.; Yang, P.-H.; Rupley, J. A. *Nature* **1980**, *284*, 572. (b) Schinkel, J. E.; Downer, N. W.; Rupley, J. A. *Biochemistry* **1985**, *24*, 352.

90.   Laane, C.; Boeren, S.; Vos, K. *Trends Biotechnol.* **1985**, *3*, 251.

91.   Klibanov, A. M. *Acc. Chem. Res.* **1990**, *23*, 114.

92.   (a) Zaks, A.; Klibanov, A. M. *J. Biol. Chem.* **1988**, *263*, 3194.   (b) Zaks, A.; Klibanov, A. M. *J. Biol. Chem.* **1988**, *263*, 8017.

93.   Burke, P. A.; Smith, S. O.; Bachovchin, W. W.; Klibanov, A. M. *J. Am. Chem. Soc.* **1989**, *111*, 8290.

94.   Kanerva, L. .; Klibanov, A. M. *J. Am. Chem. Soc.* **1989**, *111*, 6864.

95.   (a) Adams, K. A. H.; Chung, S.-H.; Klibanov, A. M. *J. Am. Chem. Soc.* **1990**, *112*, 9418.   (b) Fitzpatrick, P. A.; Steinmetz, A. C. U.; Ringe, D.; Klibanov, A. M. *Proc. Natl. Acad. Sci. USA* **1993**, *90*, 8653.

96.   Sakurai, T.; Margolin, A. L.; Russell, A. J.; Klibanov, A. M. *J. Am. Chem. Soc.* **1988**, *110*, 7236.

97.   Parida, S.; Dordick, J. S. *J. Am. Chem. Soc.* **1991**, *113*, 2253.

98.   Fitzpatrick, P. A.; Klibanov, A. M. *J. Am. Chem. Soc.* **1991**, *113*, 3166.

99.   (a) Randolph, T. W.; Clark, D. S.; Blanch, H. W.; Prausnitz, J. M. *Science* **1988**, *239*, 387.   (b) Kamat, S. V.; Beckman, E. J.; Russell, A. J. *J. Am. Chem. Soc.* **1993**, *115*, 8845.

100. Wang, Y. F.; Wong, C.-H. *J. Org. Chem.* **1988**, 53, 3127.

101. Kieboom, A. P. G. *Recl. Trav. Chim. Pays-Bas* **1988**, *107*, 347.

102. (a) Martinez, P.; Arnold, F. H. *J. Am. Chem. Soc.* **1991**, *113*, 6336. (b) Chen, K.; Arnold, F. H. *Bio/Technology* **1991**, *9*, 1073. (c) Wangikar, P. P.; Graycar, T. P.; Estell, D. A.; Clark, D. S.; Dordick, J. S. *J. Am. Chem. Soc.* **1993**, *115*, 12231.

103. (a) Wong, C.-H.; Haynie, S. L.; Whitesides, G. M. *J. Org. Chem.* **1982**, *47*, 5416. (b) Ichikawa, Y.; Shen, G.-J.; Wong, C.-H. *J. Am. Chem. Soc.* **1991**, *113*, 4698. (c) Ichikawa, Y.; Liu, J. L.-C.; Shen, G.-J.; Wong, C.-H. *J. Am. Chem. Soc.* **1991**, *113*, 6300.

104. (a) Scott, A. I. *Pure Appl. Chem.* **1993**, *65*, 1299; Scott, I. A. *Angew. Chem. Int. Ed. Engl.* **1993**, *32*, 1223. (b) Moradian, A.; Benner, S. A. *J. Am. Chem. Soc.* **1992**, *114*, 6980. (c) Fessner, W.-D.; Walter, C. *Angew. Chem. Int. Ed. Engl.* **1992**, *31*, 614. (d) Wei, L. L.; Goux, W. J. *Bioorg. Chem.* **1992**, *20*, 62.

105. Bulow, L.: Mosbach, K. *TIBTECH* **1991**, *9*, 226.

106. (a) Bailey, J. E. *Science* **1991**, *252*, 1668. (b) Stephanopoulos, G.; Sinskey, A. J. *TIBTECH* **1993**, *11*, 392.

107. Anderson, S.; Marks, C. B.; Lazarus, R.; Miller, J.; Stafford, K.; Seymour, J.; Light, D.; Rastetter, W.; Estell, D. *Science* **1985**, *230*, 144.

108. (a) Draths, K. M.; Frost, J. W. *J. Am. Chem. Soc.* **1990**, *112*, 1657. (b) Draths, K. M.; Frost, J. W. *J. Am. Chem. Soc.* **1990**, *112*, 9630.

109. Floss, H. G. *TIBTECH* **1987**, *5*, 111.

110. Martin, J. F. *TIBTECH* **1987**, *5*, 306.

111. (a) Fayerman, J. T. *Bio/Technology* **1986**, *4*, 786. (b) Fishman, S. E.; Cox, K.; Larson, J. L.; Reynolds, P. A.; Seno, E. T.; Yeh, W. K.; Frank, R. V.; Hershberger, C. L. *Proc. Natl. Acad. Sci.* **1987**, *84*, 8248. (c) Skatrud, P. L.; Tietz, A. J.; Ingolia, T. D.; Cantwell, C. A.; Fisher, D. L.; Chapman, J. L.; Queener, S. W. *Bio/Technology* **1989**, *7*, 477.

112. Mortlock, R.P. *TIBTECH* **1986**, 65.

113. Ramos, J. L.; Wasserfallen, A.; Rose, K.; Timmis, K. N. *Science* **1987**, *235*, 593.

114. (a) Muttler, M. *Angew. Chem. Int. Ed. Engl.* **1985**, *24*, 639; Degrado, W.F. *Adv. Protein Chem.* **1988**, *39*, 51; Hahn, K. S.; Klis, W. A.; Stewart, J. M. *Science* **1990**, *248*, 1544. (b) Atassi, M. Z.; Manshouri, T. *Proc. Natl. Acad. Sci. USA* **1993**, *90*, 8282.

115. (a) Ikeda, S.; Weinhouse, M. I.; Janda, K. D.; Lerner, R. A.; Danishefsky, S. J. *J. Am. Chem. Soc.* **1991**, *113*, 7763. (b) Fernholz, E.; Schloeder, D.; Liu, K. K.-C.; Bradshaw, C. W.; Huang, H.; Janda, K. D.; Lerner, R. A.; Wong, C.-H. *J. Org. Chem.* **1992**, *57*, 4756. (c) Nakayama, G. R.; Schultz, P. G. *J. Am. Chem. Soc.* **1992**, *114*, 780. (d) Janda, K. D.; Shevlin, C. G.; Lerner, R. A. *Science* **1993**, *259*, 490.

116. (a) Huse, W. D.; Sastry, L.; Iverson, S. A.; Kang, A. S.; Alting-Mees, M.; Burton, D. R.; Benkovic, S. J.; Lerner, R. A. *Science* **1989**, *246*, 1275. (b) Winter, G.; Milstein,

C. *Nature* **1991**, *349*, 293.   (c) Barbas, C. F., III; Kang, A. S.; Lerner, R. A.; Benkovic, S. J. *Proc. Natl. Acad. Sci.* **1991**, *88*, 7978.

117.   Bird, R. E.; Hardman, K. D.; Jacobson, J. W.; Johnson, S.; Kaufman, B. M.; Lee, S.-M.; Lee, T.; Pope, S. H.; Riordan, G. S.; Whitlow, M. *Science* **1988**, *242*, 423.

118.   Bowdish, K.; Tang, Y.; Hicks, J. B.; Hilvert, D. *J. Biol. Chem.* **1991**, *266*, 11901.

119.   (a) Noren, C. J.; Anthony-Cahill, S.J.; Griffith, M. C.; Schultz, P. G. *Science* **1989**, *244*, 182.   (b) Ellman, J. A.; Mendel, D.; Schultz, P.G. *Science* **1992**, *255*, 197.   (c) Bain, J.D.; Glabe, C.G.; Dix, T.A.; Chamberlin, A. R. *J. Am. Chem. Soc.* **1989**, *111*, 8013.   (d) Hecht, S. J. *Acc. Chem. Res.* **1992**, *25*, 545.   (e) Brunner, J. *Chem. Soc. Rev.* **1993**, 183.

120.   (a) Kaiser, E. T.; Lawrence, D. S. *Science* **1984**, *226*, 505.   (b) Kaiser, E. T. *Angew. Chem. Int. Ed. Engl.* **1988**, *27*, 913.

121.   (a) Chen, C. H. B.; Sigman, D. S. *Science* **1987**, *237*, 1197.   (b) Corey, D. R.; Schultz, P. G. *Science* **1987**, *238*, 1401.   (c) Pollack, S. J.; Schultz, P. G. *J. Am. Chem. Soc.* **1989**, *111*, 1929.

122.   Zuckermann, R. N.; Corey, D. R.; Schultz, P. G. *J. Am. Chem. Soc.* **1988**, *110*, 1614.

123.   Wilson, M. E.; Whitesides, G. M. *J. Am. Chem. Soc.* **1978**, *100*, 306.

124.   (a) West, J. .; Scholten, J.; Stolowich, N. J.; Hogg, J. L.; Scott, A. I.; Wong, C.-H. *J. Am. Chem. Soc.* **1988**, *110*, 3709.   (b) Zhong, Z.; Bibbs, J. A.; Yuan, W.; Wong, C.-H. *J. Am. Chem. Soc.* **1991**, *113*, 2259.

125.   (a) Nakatsuka, T.; Sasaki, T.; Kaiser, E. T. *J. Am. Chem. Soc.* **1987**, *109*, 3808.   (b) Wong, C.-H.; Schuster, M.; Wang, P.; Sears, P. *J. Am. Chem. Soc.* **1993**, *115*, 5893.   (c) Abrahmsen, L.; Tom, J.; Burnier, J.; Butcher, K. A.; Kossiakoff, A.; Wells, J. A. *Biochemistry* **1991**, *30*, 4151.

126.   Wu, Z. P.; Hilvert, D. *J. Am. Chem. Soc.* **1989**, *111*, 4513.

127.   (a) Smith, G. P. *Science* **1985**, *228*, 1315.   (b) McCafferty, J.; Griffiths, A. D.; Winter, G.; Chiswell, D. J. *Nature* **1990**, *348*, 552.   (c) Bass, S.; Greene, R.; Wells, J. A. *Proteins* **1990**, *8*, 309.   (d) McCafferty, J.; Jackson, R. H.; Chiswell, D. J. *Protein Eng.* **1991**, *4*, 286.   (e) Matthews, D. J.; Wells, J. A. *Science* **1993**, *260*, 1113.

128.   (a) Cech, T. R. *Science* **1987**, *236*, 1532.   (b) Altman, S. *Angew. Chem. Int. Ed. Engl.* **1990**, *29*, 749.   (c) Piccirilli, J. A.; McConnell, T. S.; Zaug, A. J.; Noller, H. F.; Cech, T. R. *Science* **1992**, *256*, 1420.   (d) Cech, T. R. *Current Opinion in Structural Biology* **1992**, *2*, 605.   (e) Beaudry, A. A.; Joyce, G. F. *Science* **1992**, *257*, 635.

Chapter 2. Use of Hydrolytic Enzymes: Amidases, Proteases, Esterases, Lipases, Nitrilases, Phosphatases, Epoxide Hydrolases

Hydrolytic enzymes are the biocatalysts most commonly used in organic synthesis. Of particular interest among the classes of hydrolytic enzymes are amidases, proteases, esterases and lipases; these enzymes catalyze the hydrolysis and formation of ester and amide bonds.

These reactions may or may not proceed through covalent intermediates. In general, there are four types of proteases and the mechanism of each is known (Figure 1): serine proteases, thiol

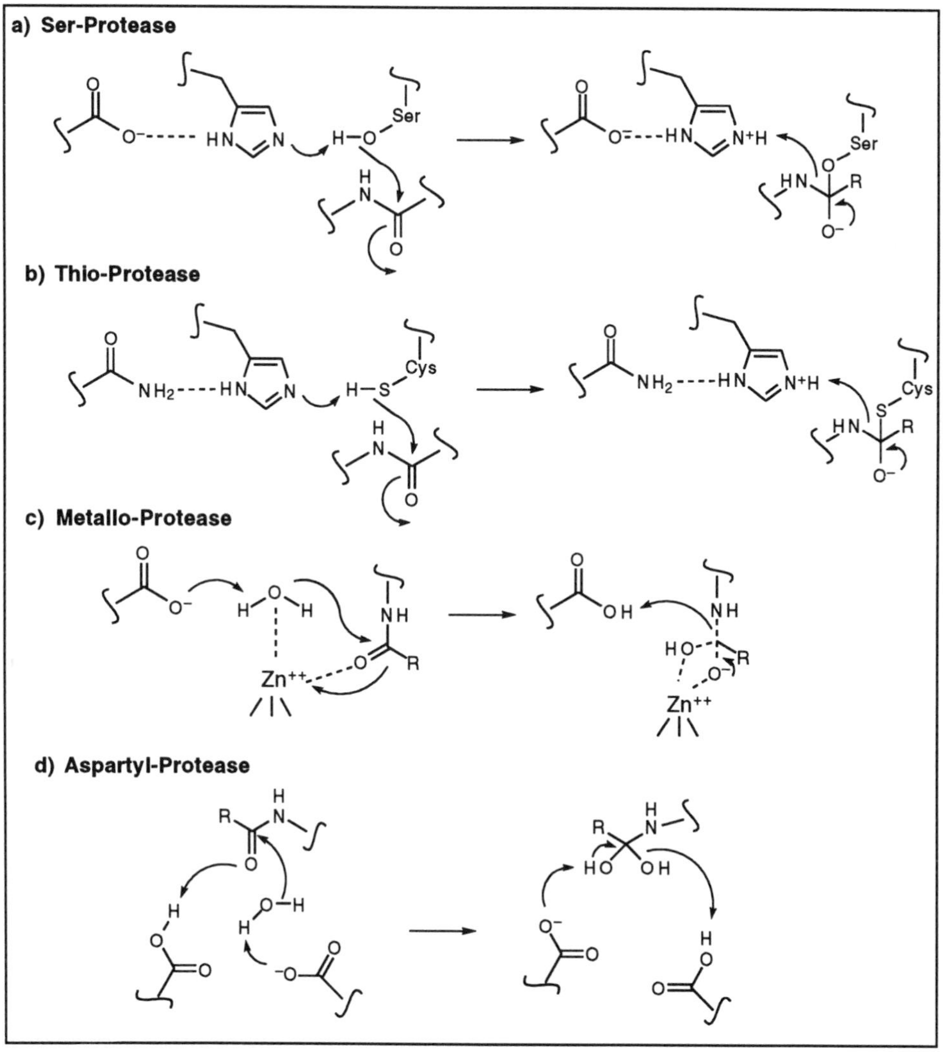

Figure 1. Mechanism of protease reactions.

proteases, metalloproteases and aspartyl proteases. The serine protease contains a catalytic triad in the active site composed of Asp, His and Ser, with Ser acting as a nucleophile. The reaction proceeds through an initial step that forms a covalent O-acyl intermediate; this intermediate is deacylated with water in a second distinct catalytic step. In the hydrolysis of peptide bonds, the acylation is often rate limiting; in the hydrolysis of esters, deacylation of the acylated enzyme is rate limiting. In addition to water, other nucleophiles—for example alcohols, amines or thiol groups—can also react with the acyl intermediate generated from an ester substrate and form new products of transacylation—esters, peptides, or thioesters. Transacylation reactions are often performed in media that contain high concentrations of organic solvents to minimize competing hydrolysis.

Examples of serine proteases include trypsin, chymotrypsin, subtilisin, α-lytic enzymes, elastase, and V8 protease. The serine-type esterases include pig liver esterase, cholesterol esterase, acetylcholine esterase, lipase, carboxypeptidase Y and penicillin acylase.

Thioproteases are similar to the serine-type proteases, except that the nucleophilic group is Cys instead of Ser. Examples of thioproteases are papain, clostripain, and cathepsin.

Metalloproteases often utilize a Zn^{++} ion as a Lewis acid that is coordinated in the catalytic step to the nucleophilic water (to increase its acidity) and to the carbonyl group of the scissile bond. A carboxylate group often serves as a base to abstract a proton from the nucleophilic water molecule. No covalent intermediate is formed. Thermolysin, acylases and carboxypeptidase A represent this type of enzyme.

Aspartyl proteases utilize a carboxylate group as a general base and a carboxylic acid as a general acid for hydrolysis. Although it is generally believed that no covalent intermediate is formed in the reaction, controversial results to the contrary have been reported. Examples of aspartyl proteases include pepsin and HIV protease.

1. Amidases

1.1 Acylases

Acylase I (aminoacylase, EC 3.5.1.1.4) from porcine kidney (PKA) and *Aspergillus sp.* (AA) are commercially available. They catalyze the enantioselective hydrolysis of N-acyl-L-amino acids.[1] A recent study of the hydrolysis of over 50 N-acyl amino acids and analogs[2] indicated that, in general, the enzyme is highly selective for L-α-amino acids, cleaving the corresponding N-acetyl, N-chloroacetyl, N-trifluoroacetyl, and N-methoxyacetyl derivatives. The N-acylated β-amino acids, amino acid esters or amides are not substrates. Representative examples of acylase I-catalyzed reactions are listed in Figure 2. N-Acyl α-methyl-α-amino acids are substrates for the

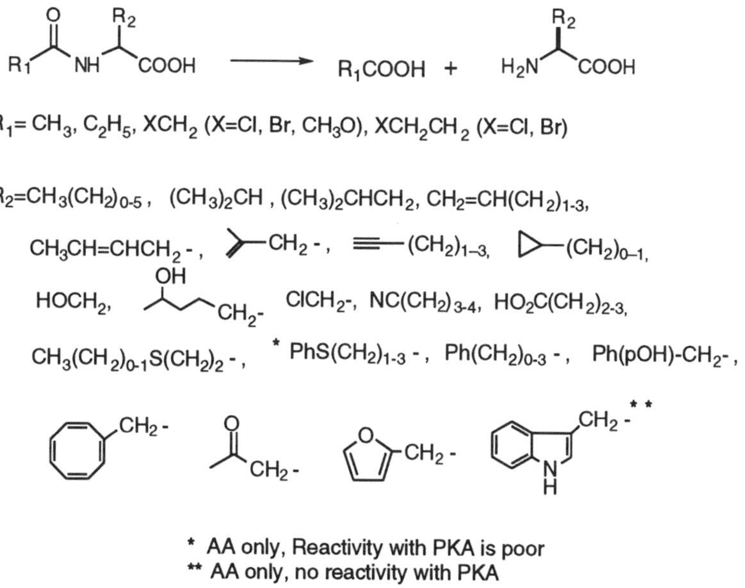

$R_1 = CH_3, C_2H_5, XCH_2$ (X=Cl, Br, CH_3O), XCH_2CH_2 (X=Cl, Br)

$R_2 = CH_3(CH_2)_{0-5}$, $(CH_3)_2CH$, $(CH_3)_2CHCH_2$, $CH_2=CH(CH_2)_{1-3}$,

$CH_3CH=CHCH_2$-, ⟩–CH_2-, ≡–$(CH_2)_{1-3}$, ▷–$(CH_2)_{0-1}$,

$HOCH_2$, (structure)CH_2-, $ClCH_2$-, $NC(CH_2)_{3-4}$, $HO_2C(CH_2)_{2-3}$,

$CH_3(CH_2)_{0-1}S(CH_2)_2$-, * $PhS(CH_2)_{1-3}$ -, $Ph(CH_2)_{0-3}$ -, $Ph(pOH)$-CH_2-,

(cyclooctatetraenyl)CH_2- , (acetyl)CH_2- , (furyl)CH_2- , (indolyl)CH_2- **

* AA only, Reactivity with PKA is poor
** AA only, no reactivity with PKA

Figure 2. Substrates for acylase I.

kidney acylase,[2,3] but not for the *Aspergillus* enzyme. The enzyme contains Zn^{++}, but Co^{++} also accelerates the enzymatic reaction. Acylase I tolerates high concentrations of organic solvent. The corresponding *O*-acyl derivatives and *N*-acyl-*N*-alkyl amino acids, including *N*-acyl proline, are generally not substrates. PKA, but not AA, accepts α-methyl-α-amino acids as weak substrates. *Aspergillus* acylase I was also used in the resolution of β-cyclooctatetraenylalanine, an analog of phenylalanine.[4]

A new bacterial acylase[5] from *Comamonas testosteroni* (DSM5416) catalyzes the hydrolysis of the L-enantiomer of acetyl and chloroacetyl proline, and the corresponding azetidine-2-carboxylic acid and pipecolic acid. Gly-Pro, Z-Gly-Pro and *N*-chloroacetyl-L-thiazolidine-4-carboxylic acid are also substrates. This enzyme is similar to that isolated from a *Pseudomonas sp.*[5b]

In addition to the L-selective acylases, D-aminoacylases has been reported.[6] The enzyme isolated from *Alcaligenes faecalis* DAI catalyzes the hydrolysis of some racemic *N*-acyl-amino acids (such as *N*-acetyl, *N*-benzyl and *N*-benzyloxycarbonyl amino acids) to D-amino acids. Neutral amino acids such as Met, and Phe are good substrates, small and polar amino acids such as Gly, Ala, Ser, Thr, Glu, Asp, Lys and Arg are not substrates. The Boc group is not accepted as substrates.[7]

1.2 Aminopeptidase

Aminopeptidases are generally Zn^{++}-containing enzymes that catalyze the hydrolysis of L-amino acid amides selectively. Some recently discovered aminopeptidases possess good properties for organic synthesis. For example, an aminopeptidase from *Pseudomonas putida* catalyzes the hydrolysis of a variety of L-amino acid amides[8a,b] and an amidase from *Mycobacterium neoaurum* (ATCC 25795) catalyzes the hydrolysis of L-α-methyl-α-amino acid amides.[8c] Another aminopeptidase from *Achromobacter sp.* selective for D-alanine amides or esters has recently been reported.[9] In the *Pseudomonas* L-aminopeptidase-catalyzed enantioselective hydrolysis of α-amino acid amides, several common and uncommon L-amino acids such as L-allylglycine and L-cyclopentylglycine have been prepared in very high ee (Figure 3a).[8b] The racemic amino acid amides used as substrates were prepared from α-methoxyglycine amides via Lewis acid-mediated coupling reactions with allylsilanes.[8b] In the enzymatic reaction, the unreacted D-amide is isolated as a Schiff base of benzaldehyde; following racemization, the racemate is again subjected to the enzymatic resolution. New amidases that catalyze the hydrolysis of the cyclic amide 2-azabicyclo-[2.2.1]hept-5-en-3-one are available as whole cell preparations[10] (Figure 3b).

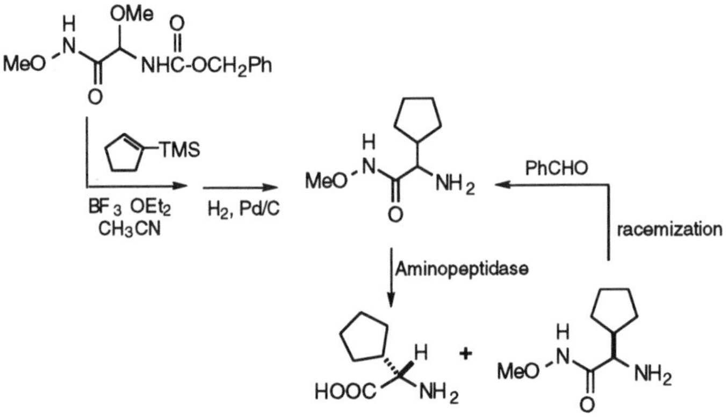

Figure 3a. Representative *Pseudomonas* aminopeptidase reaction.

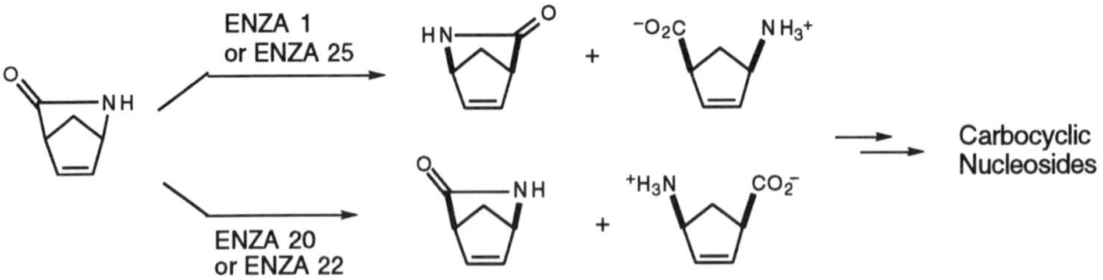

Figure 3b. New amidase reaction.

1.3 Hydantoinase

Both D- and L-amino acids can be prepared from the corresponding 5-substituted DL-hydantoins via hydantoinase-catalyzed hydrolysis (Figure 4). The advantage of this system is that the unreacted hydantoin racemizes spontaneously , and enantioselective hydrolysis of the interconverting enantiomers leads to complete conversion of a racemic mixture to one enantiomer.[11-13] The initial enzymatic product, an *N*-carbamoyl-amino acid, can be further converted to the corresponding amino acid chemically or biochemically (by enzymatically catalyzed reaction using carbamoylase from the same species).

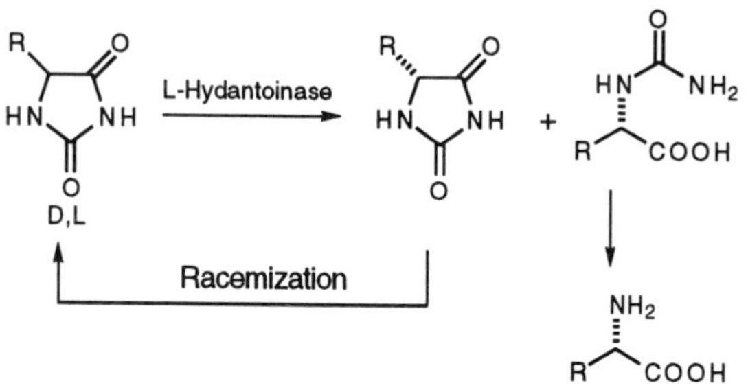

Figure 4.

1.4 Penicillin Acylase

Penicillin acylase (EC 3.5.1.11) from *E. coli* (ATCC 9637) is a serine type of enzyme catalyzing the hydrolysis of benzyl penicillin to 6-aminopenicillanic acid (6-APA); 6-APA can be converted to different acyl penicillin derivatives, a process at various times used in the pharmaceutical industry. The enzyme also catalyzes the cleavage of the *N*- or *O*-phenylacetyl protecting group from α-amino acids,[14a] β-amino acids,[14b] γ-amino acids,[14c] 1-aminoethylphosphonic acid,[14d] α-methyl amino acids,[15] peptides,[16] amines,[17] alcohols,[18] or sugars.[19] In addition to the phenylacetyl group, *p*-hydroxyphenylacetyl, furylmethyl, 2-thienylmethyl, D-α-aminobenzyl, and *n*-propoxymethyl groups can also be used as enzymatically removable protecting groups. In general, penicillin acylase accepts substrates with stereostructures related to L-amino acid. Thus, compounds with configuration related to L-amino acid are substrates (Figure 5). In peptide synthesis, the use of a phenylacetyl group as a non-

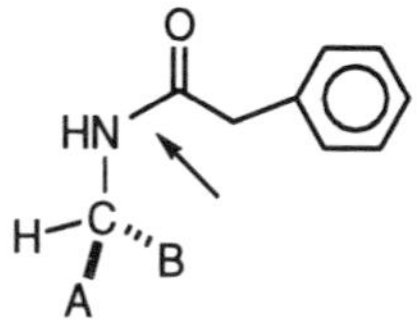

A = -CONH$_2$, -COOR, -CH$_2$CO$_2$H, -(CH$_2$)$_n$COOH (n = 0-2)
B = Alkyl groups

e.g. CH$_3$, Ph, CF$_3$, C$_2$F$_5$, C$_3$F$_7$, 2-F-Ph, 4-F-Ph
for β-amino acids (>99% ee) and
CH≡C- (E > 100), CH$_2$=C=CH- (E = 20),
CH$_2$=CH- (E = 17) for γ-amino acids

Figure 5. Preferred substrate stereoisomer for penicillin acylase-catalyzed hydrolysis.

urethane *N*-protecting group may cause racemization. Enzymatic coupling of *N*-phenylacetyl amino acids, followed by enzymatic deprotection of the phenylacetyl group with the acylase, avoids the racemization problem; the amide bonds and C-terminal ester groups are unaffected.[19,20] The enzyme was also used in the acylation of amines with methyl phenoxy acetate.[21]

1.5   Carboxypeptidases

Carboxypeptidases are exopeptidases that catalyze the release of C-terminal L-amino acid from peptides.[22-26] Three carboxypeptidases have been used in protein chemistry: carboxypeptidase A (a Zn^{++} enzyme that is specific for neutral or acidic amino acids), carboxypeptidase B (specific for basic amino acids), and carboxypeptidase Y (a serine-type enzyme from Baker's yeast that accepts a broad spectrum of substrates, including Pro). Carboxypeptidase C[25] from orange leaves has been shown to have a broad substrate specificity useful for the kinetically-controlled synthesis of peptides (see below). These carboxypeptidases may also be used in the resolution of unnatural amino acids. For example, carboxypeptidase Y catalyzes the hydrolysis of trifluoroacetyl 7-fluorotryptophan to give the corresponding L-amino acid.[26]

2.   Protease-catalyzed Peptide Synthesis

Proteases have been used in peptide synthesis[27-32] since 1938. The advantages of enzymatic peptide synthesis are freedom from racemization, minimal requirements for carboxyl activation and side-chain protection, mild reaction conditions, and high regio- and stereoselectivity. Enzyme immobilization allows the recovery of catalyst in large-scale processes. The disadvantages of protease catalysis are that the amidase activity of proteases causes a secondary hydrolysis of the growing peptides and that the substrates of proteases are often limited only to natural L-amino acids and their derivatives. Table 1 lists a number of different types of proteases, with their specificities used as catalysts in synthesis.

2.1   Strategies for Protease-Catalyzed Peptide Coupling

Three strategies are often used in protease-catalyzed peptide synthesis:[31] one is the direct reversal of the catalytic hydrolysis of peptides (i.e. a thermodynamic approach, eq. 1); the second is the aminolysis of N-protected amino acid or peptide esters (i.e. a kinetic approach, eq. 2); the third is transpeptidation (eq. 3).

The thermodynamic approach is an endergonic process, and manipulation of reaction conditions is required to increase the yield of product. The addition of water-miscible organic solvents decreases the extent of ionization of the carboxyl component, and thereby increases the concentration of substrate available for reaction.[32] The use of a biphasic system,[33] reverse micelles,[34] anhydrous media containing a minimal amount of water[35] or water mimics,[36] or the selection of appropriate N- or C-protecting groups to reduce the solubility of products are often employed to improve yields.

Table 1. Common proteases and their preferred cleavage sites.

Protease	Type	Preferred Cleavage Sites
α-chymotrypsin and subtilisins	Ser	-Trp(Tyr,Phe,Leu,Met) ↓ Xaa-
elastase	Ser	-Ala(Ser,Met,Phe) ↓ Xaa-
pepsin	Asp	-Phe(Tyr,Leu) ↓ Leu(Phe)-
thermolysin	metallo	-Phe(Gly,Asp,Leu) ↓ Leu(Phe)-
papain	Cys	-Phe(Leu,Val)-Xaa ↓ Xaa-
trypsin	Ser	-Arg(Lys) ↓ Xaa-
clostripain	Cys	-Arg ↓ Xaa-
endoprotease Lys-C (*Achromobacter*)	Ser	-Lys ↓ Xaa-
endoprotease Glu-C (V8 protease)	Ser	-Glu (Asp) ↓ Xaa-
carboxypeptidase Y	Ser	-Xaa ↓ Xaa-OH
carboxypeptidase B	metallo	-Xaa ↓ [Arg,Lys]-OH
carboxypeptidase A	metallo	-Xaa ↓ [Asp,Glu,Phe,Leu]-OH
aminopeptidase M	metallo	H_2N-Xaa ↓ Xaa-
pyroglutamate-aminopeptidase	Cys	pGlu ↓ Xaa-
cathepsin C	Cys	H_2N-Xaa-Xaa ↓ Xaa-
proline iminopeptidase	Ser	Pro ↓ Xaa-

$$RCO_2^- + \overset{+}{H_3}N\text{-}R' \rightleftharpoons RCO_2H + H_2N\text{-}R' \rightleftharpoons RCOHNR' + H_2O \qquad (1)$$

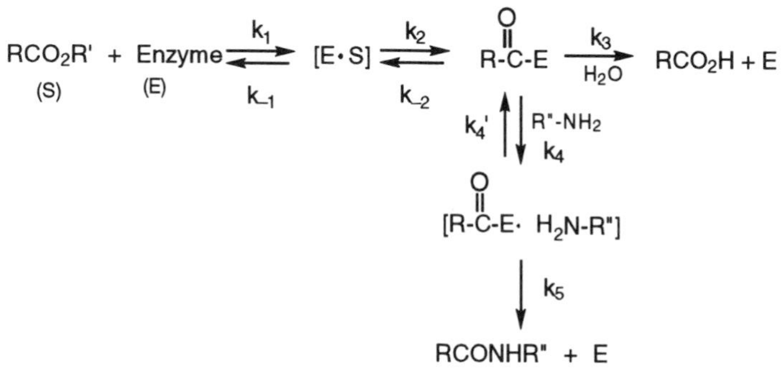

$$RCO_2R' + Enzyme \underset{k_{-1}}{\overset{k_1}{\rightleftharpoons}} [E \cdot S] \underset{k_{-2}}{\overset{k_2}{\rightleftharpoons}} R\text{-}\overset{O}{\overset{\|}{C}}\text{-}E \xrightarrow[H_2O]{k_3} RCO_2H + E \qquad (2)$$

$$k_4' \Big\Uparrow\Big\Downarrow k_4 \quad R''\text{-}NH_2$$

$$[R\text{-}\overset{O}{\overset{\|}{C}}\text{-}E \cdot H_2N\text{-}R'']$$

$$\Big\downarrow k_5$$

$$RCONHR'' + E$$

$$R\text{-}\overset{O}{\overset{\|}{C}}\text{-}NH\text{-}R' + H_2N\text{-}R'' \rightleftharpoons R\text{-}\overset{O}{\overset{\|}{C}}\text{-}NHR'' + H_2NR' \qquad (3)$$

The kinetic approach is faster, and the product yield can be improved by manipulating the reaction conditions in a manner similar to that used in the thermodynamic approach. Aminolysis, however, requires the use of esters as substrates and is limited to those enzymes (e.g. chymotrypsin, trypsin, papain, and subtilisin) that form acyl intermediates.[37] The synthesis of ester substrates for enzymatic peptide coupling can be accomplished by chemical (by solid phase or solution phase synthesis) or by enzymatic methods (as illustrated in a recent report on subtilisin catalyzed peptide segment coupling).[38a] The chemoselective esterification of the C-terminal amino acid in peptides containing acidic amino acids is, however, still a problem. Complementary to this method is the use of N-protected peptide 5(4*H*)-oxazolones (prepared by reaction of N-protected peptides with acetic anhydride) as activated acyl donors in chymotrpsyin-catalyzed segment condensation.[38b] Another method is to use water-soluble N-protecting groups to increase the substrate concentration.[38c]

2.2   Control of Amide Cleavage Activity and Enhancement of Aminolysis in Serine Protease-Catalyzed Reactions

The hydrolyses of esters and amides catalyzed by serine or thiol proteases have similar mechanisms, but different rate-determining steps. Formation of an acyl intermediate in amide hydrolysis is rate-determining and pH independent, while deacylation of the acyl intermediate in ester hydrolysis is the rate-determining step and is general base catalyzed and pH dependent. Enhancement of esterase vs. amidase activity of such proteases at higher values of pH has been utilized in the stepwise synthesis and fragment coupling of peptides via aminolysis, where hydrolysis of the growing polypeptide chain is inhibited.[39] Although esterases without amidase activities (such as lipase from *Candida*[40] or porcine pancreas[41]) can be used as catalysts for aminolysis, the rates of these reactons are generally too slow to be useful.

Serine and thiol proteases[37-38] behave similarly, as nonproteolytic esterases, when water-miscible organic solvents such as dimethylformamide (DMF), dioxane, or acetonitrile are added to the aqueous solution of enzyme. The different effect of water-miscible organic solvents on the esterase and amidase activities of trypsin has been known for some time.[42] The application of mixed aqueous-organic solvents to the kinetically controlled synthesis of peptides has , however, only recently been exploited.[37-38] Investigation of this phenomenon by reducing the water content in a subtilisin-catalyzed aminolysis revealed that the enzymatic activity decreased with decreases in water content and disappeared completely in anhydrous DMF.[37a] Although the enzyme is reasonably stable in anhydrous DMF, the aminolysis reaction with an inactive ester is too slow to be useful for practical synthesis. In the presence of 50-70% DMF, the enzyme is quite stable and active for aminolysis, while the amidase activity is insignificant.[37a] Under these conditions, peptide synthesis can be achieved in high yields, and chain elongation of thiodipeptides using chymotrypsin or subtilisin-catalyzed aminolysis with Val allyl ester has been achieved in ~70% yield.[37b]

To investigate the effect of organic solvents on catalysis further, the kinetic parameters (k_{cat} and K_m) for chymotrypsin-[43] and subtilisin-catalyzed[39] hydrolysis of ester and amide substrates were determined. Organic solvents affected both catalysis and binding. Both k_{cat} and k_{cat}/K_m for the hydrolysis of both amides and esters decreased as the content of organic solvent increased. The rate of decrease for the amide hydrolysis was, however, faster than that for the ester hydrolysis. Amides (with the exception of some hydrophobic amides) are generally more tightly bound to the enzyme than esters in the presence of organic solvents. The binding of ester substrates, however, is affected less by organic solvents. The transition-state energy for hydrolysis of both the ester and amide increases when the organic solvent is added to the solution. The aminolysis is, however, more favorable than hydrolysis. In the direction of amide cleavage, the free-energy barrier for the formation of acyl intermediate is high and rate-determining, and accounts for the irreversible nature of the aminolysis process.

Introduction of a methyl group to the ε-2N of the active-site His of chymotrypsin resulted in a significant change in the enzymatic catalysis:[44] the methylated enzyme (MeCT) favors aminolysis over hydrolysis. The effects of methylation on the enzymatic kinetics are very similar to that of organic solvents.[43] Previous studies on acylation and deacylation reactions catalyzed by MeCT indicated that a functional group with $pK_a=7$ was involved as a general base.[45] A ^{13}C-NMR study indicated the presence of an *O*-acyl intermediate[44] in the MeCT-catalyzed hydrolysis of esters. All these results suggest that the unmethylated N (δ-1) of the active-site His acts as a general base in aminolysis. This reaction may require ring-flipping of the methylated imidazole, a process first suggested by Henderson[45] and later supported by studies on solvent isotope effects,[46] proton inventories,[47] and model systems.[48] The slight change in the orientation of the active-site groups Asp, His, Ser after methylation and ring-flipping may account for the favorable ratio of aminolysis to hydrolysis for MeCT.

Subtilisin can also be converted to an effective acyl transferase for peptide synthesis in aqueous solution via modification of the active-site serine to cysteine[49] (Figure 6) or selenocysteine.[50] Although the acyl intermediate of selenosubtilisin is more reactive than acyl thiosubtilisin in aminolysis, the unacylated enzyme is very sensitive to oxygen and it requires the use of an activated ester (e.g. *p*-nitrophenyl esters) to form the acyl enzyme intermediate. Changing of Pro_{225} to Ala in thiosubtilisin improves the aminolysis activity (ligase activity)[51] by a factor of 10. Another mutant thiosubtilisin with the additional changes of $Met_{50} \rightarrow$ Phe, $Gly_{169} \rightarrow$ Ala, $Asn_{76} \rightarrow$ Asp, and $Asn_{218} \rightarrow$ Ser is thermally stable in aqueous and organic solvents,[52] and methyl esters can be used as acyl donors in aminolysis at 50 °C in aqueous solution.

Both native and thiosubtilisin accept glycosylated amino acids and peptides as donor or acceptor substrates. Incorporation of additional sugar units into glycopeptides formed using glycosyltransferases allows the synthesis of oligosaccharyl peptides.[52]

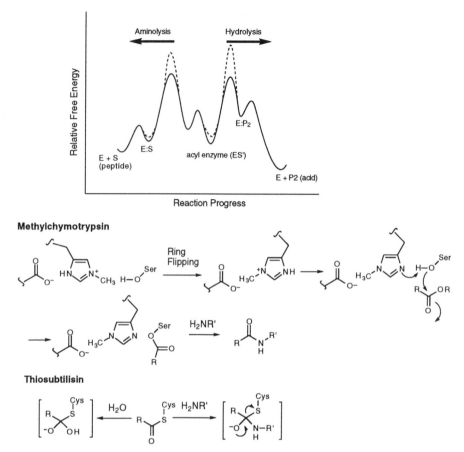

Figure 6. Free energy diagram and mechanisms of methylchymotrypsin and thiosubtilisin.

Kinetic analyses of methyl chymotrypsin and thiosubtilisin in organic solvent-mediated reactions indicated that the acyl intermediates of these modified enzymes have a higher reactivity toward the amine nucleophile than toward water.[53] The thiol protease papain is also more effective for aminolysis than hydrolysis, and has been used to catalyze the *N*-acylation of various α-hydroxy- and α-ketoamines,[54] and thus to provide a new route to peptide isosteres. It is particularly interesting that papain accepts both D- and L-isomers of some unusual amines as nucleophiles.[54]

Aminolysis was improved in alcalase-catalyzed reactions by using *N*-protected amino acid esters in high concentrations of alcohols; under this condition the ester substrate was enzymatically regenerated for aminolysis.[55] Modified trypsin, generated by active-site directed inactivation, was used as a template to condense a polypeptide and an active ester of an amino acid to form a polypeptide active ester for the use in the semi-synthesis of proteins.[56] Ligation of large peptides using proteases also provided new routes to modified polypeptides[57] such as trypsin inhibitor analogs.[57] Similarly, chymotrypsin modified at the active site His-ε$_2$N with a (terpyridine)platinum (II),[58] or trypsin with the active site His methylated at ε2-N[59] or with the active site Ser converted to Cys,[60] may be useful as ligases in aqueous solution. Enhanced aminolysis reactions in ice or at low temperatures in aqueous DMSO solution saturated with KCl[31] have also been reported.[61] High yields of enzymatic peptide synthesis can also be achieved with heterogeneous mixture of substrates formed as eutectic mixture in the presence of adjuvants such as 2-methoxyethyl acetate.[62]

Another way to avoid the problem associated with the endopeptidase activity of proteases during peptide synthesis is to use exopeptidases as catalysts. Carboxypeptidase Y, for example, has been successfully used in the stepwise synthesis of peptides, proceeding from the N to the C terminus.[23] The enzyme from Baker's yeast catalyzes both the hydrolysis of the ester group of N-protected peptide esters[20,63] and aminolysis. It also catalyzes transpeptidation, replacing the C-terminal residue with another amino acid (or its amide or *tert*-butyl ester derivative).[23] Carboxypeptidase C from orange leaves catalyzes the kinetically controlled synthesis of peptides with broad substrate specificity. A number of N-protected dipeptide and tripeptide amides have been prepared.[25] Carboxypeptidase W from wheat bran also catalyzes both the hydrolysis of the ester group of N-protected peptide esters and their aminolysis. It has no exoamidase activity, although endopeptidase activity was observed in the reaction of an N-protected dipeptide amide.[64] Carboxypeptidase Y, however, possesses both activities.

For aminolysis reaction in aqueous solution, the peptide ester used usually needs to have a good ester leaving group in order to both facilitate the formation of the acyl enzyme intermediate for reaction with the amine group (to form a peptide bond) and to minimize the hydrolysis. Good ester leaving groups in subtilisin-catalyzed reactions include *o*-chlorophenyl ester,[49] 1-hydroxypiperidine ester,[38] (prepared via reaction of 1-hydroxypiperidine with peptide attached to a solid support via an oxime ester) and esters of glycolyl-phenylalanine amide.[51] In chymotrypsin reactions, cyanomethyl esters are effective.[43,44] Nonproteases involved in peptide synthesis also hold potential in coupling reactions. These enzymes require ATP or GTP to activate the carboxyl group of an amino acid; they accept various amino acids as nucleophiles in amide bond formation (see below).[65,66]

2.3 Introduction of D- and Unnatural Amino Acids into Polypeptides

In protease-catalyzed peptide coupling, the enzyme is very specific for L-amino acids as the P1 residue (the acyl donor group). The P1' residue (the nucleophile) is more flexible, and both L- and D-amino acids are acceptable. D-Amino acids usually react about one tenth (1-5 μmol/min/mg enzyme) as fast as their L-counterparts.[67,68] Proline and other secondary amines are usually not accepted as donors or acceptors for serine proteases.[69] Peptide segments containing D- or unnatural amino acids at the position remote from the P1 or P1' site, however, can be used as substrates for enzymatic peptide segment coupling.[37,38a] Nonproteolytic enzymes such as esterases and lipases accept both D- and L-amino acid derivatives as weak acyl donors or acceptors.[40]

The enantioselectivity of enzyme catalysis can be altered by solvent engineering. The enantioselectivity of subtilisin, for example, changed dramatically on going from aqueous solution to anhydrous *tert*-amyl alcohol, as measured by the rate ratio (V_L/V_D) of the enzyme-catalyzed hydrolysis and aminolysis of *N*-acetyl-L- and D-phenylalanine chloroethyl ester.[70] Given this observation, D-amino acids could then be incorporated as the P1 residue into the *N*-terminal position of the peptide. The reaction rate, however, is very slow (i.e. <0.1 μmol/min/mg enzyme). The D-isomer selectivity of chymotrypsin in the hydrolysis of α-methyl-α-nitro carboxyl esters allowed the synthesis of pseudodipeptides containing D-α-methyl-α-amino acids at the *N*-terminal position. α-Chymotrypsin-catalyzed enantioselective aminolysis of butyl-3-(3-indolyl)-2-methyl-2-nitropropionate with L-leucine amide, for example, gave a peptide that upon catalytic hydrogenation, yielded α-methyl-D-tryptophanyl-L-leucine amide[43] (Figure 7).

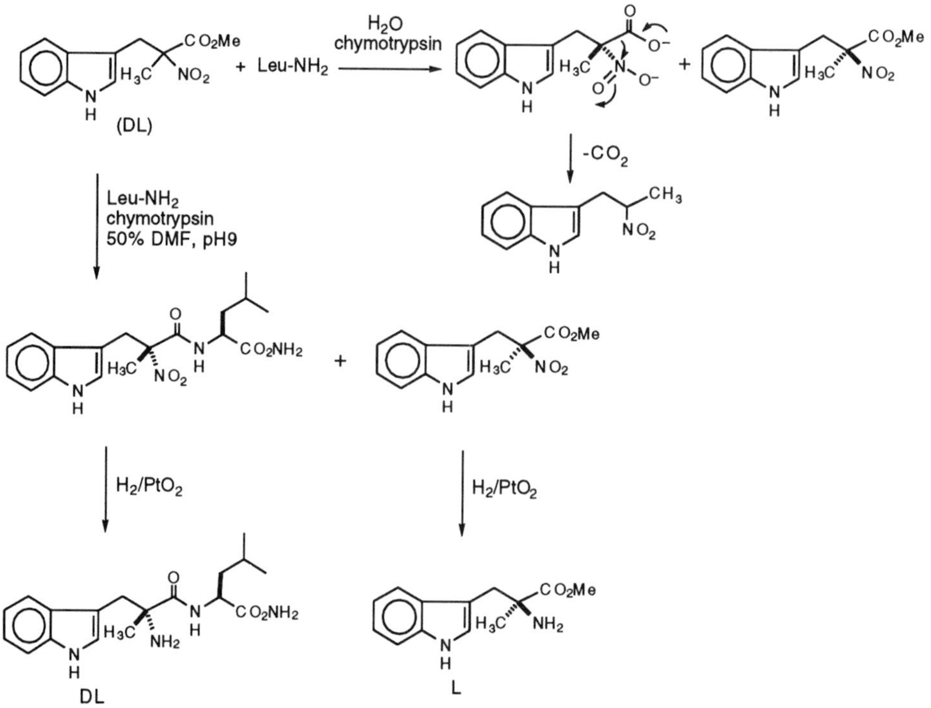

Figure 7. α-Chymotrypsin selectively accepts D-α-nitro-α-methylcarboxylic acid esters as substrates.

A new method recently developed for the site-specific incorporation of unnatural amino acids into proteins involves a chemically acylated suppressor tRNA that inserts the amino acid into a position where the codon is substituted with a nonsense codon.[71] This method currently allows μg to mg scale preparations of polypeptides.

D-Aminopeptidase from *Ochrobacterium anthropi* was used in the synthesis of poly-D-alanine[72] and D-alanyl amides[73] from D-amino acid methyl ester. The formation of Ac-L-Lys(Ac)-D-Ala-D-Ala-OMe from Ac-L-Lys(Ac)-D-Ala-D-Lac-OH and D-Ala-OMe catalyzed by muramoylpentapeptide carboxypeptidase (EC 3.4.17.8) was reported.[74]

2.4 New Enzymatic Catalysts for Peptide Synthesis

Although some proteases can be manipulated to accept D-amino acids as weak acyl donors, their utility for the synthesis of D-peptides is still quite limited. New enzymes are required for this purpose. A new thiol protease specific for D-alanine amide[72-73,75] possesses both esterase and amidase activities for certain D-amino acids and has been used in the synthesis of D-alanyl dipeptides via aminolysis. The enzyme is, however, too narrow in its specificity to be of general use. Other proteases specific for the D-configuration may exist in nature, but remain to be discovered and exploited.

The technique of site-directed mutagenesis has so far not succeeded in changing an enzyme specific for L- substrates into one specific for D-substrates. The recent developments in catalytic antibodies, however, seems to be promising in this regard. Antibodies raised against a phosphonamidate hapten catalyzed an aminolysis reaction and formed a peptide bond.[76] Phosphonate haptens were also used to elicit antibodies for bimolecular amide formation[77] and D-ester hydrolysis.[78] Applications of antibody catalysis to large-scale synthesis is still limited by the problems of product inhibition and the lack of catalytic efficiency. It may, however, be possible in the future to develop tailor-made catalytic antibodies for the synthesis of specific peptide bonds.

Understanding the structure and specificity of proteases[79] can also lead to the development of novel catalytic reactions via substrate or enzyme engineering. The specificities of trypsin,[80] α-lytic protease,[81] papain,[82] and subtilisin,[83] for example, has been altered using site-directed mutagenesis. A recent study of the mutagenesis of subtilisin at its S_1 and S_1' sites provides some interesting information in this regard.[84] The hydrophobic environment of the S_1 binding site, for which Gly_{166} is at the bottom in the wild-type enzyme, can be altered by replacing this amino acid residue with Asn or Ser to accommodate Arg as substrate. The Tyr $_{217}$→Phe change reduces the volume of the S1' pocket. The consequence of this change is a reduction in the rate of amide hydrolysis without a change in the rate of ester aminolysis. This mutant is thus good for aminolysis. The specificity of wild-type trypsin was altered with the use of Cbz-Pro *p*-guanidinophenyl ester as the donor substrate and an acceptor (Xaa) to form a Cbz-Pro-Xaa bond. The ester acts as an inverse substrate.[85] Molecular modeling of the oxyanion intermediate generated in reactions of *N*-acetyl-Tyr-OEt with different nucleophiles catalyzed by α-chymotrypsin suspended in dichloromethane containing low concentrations of water[86] explains several interesting results. For example, D-phenylalanine amide is a very effective nucleophile under these conditions, suggesting a change in protein conformation. Amino acid amides are in general better nucleophiles than the corresponding esters; this observation is attributed to the H-bonding interaction between the amide amino group of the P_1' component and the oxygen atom of the backbone carbonyl group of Phe$_{41}$. L-Lys-OBut and Arg-OBut are better substrates than the L-ornithine derivative (the amino acid with the side chain one carbon shorter than Lys) because the hydrogen bond to Cys$_{58}$ was formed with the εNH_2 group. N-Ac-D-Tyr-OEt is a poor substrate, because a high torsional strain (~8 kcal/mol) is generated in the formation of the oxyanion intermediate.

Other proteases potentially useful for synthesis of peptides include lysyl protease (which cleaves Lys-X bonds),[87] glutamyl protease (e.g. V8 protease[88] from *S. aureus*, EC 3.4.21.9 and the enzyme from *Bacillus licheniformis*,[89] ATCC 14580) and trypsin-like enzymes from microorganisms.[90a]

High-yield aminolysis reactions using glutamyl proteases and N-protected peptide benzyl or nitrobenzyl esters have been accomplished in the fragment synthesis of a polypeptide containing 29 amino acids.[89b] The proline iminopeptidase from *Bacillus brevis* catalyzes the synthesis of Pro-containing peptides (e.g. Pro-OBn + Phe → Pro-Phe, 40%).[90b] A synthetic gene coding for the poly (Asp-Phe)[91] polypeptide has been cloned and expressed in *E. coli*. The polymer may serve as a precursor to Asp-Phe-OMe.

2.5   Issues of Enzyme Stability

As proteases and other enzymatic catalysts become useful for peptide synthesis, their stability becomes a major concern. One limitation to the usefulness of most enzymes in synthesis is their intrinsic instability in many of the unnatural environments required for organic reactions. Techniques such as immobilization have been used to improve the stability of enzymes for large-scale reactions. An alternative solution to this problem of stability is to create or select a stable protein catalyst having a different amino acid sequence. More stable enzymes can be obtained by selection from thermophilic species or by site-directed mutagenesis. A subtilisin mutant derived from subtilisin BPN' by six site-specific mutations was substantially more stable than the wild-type enzyme in aqueous solution and in high concentrations of DMF, and was applied to the synthesis of peptides, chiral amino acids and sugar derivatives.[38] Another, even more DMF- and water-stable mutant, has also been developed.[92] Other serine proteases have also been engineered to improve the stability in high concentrations of DMF.[93a] In addition, mutant subtilisins with increased specific activity in high concentrations of DMF have been discovered via random mutagenesis and screening.[93b]

2.6   Synthesis of Racemization-free Peptides

The high enantio- and diastereoselectivity of protease-catalyzed peptide coupling in various conditions allows the synthesis of peptide bonds without racemization at the P_1 position. The stereospecificity of reactions of proteases has been investigated in detail at the P_2 and P_3

positions, and it has been found that reactions of subtilisin in water-organic cosolvents are highly specific for L-amino acids at the P_1 and P_2 positions.[94a] In contrast, subtilisin accepts both D- and L-amino acids at the P_3 position and positions yet further removed from the catalytic site. A new strategy based on this finding has been developed for the racemization-free synthesis of peptides that proceeds from the N to the C terminus. This procedure involved chemical coupling to form the peptide bonds, and subtilisin-catalyzed deprotection of the carboxyl ester. The enzymatic deprotection reaction only hydrolyses the ester group if there was an L-amino acid at the P_1 site; the racemized product is not hydrolyzed and can be easily separated. Repeating this procedure will therefore extend the peptide chain with no contamination of racemized products (Figure 8),

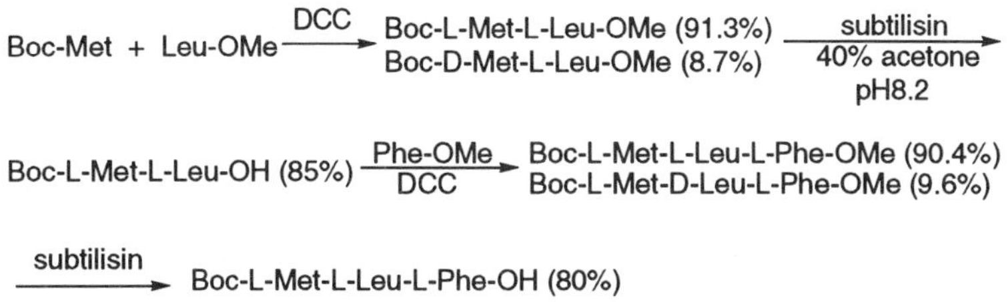

Boc-Met + Leu-OMe $\xrightarrow{\text{DCC}}$ Boc-L-Met-L-Leu-OMe (91.3%) $\xrightarrow[\text{40\% acetone} \atop \text{pH8.2}]{\text{subtilisin}}$
Boc-D-Met-L-Leu-OMe (8.7%)

Boc-L-Met-L-Leu-OH (85%) $\xrightarrow[\text{DCC}]{\text{Phe-OMe}}$ Boc-L-Met-L-Leu-L-Phe-OMe (90.4%)
Boc-L-Met-D-Leu-L-Phe-OMe (9.6%)

$\xrightarrow{\text{subtilisin}}$ Boc-L-Met-L-Leu-L-Phe-OH (80%)

Figure 8. Synthesis of racemization-free peptides: Subtilisin-catalyzed deprotection of the C-terminal ester.

provided the ester hydrolysis is enantiospecific. Similarly, stepwise synthesis of peptides from the N to C terminus using carboxypeptidase Y[23] or from the C to N terminus using papain[39] or other proteases can provide racemization-free peptides. Table 2 illustrates representative examples of enzymatic peptide coupling.[94-96]

2.7 Peptide Amides

Carboxy-terminal amidation is an important post-translational modification in the synthesis of polypeptide amides; the reaction requires two enzymes acting in sequence.[97] The first enzyme is peptidylglycine α-monooxygenase (EC 1.14.17.3) which catalyzes the formation of the α-hydroxyglycine derivative of the C-terminal glycine using a process dependent on ascorbic acid, copper and molecular oxygen. The second enzyme is peptidylamidoglycolate lyase (EC 4.3.2.5) which catalyzes the breakdown of the C-terminal α-hydroxyglycine derivative to produce the amidated peptide and glyoxylate (Figure 9). The oxygenase abstracts the Pro-*S* hydrogen from the Gly residue to form (*S*)-hydroxy-Gly; this compound is the substrate for the second enzyme. Both enzymes can be isolated from bovine pituitaries.[98] They have been cloned and expressed in mouse C-127 cells.[99] The enzymes from horse serum are commercially available (Takara Biochemical Inc., CA). Application of the enzymes in the synthesis of human growth hormone releasing factor analog GRF(1-44)-NH$_2$ has been reported.[100]

Use of Hydrolytic Enzymes

Table 2

Cbz-Tyr-Gly-Gly-Phe-OCH$_2$CN (0.1M) + Leu-NH$_2$ (0.2M)

$$\xrightarrow[\text{50\% DMF, pH 8.5, 1-2 h}]{\text{MeCT, 2mg/mL}}$$ Cbz-Tyr-Gly-Gly-Phe-Leu-NH$_2$ (99%)
Enkephalinamide

(Ref. 44)

Boc-Tyr-D-Ala-Phe-Gly-O-N◯O + Tyr-Pro-Ser(OBzl)-NH$_2$
1 mmol   3 mmol

$$\xrightarrow[\text{50\% DMF, pH 9, 10mL 1h}]{\text{subtilisin, 5}\mu\text{mol}}$$ Boc-Tyr-D-Ala-Phe-Gly-Tyr-Pro-Ser(OBzl)-NH$_2$ (82%)
Dermorphin

(Ref. 38a)

Cbz-Tyr-OMe + D-Arg-OMe
0.5M   1.5M

$$\xrightarrow[\text{50\% CH}_3\text{CN, pH 9, 1-2 h}]{\text{chymotrypsin (0.2 mM)}}$$ Cbz-Tyr-D-Arg-OMe (70%)

(Ref. 54)

Cbz-Asp(OBzl)-)Bzl + D-Ala-OPr^i
0.3M   1M

$$\xrightarrow[\text{4-methylpentan-2-one, 3mL, 2 days}]{\text{papain (200mg)-XAD-8}}$$ Cbz-Asp(OBzl)-Ala-OPr^i (86%)

(Ref. 39)

Cbz-Asp + Phe-OMe
2 mmol   4 mmol

$$\xrightarrow[\text{10 mL H}_2\text{O, pH 6-8, 40}^\circ\text{C, 3-5 h}]{\text{thermolysin, 10mg}}$$ Cbz-Asp-Phe-OMe (96%)

(Ref. 95)

Cbz-Ala-OMe (0.1M) + Gly-OPr^i (0.2M) + D-Leu-OMe (0.2M)

$$\xrightarrow[\text{60\% MeOH, pH 9.5, 10 mL, 12 h}]{\text{crude papain (150mg)}}$$ Cbz-Ala-Gly-D-Leu-OMe (69%)

(Ref. 96)

Ac-D-Phe-OEt + Phe-NH$_2$
3.1mmol   3.1mmol

$$\xrightarrow[\textit{tert}\text{-Amylalcohol, 31 mL, 45}^\circ\text{C, 3 days}]{\text{subtilisin, 100mg}}$$ Ac-D-Phe-Phe-NH$_2$ (67%)

(Ref. 56)

Cbz-Gly-Pro-Gly-Gly-Pro-Ala + Leu-Leu-Phe-NH$_2$
40 mM   40 mM

$$\xrightarrow{\text{thermolysin (3mg/mL)}}$$ Cbz-Gly-Pro-Gly-Gly-Pro-Ala-Leu-Leu-Phe-NH$_2$ (67%)
tert-Amylalcohol with 1% water,   (Ref. 36)
9% formamide, 45°C, 4 days

Porcine Insulin   Carboxypeptidase A →

A-▢   A-▢
B-▢—Arg-Lys-Ala   B-▢—Arg-Lys

10 mM

$$\xrightarrow[\text{Trypsin, (2-5}\mu\text{M), 50\% Org. Solv.}]{\text{Thr-OBu}^t\text{ (0.8 M), pH 7, <12}^\circ\text{C}}$$ A-▢
B-▢—Arg-Lys-Thr-OBu^t

(90%)   (Ref. 29a)

Thr-OBu^t, Protease from *Achromobacter lyticus* (ref 29b)

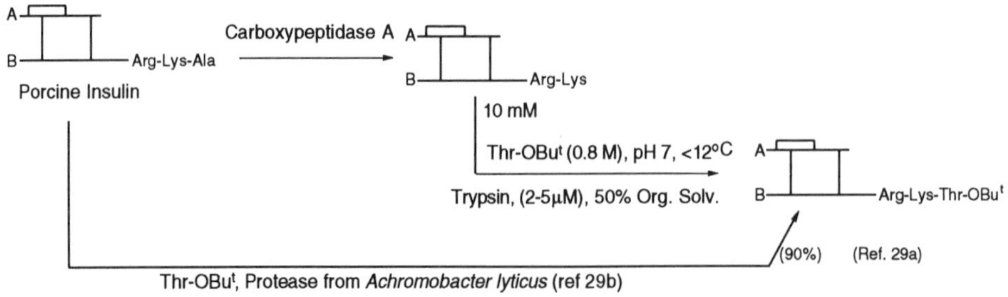

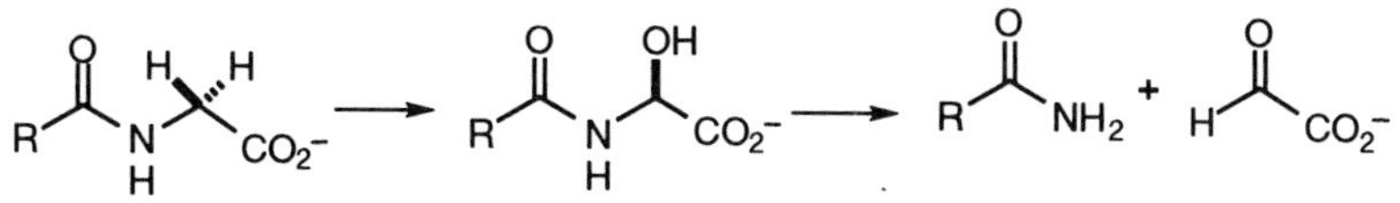

Figure 9. Enzyme-catalyzed peptide amidation.

An alternative route to peptide amides is enzymatic transpeptidation with amino acid amide.[100] None of serine proteases such as carboxypeptidase Y, however, accept prolinamide as nucleophile. The synthesis of peptide proline amide can be achieved via carboxypeptidase Y catalyzed transacylation or transpeptidation using 2-nitrobenzylamine or (2-nitrophenyl)glycinamide as the nucleophile, followed by photochemical deprotection to give peptide amides[101] (Figure 10). Another method is papain or subtilisin catalyzed aminolysis of

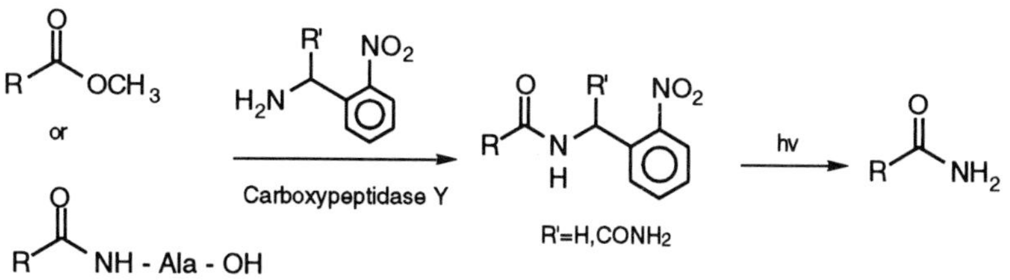

Figure 10. Synthesis of peptide amide.

peptide esters with trimethoxybenzyl amine, followed by deprotection with trifluoroacetic acid to give N-protected peptide amides.[102a] The subject of peptide amidation has recently been reviewed.[102b]

2.8   Non-Protease Catalyzed Peptide Bond Formation

In the biosynthesis of non-ribosomal peptides (e.g. peptide antibiotics and glutathione), an amino acid is activated enzymatically to aminoacyladenylate; this species then reacts with the amino group of another aminoacyl thioester to form a peptide bond (Figure 11).[65] The peptide chain is thus extended from N- to C-terminus on a multienzyme template. A typical example is the biosynthesis of δ-(L-α-aminoadipyl)-L-cysteinyl-D-valine (ACV) catalyzed by ACV synthetase,[103-105] which has recently been purified from *Streptomyces clavuligerus*[106] (Figure 12). The enzyme accepts many non-protein amino acids as well as hydroxyacids. Since ACV is a precursor of penicillins and cephalosporin, the enzymatically formed ACV analogs may be converted to the corresponding β-lactam derivatives.[105-107] Another class of non-protease enzymes potentially useful for peptide synthesis is aminoacyl-tRNA synthetases.[66,108] These enzymes catalyze the synthesis of aminoacyl-tRNA in the presence of ATP with the formation of an aminoacyl-AMP intermediate.

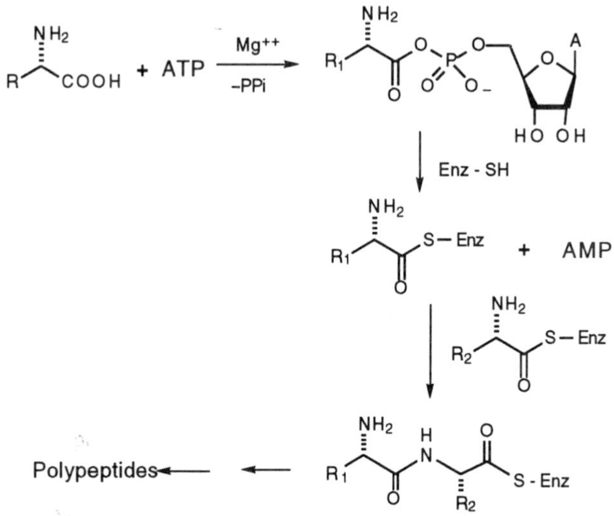

Figure 11. Peptide synthesis catalyzed by amino acid tRNA synthetase.

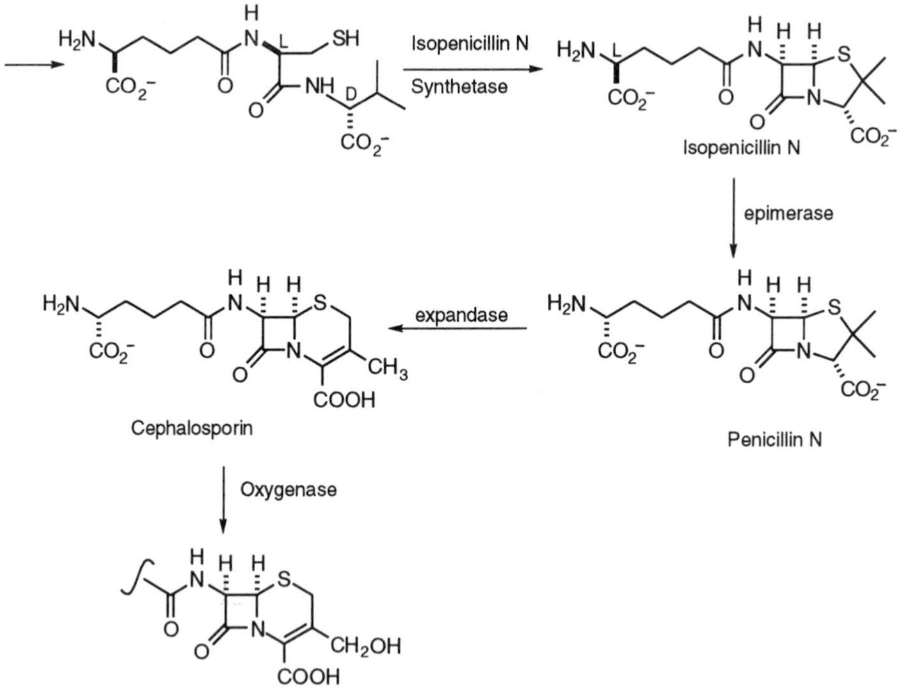

Figure 12

$$E + aa + ATP \rightarrow E \cdot aa - AMP + PPi$$

$$E \cdot aa - AMP + tRNA \rightarrow E + aa - tRNA + AMP$$

The enzymes can be utilized in peptide synthesis when tRNA is replaced with other nucleophiles such as amino acid amides. A tyrosyl-tRNA synthetase from *Bacillus stearothermophilus* has been used in the synthesis of a number of dipeptides.[109] In addition to Tyr and Leu, His and Asp were used as the acyl donor, and both D- and L-amino acid amides, including Pro-NH₂, can be used as nucleophiles. These two enzymatic processes have the advantage that no N-terminal protection is needed. For large scale process, however, regeneration of ATP is required.

2.9 Enzymatic Deprotection in Peptide Synthesis

Enzymes can be used in the selective deprotection and liberation of the α-amino group, the carboxy group and the various side chain functionalities in peptide synthesis.[20] The phenylacetyl group can be removed from dipeptides in reactions catalyzed by penicillin acylase without affecting other protecting groups.[110,111] The C-terminal ester can be hydrolyzed by carboxypeptidase Y and thermitase.[112] It is particularly interesting that thermitase also catalyzes the deprotection of C-terminal *tert*-butyl ester. The lipase from *Mucor javanicus* has been used in the deprotection of C-terminal glycopeptide heptyl ester.[113] The C-terminal amide can be deprotected with a peptide amidase from the flavedo of oranges without affecting the peptide bonds and *N*-protecting groups.[114] This enzyme accepts a variety of substrates with L-configuration at the C-terminal residue. Enzymes are also known for the deprotection of both Cbz- and Boc- groups, although the specificities are quite limited.[115]

3. Proteases that Act as Esterases
3.1 Chymotrypsin

Due to the high selectivity of proteases for L-amino acids, they have been widely used in the enantio- and regioselective transformation of amino acids and structurally related substances. Both serine- and thiol-proteases possess esterase activity and are useful for ester hydrolysis and transacylation reactions.[116] For example, α-chymotrypsin (EC 3.4.21.1) is a serine protease that catalyzes the hydrolysis of amide bonds of proteins of aromatic amino acids such as Phe, Tyr and Trp as its primary reaction. It also catalyzes the hydrolysis of various esters.[116-117] The enzyme was used in the enantioselective hydrolysis of *N*-acetyl-DL-tyrosine ethyl ester,[118] *N*-acetyl-DL-tryptophan methyl ester,[119] *N*-acetyl-phenylalanine methyl ester,[120] ring-substituted phenylalanine ethyl esters[121] (such as *m*-hydroxy, 3,4-dihydroxy, *p*-chloro- and *p*-fluoro-substituted) and other protected racemic amino acid esters[122] to give the corresponding L-enantiomers and unreacted D-enantiomers, all in very high ee. An active-site model of the enzyme was proposed by Cohen[123a] to explain the enantioselectivity. The model contains four pockets; each corresponds to one of the four groups with different sizes attached to the α-carbon of a substrate. Molecular modeling of the enzyme-substrate interactions in the calculation of enantioselectivity was recently reported.[123b] α-Methyl α-amino acid esters and α-alkenyl α-amino acid esters have also been resolved via hydrolysis by chymotrypsin to give the corresponding L-acids,[124,125] although the rates are quite slow. In the hydrolysis of α-nitro-α-methyl acid esters catalyzed by chymotrypsin, the D-isomers are preferentially hydrolyzed and the products undergo decarboxylation spontaneously.[126] The unreacted L-esters were recovered in >95% ee, and further converted to the corresponding α-methyl-α-amino acid esters via reduction. This inversion of enantioselectivity was utilized in the chymotrypsin-catalyzed aminolysis of *N*-protected amino acid or peptide esters to form a peptide bond containing D-α-methyl-α-amino acid residues.[43]

Chymotrypsin has also been used in the selective hydrolysis of other types of esters.[116] Of particular interest is the enantioselective hydrolysis of the prochiral substrates diethyl β-acetamidoglutarate[127] and diethyl β-hydroxyglutarate[128] to give the corresponding (*R*)-monoesters. The enantioselectivity can be improved (to ~93% ee) with the use of methoxymethyl protecting group for the β-NH$_2$ or β-OH group.[129] No enzymatic hydrolysis was observed when one of the two -CH$_2$CO$_2$Et groups is replaced with a -CH$_3$ group.[130] The enzyme has also been used in the hydrolysis of 5-(4*H*)-oxazolones containing various substituents in the 2- and 4-positions,[131a] *erythro*-phenylserine, and *threo*-isoserine derivatives.[131bc] The enantioselectivity was, however, very low when 5(4*H*)-oxazolones[131a] or α-hydroxyesters[131d] are used as substrates. Similar results were obtained when subtilisin was used as the catalyst. Selective hydrolysis of different ester groups in a molecule (especially the selective hydrolysis of the dihydrocinnamoyl group) has also been achieved with chymotrypsin.[132] Some representative selective hydrolyses are summarized in Figure 13.[133-136]

Chymotrypsin was also found to catalyze the formation of *N*-acetyl-L-Tyr and *N*-acetyl-L-Trp ethyl esters from the acids and ethanol containing very small amount of water.[137] The rate of enzymatic transesterifications is enhanced in the presence of crown ethers.[138] The enzyme immobilized with a mixture of polyethyleneglycol and trichlorotriazine appears to be soluble in benzene.[139]

3.2   Subtilisin, Papain and other Proteases

Subtilisin is similar to chymotrypsin in some of its reactions, it also catalyzes the enantioselective hydrolysis of amino acid esters but with a broader substrate specificity (Figure 14, also see 2.1).[38,140,141] The dibenzyl esters of L-Asp and L-Glu were hydrolyzed regioselectively at the α-position.[142] In the hydrolysis of L-aspartyl dicyclopentyl or dicyclohexyl ester, α-selectivity was observed using subtilisin; however, side chain-selectivity was observed with the use of chymotrypsin.[143] 1,2,3-Propanetricarboxylic ester and citrates were also regioselectively hydrolyzed by subtilisin at the central position.[144] The crude subtilisin "Alcalase" from NOVO (containing mainly subtilisin Carlsberg) is inexpensive, stable and useful for the enantioselective hydrolysis of amino acids[145] and peptide esters.[146] Resolution of other esters such as 2-amino-3-(2,2'-bipyridinyl) propanoic acid methyl ester[147] and the methyl ester of α-benzyl-β-sulfonamidopropionic acid[148] have been achieved with Alcalase to give the corresponding L-amino acid and unreacted D-ester, both in >93% ee. Regioselective hydrolysis of 1,6-anhydro-2,4-di-*O*-acetyl-3-azido-3-deoxy-β-D-glucopyranose at the 2-position (*Candida* lipase cleaves the 4-position) was also reported.[149] The enzyme has also been used in peptide synthesis, with anhydrous alcohols as the solvents, through aminolysis.[55] In addition to peptide synthesis (see 2.1), subtilisins have been used in organic solvents for enantio- and regio-selective transesterification or ester aminolysis. Trichloroethyl butyrate has been used as acylating reagent in anhydrous solvents for subtilisin-catalyzed acylation of sugars and related substances.[150,151]

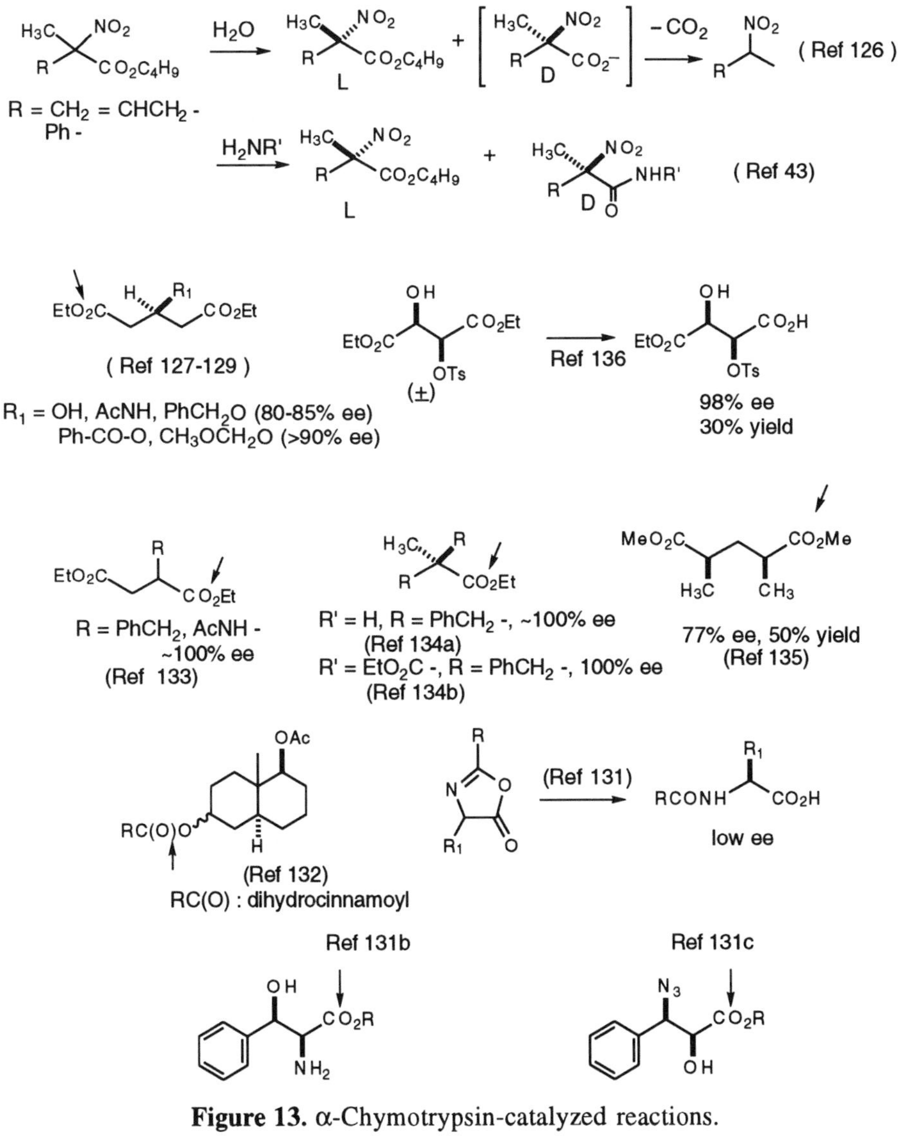

Figure 13. α-Chymotrypsin-catalyzed reactions.

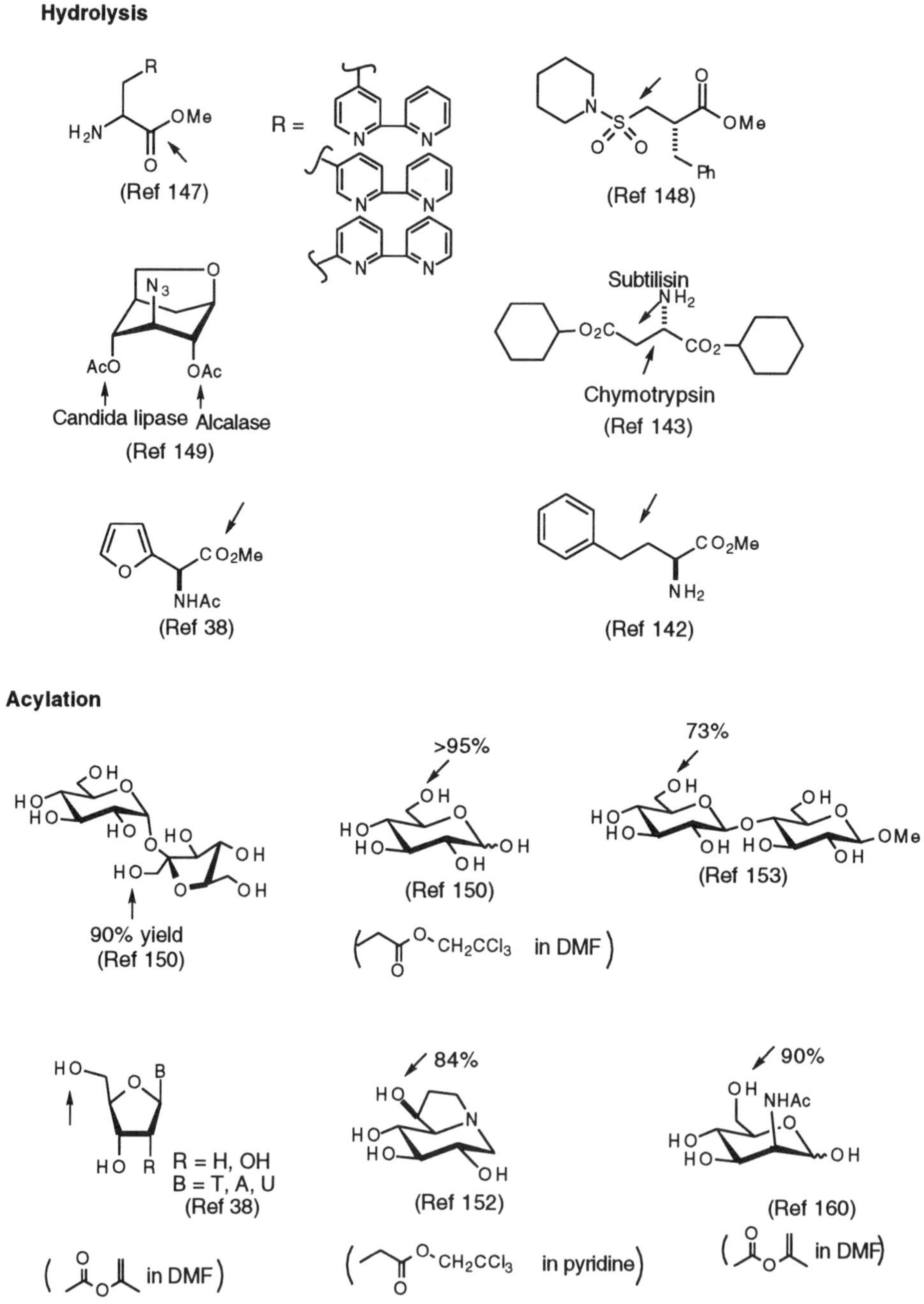

Hydrolysis

Acylation

Figure 14. Subtilisin-catalyzed reactions.

In general, a high regioselectivity for the primary hydroxyl group is observed, and several mono-, di- and trisaccharides, (glucose, sucrose, cellobiose, lactose, maltose, maltotriose), nucleosides (adenosine, uridine) and riboflavin have been selectively acylated. In the case of sucrose, with three primary hydroxyl groups, the acylation occurred selectively (~90%) at the 1-position of the fructose moiety. Acylation of monosaccharides at the primary position using *N*-acetylamino acid monochloroethyl esters or vinyl acrylate (for polymerization)[151] has also been accomplished. Similarly, a regioselective acylation of castanospermine at the 1-position[152] in pyridine, and of β-methyl lactoside at the 6-position of the Gal group in DMF[153] have also been reported. The enzymatic acylation was also found to be chemoselective, as illustrated by the enzyme-catalyzed acylation of the amine functionality of aminoalcohols using *N*-protected amino acid esters as acylating reagents in *tert*-amyl alcohol.[154] Under similar conditions, the lipase-catalyzed reactions of aminoalcohols were, however, selective for the hydroxyl group. The subtilisin-catalyzed acylation has also been extended to the enantioselective aminolysis of chiral amines, and the enantioselectivity was found to depend on the polarity of solvents used (as illustrated in the aminolysis of α-methyl benzylamine with trichloroethyl butanoate[155] to form the "*S*" amide). A higher enantioselectivity was found when more polar solvents such as 3-methyl-3-pentanol were used. Little effect of solvent on the prochiral selectivity of subtilisin Carlsberg was observed;[156] however, when 2-substituted 1,3-propanediol was used as the substrate. In contrast, the substrate specificity of subtilisin Carlsberg in the transesterification reaction of *N*-Ac-L-Ser-OEt and *N*-Ac-L-Phe-OEt with 1-propanol was quite different in different solvents.[157] The phenylalanine substrate was preferred in polar solvents, and the serine substrate was preferred in less polar solvents; and a linear relation between the substrate specificity of the enzyme-catalyzed hydrolysis in water and the solvent-to-water partition coefficients of the substrates was found. The enantioselectivity of protease-catalyzed acylation of chiral amine was markedly lower in organic solvents than in water.[158]

Subtilisin was also used in the acylation of phenylalanamide for use in the synthesis of a precursor for a chiral polymer via radical polymerization.[159] A similar reaction was performed with lipase catalyzed acrylation of chiral alcohols.[159] Although DMF is a good solvent for sugars, most enzymes, including subtilisin, are not stable in such a polar organic solvent. In addition, the transesterification reactions are generally slow and reversible, and the ee of the product may be reduced by the reverse reaction. Several DMF-stable subtilisin BPN' variants were developed using site-directed mutagenesis to improve their stability,[38] and among these, the most stable variant has a half-life of 14 days in high concentrations of DMF.[92]

The acylation reaction in DMF was also improved using isopropenyl acetate to facilitate the reaction and to avoid the problem of reaction reversibility (the released enol is tautomerized to acetone) and product inhibition. The improved enzyme and acylation reaction have been used in the selective acylation of several monosaccharides (Glc, GlcNAc, ManNAc, Man, 2-deoxy-2-fluoro-glucose).[160-161] Ethyl lactate was also used to acylate the 6 position of ManNAc, but the reaction had to be carried out in aqueous solution because the methyl ester is not activated in DMF at room temperature.[161] Both protease N and the wild-type subtilisin BPN' were also used with isopropenyl acetate in DMF for the acylation of sugars[160] and nucleosides,[38] although more enzyme and a longer reaction time were needed.

Papain was used in the hydrolytic resolution[162] of natural and unnatural amino acid derivatives such as Boc-Ser(OBzl)-OMe[163] and *N*-protected furylglycine esters.[164] The regioselective hydrolysis of the 5-ester of α,β-unsaturated α-amino-1,5-diesters was accomplished with papin while chymotrypsin was selective for the 1-ester.[165] Papain was also utilized in the enantioselective esterification of *N*-protected racemic amino acids.[166]

Enantioselective transformations with some uncommon proteases have also been reported. For example, a protease from *Aspergillus oryzae* (Sigma type XXIII) was useful for the enantioselective hydrolysis of hindered and unhindered esters. The α-trifluoromethyl mandelic acid methyl ester has been resolved with high enantioselectivity (88% ee for acid and unreacted ester at 50% conversion: E = 26).[167a] The same reaction with subtilisin gave the products with ~25% ee. Some 4,5-dihydroisoxazole-5-carboxylic acid ethyl esters were also successfully resolved, leading to the isolation of an (*S*)-acid and unreacted (*R*)-ester.[168] The protease from *Aspergillus melleus* (Seaprose S, Amano) was used in the resolution of 1,4-dihydropyridines.[167b] Another new protease-catalyzed resolution was the resolution of the anti-inflammatory agent Ketorolac via hydrolysis of its ethyl ester with a protease from *Streptomyces griseus*.[169] The unreacted (*R*)-substrate was racemized *in situ*, leading to a high yield of the (*S*)-acid. Other regio- and enantioselective transformations based on penicillin acylase for the hydrolysis of phenylacetyl group were also reported.[170,171] The enzyme was also useful as a deprotecting catalyst.[172] The selectivity of the enzyme is generally consistent with that found in the hydrolysis of *N*-phenylacetyl amino acids (i.e. L-selective) (see Figure 15 for different protease reactions).

Thermitase from the thermophilic microorganism *Thermoactinomyces vulgaris* has a high ratio of esterase to peptidase activities[173] and its selectivity for α-ester hydrolysis in the presence of β- and γ-esters has been used in the synthesis of Boc-Asp(OMe)-OH and 2-Phe-Asp(OMe)-OH. The enzyme has recently been used in the selective deprotection of methyl, ethyl, benzyl[174] and *tert*-butyl esters of *N*-protected amino acids and peptides[112] (e.g. Z-Ser-Ala-tBu, Z-Gly-Val-Ala-tBu).

4. Acetylcholine Esterase

Acetylcholine esterase (EC 3.1.1.7) is a serine type of esterase containing the Glu-Ser-His catalytic triad.[175] The structure of the enzyme from *Torpedo Californica* has recently been solved to 2.8Å.[176] The enzyme from electric eel is commercially available, and has been used in the enantioselective hydrolysis of acetyl-DL-carnitine to D-carnitine,[177] and esters of meso-diols to monoesters[178,179] including 3*R*-acetoxy-5*S*-hydroxycyclopent-1-ene, 1*R*-acetoxy-2*R*,3*R*,4*S*-trihydroxy-2,3-isopropylidene cyclopentane, and 6*S*-acetoxy-4-cycloheptene-1*S*,3*R*-diol (Figure 16).

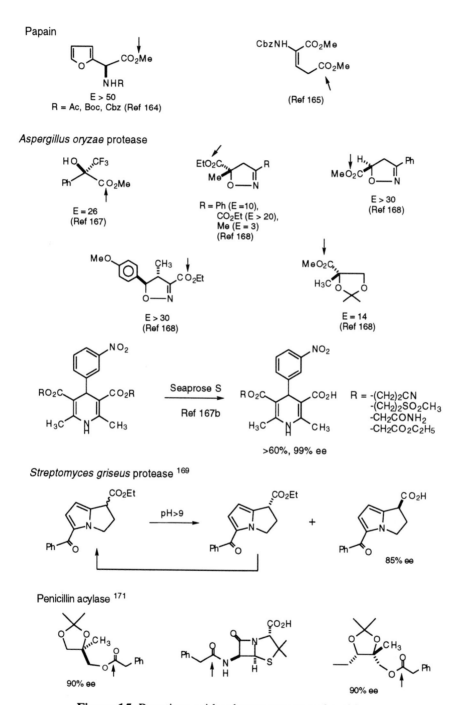

Figure 15. Reactions with other proteases and amidases.

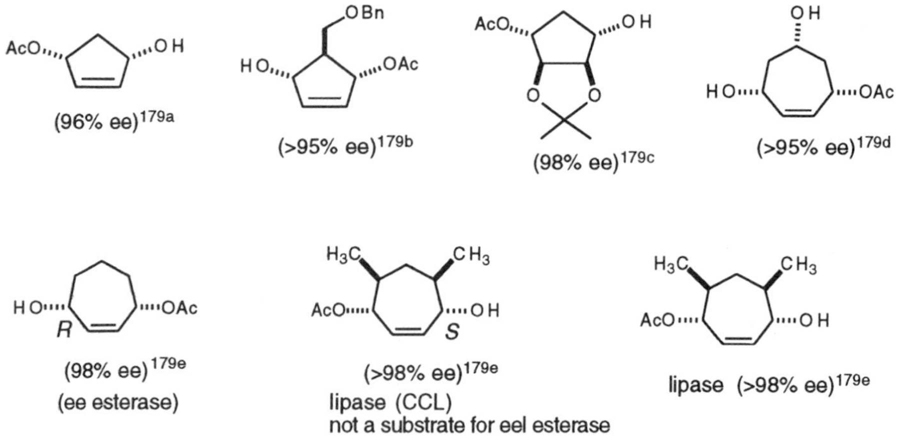

Figure 16. Acetylcholine esterase reaction (Products from the hydrolysis of meso-diesters).

5. Pig Liver Esterase

Pig liver esterase (EC 3.1.1.1) is a serine type of esterase[180] that catalyzes the stereoselective hydrolysis of a wide variety of esters.[181-184] The commercially available enzyme contains several isozymes; they behave, however, similarly in terms of stereoselectivity.[185] Of several models developed for interpreting and predicting the specificity of pig liver esterase,[184,186] the cubic-space active-site model reported by Jones[184] seems most broadly useful, and has been applied to a number of prochiral, cyclic and acyclic meso esters, and racemic esters (Figure 17). The enzyme also catalyzes the hydrolysis of lactams,[187a] lactones,[187b] cyclic carbonates,[187c] phosphonates[188] and dimethyl aziridine 2,3-dicarboxylates.[183b] The enzymatic hydrolysis of γ-phenyl-γ-butyrolactone is selective for the S-enantiomer (>94%ee), but no selectivity was observed in the hydrolysis of the corresponding acyclic ester (e.g. the γ-phenyl-γ-hydroxy substituted methyl butanoate) suggesting that the acylation step shows the selectvity.[187b] A high enantioselectivity (92%ee) was observed for the *meso*-aziridine substrate[183b] without protection of the N group. A low enantioselectivity was, however, observed in the enzymatic hydrolysis of racemic or N-protected *meso*-aziridine derivatives. In general, addition of dimethyl sulfoxide (10-20%) to the hydrolysis mixture enhances the stereoselectivity, and temperature has little effect.[189] In the kinetic resolution of trans-1,2-diacetoxyhexanediol with pig liver esterase, the overall enantioselectivity was enhanced if the rates for both sequential resolution steps were equal.[190] Pig liver esterase was also used to catalyze enantioselective transesterifications of chiral alcohols in a biphasic organic-aqueous system having high volume fractions of the organic phase.[191]

Figure 17: Products from pig liver esterase-catalyzed enantioselective hydrolysis of esters

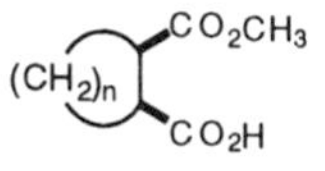

$n = 1-2, >97\%$ ee[1,2]

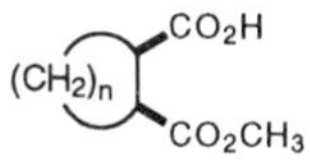

$n = 3, 17\%$ ee[1]
$n = 4, >97\%$ ee[1,2]

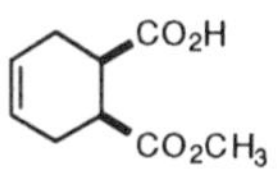

$>97\%$ ee[2,3,4,5]

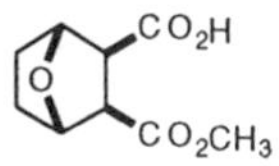

98% ee[6]

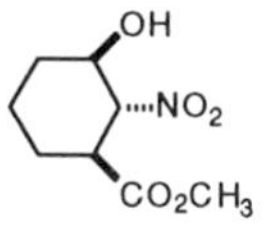

98% ee[7]

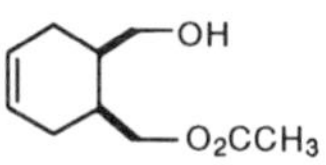

96% ee[8]

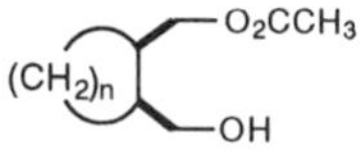

$n = 1, 44\%$ ee[9]
$n = 2, 0\%$ ee[9]

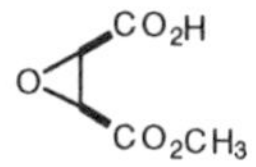

20% ee[10]

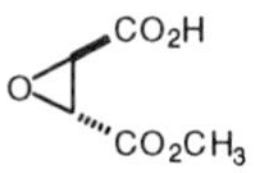

99% ee[11]

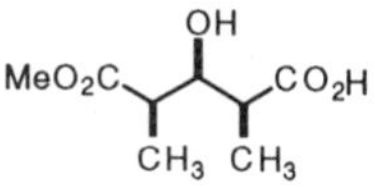

98% ee[12]

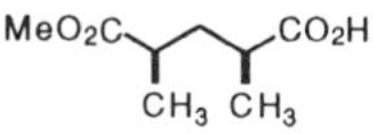

60% ee[12]

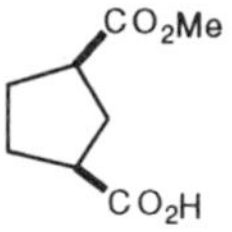

20% ee[13]

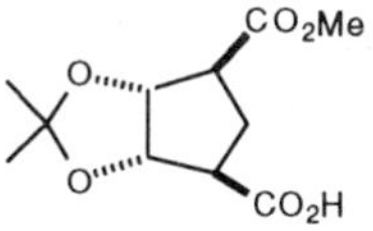

60% ee[14]

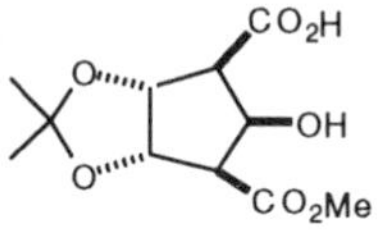

$>95\%$ ee[15]

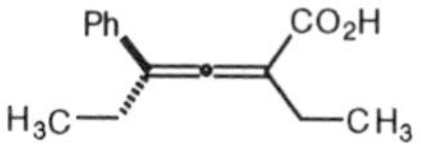

93% ee[16]

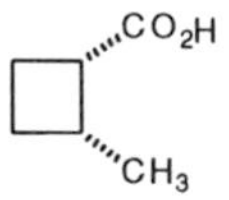

$>97\%$ ee[17]

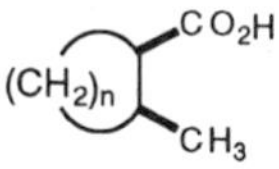

$n = 3, 22\%$ ee[17]
$n = 4, >97\%$ ee[17]

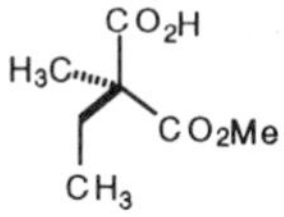

73% ee[18]

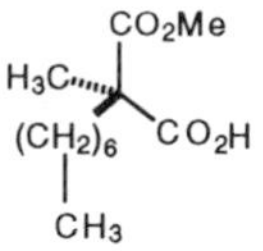

88% ee[18]

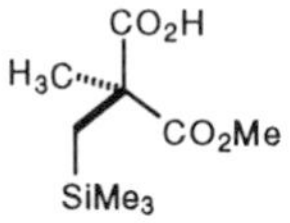

98% ee[19]

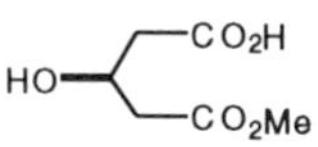

12% ee[3]

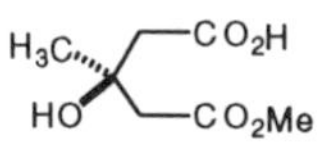

99% ee[3,20]

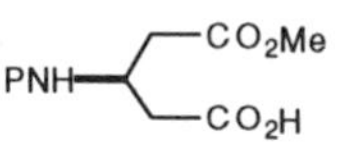

P=Ac, Cbz, 93% ee[21]

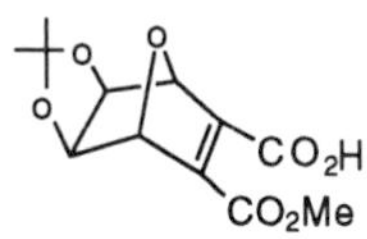

85% ee[22]

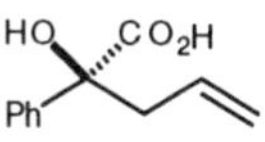

94% ee[23]

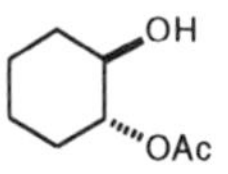

99% ee[24]

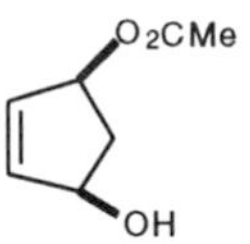

86% ee[25]

Figure 17. continue

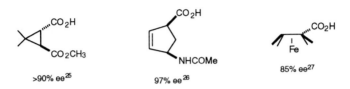

>90% ee[25]   97% ee[26]   85% ee[27]

Active-site model for PLE

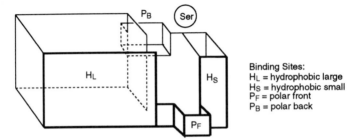

Binding Sites:
H_L = hydrophobic large
H_S = hydrophobic small
P_F = polar front
P_B = polar back

1. Sabbioni, G.; Shea, M.L.; Jones, J.B. *J. Chem. Soc., Chem. Commun.* **1984**, 236.
2. Schneider, M.; Engel, N.; Honicke, P.; Heinemann, G.; Gorisch, H. *Angew. Chem. Int. Ed. Engl.* **1984**, *23*, 67.
3. Mohr, P.; Waespe-Sarcevic, N.; Tamm, C.; Gawronska, K.; Gawronski, J.K. *Helv. Chim. Acta* **1983**, *66*, 2501.
4. Gais, H.-J.; Lukas, K.L. *Angew. Chem. Int. Ed. Engl.* **1984**, *23*, 142.
5. Kobayashi, S.; Kamiyama, K.; Imori, T.; Ohno, M. *Tetrahedron Lett.* **1984**, *25*, 2557.
6. Bloch, R.; Guibe-Jampel, E.; Girard, C. *Tetrahedron Lett.* **1985**, *26*, 4087.
7. Seebach, D.; Eberle, M. *Chimia* **1986**, *40*, 315.
8. Guanti, G.; Banfi, L.; Narisano, E.; Riva, R.; Thea, S. *Tetrahedron Lett.* **1986**, *27*, 4639.
9. Laumen, L.; Schneider, M. *Tetrahedron Lett.* **1985**, *26*, 2073.
10. Habich, D.; Hartwig, W. *Tetrahedron Lett.* **1987**, *28*, 781.
11. Mohr, P.; Rosslein, L.; Tamm, C. *Helv. Chim. Acta* **1987**, *70*, 142.
12. Chen, C.-S.; Fujimoto, Y.; Sih, C.J. *J. Am. Chem. Soc.* **1981**, *103*, 3580.
13. Jones, J.B.; Hinks, R.S.; Hultin, P.G. *Can. J. Chem.* **1985**, *63*, 452.
14. Hultin, P.G.; Mueseler, F.-J.; Jones, J.B. *J. Org. Chem.* **1991**, *56*, 5375.
15. Zemlicka, J.; Craine, L.E.; Heeg, M.J.; Oliver, J.P. *J. Org. Chem.* **1988**, *53*, 937.
16. Ramaswamy, S.; Hui, R.A.H.F.; Jones, J.B. *J. Chem. Soc., Chem. Commun.* **1986**, 1545.
17. Toone, E.J.; Jones, J.B. *Tetrahedron: Asymmetry* **1991**, *2*, 207.
18. Bjorkling, F.; Boutelje, J.; Gatenbeck, S.; Hult, K.; Norin, T.; Szmulik, P. *Tetrahedron* **1985**, *41*, 1347.
19. De Jeso, B.; Belair, N.; Deleuze, H.; Rascle, M.-C.; Maillard, B. *Tetrahedron Lett.* **1990**, *31*, 653.
20. Huang, F.-C.; Lee, L.F.H.; Mittal, R.S.D.; Ravikumar, P.R.; Chan, J.A.; Sih, C.J.; Caspi, E.; Eck, C.R. *J. Am. Chem. Soc.* **1975**, *97*, 4144.
21. Wang, Y.-F.; Izawa, T.; Kobayashi, S.; Ohno, M. *J. Am. Chem. Soc.* **1982**, *104*, 6465.
22. Ho, Y.; Shibata, T.; Arita, M.; Sawai, H.; Ohno, M. *J. Am. Chem. Soc.* **1981**, *103*, 6739.
23. Moorlag, H.; Kellogg, R.M.; Kloosterman, M.; Kaptein, B.; Kamphuis, J.; Shoemaker, H.E. *J. Org. Chem.* **1990**, *55*, 5878.
24. Caron, G.; Kazlauskas, R.J. *J. Org. Chem.* **1991**, *56*, 7251. If the diacetate is mixed in the H_2O-hexane biphasic system to slow down the first hydrolysis, the enantioselectivity is enhanced when the rates for the two steps are equal $E = (1 + E_1E_2)/2$.
25. Schneider, M.; Laumen, K. *Tetrahedron Lett.* **1984**, *25*, 5875; Laumen, K.; Reimerdes, E.H.; Schneider, M. *Tetrahedron Lett.* **1985**, *26*, 407. Using the lipases from *Candida* and *Rhizopus* species gave enantiomeric product with ~50% ee, and the electric eel esterase gave >98% ee of the enantiomer.
26. Sicsic, S.; Ikbal, M.; LeGoffic, F. *Tetrahedron Lett.* **1987**, *28*, 1887.
27. Alcock, N.W.; Crout, D.H.G.; Henderson, C.M.; Thomas, S.E. *J. Chem. Soc., Chem. Commun.* **1988**, 746.

6. Phospholipases

Phospholipases catalyze the hydrolysis of phospholipids. Four different types of enzymes with different regio- and stereoselectivity have been identified; their cleavage sites are indicated in Figure 18. The enzymatic catalysis generally occurs at the interface of the aggregated

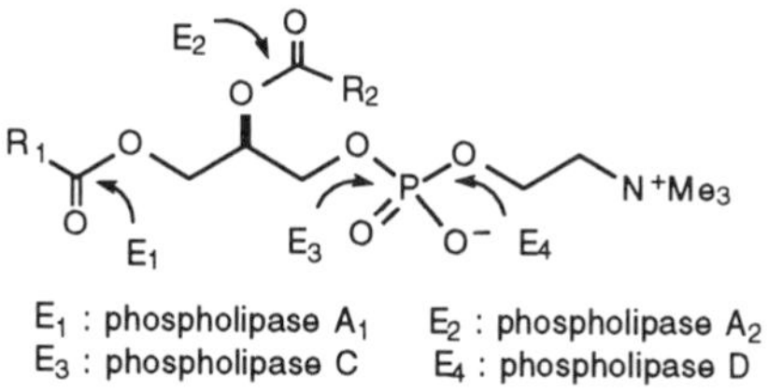

E$_1$: phospholipase A$_1$ E$_2$: phospholipase A$_2$
E$_3$: phospholipase C   E$_4$: phospholipase D

Figure 18. Phospholipase-catalyzed hydrolysis.

substrates.[192-193] Of many phospholipases known, phospholipase A$_2$ from Cobra venom is the best studied.[192] The enzyme contains a His residue in the active site as a general base: the δ1-N abstracts a proton from the bound water to facilitate its attack on the substrate carbonyl group; this carbonyl group is polarized by the bound Ca^{++} ion, a Lewis acid. The ε_2-N of the active-site His$_{48}$ has a H-bonding interaction with the carboxylate group of Asp$_{99}$. The enzyme-substrate interaction facilitates the diffusion of the substrate from the interfacial binding surface to the catalytic site, rather than facilitating a conformational change in the enzyme (as suggested in the case of lipase).

A number of phospholipids chiral at phosphorus have been prepared to study the stereospecificity of phospholipases.[194] Phospholipids and 2-thiophospholipids treated with phospholipase A$_2$ have been used in the preparation of phospholipid analogs.[195]

Phospholipase D from *Streptomyces sp.* was used to catalyze the exchange of choline from a phospholipid with other primary alcohols such as nucleosides.[196] The 1-*O*-alkyl derivatives of phospholipids can also be converted to other analogs by replacing the choline moiety with alcohols, including 3-(2-hydroxyethyl)indole, 2-(hydroxymethyl)-1,4-benzodioxan, 2-(2-thienyl)-ethanol, nucleosides[197-198] and glycerol.[199] The chemo-, regio- and enantio-selectivities of phospholipase D from *Streptomyces sp.* (Sigma, P-4912) have been investigated, and the results suggest that the primary hydroxyl group of aza-sugars, nucleosides, serine- and ω-hydroxyacid-containing peptides is acceptable for the transphosphatidylation reaction.[196b] Amine and thiol groups are, however, not reactive.

7. Cholesterol Esterase

Bovine pancreatic cholesterol esterase (EC 3.1.1.13) catalyzes the hydrolysis of cholesteryl and other esters via a serine esterase mechanism.[200] It also catalyzes the enantioselective hydrolysis of binaphthol,[201] (Figure 19) spirobindanol,[201] and protected myoinositol diesters[202] and secondary alcohol esters.[203] The reactions proceed through a two-step resolution, which can yield products with higher ee than one single-step resolution can.[201]

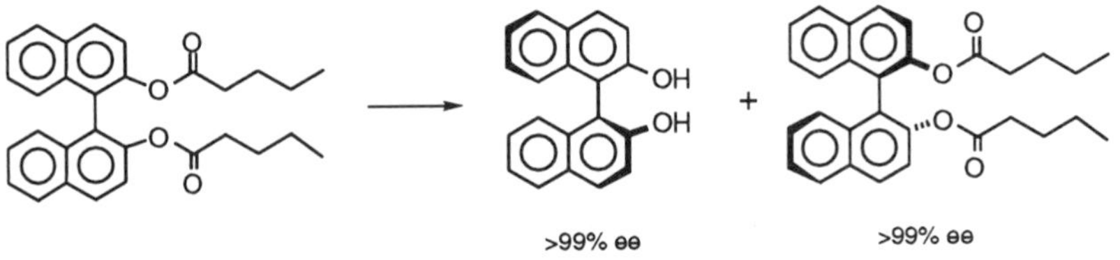

Figure 19. Cholesterol esterase reaction.

8. Lipases

Lipases are serine hydrolases that catalyze the hydrolysis of lipids of fatty acids and glycerol at the lipid-water interface .[204] Unlike esterases, which show a normal Michaelis-Menten activity, lipases display little activity in aqueous solutions with soluble substrates. A sharp increase in activity is observed when the substrate concentration is increased beyond its critical micellular concentration[205-207] (Figure 20). The increase in lipase activity at the lipid-water

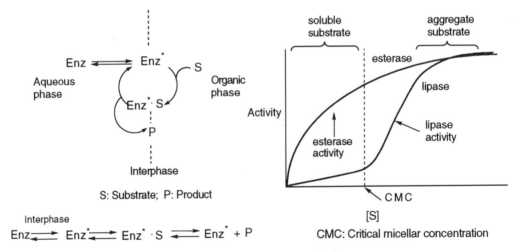

Figure 20. Interfacial catalysis.

interface led to the suggestion that soluble lipases might undergo conformational change at the oil-water interface,[208] before substrate binding takes place. This conformational change at the interface is supported by the X-ray structures of human[209] and *Mucor miehei*[210] lipases, phospholipase A$_2$,[211] and their complexes with inhibitors. In both lipases the active centers contain structurally analogous Asp-His-Ser triads, which are buried completely beneath a short helical lid. When the enzymes and inhibitors (diethyl *p*-nitrophenyl phosphate for *M. miehei* lipase) form a complex, active site of the the enzyme is exposed by the movement of the lid. The structure of the enzyme in this complex is believed to be equivalent to the activated state generated by the oil-water interface. The lipase from *Geotrichum candidum* contains a catalytic triad consisting of Glu-His-Ser sequence,[212] and the *humicola lanuginosa* lipase requires Asp-His-Tyr for activity.[213]

Several types of reactions catalyzed by lipases have been applied to synthesis (Table 3):

Table 3. Microbial Lipases (EC 3.1.1.3) that are commonly used in organic synthesis.

Source	Trade name	Suppliers
Alcaligenes sp.	Amano lipase PL	Amano
Achromobacter sp.	-	Meito Sangyo
Aspergillus niger	Amano A	Amano
Bacillus subtilis	-	Towa Koso
Candida cylindracea[a] (CCL)	Lipase OF	Sigma,
	Amano Lipase AY	Boehringer Mannheim,
		Amano, Meito Sangyo
Candida lypolytica	-	Amano, Fluka
Candida antarctica	SP-435[b]	NOVO
Chromobacterium viscosum (CVL)	Amano lipase LP	Sigma, Toyo Jozo
Geotrichum candidum (GCL)	Amano lipase GC	Sigma, Amano
Humicola lanuginosa	Amano lipase R-10	Amano
Mucor miehei (MML)	Amano MAP-10	Amano, NOVO, Fluka
	Lipozyme	
Penicillium camemberti	-	Rhone-Poulenc
Penicillium roqueforti	-	Fluka
Phycomyces nitens	-	Takeda Yakuhin
Porcine pancreas (PPL)	-	Sigma, Amano, Fluka
		Boehringer Mannheim
Pseudomonas cepacia[c]	Amano, P. PS.	Amano, Fluka
	PS30, LP80	
	SAM-I	
Pseudomonas sp.	Amano AK, K-10,	Amano
	SAM-II	
Pseudomonas aeruginosa	LPL, PAL	Amano
Rhizopus arrhizus	-	Sigma, Boehringer-Mannheim, Fluka
Rhizopus delemar	-	Sigma, Amano, Tanabe, Fluka
Rhizopus japanicus	Amano lipase FAP	Amano, Nagase Sangyo, Osaka Saiken, Fluka
Rhizopus niveus	Amano lipase N	Amano, Fluka
Rhizopus oryzae	-	Amano
Wheat germ	-	Sigma, Fluka

[a]Now called *Candida rugosa*.

[b]The acrylic resin supported lipase which was produced by a host organism, *Aspergillus oryzae*, after transfer of the gene coding for lipase B from *Candida antarctica*.

[c]Formerly called *Pseudomonas fluorescens*.

Use of Hydrolytic Enzymes

hydrolysis of esters, esterification (acid and alcohol), transesterification (alcohol and ester), interesterification (ester and acid), and transfer of acyl groups from esters to other nucleophiles such as amines, thiols, and hydroperoxide . The enantioselectivity of lipase-catalyzed reactions in aqueous solutions, water-organic solvent mixtures, and in anhydrous organic solvents follows the classical homocompetitive equation.[214-215]

8.1 From Reversible to Irreversible Reactions in Organic Solvents

Many substrates for lipases are poorly soluble in water. Reactions are often carried out in the presence of organic solvents, or using an excess of suspended substrates with vigorous stirring in order to create interface catalysis to facilitate the reaction. Addition of fine glass beads or adsorption of enzymes to supports such as XAD-8 also facilitates the reaction. Esterifications and transesterifications are much slower than hydrolysis, and sometimes the products are difficult to separate. In addition, due to the reversible nature of these reactions in organic solvents, and since the same stereoisomer is used in both directions, the enantiomeric excess of the desired product obtained decreases as the reverse reaction proceeds.

Product inhibition caused by the released alcohol in the transesterification is anotherproblem. As shown in Figure 21 for a transesterification reaction, if the D isomer is a

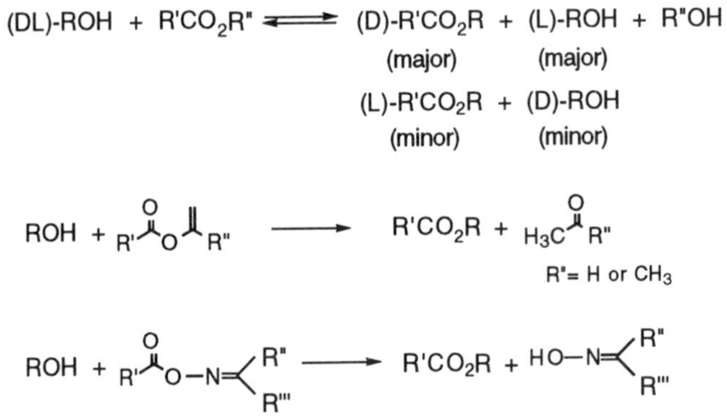

$$(DL)-ROH + R'CO_2R'' \rightleftharpoons (D)-R'CO_2R + (L)-ROH + R''OH$$

$$\text{(major)} \qquad \text{(major)}$$

$$(L)-R'CO_2R + (D)-ROH$$

$$\text{(minor)} \qquad \text{(minor)}$$

Figure 21. Reversible and irreversible transesterifications.

better substrate for the enzyme than the L isomer, accumulation of the D ester and the unreacted L alcohol will be observed. In the reverse reaction, however, the D ester should be a better substrate than the L ester based on the principle of microscopic reversibility. The enantiomeric excesses of both the D ester and the L alcohol will therefore decrease progressively as the extent of the reverse reaction increases. This problem has been clearly illustrated in the kinetic resolution of menthol.[215] The same situation occurs in lipase-catalyzed transesterification reactions. One way to avoid part of these problems is to use activated esters as acylating reagents to facilitate the reaction and to make it kinetically irreversible. Several activated esters such as chloroethyl or trifluoroethyl esters,[216] cyanomethyl esters,[44] enol esters,[217,218] anhydrides,[219] thioesters,[220] oxime esters,[221-222] and vinyl carbonates[223] have been used in enzyme-catalyzed reactions, and their rates of catalysis are about one to two orders of magnitude faster than those of non-activated methyl or ethyl esters. Cyanomethyl esters gave formaldehyde cyanohydrin (a toxin). Trichloroethyl esters are less expensive than trifluoroethyl esters, but trichloroethanol is difficult to remove (bp = 151 °C). 2-Chloroethanol is easier to remove (bp = 130 °C), but the degree of activation is limited. The enol esters liberate unstable enols which tautomerize instantaneously to aldehydes or ketones thus making the reaction completely irreversible with no product inhibition. The acetaldehyde liberated from reactions with vinyl esters may inactivate the enzyme due to the formation of a Schiff base with Lys residues. This problem may be avioded by immobilization of the enzyme via reaction of the Lys residues with an epoxy-derivatives on a support.[224] Product inhibition caused by oxime[225] and trifluoroethanol[226] has also been observed.

For transesterifications using oxime esters, the esters of acetone and diacetone oxime (prepared from acyl chloride and oxime) are commonly used.[227] Oxime carbonates are also accepted by lipases;[228-230] this reaction provides a useful enzymatic method for the selective introduction of an alkoxy carbonyl group (e.g. the benzyloxycarbonyl group) into hydroxy compounds.

A more detailed study on the enzymatic acylation[230] and alkoxycarbonylation (with oxime carbonate)[229] of nucleosides, 2'-deoxynucleosides and sugars (L-sorbose and L-arabinose)[231] indicates that *Candida antarctica* lipases (SP435 and SP435A from NOVO) showed high regioselectivity toward the primary hydroxy group. The lipases from *Candida cylindracea* (CCL), *Pseudomonas cepacea* (PSL) and porcine pancrease (PPL) are selective for the 3'-OH group in reaction with 2'-deoxynucleosides such as thymidine (Figure 22). The solvent used was pyridine,

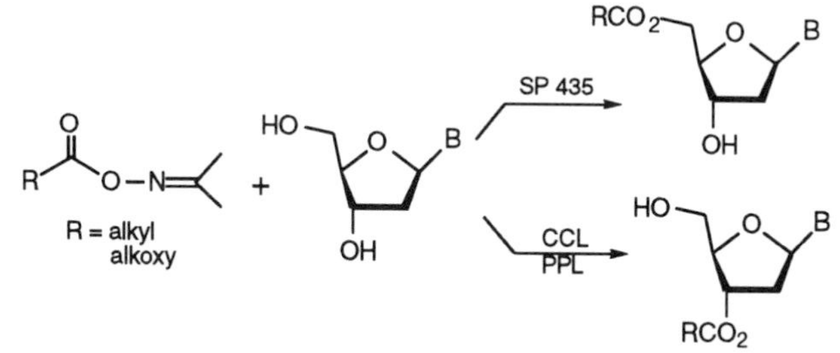

Figure 22. Use of oxime-esters in acylation.

THF or dioxane. Different acylation reagents such as enol esters, acid anhydrides, trifluoroethyl esters were also used for comparison with the oxime esters in the lipase reaction with deoxynucleosides.[230] The results indicate that enol and oxime esters give higher yield with higher regioselectivities. Trifluoroethyl esters also gave good regioselectivity, but with low conversion. Anhydrides showed different behavior: complete regioselectivity for the 5'-OH was observed with PSL-catalyzed acylation of thymidine, but a mixture was obtained with 2'-deoxyadenosine.

Although the regioselectivity is generally high, reaction with cytidine and guanosine gave *N*-acyl byproducts. Of these activated esters, the enol esters are the most often used. It is also worth noting that enzyme-catalyzed transesterification (or esterification) and hydrolysis are often complementary in their enantioselectivities.

The following section further describes the applications of lipase-catalyzed reactions in aqueous and organic solvents.

8.2 Porcine Pancreatic Lipase (PPL)

The enzyme PPL from commercial sources used in organic synthesis is usually not pure and may contain other enzymes such as chymotrypsin and cholesterol esterase. The enzyme is selective for esters of primary alcohols, not secondary alcohols, as indicated in one case where the enzyme activity for secondary alcohol ester substrates is totally abolished when pure PPL is used.[232] The enantioselective hydrolysis of meso-diacetates by PPL is complementary to the pig liver esterase-catalyzed hydrolysis of the corresponding meso-1,2-dicarboxylates,[233-234] (Figure 23) and the enantioselectivity can be improved with modification of the substrate.[233-234] Hydrolysis of prochiral 1,3-diol derivatives is also a useful application of PPL. The substituent on position 2 influences the enantioselectivity, and a reversal of enantioselectivity was observed when a (*E*)-substituent was changed to a (*Z*)-substituent, or when a phenyl substituent was changed to a benzyloxy substituent.[235] The 2-phenyl derivative was not a substrate of pure PPL.[236] An improvement of enantioselectivity was also observed when the acetate group was changed to a propionate group in the 2-benzyloxy case.[237]

Figure 23: PPL-catalyzed enantioselective reactions

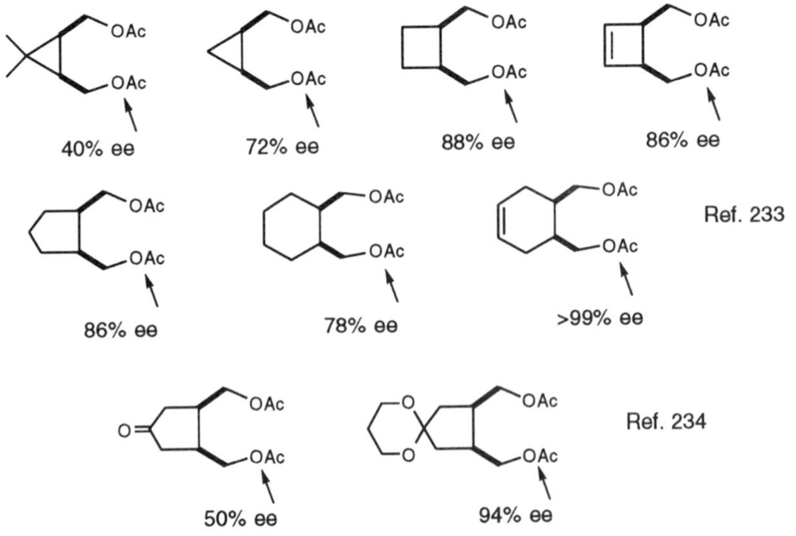

40% ee   72% ee   88% ee   86% ee

86% ee   78% ee   >99% ee   Ref. 233

50% ee   94% ee   Ref. 234

Figure 23: continue

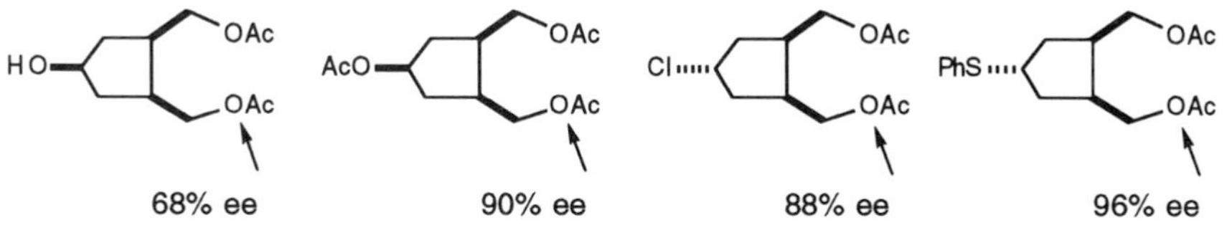

68% ee   90% ee   88% ee   96% ee   Ref 234

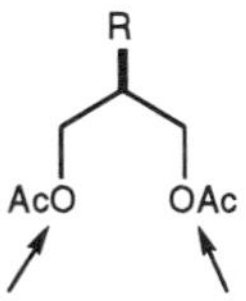

Ref 236, 237

R	% ee
(Z)-n-C_5H_{11}CH=CH-	53
PhCH$_2$O-	90

Ref 236

R	% ee
n-C_7CH_{15}-	70
(E)-n-C_5H_{11}CH=CH-	95
(E)-$(CH_3)_2$CHCH=CH-	97
Ph	99

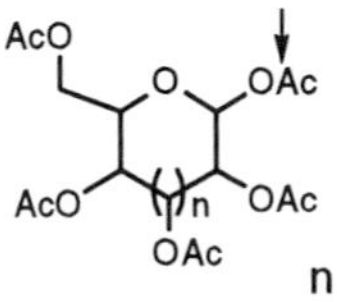

n = 0,1
Ref 253

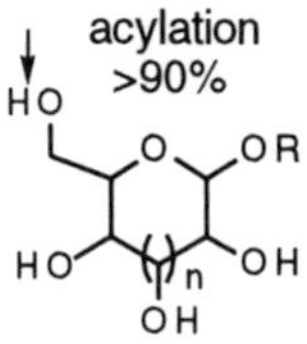

acylation
>90%

R = H, CH$_3$
Ref 253, 255

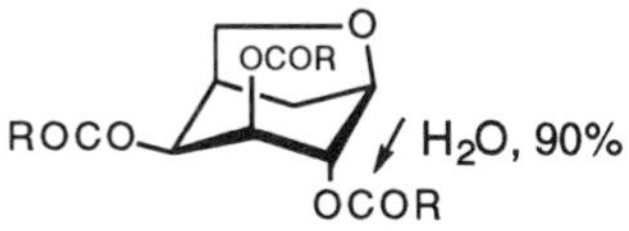

R = $CH_3(CH_2)_2$-
Ref 259

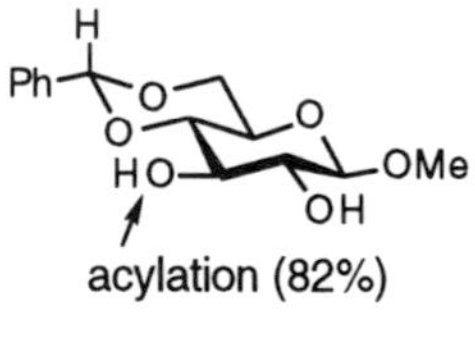

acylation (82%)

Ref 258

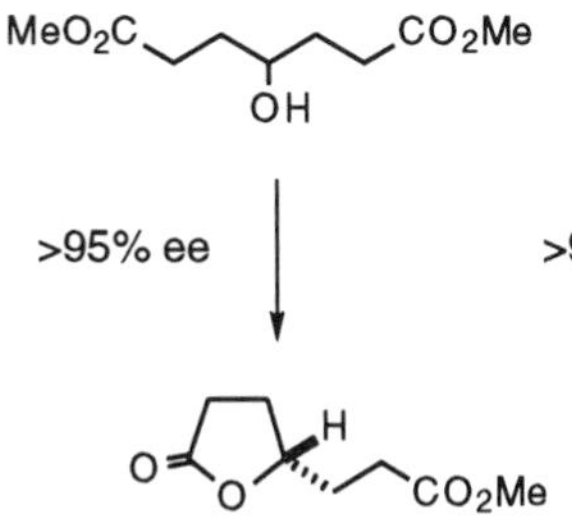

>95% ee

>95% ee

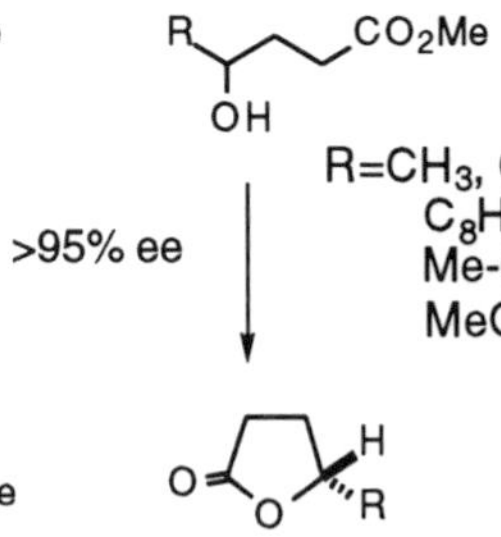

R=CH_3, C_2H_5,
C_8H_{17}, C_6H_5,
Me-C_6H_4,
MeO-C_6H_4

Ref 263c

Ref 263c

Figure 23. continue

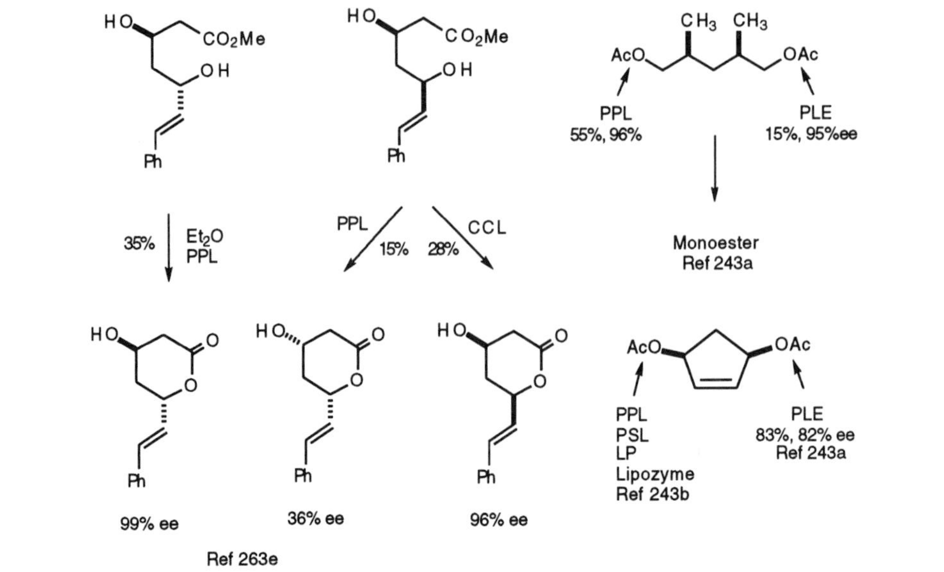

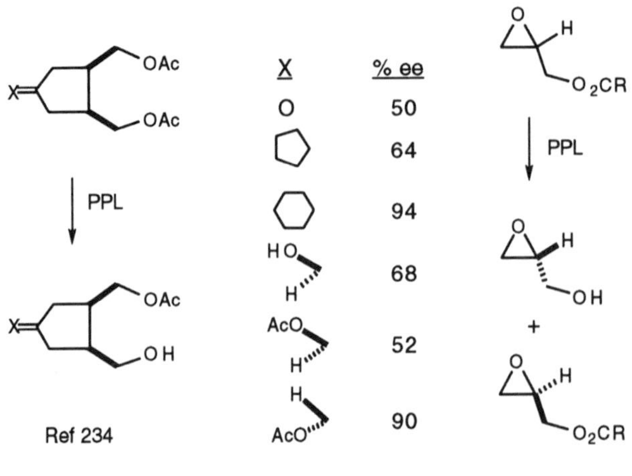

99% ee

36% ee

96% ee

Ref 263e

PPL 55%, 96%

PLE 15%, 95%ee

Monoester
Ref 243a

PPL
PSL
LP
Lipozyme
Ref 243b

PLE
83%, 82% ee
Ref 243a

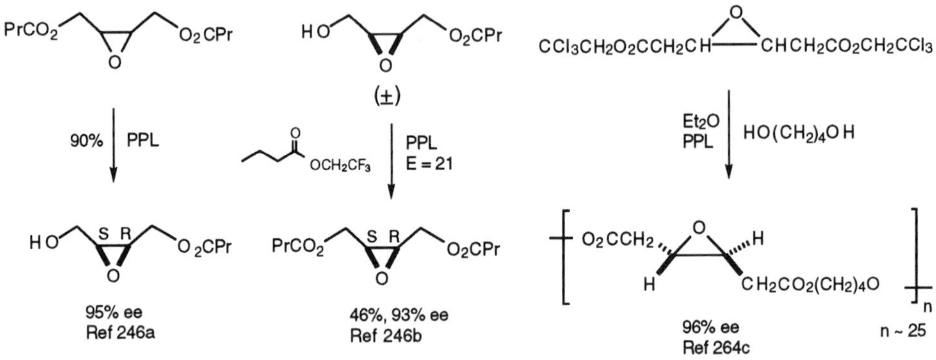

95% ee
Ref 246a

46%, 93% ee
Ref 246b

96% ee
n ~ 25
Ref 264c

X	% ee
O	50
(cyclopentane)	64
(cyclohexane)	94
HO,,,	68
AcO,,,	52
AcO,,,	90

R	% ee
CH_3	53
C_2H_5	88
$n\text{-}C_3H_7$	92
$n\text{-}C_4H_9$	96

Ref 238

Ref 234

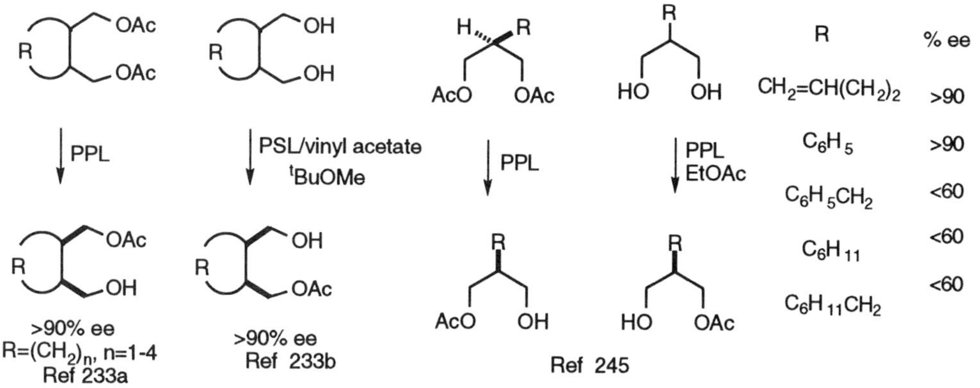

Figure 23. continue

A typical application of PPL to kinetic resolution is the enzyme-catalyzed enantioselective hydrolysis of epoxy alcohol esters.[238] The enantioselectivity depended on the size of the acyl group, and the propionate derivatives were optimal. Based on more than one hundred esters of primary alcohols surveyed (all of which gave hydrolysis products with >85% ee) an active-site model was recently developed by Jones[239] (Figure 24). The model contains a region to accommodate a small and polar group, and a region for large and hydrophobic group. This model is similar to that of Guanti,[240] but enantiomeric to that of Seebach.[241] Another model similar to that of Jones suggested a new explanation of the nature of the hydrophobic and hydrophilic substituents.[242]

PPL also possesses high regio- and chemo-selectivities. Selective removal of acyl groups of peracylated sugars is of interest in carbohydrate synthesis. Early work on the use of wheat germ lipase[251] or *Aspergillus niger* lipase[252] in the hydrolysis of glucose peracetate showed a selective cleavage of the 1-O-acetyl group at 20% conversion. When all the starting material had been consumed, several products were obtained due to the cleavage of other positions. Among the lipases tested for the hydrolysis of sugar peracetates, PPL was shown to be a better catalyst and the 1-O-acetyl group of glucose, galactose, mannose, N-acetylglucosamine, N-acetylmannosamine, L-fucose and rhamnose were selectively cleaved in 10% aqueous DMF to give the 1-OH derivative in high yields (55-96%).[253] For the hydrolysis of peracylated methyl glycosides (pentoses and hexoses) the acyl group of the primary position was selectively hydrolyzed: the lipase from *Candida cylindracea* was a better catalyst, especially for the removal of the pentanoyl group.[217a,254] The regioselective acetylation of glucose at the 6-position with PPL was also described.[217a,254] Likewise, regioselective acylation of free sugars (including glucose, mannose and galactose) with trichloroethyl esters in pyridine using PPL gave the corresponding 6-O-acyl (e.g. butyryl, acetyl, capryl and lauryl) sugars in 50-90% isolated

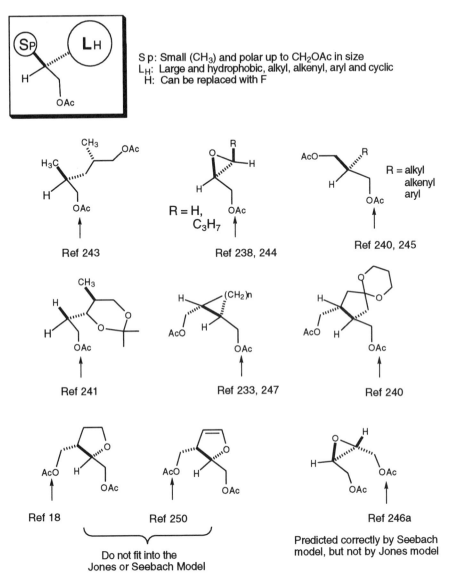

Figure 24. Active-site model for PPL and representative enantioselective hydrolyses explained by the model.

yields.[255] When the 6-OH group of hexoses is acylated, the PPL-catalyzed acylation of the secondary OH groups is not very selective[256,257] (the lipase from *Chromobacterium viscosum* or *Aspergillus niger* is selective for 3-position).[256] Exceptions occur in the case of 6-deoxysugars such as α-methyl-L-fucoside and L-rhamnoside, where selective acylation occurs at the 4-position, and at the 2-position for the corresponding D-enantiomers.[257] A highly selective acylation at the 3-position was also observed in the PPL-catalyzed acetylation (with vinyl acetate) of methyl 4,6-*O*-benzylidene β-D-glucopyranoside (82%), while almost no acylation was observed for the α-isomer.[258] Using the lipase from *Pseudomonas fluorescens*, the 2-position of the α-isomer and the 3-position of the β-glycoside were acylated.[258] Other representative PPL-catalyzed regioselective hydrolyses include the selective hydrolysis on position 2 of 1,6-anhydro-2,3,4-tri-O-butanoyl-galactopyranose[259] and the selective hydrolysis on the β-position of a racemic dimethyl α-methylsuccinate to give the "S" product[260] (α-chymotrypsin cleaved the α-position).[261]

PPL was also used in organic solvents for the formation of cyclic amides,[262] optically active lactones,[263] polyesters[264] and for asymmetric acylation of acylic (Diene) Fe(CO)3 series bearing CHO and CH2OH groups.[265] Some other PPL reactions are listed in Figure 25.

Figure 25. Other representative PPL-catalyzed reaction.

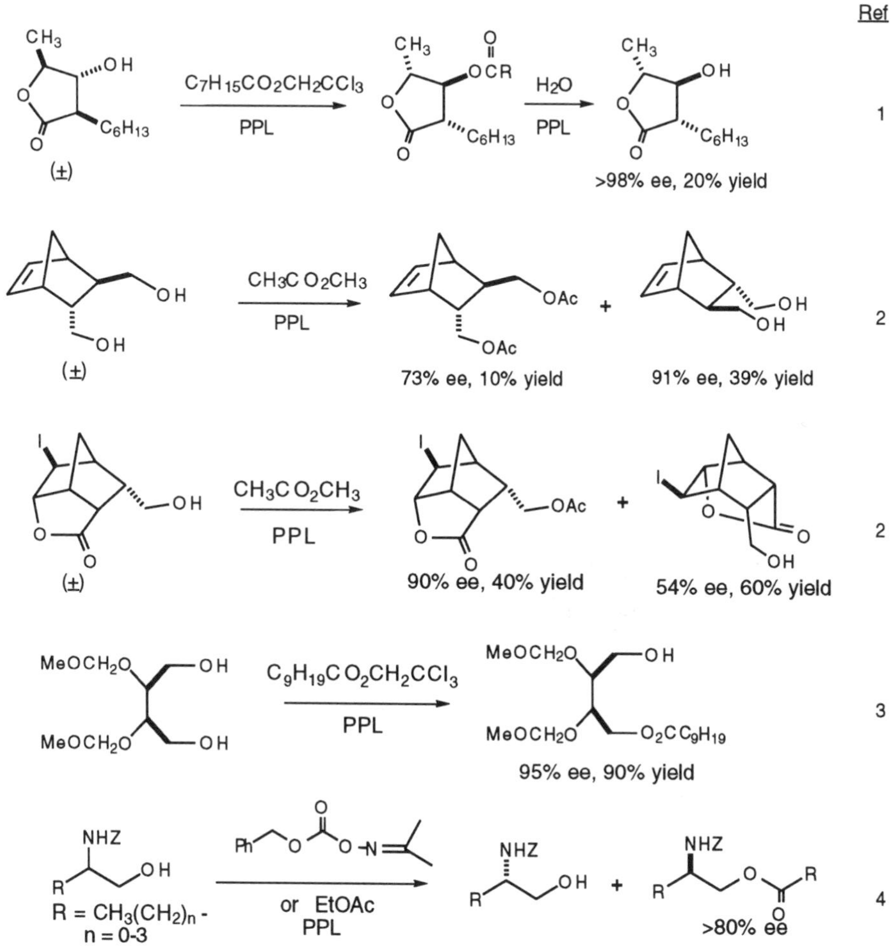

Figure 25: continue

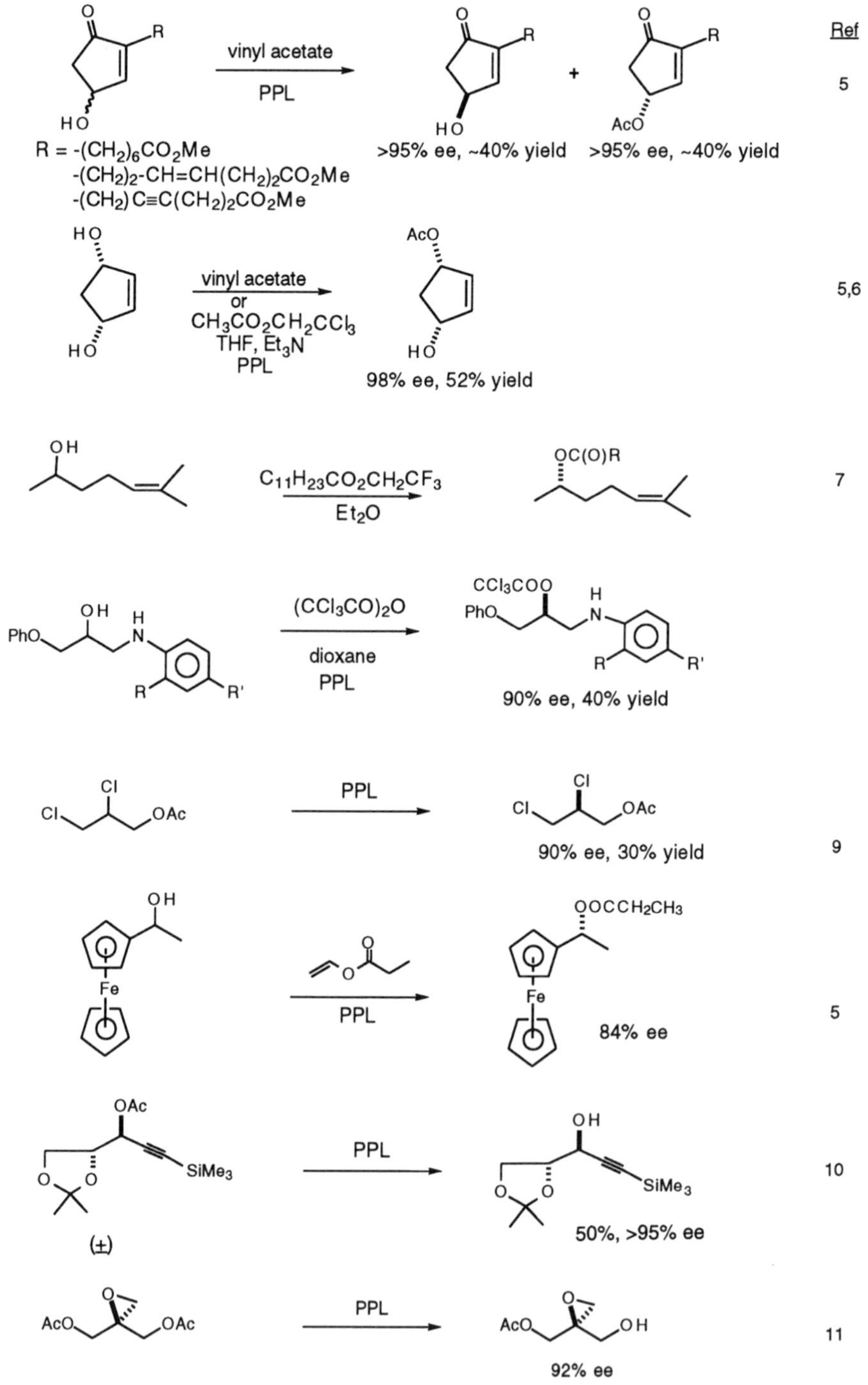

			Ref
	vinyl acetate / PPL		5
R = -(CH₂)₆CO₂Me		>95% ee, ~40% yield >95% ee, ~40% yield	

$R = -(CH_2)_6CO_2Me$
$-(CH_2)_2-CH=CH(CH_2)_2CO_2Me$
$-(CH_2)C\equiv C(CH_2)_2CO_2Me$

vinyl acetate
or
$CH_3CO_2CH_2CCl_3$
THF, Et₃N
PPL

98% ee, 52% yield   5,6

$C_{11}H_{23}CO_2CH_2CF_3$
Et₂O   7

(CCl₃CO)₂O
dioxane
PPL

90% ee, 40% yield

PPL

90% ee, 30% yield   9

PPL

84% ee   5

PPL

50%, >95% ee   10

PPL

92% ee   11

Figure 25: continue

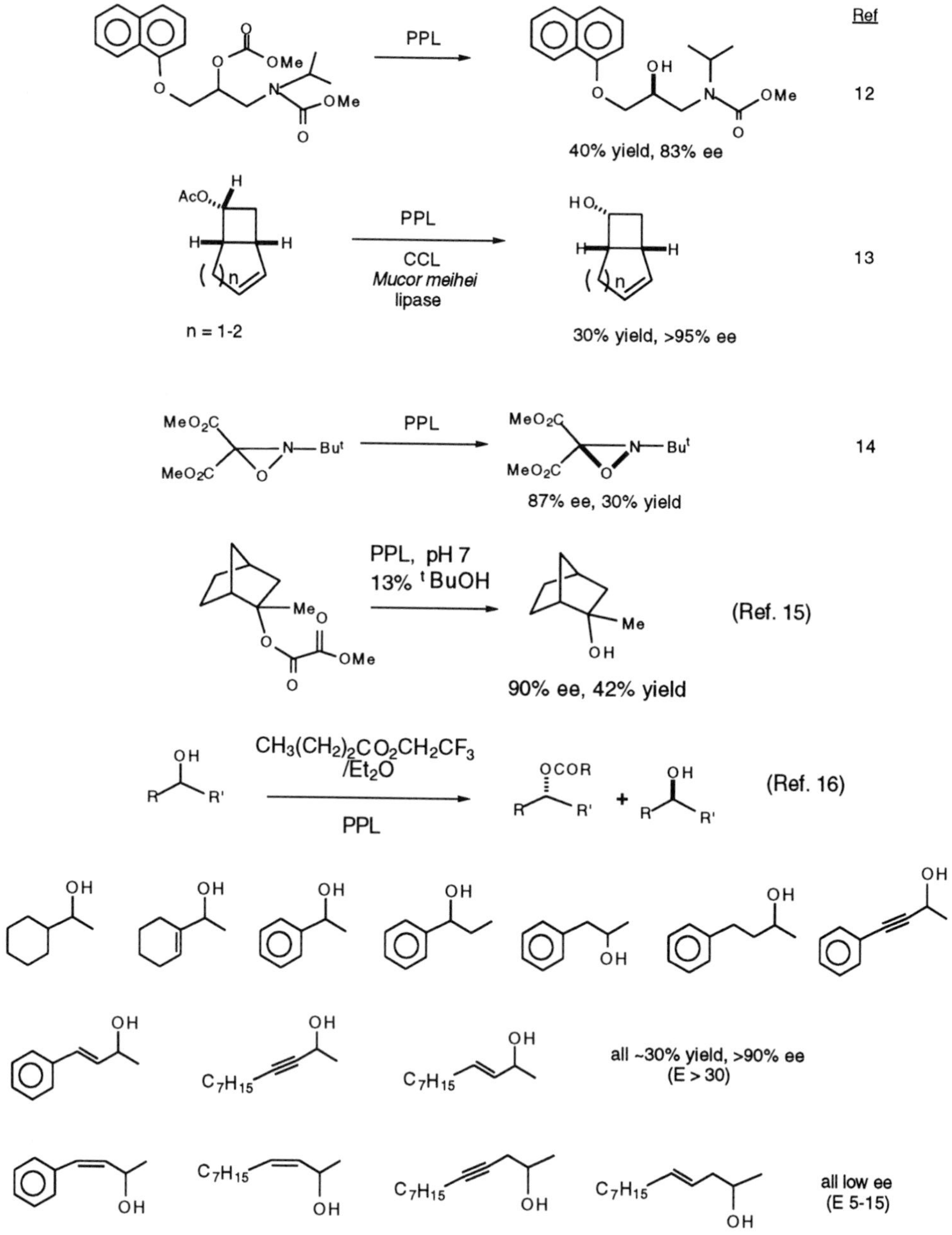

	Ref

PPL → 40% yield, 83% ee — 12

PPL / CCL / *Mucor meihei* lipase, n = 1-2 → 30% yield, >95% ee — 13

PPL → 87% ee, 30% yield — 14

PPL, pH 7 / 13% ^{t}BuOH → 90% ee, 42% yield — (Ref. 15)

$CH_3(CH_2)_2CO_2CH_2CF_3$ /Et$_2$O / PPL → (Ref. 16)

all ~30% yield, >90% ee (E > 30)

C_7H_{15}

all low ee (E 5-15)

C_7H_{15}

Figure 25: continue

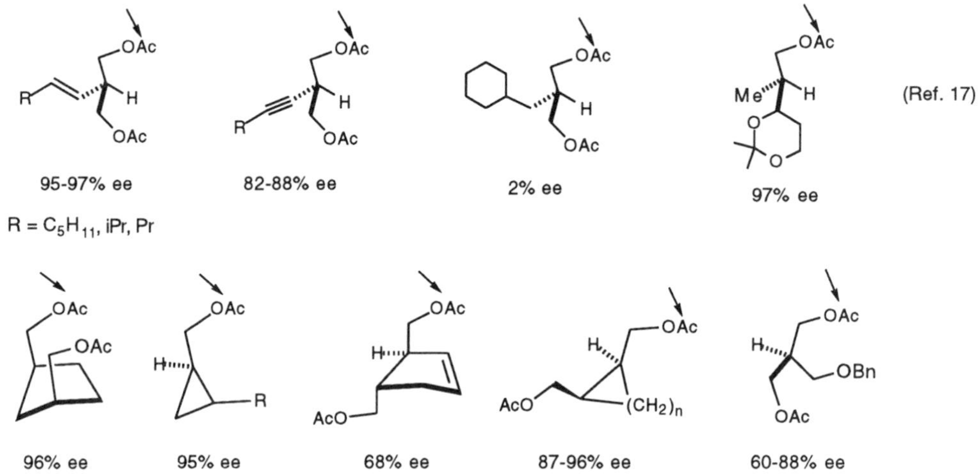

95-97% ee   82-88% ee   2% ee   97% ee   (Ref. 17)

R = C$_5$H$_{11}$, iPr, Pr

96% ee   95% ee   68% ee   87-96% ee   60-88% ee

1.   Nishida, T.; Nihira, T.; Yamada, Y. *Tetrahedron* **1991**, *47*, 6623.
2.   Janssen, A.J.M.; Klunder, A.J.H.; Zwanenburg, B. *Tetrahedron* **1991**, *47*, 5513.
3.   Bestmann, H.J.; Philipp, U.C. *Angew. Chem. Int. Ed. Engl.* **1991**, *30*, 86.
4.   Fernandez, S.; Brieva, R.; Rebolledo, F.; Gotor, V. *J. Chem. Soc., Perkin Trans. 1* **1992**, 2885.
5.   Babiak, K.A.; Ng, J.S.; Dygos, J.H.; Weyker, C.L.; Wang, Y.-F.; Wong, C.-H. *J. Org. Chem.* **1990**, *55*, 3377.
6.   Theil, F.; Ballschuh, S.; Schick, H.; Haupt, M.; Hafner, B.; Schwarz, S. *Synthesis* **1988**, 540.
7.   Stokes, T.M.; Oehlschlager, A.C. *Tetrahedron Lett.* **1987**, *28*, 2091; A lower ee (88%) was obtained with trichloroethyl butyrate: Belan, A.; Bolte, J.; Fauve, A.; Gourey, J.G.; Veschambre, H. *J. Org. Chem.* **1987**, *52*, 256 and with vinyl acetate: Wang, Y.-F.; Lalonde, J.J.; Momongan, M.; Bergbreiter, D.E.; Wong, C.-H. *J. Am. Chem. Soc.* **1988**, *110*, 7200.
8.   Kamal, A.; Rao, M.W. *Tetrahedron: Asymmetry* **1991**, *2*, 751.
9.   Iriuchijima, S.; Keiya, A.; Kojima, N. *Agric. Biol. Chem.* **1982**, *46*, 1593.
10.   Mulzer, J.; Greifenberg, S.; Beckstett, A.; Gottwald, M. *Liebigs Ann. Chem.* **1992**, *11*, 1131.
11.   Sen, Y.-B.; Kho, Y.-H. *Tetrahedron Lett.* **1992**, *33*, 7015.
12.   Shieh, W.-R.; Gou, D.-M.; Chen, C.-S. *J. Chem. Soc., Chem. Commun.* **1991**, 651.
13.   Cotterill, I.C.; Macfarlane, E.L.A.; Roberts, S.M. *J. Chem. Soc., Perkin Trans. 1* **1988**, 3387.
14.   Bucciarelli, M.; Forni, A.; Moretti, I.; Prati, F. *J. Chem. Soc., Chem. Commun.* **1988**, 1614.
15.   a) Brackenridge, I.; McCague, R.; Roberts, S.M.; Turner, N.J. *J. Chem. Soc., Perkin Trans .1*, **1993**, 1093. b) Vanttinen, E.; Kanerva, L.T. *Tetrahedron: Asymmetry* **1992**, *3*, 1529.
16.   Morgan, B.; Oehlschlager, A.C.; Stokes, T.M. *J. Org. Chem.* **1992**, *57*, 3231.
17.   Guanti, G.; Banfi, L.; Narisano, E. *J. Org. Chem.* **1992**, *57*, 1540.

8.3   *Pseudomonas sp.* Lipases

The lipases isolated from different *Pseudomonas* species (PSL) are highly selective, especially for the hydrolysis of the esters of secondary alcohols, and for the corresponding reverse reactions.[266] Although these lipases seem to accept less bulky substrates than do *Candida* lipases, they possess exceptionally high selectivity on narrow open-chain substrates with chiral center located both near and remote from the reaction center (Figure 26).[267] The enzymatic

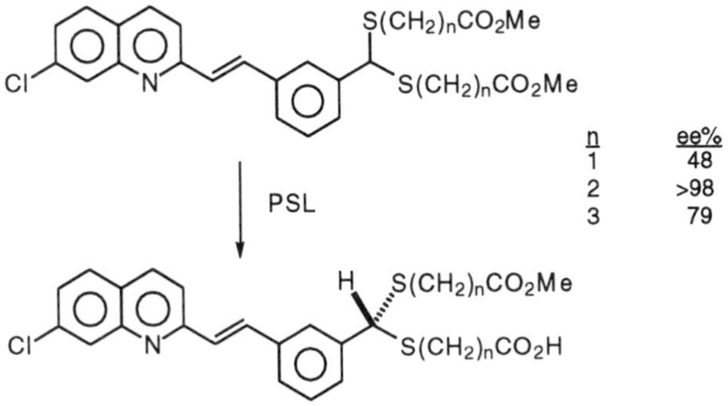

n	ee%
1	48
2	>98
3	79

Figure 26. Enantioselective hydrolysis with PSL of esters with remote prochiral centers.

hydrolysis and transesterification are enantiocomplementary, and allow the preparation of both enantiomers (Figure 27).[218a,268] All commercially available pure or crude *Pseudomonas sp.* lipases possess a stereochemical preference for the (*R*)-configuration at the reaction center of secondary alcohols. This stereochemical preference has led to the development of an active-site model (Figure 28)[269] which is also found in the transformations catalyzed by cholesterol esterase and *Candida cylindracea* lipase.

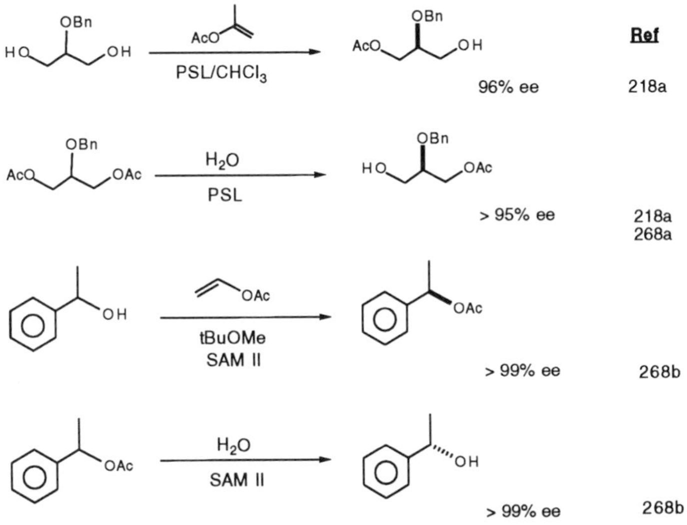

Figure 27. PSL-catalyzed enanticomplementary hydrolysis and transesterification.

In the resolution of α- and β-hydroxyaldehyde acetals, it was found that a significant enhancement in the selectivity was achieved with the use of thioacetals instead of *O*-acetals[270] (Figure 29). A similar situation was observed in the resolution of β-hydroxynitrile esters.[271] The enantioselectivity was improved when the acyl moiety contained a sulfur atom. A high enantioselectivity was also observed in the PSL catalyzed hydrolysis and transesterification of various methyl sulfinylalkanoates.[272]

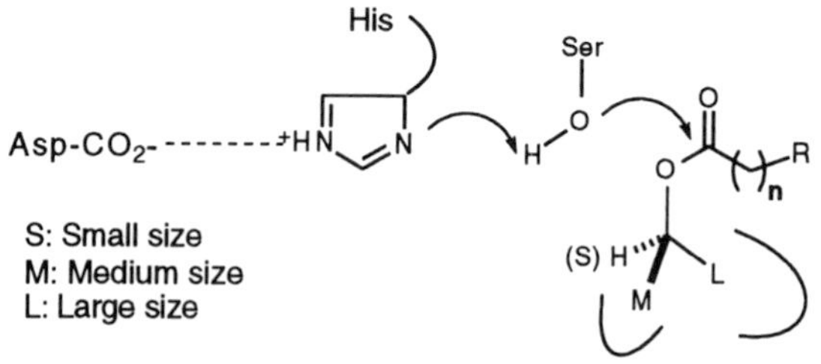

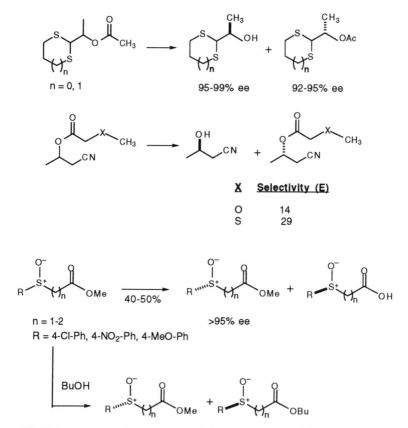

Figure 28. Acitve-site model: Preferred ester substrates for *Pseudomonas* lipase, Cholesterol esterase and *Candida cylindradea* lipase.

Figure 29. Enhancement of enantioselectivity using *S*-containing substrates for PSL.

In the resolution of hydroxyalkanoic acids, two strategies can be employed: hydrolysis of the *O*-acetate of the *tert*-butyl esters[273] or *O*-acetylation of the free hydroxyalkanoic acids[274] (Figure 30). A lower ee was observed when the carboxylic esters were used as substrates for acylation. In the resolution of chiral cyanohydrins with PSL-catalyzed hydrolysis of the esters, the enantioselectivity is very high for aromatic cyanohydrins.[275] The resolved cyanohydrins,

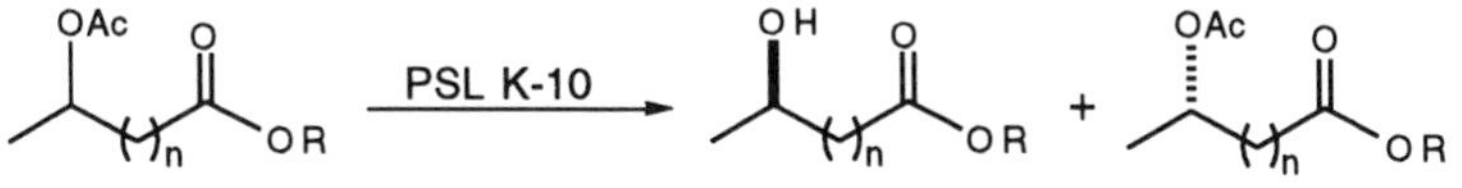

n	R	% conversion	ee alcohol	E
1	H	17	50%	3
1	t-Butyl	38	>99%	>100
5	H	26	45%	3
5	t-Butyl	28	>99%	>100
10	H	16	64%	3
10	t-Butyl	28	99%	>100

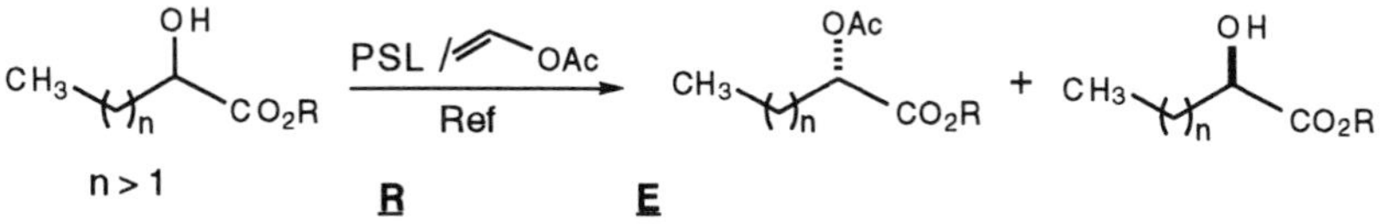

R	E
H	37
CH₃	9
C₆H₅CH₂-	2

Figure 30.

however, tend to racemize in aqueous solution. Although the extent of racemization can be decreased when the resolution is performed at low pH or in organic solvents,[276] the isolation of resolved free cyanohydrins still represents a significant problem.

An improvement of this procedure is the use of acetone cyanohydrin for transhydrocyanation with an aldehyde to form a cyanohydrin, which is subsequently resolved *in situ* via transesterification using isopropenyl acetate (Figure 31).[277] The reversible nature of

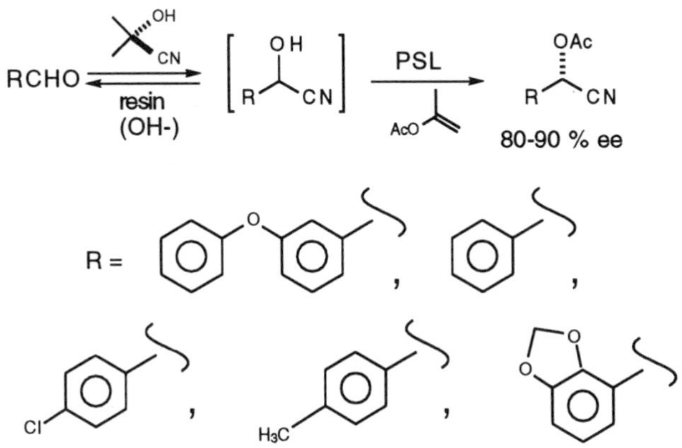

Figure 31.

transhydrocyanation allows a high-yield preparation of a single enantiomer of a cyanohydrin acetate starting from the aldehyde. The rate of this process is, however, quite slow compared to the resolution of racemic cyanohydrin, as described previously, and the asymmetric cyanohydrin formation based on oxynitrilase (see Chapter 4).

In addition to the enantioselective transformation of many primary and secondary chiral and prochiral alcohols, the acyl transfer capability of PSL has also been used in the enantioselective acylation of hydroperoxides[278] and organometallic alcohols[279] (Figure 32), the

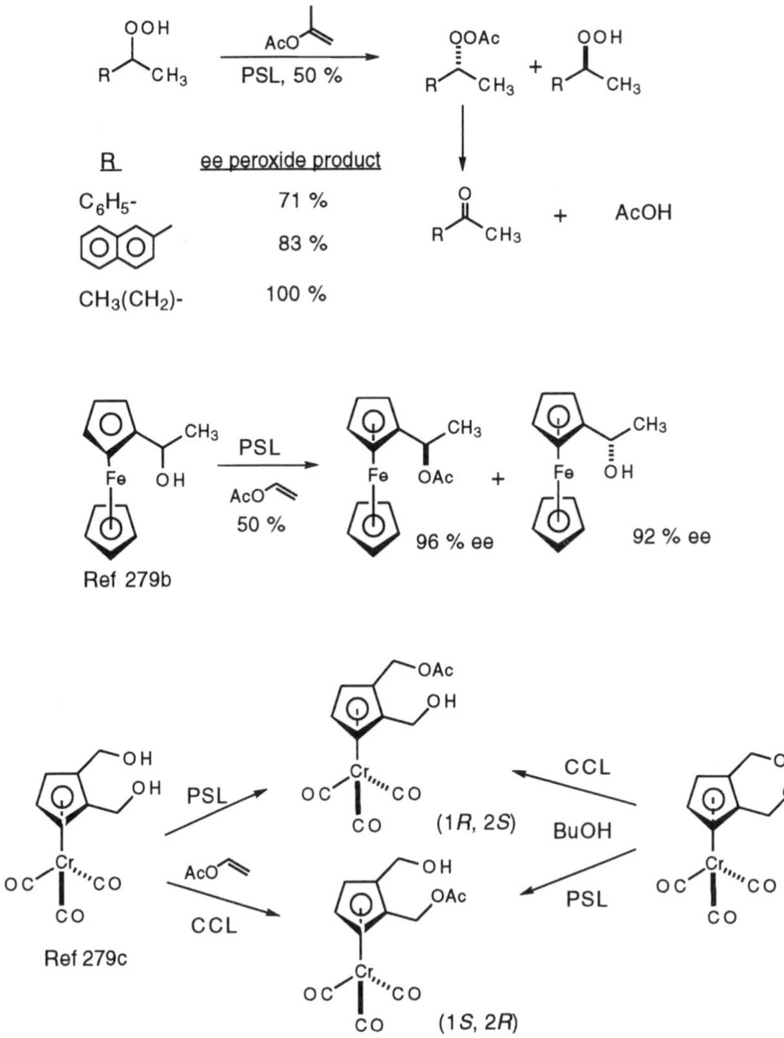

R	ee peroxide product
C_6H_5-	71 %
(naphthyl)	83 %
$CH_3(CH_2)$-	100 %

Ref 279b

Ref 279c

Figure 32.

chemoselective hydrolysis of thioesters in the presence of *O*-esters,[280] the acylation of ε-NH_2 vs. α-NH_2 and β-OH vs. α-NH_2 of dipeptides,[281a] the enantioselective acylation of amines with ethyl acetate,[281b] the regioselective hydrolysis or acylation of glycals[282] (Figure 33), sugars,[283]

Ref

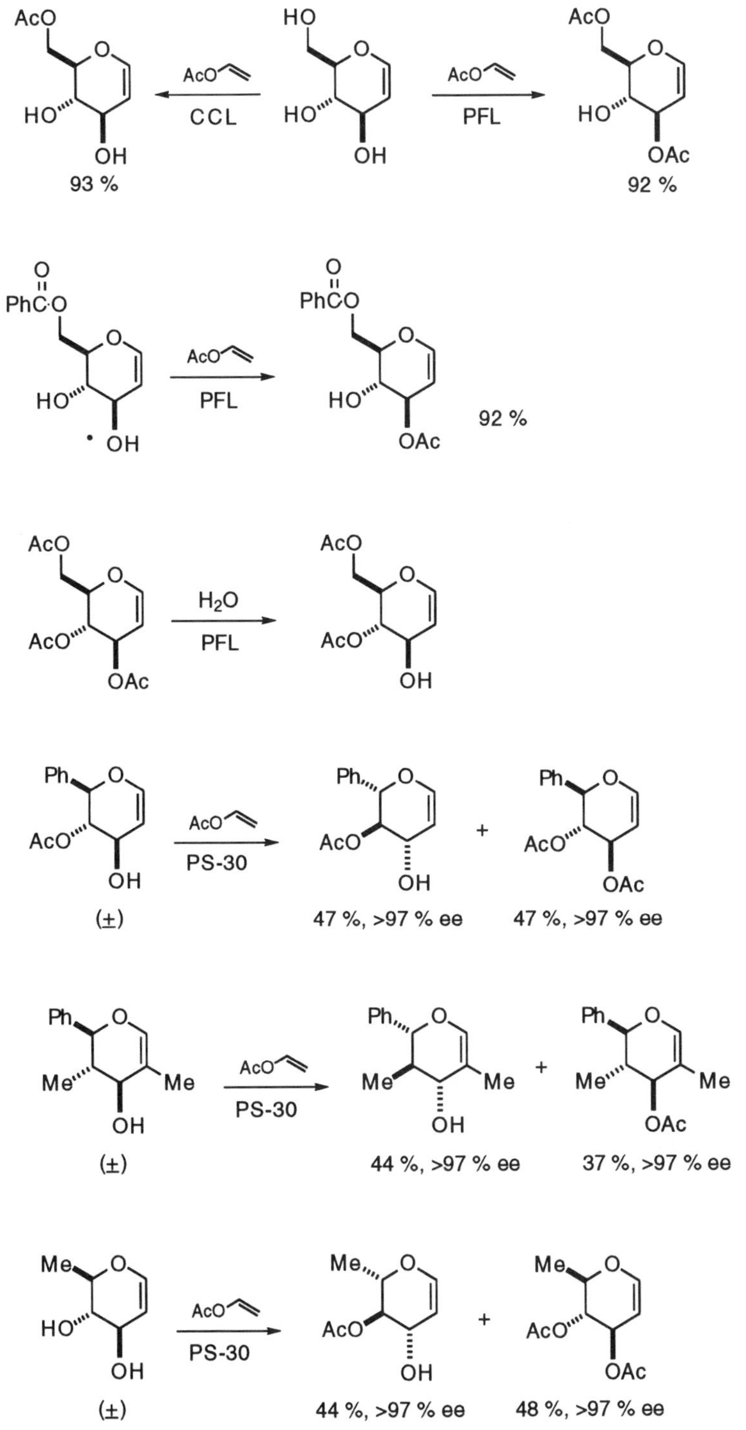

282a

282a

282a

282b

282b

282b

Figure 33.

and nucleosides (PFL is selective for 3-OH and subtilisin for 5-OH),[284] and lactonization.[285] In reaction with 6-deoxysugars such as rhamnose and fucose, the 2-OH group of the D-sugars is selectively acylated by PPL, while the 4-OH group of L-sugars is acylated with PSL.[283] In most cases, the enzyme is used as free form. The efficiency and stability of the enzyme can be improved when immobilized on XAD-8,[286ab] Celite,[286c] gelatin,[286d] Hyflo Super Gel[286e] or when it is used as a lipid-lipase aggregate.[286f] The low cost and high selectivity and stability of PSL make this enzyme a very useful reagent for organic synthesis. In addition to the reactions described in this section, numerous other transformations based on PSL have been reported[266,287] (Figure 34).

Figure 34. Other Selective examples of PSL catalyzed reactions.

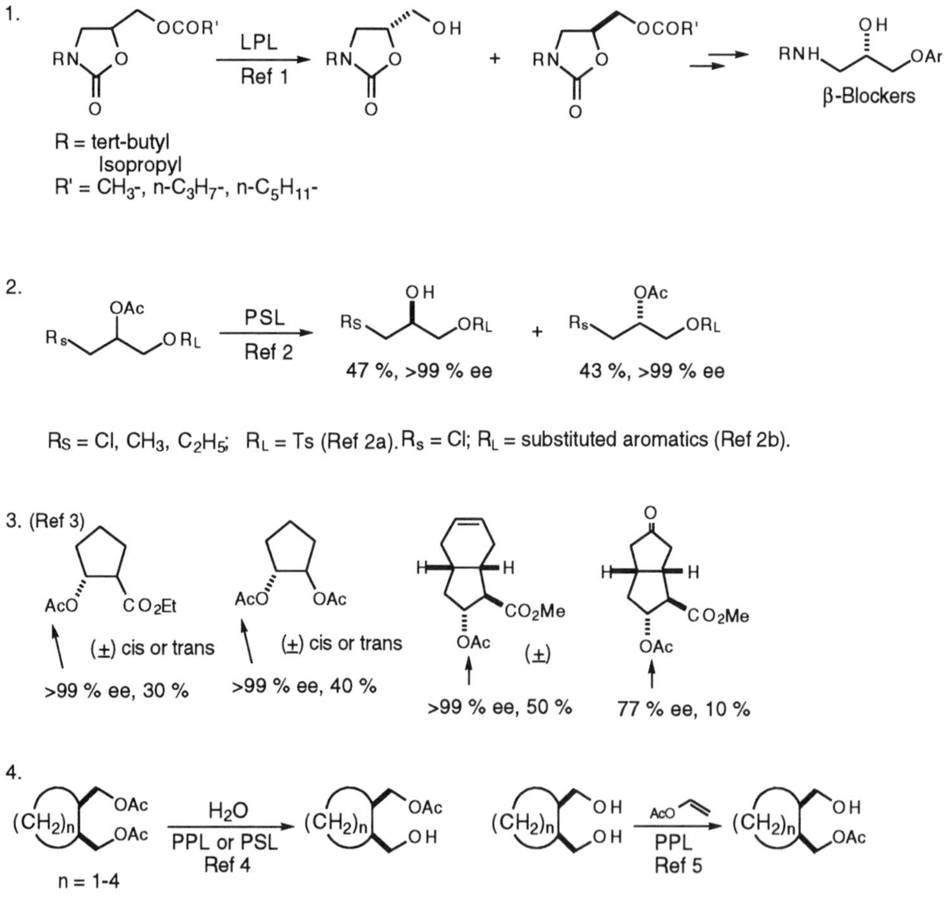

Figure 34: continue

5.

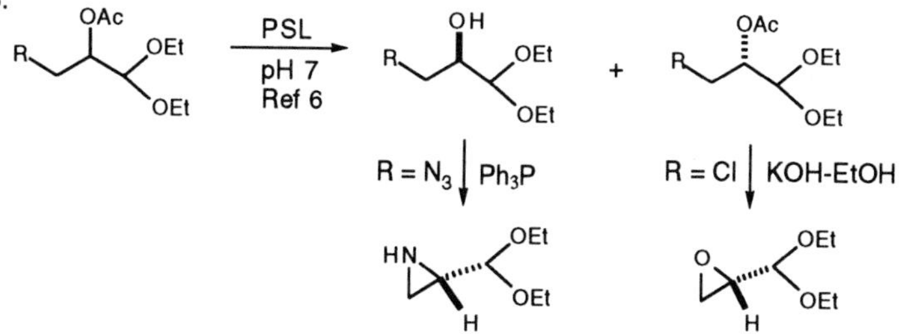

PSL, pH 7, Ref 6

R = N$_3$ | Ph$_3$P

R = Cl | KOH-EtOH

6.

lipase AK
Ref 7b, 7c

32 %, >95 % ee   47 %, >95 % ee

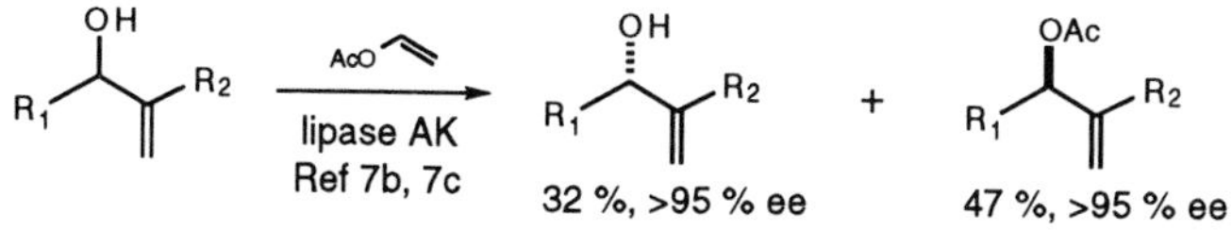

R$_1$ = Me, R$_2$ = Ph (Ref 7a); R$_1$ = Me, Pr; R$_2$ = -$\overset{O}{\overset{\|}{C}}$-OBu, -$\overset{O}{\overset{\|}{C}}$-OCH$_2$CH=CH$_2$, -COBu (Ref 7b

Other compounds prepared:

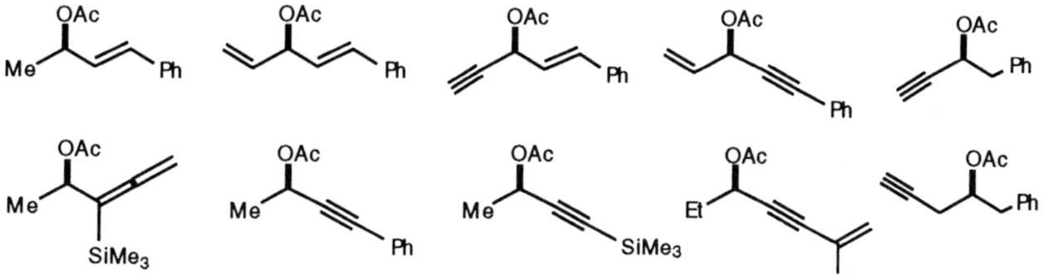

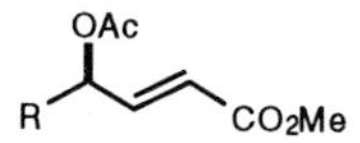

(R = Me, Et, Pr, Ref 7c)

7.

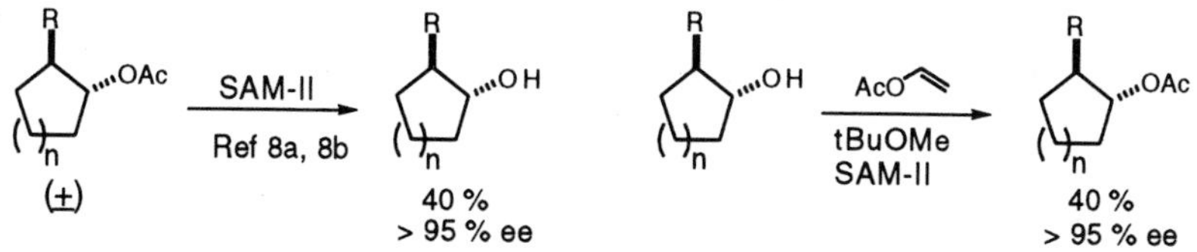

SAM-II
Ref 8a, 8b

40 %
> 95 % ee

AcO, tBuOMe, SAM-II

40 %
> 95 % ee

R = OAc, n = 1-4 (Ref 8a); R = Ph, PhCH$_2$-; PhO-, PhCH$_2$O, n = 2 (Ref 8b)

8.

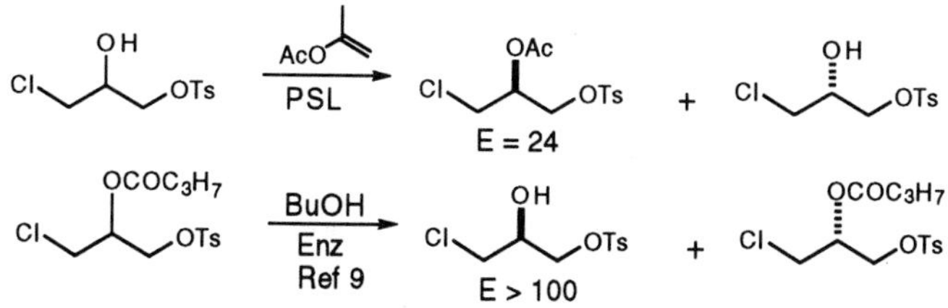

AcO, PSL

E = 24

BuOH
Enz
Ref 9

E > 100

Enz: lipase from *Pseudomonas* (R = ClCH$_2$-), *Candida* (R = Ph-), *Humicola* (R = PhOCH$_2$-)

Figure 34: continue

9.

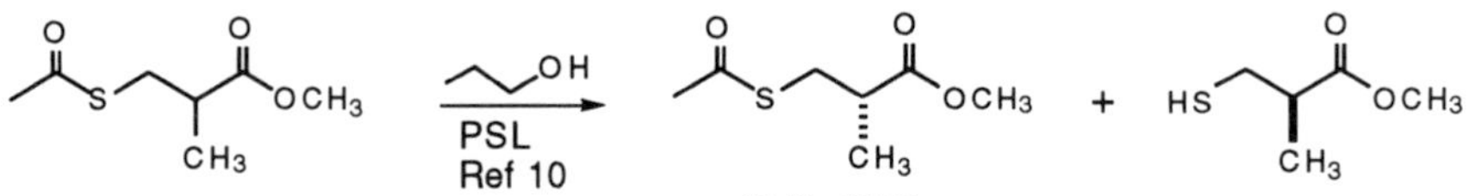

PSL
Ref 10

40 %, 93 % ee

10.

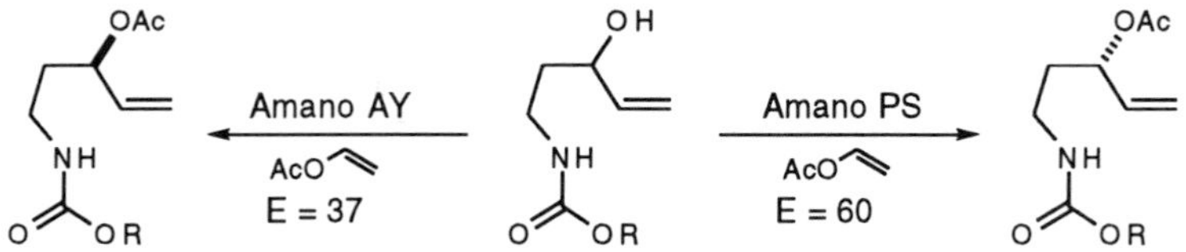

Amano AY

AcO
E = 37

Amano PS

AcO
E = 60

Ref 11

11.

AcO

P-30
Ref 12

> 95 % ee
95 %

L-glucose

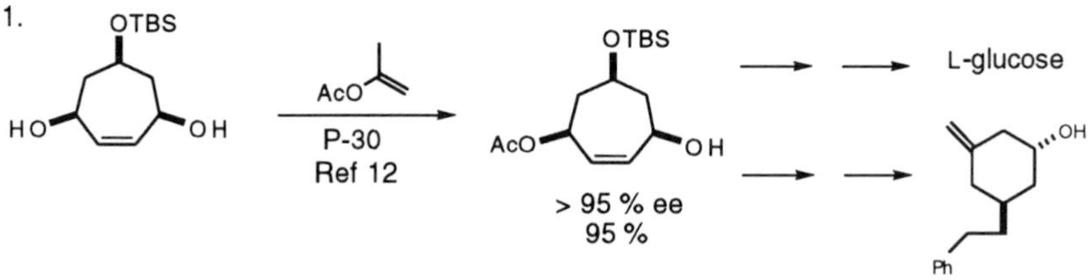

12. Other PSL products via acetylation with vinyl acetate:

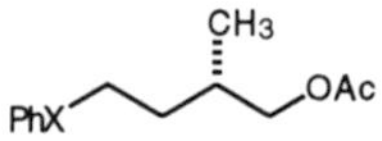

X = Se, S, SO₂ [13a]
40 %, > 98 % ee

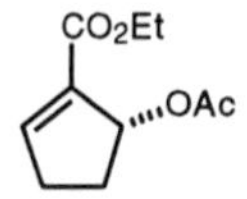

> 98 % ee [13b]

> 98 % ee [13c]
R = C₆H₅CH₂-, C₉H₁₉-

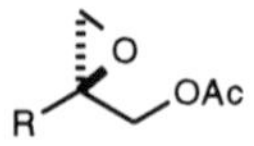

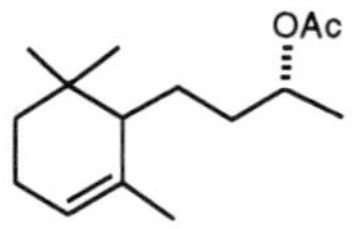

94 % ee [13d]

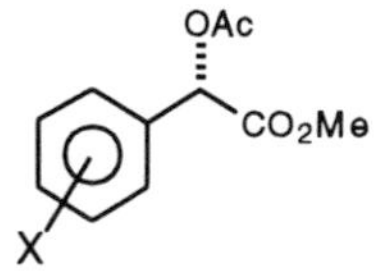

> 95 % ee [13e]
X = H, F, Cl, Br, MeO

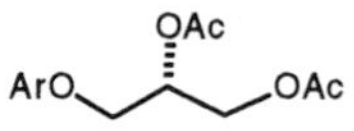

From diol
Ar: Aromatics
80-95 % ee [13f]

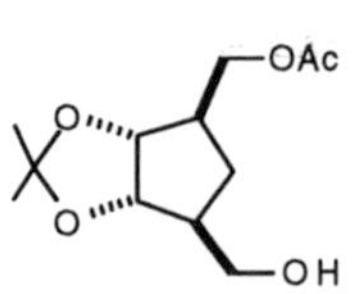

> 99 % ee [13g]

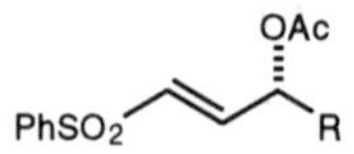

R = Me, Et, C₆H₁₃, C₁₀H₂₁
> 98 % ee [13h]

> 97 % ee [13i]

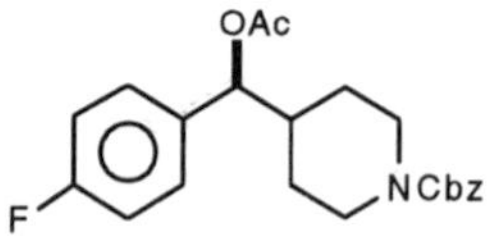

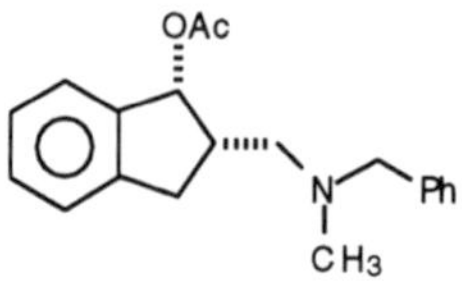

48 %, 98 % ee [13j]

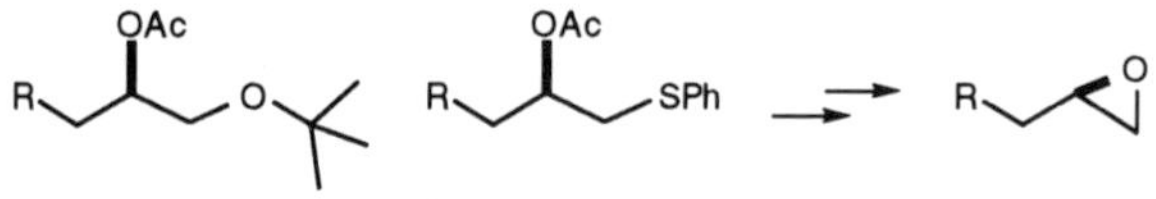

R = H, CH₃; > 95 % ee [13k]

Figure 34: continue

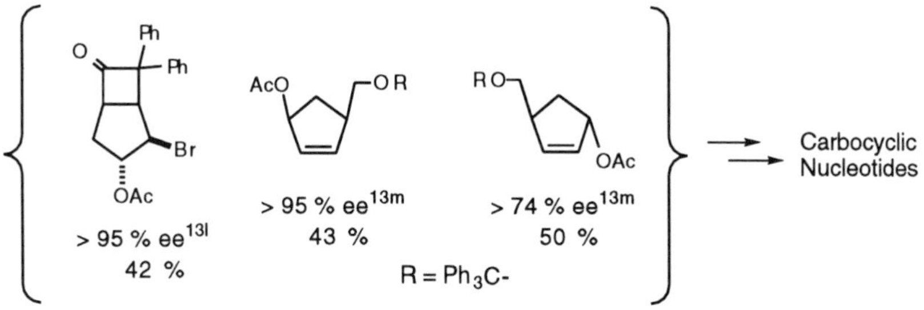

> 95 % ee[13l]
42 %

> 95 % ee[13m]
43 %

> 74 % ee[13m]
50 %

R = Ph₃C-

Carbocyclic Nucleotides

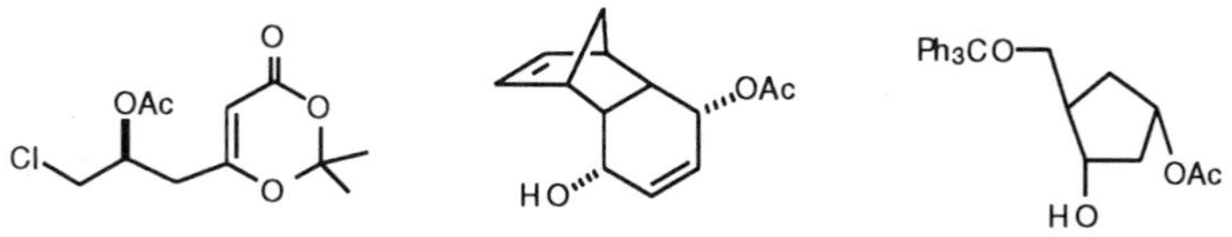

> 98 % ee[13n]

87 %, 100 % ee[13o]

43 %, 98 % ee[13p]

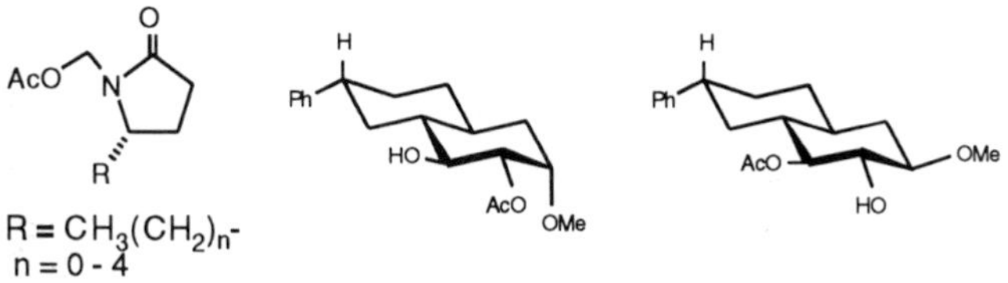

R = CH₃(CH₂)ₙ-
n = 0 - 4
> 98 % ee[13q]

> 90 % ee[13r]

50 %, 99 % ee[13s]

Diltiazem

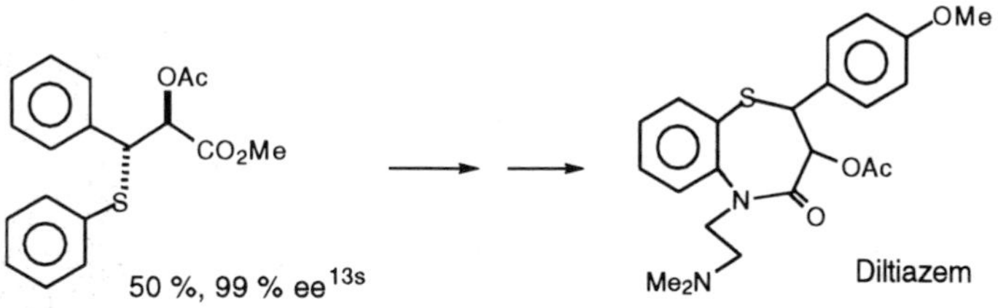

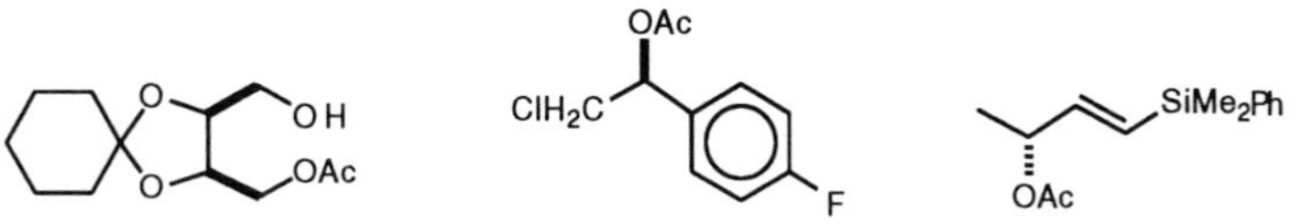

> 99 % ee, 78 % yield [13t]

> 97 % ee[13u]

30 %, 95 % ee[13v]

13.

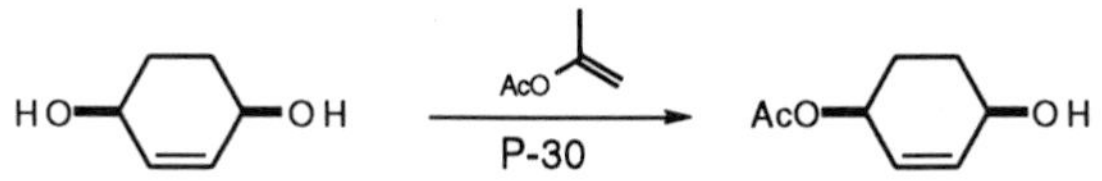

Ref 14

P-30

51 %, 95 % ee

H₂O
P-30

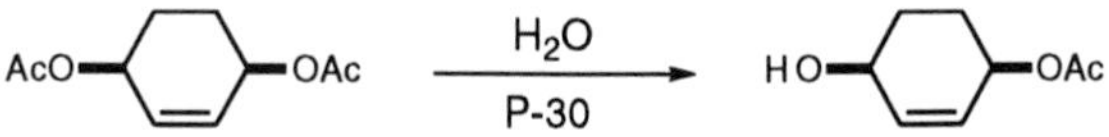

Figure 34: continue

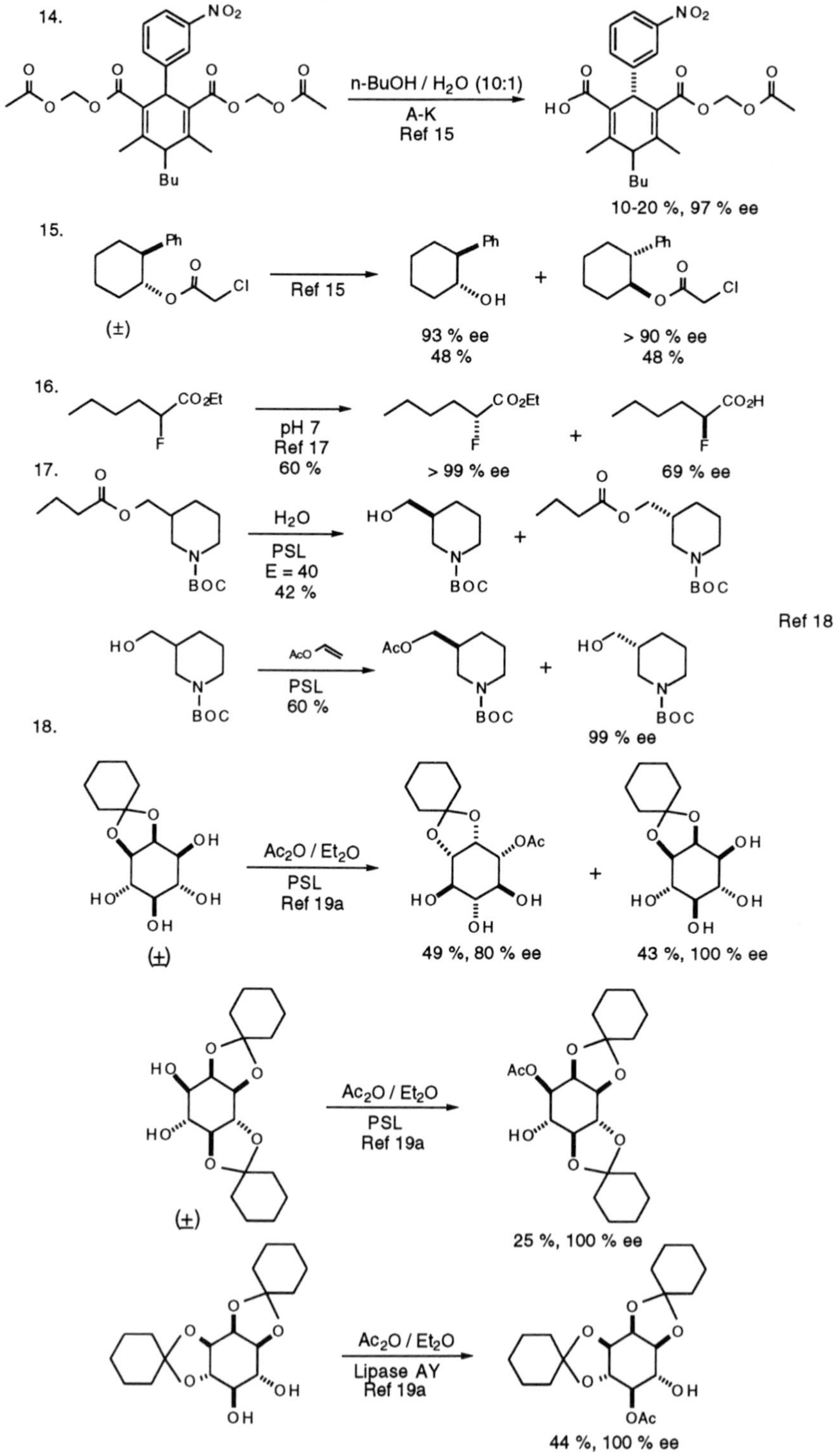

14.

n-BuOH / H₂O (10:1)
A-K
Ref 15

10-20 %, 97 % ee

15.

(±)

Ref 15

93 % ee
48 %

+

> 90 % ee
48 %

16.

pH 7
Ref 17
60 %

> 99 % ee

+

69 % ee

17.

H₂O
PSL
E = 40
42 %

+

Ref 18

AcO⟋

PSL
60 %

+

99 % ee

18.

(±)

Ac₂O / Et₂O
PSL
Ref 19a

49 %, 80 % ee

+

43 %, 100 % ee

(±)

Ac₂O / Et₂O
PSL
Ref 19a

25 %, 100 % ee

Ac₂O / Et₂O
Lipase AY
Ref 19a

44 %, 100 % ee

Figure 34: continue

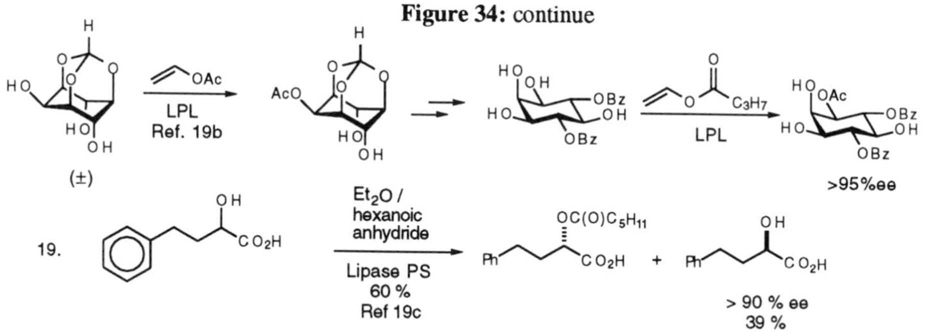

19.

1. Hamaguchi, S.; Asada, M.; Hasegawa, J.; Watanabe, K. *Agric. Biol. Chem.* **1984**, *48*, 2331; Kan, K.; Miyama, A.; Hamaguchi, S.; Ohashi, T.; Watanabe, K. *Agric. Biol. Chem.* **1985**, *49*, 207; also **1985**, *49*, 1669.
2. a) Hamaguchi, S.; Ohashi, T.; Watanabe, K. *Agric. Biol. Chem.* **1986**, *50*, 375; also **1986**, *50*, 1629. b) Ader, U.; Schneider, M.P. *Tetrahedron: Asymmetry* **1992**, *3*, 521.
3. Xie, Z.F.; Suemune, H.; Sakai, K. *J. Chem. Soc., Chem. Commun.* **1987**, 838.
4. Laumen, K.; Schneider, M. *Tetrahedron Lett.* **1985**, *26*, 2073; Kasel, W.; Hultin, P.G.; Jones, J.B. *J. Chem. Soc., Chem. Commun.* **1985**, 1563.
5. Ader, U.; Breitgoff, D.; Klein, P.; Laumen, K.E.; Schneider, M.P. *Tetrahedron Lett.* **1989**, *30*, 1793.
6. Pederson, R.L.; Liu, K.K.-C.; Rutan, J.F.; Chen, L.; Wong, C.-H. *J. Org. Chem.* **1990**, *55*, 4897.
7. a) Burgess, K.; Jennings, L.D. *J. Am. Chem. Soc.* **1991**, *113*, 6129. b) Burgess, K.; Jennings, L.D. *J. Org. Chem.* **1990**, *55*, 1138. c) Burgess, K.; Cassidy, J.; Henderson, I. *J. Org. Chem.* **1991**, *56*, 2050.
8. a) Seemayer, R.; Schneider, M.P. *J. Chem. Soc., Chem. Commun.* **1991**, 49. b) Laumen, K.; Seemayer, R.; Schneider, M.P. *J. Chem. Soc., Chem. Commun.* **1990**, 49.
9. Chen, C.-S.; Liu, Y.-C. *Tetrahedron Lett.* **1989**, *30*, 7165.
10. Bianchi, D.; Cesti, P. *J. Org. Chem.* **1990**, *55*, 5657.
11. Takahata, H.; Uchida, Y.; Momose, T. *Tetrahedron Lett.* **1992**, *33*, 3331.
12. Johnson, C.R.; Golebiowski, A.; McGill, T.K.; Steensma, D.H. *Tetrahedron Lett.* **1991**, *32*, 2597; Johnson, C.R.; Golebiowski, A.; Steensma, D.H. *J. Am. Chem. Soc.* **1992**, *114*, 9414; also *Tetrahedron Lett.* **1991**, *32*, 3931; Johnson, C.R.; Senanayake, C.H. *J. Org. Chem.* **1989**, *54*, 736.
13. a) Ferraboschi, P.; Grisenti, P.; Manzocchi, A.; Santaniello, E. *J. Org. Chem.* **1990**, *55*, 6214; also *Synlett* **1990**, 545. b) Takano, S.; Yamane, T.; Takahashi, M.; Ogasawara, K. *Synlett* **1992**, 410. c) Ferraboschi, P.; Brembilla, D.; Grisenti, P.; Santaniello, E. *J. Org. Chem.* **1991**, *56*, 5478. d) Sugai, T.; Yokochi, T.; Watanabe, N.; Ohta, H. *Tetrahedron* **1991**, *47*, 7227. e) Miyazawa, T.; Kurita, S.; Ueji, S.; Yamada, T.; Kuwata, S. *J. Chem. Soc., Perkin Trans. 1* **1992**, 2253. f) Theil, F.; Weidner, J.; Ballschuh, S.; Kunath, A.; Schick, H. *Tetrahedron Lett.* **1993**, *34*, 305. g) Tanaka, M.; Yoshioka, M.; Sakai, K. *J. Chem. Soc., Chem. Commun.* **1992**, 1454. h) Carretero, J.C.; Dominguez, E. *J. Org. Chem.* **1992**, *57*, 3867. i) Nieduzak, T.R.; Margolin, A.L. *Tetrahedron: Asymmetry* **1991**, *2*, 113. j) Cregge, R.J.; Wagner, E.R.; Freeman, J.; Margolin, A.L. *J. Org. Chem.* **1990**, *55*, 4237. k) Goergens, U.; Schneider, M.P. *J. Chem. Soc., Chem. Commun.* **1991**, 1064. l) Cotterilll, I.C.; Roberts, S.M. *J. Chem. Soc., Perkin Trans. 1* **1992**, 2585. m) Roberts, S.M.; Shoberu, K.A. *J. Chem. Soc., Perkin Trans 1* **1991**, 2605. n) Sasaki, J.-I.; Sakoda, H.; Sugita, Y.; Sato, M.; Kaneko, C. *Tetrahedron Lett.* **1991**, *2*, 343. o) Takano, S.; Moriya, M.; Higashi, Y.; Ogasawara, K. *J. Chem. Soc., Chem. Commun.* **1993**, 177. p) Henly, R.; Elie, C.J.J.; Buser, H.P.; Ramos, G.; Moser, H.E. *Tetrahedron Lett.* **1993**, *34*, 2923. q) Jouglet, B.; Rousseau, G. *Tetrahedron Lett.* **1993**, *34*, 2307. r) Chinn, M.J.; Lacazio, G.; Spackman, D.G.; Turner, N.-J. Roberts, S.M. *J. Chem. Soc., Perkin Trans 1* **1992**, 661. s) Kanerva, L.T.; Sundholm, O. *J. Chem. Soc., Perkin Trans. 1* **1993**, 1385. t) Gais, H.-J.; Hemmerle, H.; Kossek, S. *Synthesis* **1992**, 169. u) Nieduzak, T.R.; Margolin, A.L. *Tetrahedron: Asymmetry* **1991**, *2*, 113. v) Sparks, M.A.; Panek, J.S. *Tetrahedron Lett.* **1991**, *32*, 4085.
14. Harris, K.J.; Gu, Q.-M.; Shih, Y.-E.; Girdaukas, G.; Sih, C.J. *Tetrahedron Lett.* **1991**, *32*, 3941.
15. Holdgrun, X.K.; Sih, C.J. *Tetrahedron Lett.* **1991**, *32*, 3465.
16. Schwartz, A.; Madan, P.; Whitesell, J.K.; Lawrence, R.M. *Org. Syn.* **1990**, *69*, 1.
17. Kalaritis, P.; Regenye, R.W. *Org. Syn.* **1990**, *69*, 10.
18. Ling, L.; Ozaki, S. *Tetrahedron Lett.* **1993**, *34*, 2501.
19. a) Ling, L.; Watanabe, Y.; Akiyama, T.; Ozaki, S. *Tetrahedron Lett.* **1992**, *33*, 1911. b) Andersch, P.; Schneider, M. P. *Tetrahedron: Asymmetry* **1993**, *4*, 2135. c) Sugai, T.; Ohta, H. *Agric. Biol. Chem.* **1991**, *55*, 293.

8.4 *Candida sp.* Lipases

The lipases from the yeast *Candida lipolytica, Candida antarctica, Candida rugosa,* and *Candida cylindracea* (CCL) are commonly used in organic synthesis. The recombinant form of *C. antarctica* lipase (SP435) is available from NOVO. The *C. lipolytica* lipase preparation also contains a small portion (0.3-0.7 mg/g) of esterases that catalyze the selective hydrolysis of hindered esters such as α-methyl-α-amino acid esters and tertiary α-substituted carboxylic acid esters.[288] The enzyme CCL was also used in organic solvents for the resolution of open-chain α-halogenated acids via esterification, and for the resolution of racemic open-chain secondary alcohols via transesterification.[289] This enzyme was also applied to the resolution of menthol in aqueous and organic solvents.[290] The enzyme accommodates bulky esters in its active site.[291-297] For transesterifications the length of the alkyl chain of the acyl group influences the efficiency and selectivity, and a clear minimum was observed for transesterification or esterification of six-carbon acyl moieties.[298] The enzyme showed better enantioselectivity in hydrophobic organic solvents than that in hydrophilic solvents.[299] High selectivity of the enzymes for bulky substrates can be seen in a number of CCL-catalyzed enantioselective transformations. Cyclohexyl enol esters can be hydrolyzed, and the enol products are selectively protonated to give ketoester[292] (Figure 35). Racemic 2,3-dihydroxy-2-methyl carboxylates,[293] cyclohexane 1,2,3-triol esters,[294]

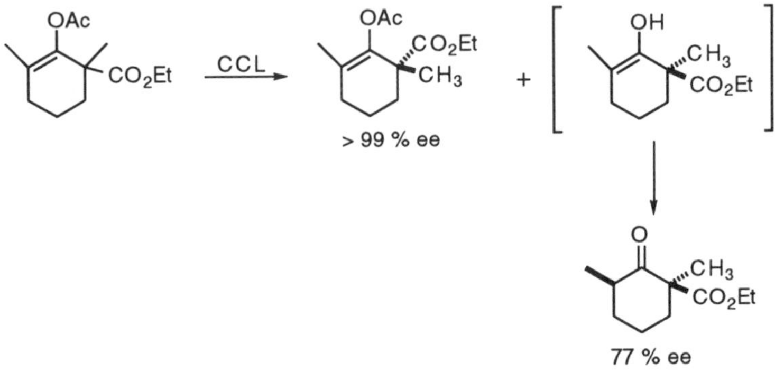

Figure 35. CCL-catalyzed hydrolysis of enol ester.

α-aryl[295] and aryloxypropionic acids (Figure 36) and bicyclic secondary alcohols[291] can be resolved. The results obtained from twenty-five norbornane-type alcohol substrates led to the development of a model[300] (Figure 37) for rationalization and prediction of the enzymatic reactions.

The crude preparation of CCL contains two isomeric forms of the enzyme, both possessing the same stereochemical preference, but with different degrees of enantioselectivity.[301] The more selective form can be separated. Alternatively, the mixture can be treated with the surface-active agent deoxycholate and an organic solvent system (ethanol/ether) to transform the less selective isomer to the more selective one.[301] The enantioselectivity of CCL can also be enhanced in the presence of noncompetitive inhibitors such as chiral amines, as illustrated in the enantioselective hydrolysis of 2-arylpropionic acid and 2-aryloxypropionic acid esters.[302] The enantioselectivity of CCL is also influenced by organic solvents. In the esterification of racemic 2-(4-chlorophenoxy)propionic acid with *n*-butanol, the *R*-enantiomer is preferred in hexane, isooctane, or toluene in the presence of water, and the *S*-enantiomer is preferred in CH$_2$Cl$_2$, acetone, *n*-butanol and THF.[303] In the resolution of 2-chloropropionic acid, an alcohol having substituents both at the α- and the β-carbon preferentially esterified the *S*-acid in CCL-catalyzed reactions, while a straight chain alcohol preferentially esterified the *R*-acid.[304] The enantioselectivities of these reactions are similar.

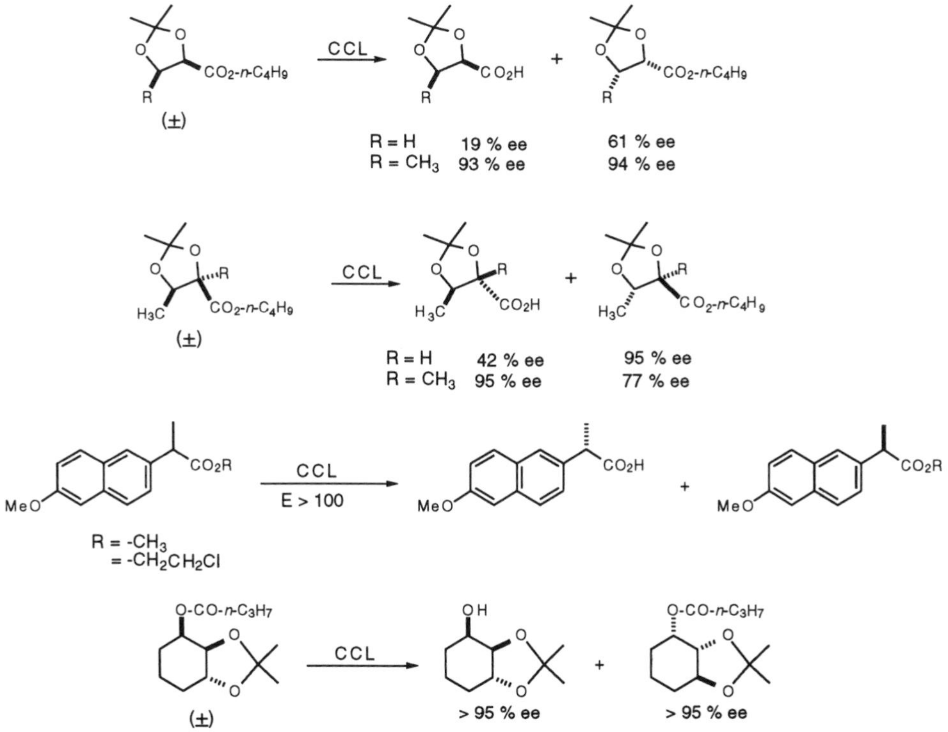

R = H   19 % ee   61 % ee
R = CH$_3$ 93 % ee   94 % ee

R = H   42 % ee   95 % ee
R = CH$_3$ 95 % ee   77 % ee

R = -CH$_3$
  = -CH$_2$CH$_2$Cl

> 95 % ee   > 95 % ee

Figure 36. CCL-catalyzed reactions with bulky substrates.

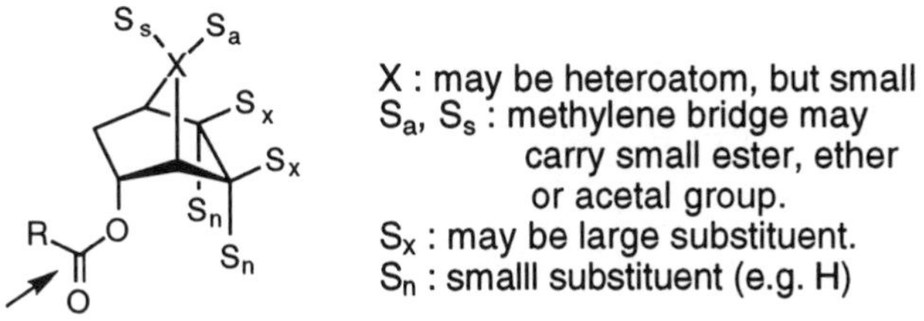

X : may be heteroatom, but small
S$_a$, S$_s$: methylene bridge may
          carry small ester, ether
          or acetal group.
S$_x$: may be large substituent.
S$_n$: smalll substituent (e.g. H)

Figure 37. Bicyclic substrate model for CCL.

CCL also exhibits a high regioselectivity in reactions with sugars and polyhydroxylated compounds. The CCL-catalyzed hydrolysis of peracylated α-methyl glucoside occurred regioselectively at the 6-position and the pentanoyl group was found to be the optimal acyl group[217a] (92% isolated yield). Regioselective acylation of β-glucoside at 6-position was also reported.[256,305] Further acylation of the 6-*O*-acyl product however gave a mixture of products, indicating a poor regioselectivity for the secondary hydroxyl groups. Both *Aspergillus niger* and *Chromobacterium viscosum* lipase selectively acylate the 3-position, and are the basis for a high yield synthesis of 3,6-diacyl glucoside.[256] The enzyme CCL also selectively hydrolyzes the 4-*O*-acetyl group of 1,6-anhydro-2,4-di-*O*-acetyl-3-azido-3-deoxy-β-D-glucose,[306] while Alcalase is selective for the 2-position. Treatment of sucrose peracetate with CCL resulted in a selective hydrolysis of the 6-acetyl group of the glucose moiety while Alcalase or protease N is selective for the 1-acetyl group of the fructose moiety.[307]

Selective acylation (~80%) of the 3α-OH group of bile acid derivatives with CCL and trichloroethyl butanoate has also been accomplished.[296] In the acylation of other polyols, CCL is selective for the less hindered primary hydroxy, or the benzylic hydroxyl group.[308] Hydrogen peroxide is also a substrate for CCL and fatty acids have been converted to peracids for use in epoxidation[309] (Figure 38). Chiral amines can also be resolved in diisopropyl ether via CCL-

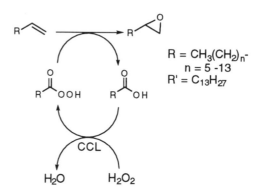

$R = CH_3(CH_2)_n-$
$n = 5 - 13$
$R' = C_{13}H_{27}$

Figure 38. CCL-catalyzed synthesis of peracids.

catalyzed acylation with the methyl esters of β-furyl- or β-phenyl propionate, and the enzyme from *C. antartica* (SP435) appears to be more selective than CCL.[310] Other *Candida* lipase reactions include the enantioselective hydrolyses of α-fluoromalonates (with CCL),[311a] cyanohydrin esters (with CCL),[311b] carboxylic acid esters containing a chiral sulfoxide (with *C. rugosa*),[311c] 2-*O*-acyl derivatives of chiral glycerols (with *C. antarctica*),[311d] polymerization of 10-hydroxydecanoic acid (with CCL),[311e] and benzyloxycarbonylation of α-, *xylo*-, anhydro- and arabino-nucleosides at the 5'-OH group (with *C. antarctica*),[311f] and many other transformations.[311g] Some representative transformations are shown in Figure 39.

Figure 39: Other CCL-catalyzed reactions

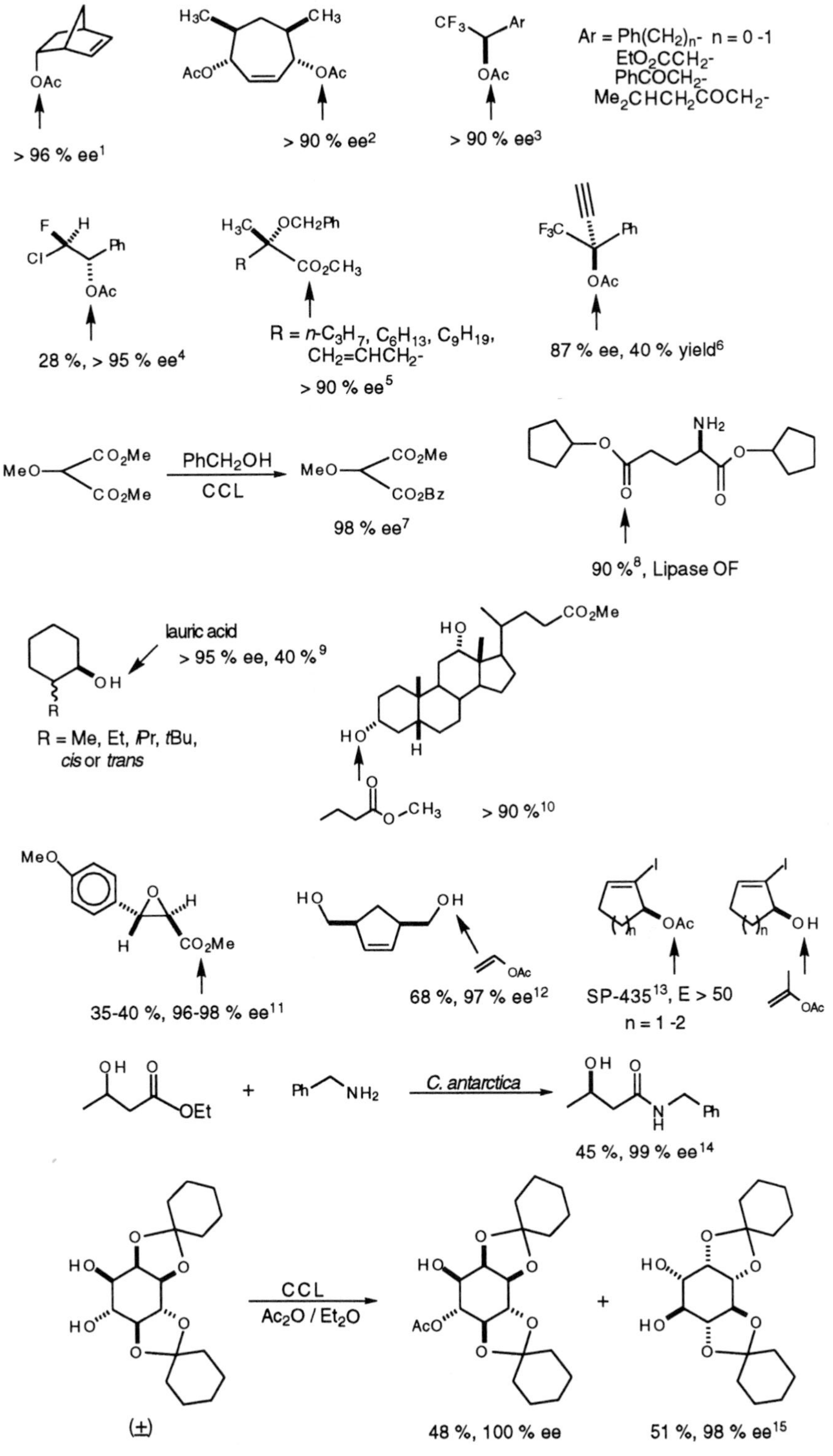

> 96 % ee[1]

> 90 % ee[2]

> 90 % ee[3]

Ar = Ph(CH$_2$)$_n$- n = 0 -1
EtO$_2$CCH$_2$-
PhCOCH$_2$-
Me$_2$CHCH$_2$COCH$_2$-

28 %, > 95 % ee[4]

R = n-C$_3$H$_7$, C$_6$H$_{13}$, C$_9$H$_{19}$,
CH$_2$=CHCH$_2$-

> 90 % ee[5]

87 % ee, 40 % yield[6]

98 % ee[7]

90 %[8], Lipase OF

lauric acid
> 95 % ee, 40 %[9]

R = Me, Et, iPr, tBu,
cis or *trans*

> 90 %[10]

35-40 %, 96-98 % ee[11]

68 %, 97 % ee[12]

SP-435[13], E > 50
n = 1 -2

C. antarctica

45 %, 99 % ee[14]

CCL
Ac$_2$O / Et$_2$O

(±)

48 %, 100 % ee

51 %, 98 % ee[15]

Figure 39: continue

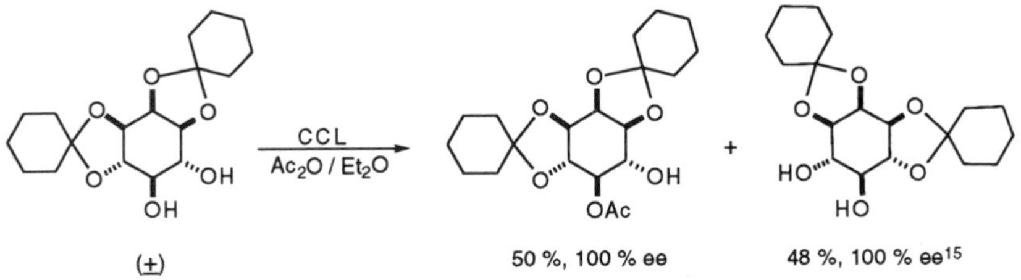

CCL
Ac$_2$O / Et$_2$O

+

(±)   50 %, 100 % ee   48 %, 100 % ee[15]

1.   Eichberger, G.; Penn, G.; Faber, K.; Griengl, H. *Tetrahedron Lett.* **1986**, *27*, 2843.
2.   Pearson, A.J.; Bansal, H.S.; Lai, Y.-S. *J. Chem. Soc., Chem. Commun.* **1987**, 519.
3.   Lin, J.T.; Yamazaki, T.; Kitazume, T. *J. Org. Chem.* **1987**, *52*, 3211.
4.   Yamazaki, T.; Ichikana, S.; Kitazume, T. *J. Chem. Soc., Chem. Commun.* **1989**, 253.
5.   Sugai, T.; Kakeya, H.; Ohta, H. *J. Org. Chem.* **1990**, *55*, 4643.
6.   O'Hagan, D.; Zaidi, N.A. *J. Chem. Soc., Perkin Trans. 1* **1992**, 947.
7.   Gutman, A.L.; Shapira, M.; Boltanski, A. *J. Org. Chem.* **1992**, *57*, 1063.
8.   Wu, S.-H.; Chu, F.-Y.; Chang, C.-H.; Wang, K.-T. *Tetrahedron Lett.* **1991**, *32*, 3529.
9.   Langrand, G.; Secchi, M.; Buono, G.; Baratti, J.; Triantaphylides, C. *Tetrahedron Lett.* **1985**, *26*, 1857.
10.   Riva, S.; Bovara, R.; Ottolina, G.; Secundo, F.; Carrea, G. *J. Org. Chem.* **1989**, *54*, 3161.
11.   Gentile, A.; Giordano, C.; Fuganti, C.; Ghirotto, L.; Servi, S. *J. Org. Chem.* **1992**, *57*, 6635.
12.   Mekrami, M.; Sicsic, S. *Tetrahedron: Asymmetry* **1992**, *3*, 431.
13.   Johnson, C.R.; Sakaguchi, H. *Synlett* **1992**, 813.
14.   Garcia, M.J.; Rebolledo, F.; Gotor, V. *Tetrahedron: Asymmetry* **1992**, *3*, 1519.
15.   Ling, L.; Watanabe, Y.; Akiyama, T.; Ozaki, S. *Tetrahedron Lett.* **1992**, *33*, 194.

8.5 Lipases and Esterases from Other Species, Case Study

The lipases from *Mucor* species such as *M. miehei* and *M. javanicus* have recently been used in synthesis.[312-313] The former is available from Amano, or from NOVO as an immobilized form called Lipozyme. The latter is available from Amano as lipase M. Both enzymes seem to possess a stereochemical preference similar to that of *Pseudomonas* lipases, but opposite to that of pig liver esterase and protease N. The three-dimensional structure of *M. Miehei* has been determined and contains an Asp-His-Ser catalytic triad.[210] A new process for the increase of enantiomeric excess via Lipozyme-catalyzed interesterification is to have the chiral alcohol moiety that is to be resolved enters the active-site twice[314] (Figure 40). An ester of a chiral alcohol, for

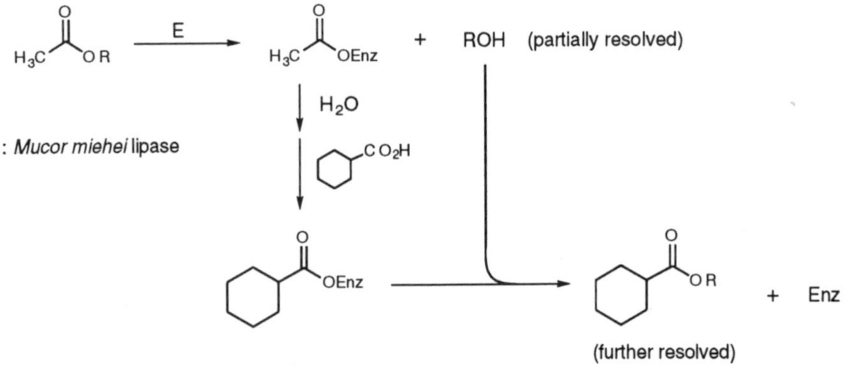

Figure 40. Kinetic double resolution.

example, may react with cyclohexylcarboxylic acid in the presence of a small amount of water to give a new cyclohexylcarboxylic acid ester of the resolved alcohol.

A kinetic resolution for the preparation of the antiinflammatory agent Ketorolac was achieved with *M. miehei* lipase catalyzed hydrolysis of the racemic methyl ester (Figure 41).[315]

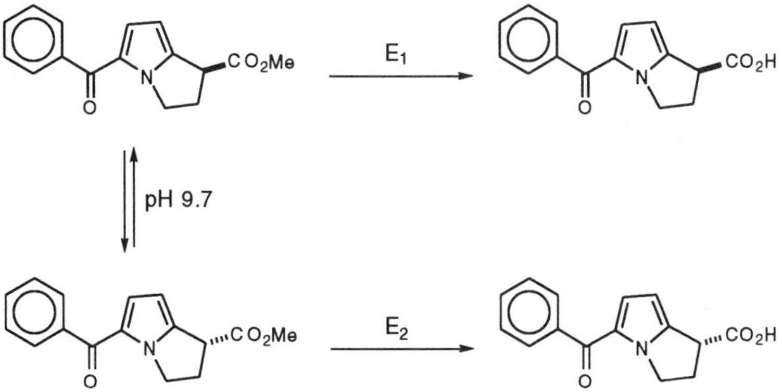

E$_1$: *Streptomyces griseus* protease (sigma type XXI), E > 100
Aspergillus saitoi protease (sigma type XXI), E > 100
Bacillus subtilis protease (Amano protease N), E > 100

E2 : *Mucor miehei* lipase, E = 101
Pseudomonas lipase, E = 65
C. cylindracea lipase, E = 23

Figure 41. Kinetic resolution coupled with racemization *in situ.*

Since the ester is easily racemized at pH 9.7, both enantiomers are eventually converted to the (*R*)-product. Enantiocomplementary to this process is the use of protease N and some other microbial proteases to produce the (*S*)-enantiomer. *M. miehei* is also very selective for enantiotopic groups. *cis*-1,4-Diacyl-2-cyclopentenediol diacetate, for example, has been selectively hydrolyzed at the (*S*)-position to give the monoester in 97% yield and 97% ee.[316a] This high degree of enantioselectivity is similar to that based on the acetylcholine esterase reaction.[179a,316b] The lipases PPL, and PSL, and *Chromobacterium viscosum* lipase, all show the same enantioselectivity and give the monoester in >80% yield and >90% ee. Reaction with CCL, however, gave a lower enantioselectivity.[316c] To prepare the enantiomeric monoester, pig liver esterase was used to hydrolyze the (4*S*)-center selectively. The selectivity was, however, not very high (83% yield, 81% ee).[316d] Alternatively, transesterification of the *cis*-diol with an ester (enol esters or other active ester) catalyzed by PPL gave the (1*R*,4*S*)-monoester[317] (Figure 42). An active site model of *M. miehei* lipase (MML) has been reported.[318] The polyesterification of adipic acid and butane-1,4-diol by Lipozyme using a two-chamber reaction and molecular sieves as a dehydrating agent gave polymeric products containing 20 or more repeating units.[319] The hydrolysis of *bis*-(hydroxymethyl)-tetrahydrofuran dibutyrates with *M. javanicus* lipase (MJL) occurs at the (*R*)-center to give the monoester in >97% ee[320] (Figure 43). Other lipases such as

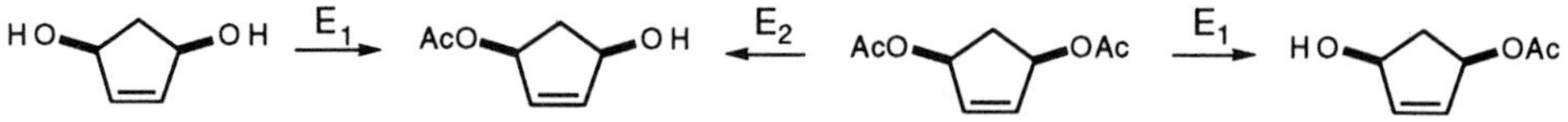

E$_1$ = PPL, PSL, MML, CVL, electric eel acetylcholine esterase
E$_2$ = PLE

Figure 42.

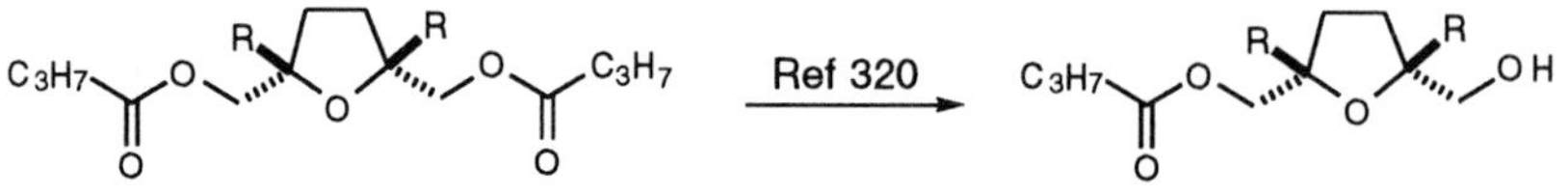

(2S, 5R)-monoester

Lipase	R	ee%
MJL	H, CH$_3$	>97
CCL	H	12
PSL	H	81
PPL	H	85

Figure 43.

PPL and CCL gave the same product with lower ee. An opposite stereoselectivity was observed with the use of PLE. The enzyme MJL has also been used in the selective deprotection of the heptyl ester of *N*-Cbz-(or Fmoc)-*O*-glycosylated amino acids wherein the sugar moiety is an α- or β-linked 2-azido-3,4,6-triacetyl gluco- (or galactosyl) group.[321]

In another application, *Mucor miehei* lipase (Amano, MAP-10) was used in the resolution of methyl *trans*-β-phenylglycidate via transesterification in hexane/isobutyl alcohol (1:1).[322] The unreacted substrate (2*R*, 3*S*, 42%) and product (2*S*, 3*R*, 43%) were obtained in 95% ee (Figure 44). Both enantiomers were converted to *N*-benzoyl-(2*R*,3*S*)-3-phenylisoserine, the C-13 side

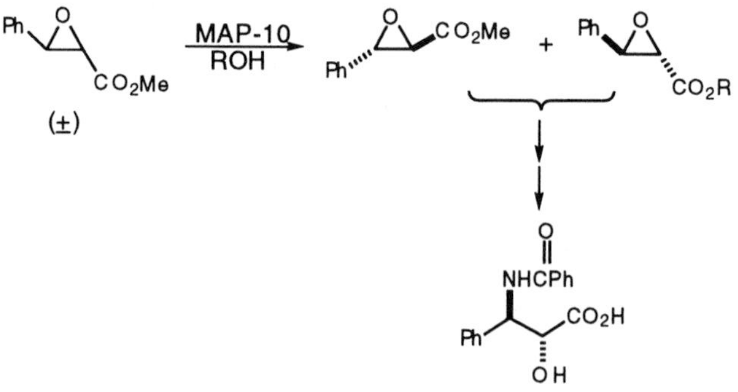

Figure 44.

chain of the antitumor agent Taxol. The hindered isobutyl alcohol was used to avoid the reverse transesterification. Honig et al. utilized *Pseudomonas* lipase P-30 to catalyze the enantioselective hydrolysis of 3-azido-2-(butanoyloxy)-3-phenylpropionic esters to obtain the unreacted (2*R*,3*S*)-

isomer.[323] Another interesting approach to the synthesis of the C-13 side chain of Taxol is based on the lipase-catalyzed enantioselective transformation of 3-hydroxy-4-phenyl β-lactam derivatives[324] (Figure 45). *Pseudomonas* lipase (P-30) was found to be highly selective (E>100)

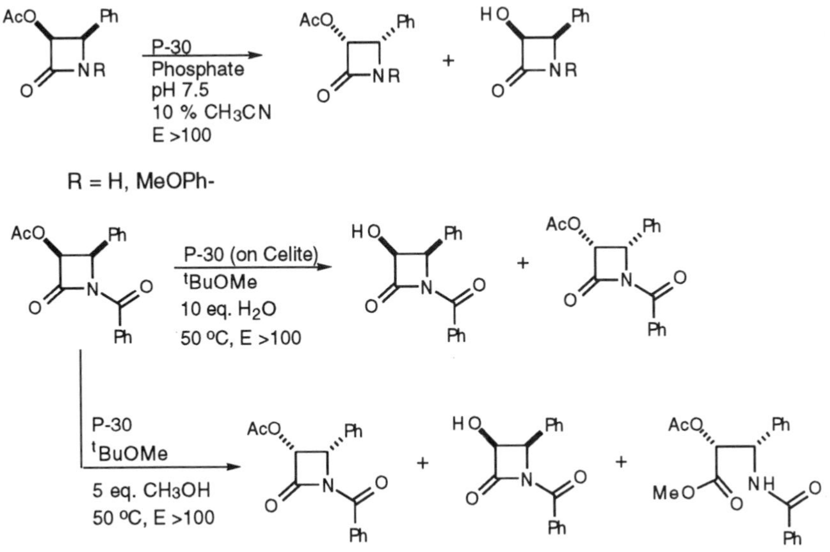

Figure 45.

for the hydrolysis of the 3-acetoxy derivatives, with the ring nitrogen free or protected with benzoyl or *p*-methoxy-phenyl group. The best result was to carry out the hydrolysis in phosphate buffer (pH 7.5) containing 10% acetonitrile. Addition of acetonitrile was found to increase the rate and enantioselectivity. The *N*-benzoyl lactam derivative was not stable in aqueous solution. *t*-Butyl methyl ether containing 10 equivalents of water was therefore used as solvent, and the reaction was carried out at 50 °C to improve the rate. When water was replaced with CH_3OH, the lactam ring of the *N*-benzoyl derivative was cleaved enzymatically, in addition to the methanolysis of the *O*-acetoxy group. The enantioselectivities of these two reactions were, however, different. The (3*S*,4*R*)-enantiomer of the lactam was obtained, while the (2*R*,3*S*)-enantiomer of the ring-opening product was generated. Enantioselective acetylation of lactams with vinyl acetate in MeOBu^t was also accomplished with high enantiomeric excess.[324]

With *N*-α-benzoyl-L-lysinol as the substrate for acylation with trifluoroethyl butyrate, *M. miehei* lipase prefers the NH_2 group and PSL the OH group in several solvents tested.[325] With lipid substrates, *M. miehei* and *P. fluorescens* lipase do not display high 1,3-specificity, while *Rhizopus delemar* lipase is 1,3-specific.[326] The lipase from *Mucor javanicus* (lipase M from Amano) catalyzes the selective deprotection of the heptyl ester of *O*-glycosylated amino acid esters, and the lipase from *Rhizopus nives* (lipase N from Amano) catalyzes the selective deprotection of *N*-protected dipeptide heptyl esters.[321] The lipase from *Aspergillus sp.* seems to

Use of Hydrolytic Enzymes

be quite selective for the sterically demanding α-substituted-β-acyloxy esters[326a,b] (Figure 46),

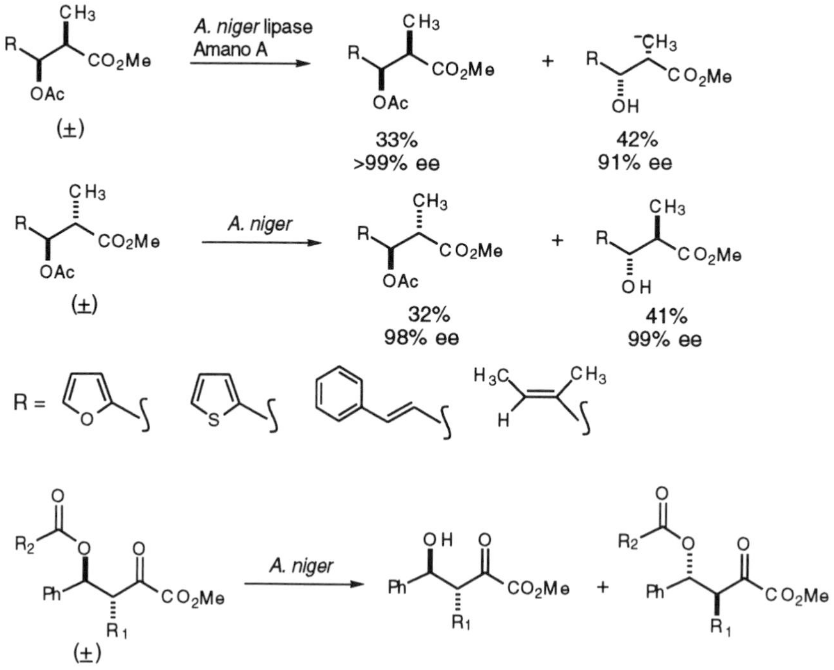

R_1	R_2	Selectivity
Me	Me	80
Et	Me	20
MeS-	Me	30
MeS-	MeSCH$_2$-	180

Figure 46. *A. niger* lipase-catalyzed resolution.

and the selectivity can be further improved by introduction of a thioacetate as the acyl moiety. The *anti*-substrates are better resolved than the *syn*-substrates. In the hydrolysis of peracetylated methyl glycosides of gluco-, manno- and galacto-series, the *Aspergillus* lipase regioselectively cleaves the 2-acetyl group.[326c] When the peracetates of some disaccharides were used as substrates both 1- and 2-positions were hydrolyzed.[326d]

The lipase of *A. niger* catalyzes highly enantioselective hydrolysis of 4-substituted-2-phenyloxazolin-5-ones to give (R)-N-benzoyl-α-amino acids, while PPL exhibits an opposite stereoselectivity in converting the same substrates to the (S)-enantiomers of the product.[327] Since the substrates are interconvertible (due to rapid racemization under the reaction condition; pH 7.6), the enzymatic process in principle can transform both enantiomeric substrates to a single enantiomeric product.[328] The best of the substrates tested was the 4-benzyl derivative which gave (R)-phenylalanine in 99% ee.[327a] Enantioselective alcoholysis of 5(4H)-oxazolone derivatives in organic solvents was also investigated and the best result was that based on *M. miehei* lipase-catalyzed solvolysis of 2-phenyl-4-methyloxazolin-5-one to (*S*)-butyl *N*-benzoylalanine (45% yield, 57% ee).[329] A detailed investigation of lipase-catalyzed solvolysis of 5(4*H*)-oxazolone derivatives was carried out with PSL (AK, K-10, P-30). In *tert*-butyl methyl ether at 50 °C in the presence of 5 equivalents of methanol (for methanolysis) or 5 equivalents of water (for hydrolysis), the enantioselectivity ranged from 66 to 95% ee for the (*S*)-enantiomer and appeared to improve with a larger C-4 substituent (a poor selectivity was found for the 4-methyl derivative).[327b] To improve the enantioselectivity, the methanolysis product was subjected to a second enzymatic enantioselective hydrolysis. Both prozyme 6 and protease N (from Amano) were selective for the (*S*)-ester, and gave the (*S*)-*N*-benzoyl-amino acids in very high ee.[327b] Prozyme 6 and protease 2A (from Amano) also showed a high (*S*)-stereochemical preference toward 5(4*H*)-thiazolin-5-ones, and gave (*S*)-*N*-(benzoylthio)amino acids in very high ee.[327b] The reaction rate was improved in the presence of 10% acetonitrile (Figure 47). This new enzymatic reaction appears to be better than the corresponding process based on chymotrypsin.[131,328]

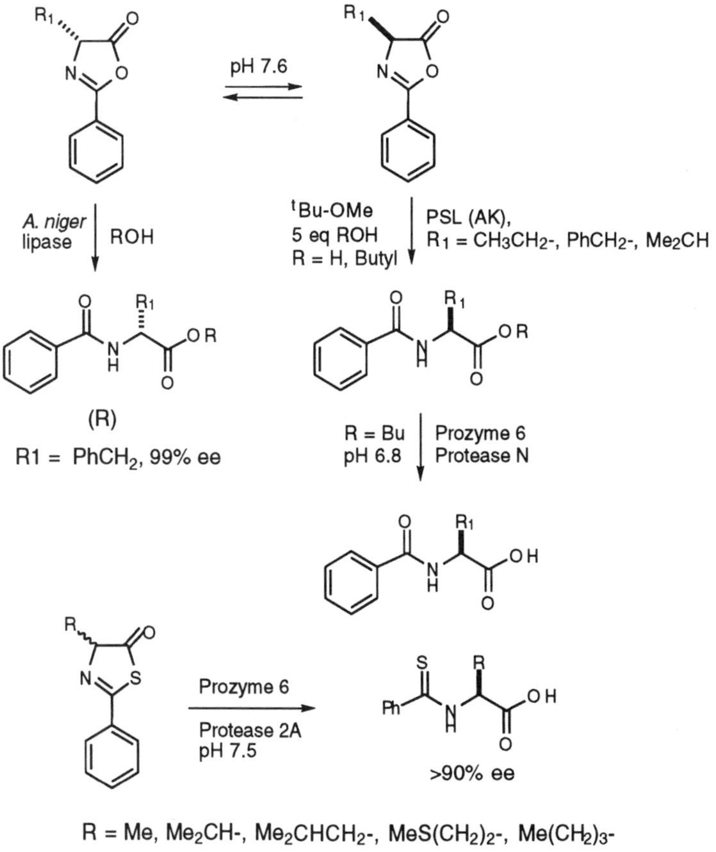

R = Me, Me₂CH-, Me₂CHCH₂-, MeS(CH₂)₂-, Me(CH₂)₃-

Figure 47. Enantioselective solvolysis of 5(4*H*)oxazolones and thiazolones.

The lipase from *Chromobacterium viscosum* is another useful enzyme. It was used in the regioselective acylation of polyhydroxy compounds[330] and enantioselective resolution of racemic hydroxy compounds such as cyanohydrins[331] (Figure 48). The regioselectivity is different from that of subtilisin, and the combination of both enzymatic reactions provides a useful route to selectively acylated sugars and sterols.

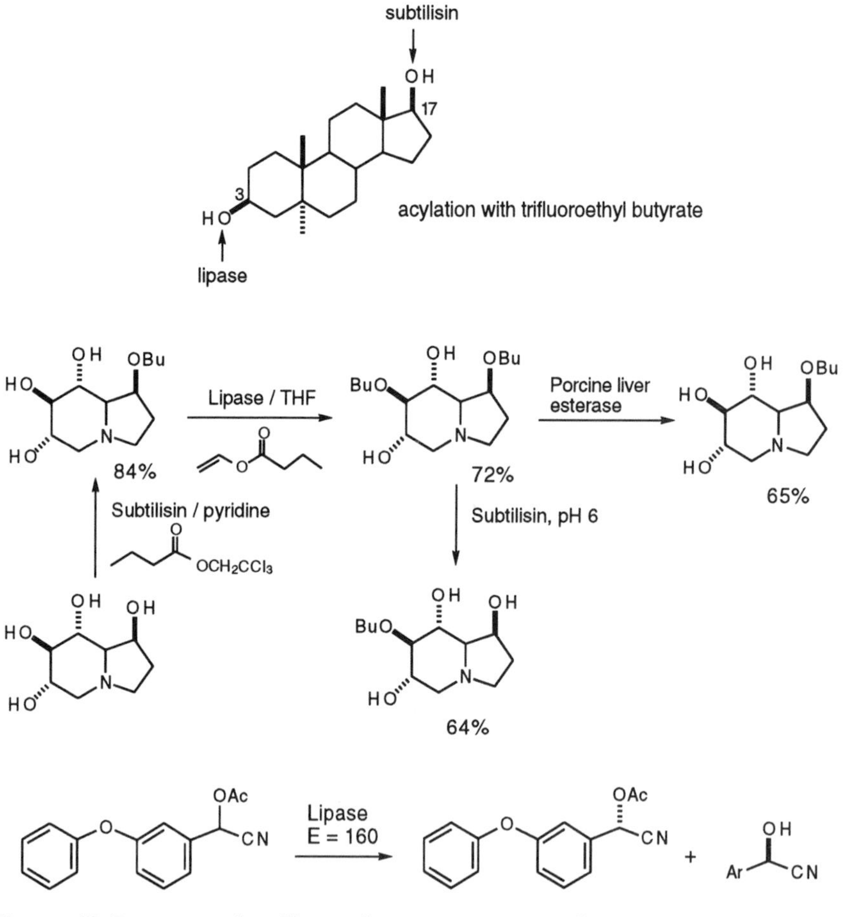

Figure 48. Representative *Chromobacterium viscosum* lipase catalyzed reactions.

For resolution of the cyanohydrin α-cyano-3-phenoxybenzyl alcohol (an important precursor to an insectiside), the lipase from *Arthrobacter* species is better (E >1000) than *C. viscosum* lipase and PSL (E=88).[331] Another useful application of *Arthrobacter* lipase is the resolution of the alcohol moiety (*S* configuration) of the pyrethroid insecticides where the mixture of the resolved alcohol (*R*) and unreacted acetate (*S*) was converted to the (*S*)-alcohol via inversion without separation (Figure 49).[332]

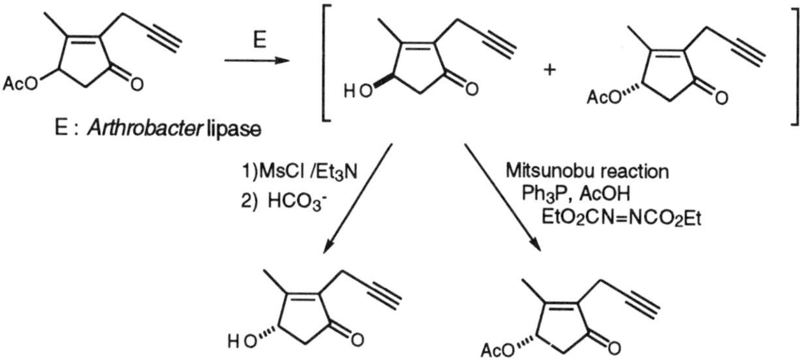

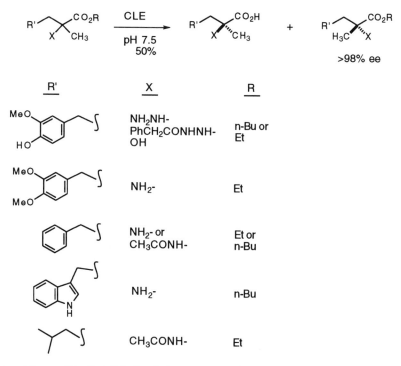

Figure 49. Enzymatic resolution and inversion.

The esterase isolated from *Candida lipolytica* lipase (Amano) was highly selective for the (S)-enantiomer of α-substituted carboxylic acid esters in hydrolysis, and provided a very useful route to α-methyl-α-amino acids and related compounds (Figure 50).[288] The esterase from rabbit liver (Sigma E9636) was found to be better than other esterases for the resolution of 2-arylpropanoic acid,[333] in both transesterification and hydrolysis (Figure 51).

R'	X	R
MeO / HO (aryl)	NH$_2$NH- PhCH$_2$CONHNH- OH	n-Bu or Et
MeO / MeO (aryl)	NH$_2$-	Et
(phenyl)	NH$_2$- or CH$_3$CONH-	Et or n-Bu
(indole)	NH$_2$-	n-Bu
(isobutyl)	CH$_3$CONH-	Et

Figure 50. *Candida lipolytica* esterase (CLE) catalyzed resolution.

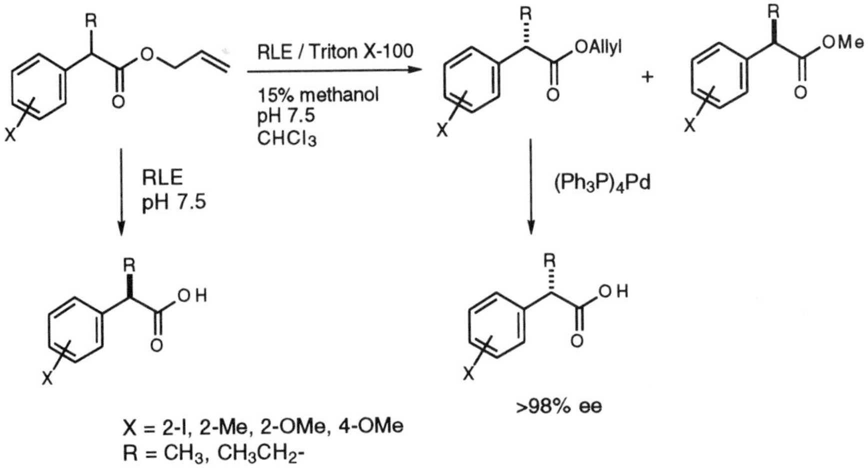

X = 2-I, 2-Me, 2-OMe, 4-OMe
R = CH$_3$, CH$_3$CH$_2$-

Figure 51. Rabbit liver esterase (RLE)-catalyzed resolution of 2-arylpropanoic acid.

The lipase from *Geotrichum candidum* (GCL) is a serine hydrolase with a catalytic triad consisting of the Glu-His-Ser sequence.[334] It was better than other lipases in the resolution of β-acyloxy fatty acid esters with chain lengths of C$_{10}$ to C$_{18}$ (Figure 52).[335] Figure 53 lists some other lipase-catalyzed reactions.

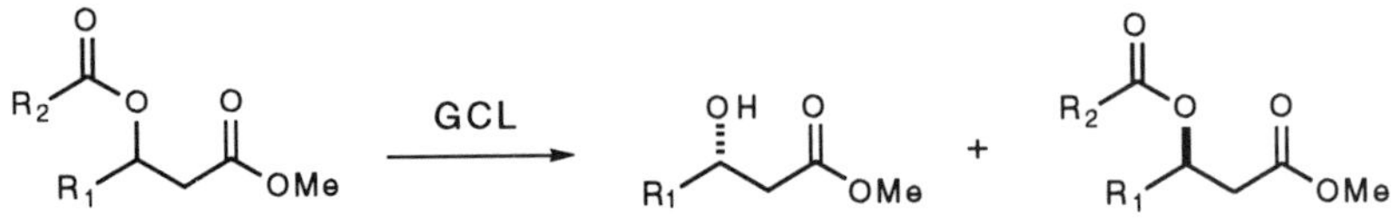

R$_1$	R$_2$	E
n-C$_7$H$_{15}$-	n-C$_3$H$_7$-	11
n-C$_{11}$H$_{23}$-	n-C$_3$H$_7$-	20
n-C$_{15}$H$_{31}$-	n-C$_3$H$_7$-	20
(n-C$_4$H$_9$)$_2$CH(CH$_2$)$_4$-	n-C$_3$H$_7$-	62
n-C$_{11}$H$_{23}$-	ClCH$_2$-	3
n-C$_{11}$H$_{23}$-	n-C$_7$H$_{15}$-	5

Figure 52. *Geotrichum candidum* lipase (GCL)-catalyzed resolution of β-hydroxy fatty acids.

It appears that lipases, proteases and other esterolytic enzymes[266] are, at present, the most extensively used classes of enzymes in organic synthesis. The rate of lipase-catalyzed hydrolysis can be improved by performing the reaction in hydrophobic organic solvents containing a small amount of water.[327,336] The chemo- and enantioselectivity can be manipulated by solvents with different log P values (P represents the partition between octanol and water)[337] or by solvent dielectric constants.[338] The most effective way of performing highly enantioselective transformations is, however, to design substrates with appropriate protecting groups so that the

Figure 53: Other lipase or esterase-catalyzed reaction

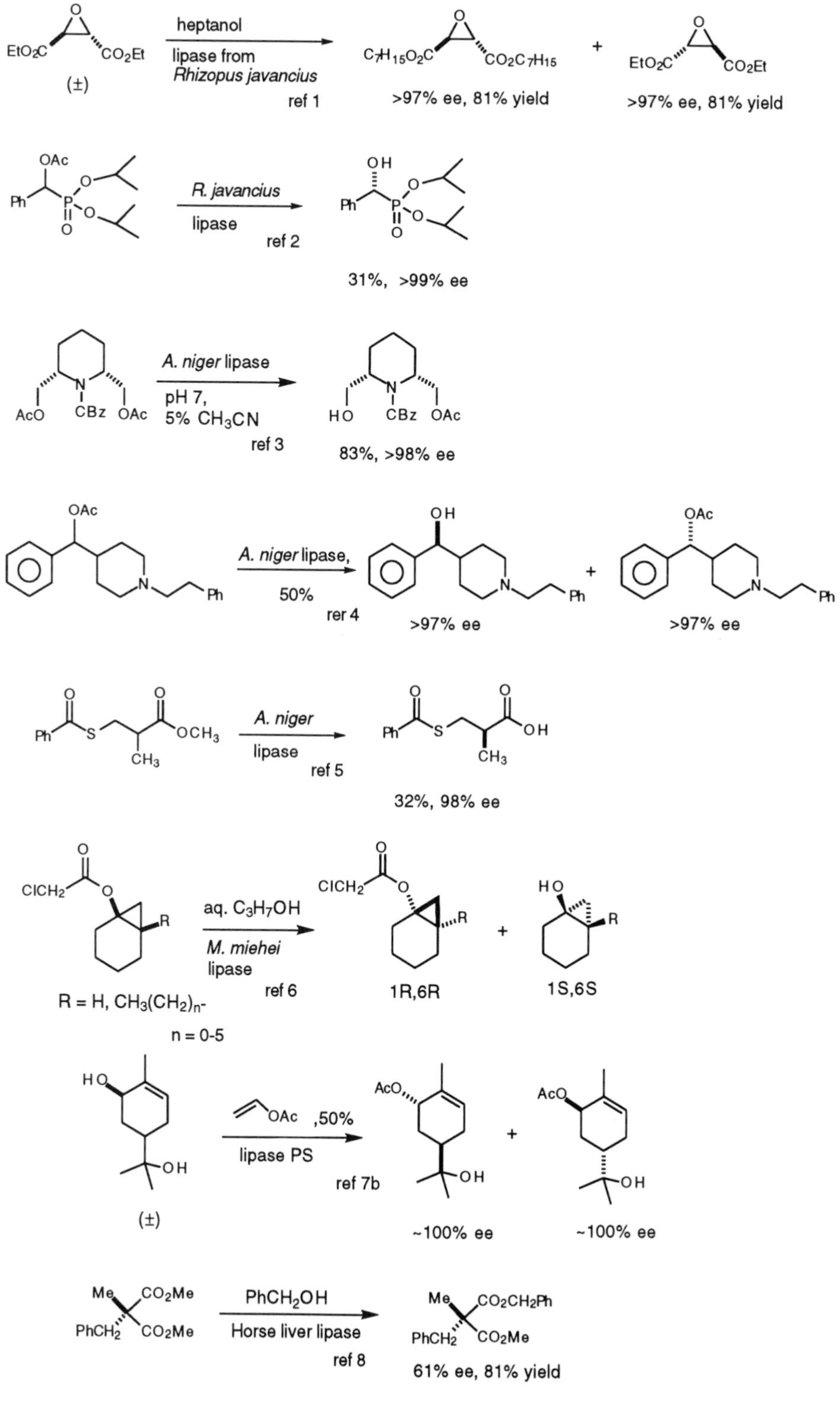
EtO2C
CO2Et
(±)
heptanol
lipase from
Rhizopus javancius
ref 1
C7H15O2C
CO2C7H15
>97% ee, 81% yield
+
EtO2C
CO2Et
>97% ee, 81% yield
OAc
Ph
R. javancius
lipase
ref 2
OH
31%, >99% ee
A. niger lipase
pH 7,
5% CH3CN
ref 3
AcO
CBz
OAc
HO
83%, >98% ee
A. niger lipase,
50%
rer 4
>97% ee
>97% ee
OCH3
CH3
A. niger
lipase
ref 5
OH
32%, 98% ee
ClCH2
aq. C3H7OH
M. miehei
lipase
ref 6
R = H, CH3(CH2)n-
n = 0-5
1R,6R
1S,6S
OAc ,50%
lipase PS
ref 7b
(±)
~100% ee
~100% ee
Me
CO2Me
PhCH2
PhCH2OH
Horse liver lipase
ref 8
CO2CH2Ph
61% ee, 81% yield

Figure 53: continue

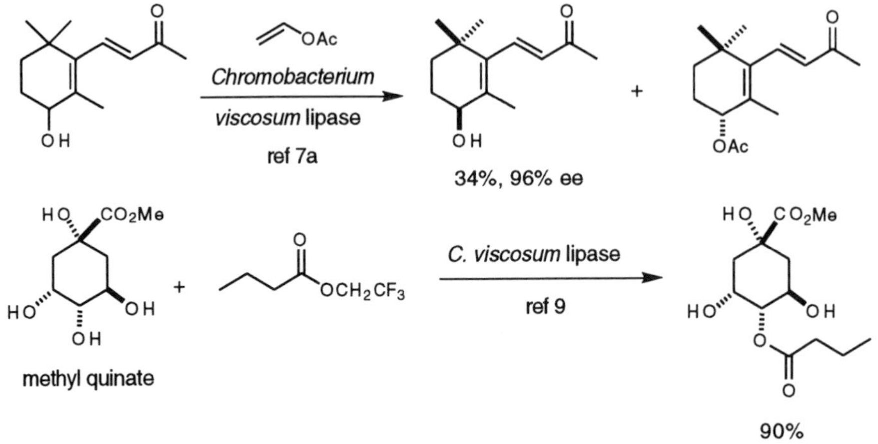

34%, 96% ee

methyl quinate

90%

1. Crout, D.H.G.; Gaudet, V.S.B.; Hallinan, K.O. *J. Chem. Soc., Perkin Trans. 1* **1993**, 805.
2. Li, Y.-F.; Hammerschmidt, F. *Tetrahedron: Asymmetry* **1993**, *4*, 109.
3. Chenevert, R.; Dickman, M. *Tetrahedron: Asymmetry* **1992**, *3*, 1021.
4. Nieduzak, T.R.; Carr, A.A. *Tetrahedron: Asymmetry* **1990**, *1*, 535.
5. Gu, Q.-M.; Reddy, D.R.; Sih, C.J. *Tetrahedron Lett.* **1986**, *27*, 5203.
6. Barnier, J.-P.; Blanco, L.; Rousseau, G.; Guibe-Jampel, E.; Fresse, I. *J. Org. Chem.* **1993**, *58*, 1570.
7. a) Kakeya, H.; Sugai, T.; Ohta, H. *Agric. Biol. Chem.* **1991**, *55*, 1873. b) Bovara, R.; Carrea, G.; Ferrara, L.; Riva, S. *Tetrahedron: Asymmetry* **1991**, *2*, 931.
8. Gutman, A.L.; Shapira, M. *J. Chem. Soc., Chem. Commun.* **1991**, 1467.
9. Danieli, B.; Bellis, P.D.; Barzaghi, L.; Carrea, G.; Ottolina, G.; Riva, S. *Helv. Chim. Acta* **1992**, *75*, 1297.

enzymes can distinguish between enantiotopic or diastereotopic groups to achieve high enantioselection.

Some other novel enzymatic transformations with hydrolases include the asymmetric opening of epoxides by amine or azide catalyzed by lipases, enantioselective ester hydrolysis of *N*-acetyl-DL-amino acid methyl esters catalyzed by carbonic anhydrase[339c] (selective for D isomer) and carbon–carbon bond formation mediated by modified papain.[339d] Chemical and biological techniques are also available to modify enzyme stability and specificity. Selection of new protease substrates by monovalent phage display is now possible,[339e] and detection of catalytic species involved in a coupling reaction between a tagged (e.g. biotin-linked substrates) and a surface-bound molecule is also available for finding appropriate condition in peptide synthesis.[339f]

9. Nitrile Hydrolysis Enzymes

Microbial hydrolysis of nitriles may be catalyzed by nitrilase to give the corresponding acids, or by nitrile hydratase to give amides. The amides may be hydrolyzed to acids in reactions catalyzed by amidases[340-341] (Figure 54). An early proposed enzymatic synthesis of acrylamide

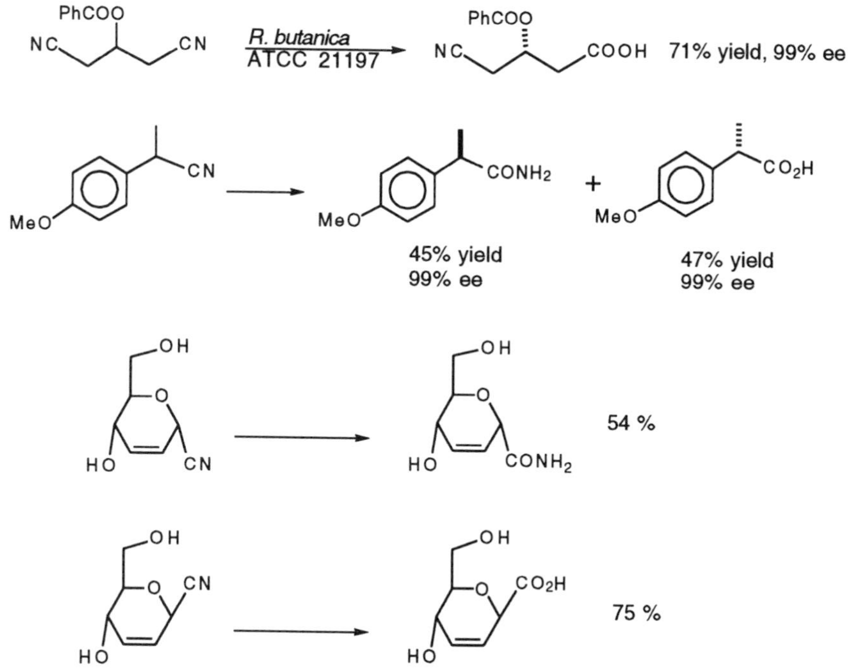

Figure 54: Nitrile hydrolysis enzyme

from acrylnitrile[342-343] has led to the industrial production of acrylamide (400 g/L) based on the resting cells of a *Pseudomonas species*.[341,343] The enzyme has been purified and characterized.[340] It contains a ferric ion and a cofactor pyrroloquinoline quinone (PQQ).[344] Both are involved in the hydration of nitriles. Several other nitrile hydratases containing cobalt have also been isolated.[340] The immobilized *Rhodococcus sp.* from NOVO (SP409) contains both nitrile hydratase (RCN to $RCONH_2$) and amidase ($RCONH_2$ to RCO_2H) activity (but not nitrilase activity, RCN to RCO_2H), and has been used in the chemoselective hydrolysis of nitriles.[345,346] One particularly interesting application of nitrile hydrolysis employs the enzyme from *Rhodococcus butanica* (ATCC 21197). The microbial transformations of α-arylpropionitriles to (*R*)-amides and (*S*)-carboxylic acids, and of 3-benzoyloxy glutaronitrile to an (*S*)-acid with very high ee (>95%) have been reported.[347,348] Several prochiral nitriles were hydrolyzed with SP409 to give the corresponding (*S*)-nitrile-carboxylic acids with ee's ranging from 22-84%.[349a] The amidase activity was found to be (*S*)-selective.[349b] Hydrolysis of aldono- and aldurononitriles to the corresponding amides and/or carboxylic acids was also reported.[349c] Studies on the electronic substituent effect in the SP409-catalyzed hydrolysis of *para*-substituted benzyl cyanides revealed a Hammett-type linear free energy correlation (i.e. the rate is faster with electron-withdrawing substituents).[349d] The nitrilase gene has been cloned and expressed in *E. coli*.[350]

10. Epoxide Hydrolase

Various epoxide hydrolases, particularly that from rabbit liver microsome and microbial sources, have been used in selective hydrolysis of epoxides.[351-356] Synthesis of D-,[356a] L-[356b] and meso-tartaric acid[356c] from the precursor epoxides catalyzed by microbial epoxide hydrolases still represent the most useful reactions of epoxide hydrolases. The microsomal epoxide hydrolases prefer cyclic *cis*-meso-epoxides as substrates, attacking the (*S*)-center to give the (*R,R*)-diol.[354a,b] The microsomal epoxide hydrolase catalysis involves an ester intermediate, formed by the reaction of a carboxylate group of the enzyme and the epoxide.[354c]

11. Phosphatase

Phosphatases (EC 3.1.3.1) have been used as deprotecting agents in the hydrolysis of phosphates to alcohols. A particularly useful phosphatase is the commercially available potato acid phosphotase.[357] 5'-Ribonucleotide phosphatase (5'-ribonucleotidase) has been used in the enantioselective hydrolysis of racemic carbocyclic ribonucleotides to prepare (-)-aristeromycin[358a] and a 2'-fluoroguanosine derivative.[358b] Wheat germ acid phosphatase catalyzes the selective hydrolysis of L-*O*-phosphothreonine from the racemic mixture.[359] Calf intestine alkaline phosphatase catalyzes the transfer of a phosphoryl group from pyrophosphate to various alcohols to form monoalkyl phosphates[360a] and to decompose nucleoside phosphates selectively in the presence of sugar nucleotides.[360b] DL-Glycerol 1-phosphate has been prepared from glycerol and alkaline phosphatase.[361]

References

1. Greenstein, J. P.; Winitz, M. *Chemistry of the Amino Acids*; Wiley: New York, **1961**, 2, 1753; Greenstein, J.P. *Method. Enzymol.* **1957**, *3*, 554; Chibata, I.; Tosa, T.; Sato, T.; Mori, T. *Method. Enzymol.* **1976**, *44*, 746.

2. Chenault, H. K.; Dahmer, J.; Whitesides, G. M. *J. Am. Chem. Soc.* **1989**, *111*, 6354.

3. Keller, J. W.; Hamilton, B. J. *Tetrahedron Lett.* **1986**, *27*, 1249.

4. Pirrung, M. C.; Krishnamurthy, N. *J. Org. Chem.* **1993**, *58*, 954.

5. a) Groeger, U.; Drauz, K.; Klenk, H. *Angew. Chem. Int. Ed. Engl.* **1990**, *29*, 417; **1992**, *31*, 195. b) Kikuchi, M.; Koshiyama, I.; Fukushima, D. *Biochim. Biophys. Acta* **1983**, *744*, 180.

6. Kameda, Y.; Toyoura, E.; Kimura, Y. *Nature* **1958**, *181*, 1225; Kubo, K.; Ishikura, T.; Fukagawa, Y. *J. Antibiot.* **1980**, *33*, 550; Yang, Y.B.; Lin, C. S.; Tseng, C. P.; Wang, Y. J.; Tsai, Y. C. *Appl. Environ. Microbiol.* **1991**, *57*, 1259.

7. Chen, H.-P.; Wu,, S.-H.; Tsai, Y.-C.; Yang, Y.-B.; Wang, K.-T. *Bioorg. Med. Chem. Lett.* **1992**, *2*, 697.

8. a) Schoemaker, H. E.; Boesten, W. H. J.; Kaptein, B.; Hermes, H. F. M.; Sonke, T.; Broxterman, Q.B.; van den Tweel, W. J. J.; Kamphuis, J. *Pure Appl. Chem.* **1992**, *64*, 1171. b) Roos, E.C.; Mooiweer, H. H.; Heimstra, H.; Speckamp, W. N.; Kaptein, B.; Boesten, W. H. J.; Kamphuis, J. *J. Org. Chem.* **1992**, *57*, 6769. c) Kaptein, B.; Boesten, W. H. J.; Broxterman, Q. B.; Peters, P. J. H.; Schoemaker, H. E.; Kamphuis, J. *Tetrahedron: Asymmetry* **1993**, *4*, 1113.

9. Asano, Y.; Nakazawa, A.; Kato, Y.; Kondo, K. *Angew. Chem. Int. Ed. Engl.* **1989**, *28*, 450.

10. Taylor, S. J. C.; McCague, R.; Wisdom, R. et al. *Tetrahedron: Asymmetry* **1993**, *4*, 1117.

11. Yokozeki, K.; Nakamori, S.; Eguchi, C.; Yamada, K.; Mitsugi, K. *Agric. Biol. Chem.* **1987**, *51*, 355. (*Pseudomonas sp.* for D-*p*-hydroxyphenylglycine); Yokozeki, K.; Sano, K.; Eguchi, C.; Yamada, K.; Mitsugi, K. *Agric. Biol. Chem.* **1987**, *51*, 363 (L-Trp); Yamashiro, A.; Yokozeki, K.; Kano, H.; Kubota, K. *Agric. Biol. Chem.* **1988**, *52*, 2851 (Other L-amino acids).

12. Syldatk, C.; Cotoras, D.; Dombach, G.; Grob, C.; Kallwab, H.; Wagner, F. *Biotech. Lett.* **1987**, *9*, 25 (L-Trp).

13. Drauz, K.; Kottenhahn, M.; Makryaleas, K.; Klenk, H.; Bernd, M. *Angew. Chem. Int. Ed. Engl.* **1991**, *30*, 712 (*Agrobacterium sp.* for D-citruline).

14. a) Cole, M. *Biochem. J.* **1969**, *115*, 733; Andersson, E.; Mattiasson, B.; Hahn-Hagerdal, B. *Enz. Microb. Tech.* **1984**, *6*, 301. b) Soloshonok, V. A.; Svedas, V. K.; Kukhar, V. P.; Kirilenko, A. G.; Rybakova, A. V.; Solodenko, V. A.; Fokina, N. A.;

Kogut, O. V.; Galaev, I. Y.; Kozlova, E. V.; Shishkina, I. P.; Galushko, V. S. *Synlett* **1993**, 339. c) Margolin, A. L. *Tetrahedron Lett.* **1993**, *34*, 1239. d) Solodenko, V. A.; Belik, M. Y.; Galushko, S. V.; Kukhar, V. P.; Kozlova, E. V.; Mironenko, D. A.; Svedas, V. K. *Tetrahedron: Asymmetry* **1993**, *4*, 1965.

15.   Rossi, D.; Calcagni, A. *Experientia* **1985**, *41*, 35.

16.   a) Fuganti, C.; Grasselli, P. *Tetrahedron Lett.* **1986**, *27*, 3191. (Synthesis of phenyl-acetyl-L-Asp-L-Phe-OMe). b) Fuganti, C.; Grasseli, P.; Seneci, P. F.; Servi, S. *Tetrahedron Lett.* **1986**, *27*, 2601.

17.   Romeo, A.; Lucente, G.; Rossie, D.; Zanotti, G. *Tetrahedron Lett.* **1971**, *21*, 1799.

18.   Fuganti, C.; Grasselli, P.; Servi, S.; Lazzarini, A.; Casati, P. *J. Chem. Soc., Chem. Commun.* **1987**, 538.

19.   Waldmann, H. *Liebigs Ann. Chem.* **1988**, 1175; Waldmann, H. *Tetrahedron Lett.* **1988**, *29*, 1131.

20.   Waldmann, H. *Kontakte* **1991**, *2*, 33.

21.   Zmijewski, M. J.; Briggs, B. S.; Thompson, A. R.; Wright, I. G. *Tetrahedron Lett.* **1991**, *32*, 1621.

22.   Hayashi, R.; Moore, S.; Stein, W. H. *J. Biol. Chem.* **1973**, *248*, 2296; Hayashi, R.; Bai, Y.; Hata, T. *J. Biochem.* **1975**, *77*, 69; Bai, Y.; Hayashi, R.; Hata, T. *J. Biochem.* **1975**, *77*, 81; Bai, Y.; Hayashi, R.; Hata, T. *J. Biochem.* **1975**, *78*, 617.

23.   Widmer, F.; Johansen, J. T. *Carlsberg, Res. Commun.* **1979**, *44*, 37; Widmer, F.; Breddam, K.; Johansen, J. T. *Carlsberg Res. Commun.* **1980**, *45*, 453.

24.   Breddam, K.; Johansen, J. T.; Ottesen, M. *Carlsberg. Res. Commun.* **1984**, *49*, 457; Berne, P.-F.; Blanquet, S.; Schmitter, J.-M. *J. Am. Chem. Soc.* **1992**, *114*, 2603.

25.   Steinke, D.; Schwarz, A.; Wandrey, C.; Kula, M. R. *Enz. Microb. Tech.* **1991**, *13*, 262.

26.   Lee, M.; Phillips, R. S. *Bioorg. & Med. Chem. Lett.* **1991**, *1*, 477.

27.   Bergmann, M.; Fraenkel-Conrat, H. *J. Biol. Chem.* **1938**, *124*, 1.; Fruton, J. S. *Adv. Enzymol.* **1982**, *53*, 239.

28.   Jakubke, H.-D.; Kuhl, P.; Konnecke, A. *Angew. Chem. Int. Ed. Engl.* **1985**, *24*, 85; Schellenberger, V.; Jakubke, H.-D. *Angew. Chem. Int. Ed. Engl.* **1991**, *30*, 1437.

29.   a) Morihara, K. *TIBTECH* **1987**, *5*, 164. b) Morihara, K.; Oka, T.; Tsuzuki, H.; Tochino, Y.; Kanaya, T. *Biochem. Biophys. Res. Commun.* **1980**, *92*, 396.

30.   For a representative monograph, see Kullmann, W. *Enzymatic Peptide Synthesis*, CRC Press, **1987**.

31.   Wong, C.-H.; Wang, K.-T. *Experientia* **1991**, *47*, 1123.

32.   Homandberg, G. A.; Mattis, J. A.; Laskowski, M., Jr. *Biochemistry* **1978**, *17*, 5220.

33.   Martinek, K.; Semenov, A.; Berezin, I. V. *Biochim. Biophys. Acta* **1981**, *658*, 76.

34. Luthi, P.; Luisi, P. L. Enzymatic synthesis of hydrocarbon-soluble peptides with reverse micelles. *J. Am. Chem. Soc.* **1984**, *106*, 7285.

35. Oyama, K.; Nishimura, S.; Sonaka, Y.; Kihara, K.; Hashimoto, T. *J. Org. Chem.* **1981**, *46*, 5242.

36. Kitaguchi, H.; Klibanov, A. M. *J. Am. Chem. Soc.* **1989**, *111*, 9272.

37. a) Barbas, C. F., III; Matos, J. R.; West, J. B.; Wong, C.-H. *J. Am. Chem. Soc.* **1988**, *110*, 5162. b) Unverzagt, C.; Geyer, A.; Kessler, H. *Angew. Chem. Int. Ed. Engl.* **1992**, *31*, 1229. c) A representative large-scale synthesis of the dipeptide Kyotorphin: Herrmann, G.; Schwarz, A.; Wandrey, C.; Kula, M.-R.; Knaup, G.; Drauz, K. H.; Berndt, H. *Biotech. Appl. Biochem.* **1991**, *13*, 346.

38. a) Wong, C.-H.; Chen, S.-T.; Hennen, W.J.; Bibbs, J.A.; Wang, Y.-F.; Liu, J.L.-C.; Pantoliano, M. W.; Whitlow, M.; Bryan, P. N. *J. Am. Chem. Soc.* **1990**, *112*, 945. b) Hwang, B. K.; Gu, Q.-M.; Sih, C. J. *J. Am. Chem. Soc.* **1993**, *115*, 7912. c) Fischer, A.; Schwarz, A.; Wandrey, C.; Bommarius, A. S.; Knaup, G.; Drauz, K. *Biomed. Biochim. Acta* **1991**, *50*, 169.

39. Barbas, C. F., III; Wong, C.-H. *J. Chem. Soc., Chem. Commun.* **1987**, 532.

40. West, J. B.; Wong, C.-H. *Tetrahedron Lett.* **1987**, *28*, 1629.; Yu, V.M.; Zapevalova, N. P.; Gorbubova, E. Y. *Int. J. Pept. Protein Res.* **1984**, *23*, 528; Whitaker, J. R.; Bender, M. L. *J. Am. Chem. Soc.* **1965**, *87*, 2728.

41. Margolin, A. L.; Klibanov, A. M. *J. Am. Chem. Soc.* **1987**, *109*, 3802; Kawashiro, K.; Kaiso, K.; Minato, D.; Sugiyama, S.; Hayashi, H. *Tetrahedron* **1993**, *49*, 4541.

42. Coletti-Previero, M.-A.; Previero, A.; Zuckerkandl, E. *J. Mol. Biol.* **1969**, *39*, 493.

43. West, J. B.; Hennen, W.J.; Lalonde, J. L.; Bibbs, J. A.; Zhong, Z.; Meyer, E. F.; Wong, C.-H. *J. Am. Chem. Soc.* **1990**, *112*, 5313.

44. West, J. B.; Scholten, J.; Stolowich, N. J.; Hogg, J. L.; Scott, A. I.; Wong, C.-H. *J. Am. Chem. Soc.* **1988**, *110*, 3709.

45. Henderson, R. *Biochem. J.* **1971**, *124*, 13.

46. Byers, L. D.; Koshland, D. E., Jr. *Bioorg. Chem.* **1978**, *7*, 15.

47. Scholten, J. D.; Hogg, J. L.; Raushel, F. M. *J. Am. Chem. Soc.* **1988**, *110*, 8246.

48. Rebek, J., Jr. *Chemtracts - Org. Chem.* **1989**, *2*, 337.

49. Nakatsuka, T.; Sasaki, T.; Kaiser, E. T. *J. Am. Chem. Soc.* **1987**, *109*, 3808.

50. Wu, Z.-P.; Hilvert, D. *J. Am. Chem. Soc.* **1989**, *111*, 4513.

51. Abrahmsen, L.; Tom, J.; Burnier, J.; Butcher, K. A.; Kossiakoff, A.; Wells, J. A. *Biochemistry* **1991**, *30*, 4151.

52. Wong, C.-H.; Schuster, M.; Wang, P.; Sears, P. *J. Am. Chem. Soc.* 1993, 115, 5893.

53. Wong, C.-H. *TIBTECH* **1992**, *10*, 378.

54. Schuster, M.; Munoz, B.; Yuan, W.; Wong, C.-H. *Tetrahedron Lett.* **1993**, *34*, 1247.

55. Chen, S.-T.; Hsiao, S.-C.; Wang, K.-T. *Bioorg. Med. Chem. Lett.* **1991**, *1*, 445; Chen, S.-T.; Chen, S.-Y.; Wang, K.-T. *J. Org. Chem.* **1992**, *57*, 6960.

56. Rose, K.; Herrero, C.; Proudfoot, A. E. I.; Offord, R. E.; Wallace, C. J. A. *Biochem. J.* **1988**, *249*, 83.

57. Laskowski, M., Jr. in *Semisynthetic Peptides and Proteins* (R.E. Offord and C.D. Bello, eds.) Academic Press, London 1978, p255.

58. Brothers, H. M., II; Kostic, N. M. *Biochemistry* **1990**, *29*, 7468.

59. Magnotti, R. A., Jr. *Biochim. Biophys. Acta* **1987**, *915*, 46.

60. Yokosawa, H.; Ojima, S.; Ishii, S.-I. *J. Biochem.* **1977**, *82*, 869.

61. Schuster, M.; Aaviksaar, A.; Jakubke, H.-D. *Tetrahedron Lett.* **1992**, *33*, 2799.

62. Gill, I.; Vulfson, E. N. *J. Am. Chem. Soc.* **1993**, *115*, 3348.

63. Royer, G. P. *Method. Enzymol.* **1987**, *136*, 157.

64. Shima, H.; Fukuda, M.; Tanabe, K.; Ito, T.; Kunigi, S. *Bull. Chem. Soc. Jpn.* **1987**, *60*, 1403.

65. Lipmann, F. *Acc. Chem. Res.* **1973**, *6*, 361.

66. Schimmel, P. *Ann. Rev. Biochem.* **1987**, *56*, 125.

67. Petkov, D. D.; Stoineva, I. B. *Tetrahedron Lett.* **1984**, *25*, 3751.

68. West, J. B.; Wong, C.-H. *J. Org. Chem.* **1986**, *51*, 2728.

69. Bizzozero, S. A.; Butler, H. *Bioorg. Chem.* **1981**, *10*, 46.

70. Margolin, A. L.; Tai, D.-F.; Klibanov, A. M. *J. Am. Chem. Soc.* **1987**, *109*, 7885.

71. Noren, C. J.; Anthony-Cahilll, S. J.; Griffith, M. C.; Schultz, P. G. *Science* **1989**, *244*, 182.

72. Kato, Y.; Asano, Y.; Nakayama, A.; Kondo, K. *Biocatalysis* **1990**, *3*, 207.

73. Kato, Y.; Asano, Y.; Nakazawa, A.; Kondo, K. *Tetrahedron* **1989**, *45*, 5743.

74. Ekberg, B.; Lindbladh, C.; Kempe, M.; Mosbach, K. *Tetrahedron Lett.* **1989**, *30*, 583.

75. Asano, Y.; Nakazawa, A.; Kato, Y.; Kondo, K. *Angew. Chem. Int. Ed. Engl.* **1989**, *28*, 450.

76. Janda, K. D.; Lerner, R. A.; Tramontano, A. *J. Am. Chem. Soc.* **1988**, *110*, 4835.

77. Benkovic, S. J.; Napper, A. D.; Lerner, R. A. *Proc. Natl. Acad. Sci.* **1988**, *85*, 5355.

78. Pollack, S. J.; Hsiun, P.; Schultz, P. G. *J. Am. Chem. Soc.* **1989**, *111*, 5961.

79. For chymotrypsin, Schellenberger, V.; Braune, K.; Hofmann, H.-J.; Jakubke, H.-D. *Eur. J. Biochem.* **1991**, *199*, 623; for subtilisin, Philipp, M.; Bender, M. L. *Mol. Cell. Biochem.* **1983**, *51*, 5.

80. Craik, C. S.; Largman, C.; Fletcher, T.; Roczniak, S.; Barr, P. J.; Fletterick, R.; Rutter, W. J. *Science* **1985**, *228*, 291; Graf, L.; Jancso, A.; Szilagyi, L.; Hegyi, G.; Pinter, K.; Naray-Szabo, G.; Hepp, J.; Medzihradszky, K.; Rutter, W. J. *Proc. Natl. Acad. Sci. USA* **1988**, *85*, 4961.

81. Bone, R.; Silen, J. L.; Agard, D. A. *Nature* **1989**, *339*, 191.

82. Khouri, H. E.; Vernet, T.; Menard, R.; Parlati, F.; Laflamme, P.; Tessier, D. C.; Gour-Salin, B.; Thomas, D. Y.; Storer, A. C. *Biochemistry* **1991**, *30*, 8929.

83. Fersht, A.; Winter, G. *TIBS* **1992**, 292; Wells, J. A.; Estell, D. A. *TIBS* **1988**, 291; Pantoliano, M. W. *Current Opinion in Structural Biology* **1992**, *2*, 559.

84. Bonneau, P. R.; Graycar, T. P.; Estell, D. A.; Jones, J. B. *J. Am. Chem. Soc.* **1991**, *113*, 1026.

85. Schellenberger, V.; Schellenberger, U.; Jakubke, H.-D.; Zapevalova, N. P.; Mitin, Y. V. *Biotech. Bioeng.* **1991**, *38*, 319.

86. Ricca, J.-M.; Crout, D. H. G. *J. Chem. Soc. Perkin Trans. 1* **1989**, 2126; Crout, D. H. G.; MacManus, D. A.; Ricca, J.-M.; Singh, S.; Critchley, P.; Gibson, W. T. *Pure Appl. Chem.* **1992**, *64*, 1079.

87. Masaki, T.; Fujihashi, T.; Nakamura, K.; Soejima, M. *Biochim. Biophys. Acta* **1981**, *660*, 51.

88. Drapeau, G. R.; Boily, Y.; Houmard, J. *J. Biol. Chem.* **1992**, *247*, 6720; Yoshikawa, K.; Tsuzuki, H.; Fujiwara, T.; Nakamura, E.; Iwamoto, H.; Matsumoto, K.; Shin, M.; Yoshida, N.; Teraoka, H. *Biochim. Biophys. Acta* **1992**, *1121*, 221; Cerovsky, V. *Tetrahedron Lett.* **1991**, *32*, 3421; Sorensen, S. B.; Sorensen, T. L.; Breddam, K. *FEBS* **1991**, *294*, 195.

89. a) Sakudo, S.; Kikuchi, N.; Kitadokoro, K.; Fujiwara, T.; Nakamura, E.; Okamoto, H.; Shin, M.; Tamaki, M.; Teraoka, H.; Tsuzuki, H.; Yoshida, N. *J. Biol. Chem.* **1992**, *267*, 23782; Breddam, K.; Meldal, M. *Eur. J. Biochem.* 1992, 206, 103; Rolland-Fulcrand, V.; Breddam, K. *Biocatalysis* **1993**, *7*, 75. b) Bongers, J.; Lambros, T.; Liu, W.; Ahmad, M.; Campbell, R. M.; Felix, A. M.; Heimer, E. P. *J. Med. Chem.* **1992**, *35*, 3934.

90. a) Morihara, K.; Tsuzuki, H. *Arch. Biochem. Biophys.* **1968**, *126*, 971; Narahashi, Y.; Fukunaga, J. *J. Biochem.* **1969**, *66*, 743. b) Ohshiro, T.; Mochida, K.; Uwajima, T. *Biotechnol. Lett.* **1992**, *14*, 175.

91. Doel, M.T.; Eaton, M.; Cook, E.A.; Lewis, H.; Patel, T.; Carey, N.H. *Nucleic Acid Res.* **1980**, *8*, 4575.

92. Zhong, Z.; Liu, J. L.-C.; Dinterman, L. M.; Finkelman, M. A.; Mueller, W. T.; Rollence, M. L.; Whitlow, M.; Wong, C.-H. *J. Am. Chem. Soc.* **1991**, *113*, 683.

93. a) Arnold, F.H. *TIBTECH* **1990**, *8*, 244; b) Chen, K.; Arnold, F.H. *Bio/Technology* **1991**, *9*, 1073. c) Arnold, F. H. *The FASEB Journal* **1993**, *7*, 748.

94. a) Chen, W.-T.; Wu, S.-H.; Wang, K.-T. *Int. J. Peptide Protein Res.* **1991**, *37*, 347. b) Chen, S.-T.; Wang, K.-T. *J. Org. Chem.* **1988**, *53*, 4589.

95. Isowa, Y.; Ohmori, M.; Ichikawa, T.; Mori, K.; Nonaka, Y.; Kihara, K.I.; Oyama, K. *Tetrahedron Lett.* **1979**, *20*, 2611.

96. Barbas, C. F., III; Wong, C.-H. *Tetrahedron Lett.* **1988**, *29*, 2907.

97. Ping, D.; Katopodis, A. G.; May, S. W. *J. Am. Chem. Soc.* **1992**, *114*, 3998; Mark, T. M.; Cheng, H.; Vederas, J. C. *J. Am. Chem. Soc.* **1992**, *114*, 2270.

98. Katopodis, A. G.; Ping, D.; May, S. W. *Biochemistry* **1991**, *30*, 6189.

99. Beaudry, G. A.; Mehta, N. M.; Ray, M. L.; Bertelsen, A. H. *J. Biol. Chem.* **1990**, *265*, 17694.

100. Bongers, J.; Offord, R. E.; Felix, A. M.; Lambros, T.; Liu, W.; Ahmad, M.; Campbell, R. M.; Heimer, E. P. *Biomed. Biochim. Acta* **1991**, *50*, S157.

101. Henriksen, D. B.; Breddam, K.; Moller, J.; Buchardt, O. *J. Am. Chem. Soc.* **1992**, *114*, 1876.

102. a) Green, J.; Margolin, A. L. *Tetrahedron Lett.* **1992**, *33*, 7759. b) Hendriksen, D. B.; Breddam, K.; Buchardt, O. *Int. J. Peptide Protein Res.* **1993**, *41*, 169.

103. Wolfe, S.; Demain, A. L.; Jensen, S. E.; Westlake, D. W. S. *Science* **1984**, *226*, 1386.

104. van Liempt, H.; von Doehren, H.; Kleinkauf, H. *J. Biol. Chem.* **1989**, *264*, 3680.

105. Banko, G.; Demain, A. L.; Wolfe, S. *J. Am. Chem. Soc.* **1987**, *109*, 2858.

106. Schwecke, T.; Aharonowitz, Y.; Palissa, H.; von Dohren, H.; Kleinkauf, H.; van Liempt, H. *Eur. J. Biochem.* **1992**, *205*, 687.

107. Kleinkauf, H.; von Dohren, H. *Annu. Rev. Microbiol.* **1987**, *41*, 259.

108. Fersht, A. R. *Biochemistry* **1987**, *26*, 8031.

109. Nakajima, H.; Kitabatake, S.; Tsurutani, R.; Yamamoto, K.; Tomioka, I.; Imahori, K. *Int. J. Peptide Protein Res.* **1986**, *28*, 179.

110. Stoineva, I. B.; Galunsky, B. P.; Lazanov, V. S.; Ivanov, I. P.; Petkov, D. D. *Tetrahedron* **1992**, *48*, 1115.

111. Fuganti, C.; Grasselli, P.; Casati, P. *Tetrahedron Lett.* **1986**, *22*, 3191.

112. Schultz, M.; Hermann, P.; Kunz, H. *Synlett* **1992**, 37.

113. Braun, P.; Waldmann, H.; Vogt, W.; Kunz, H. *Synlett* **1990**, 105; Braun, P.; Waldmann, H.; Kunz, H. *Synlett* **1992**, 39.

114. Steinke, D.; Kula, M.-R. *Angew. Chem. Int. Ed. Engl.* **1990**, *29*, 1139.

115. Matsumura, E.; Shin, T.; Murao, S.; Sakaguchi, M.; Kawano, T. *Agric. Biol. Chem.* **1985**, *49*, 3643.

116. Jones, J.B.; Sih, C.-J.; Perlmann, D. *Tech. Chem.* **1976**, *10*, 107.

117. Schubert, W.C. *J. Mol. Biol.* **1972**, *67*, 151; Bender, M. L.; Killhefer, J. V. *Crit. Rev. Biochem.* **1973**, *1*, 149.

118. Niemann, C.; Thomas, D. W.; MacAllister, R. V. *J. Am. Chem. Soc.* **1951**, *73*, 1548.

119. Niemann, C.; Huang, H. T. *J. Am. Chem. Soc.* **1951**, *73*, 1541.

120. Clement, G. E.; Potter, R. *J. Chem. Educ.* **1971**, *48*, 695.

121. Tong, J. H.; Petitclerc, C.; D'Lorio, A.; Benoiton, N. L. *Can. J. Biochem.* **1971**, *49*, 877.

122. Berger, A.; Smolarsky, M.; Kurn, N.; Bosshard, H. R. *J. Org. Chem.* **1973**, *38*, 457.

123. a) Cohen, S. G. *Trans. N.Y. Acad. Sci.* **1969**, *31*, 705. b) Norin, M.; Hult, K. *Biocatalysis* **1993**, *7*, 131.

124. Anantharamaiah, G. M.; Roeske, R. W. *Tetrahedron Lett.* **1982**, *23*, 3335.

125. Schricker, B.; Thirring, K.; Berner, H. *Bioorg. Med. Chem. Lett.* **1992**, *12*, 387.

126. Lalonde, J. J.; Bergbreiter, D. E.; Wong, C.-H. *J. Org. Chem.* **1988**, *53*, 2523.

127. Cohen, S. G.; Khedouri, E. *J. Am. Chem. Soc.* **1961**, *83*, 1093.

128. Cohen, S. G.; Khedouri, E. *J. Am. Chem. Soc.* **1961**, *83*, 4228.

129. Roy, R.; Rey, A. W. *Tetrahedron Lett.* **1987**, *28*, 4935.

130. Cohen, S. G.; Sprinzak, Y.; Khodouri, E. *J. Am. Chem. Soc.* **1961**, *83*, 4225.

131. a) Daffe, V.; Fastrez, J. *J. Am. Chem. Soc.* **1980**, *102*, 3601. b) Chenevert, R.; Letourneau, M. *Chem. Lett.* **1986**, 1151. c) Honig, H.; Seufer-Wasserthal, P.; Weber, H. *Tetrahedron* **1990**, *46*, 3841. d) Ricca, J.-M.; Crout, D. H. G. *J. Chem. Soc., Perkin Trans. 1* **1993**, 1225.

132. Lin, Y. Y.; Jones, J. B. *J. Org. Chem.* **1973**, *38*, 3575; Tamura, M.; Kinomura, K.; Tada, M.; Nakatsuka, T.; Okai, H.; Fukui, S. *Agric. Biol. Chem.* **1985**, *49*, 2011; Bender, M. L.; Kezdy, F. J.; Gunter, C. R. *J. Am. Chem. Soc.* **1964**, *86*, 3714. d) Ricca, J.-M.; Crout, D. H. G. *J. Chem. Soc., Perkin Trans. 1* **1993**, 1225.

133. Cohen, S. G.; Crossley, J.; Khedouri, E. *Biochemistry* **1963**, *2*, 820.

134. a) Cohen, S. G.; Milovanovic, A. *J. Am. Chem. Soc.* **1968**, *90*, 3495. b) Luyten, M.; Muller, S.; Herzog, B.; Keese, R. *Helv. Chim. Acta* **1987**, *70*, 1250.

135. Schregenberger, C.; Seebach, D. *Liebigs Ann. Chem.* **1986**, 2081.

136. Crout, D. H. G.; Gaudet, V. S. B.; Hallinan, K. O. *J. Chem. Soc., Perkin Trans. 1* **1993**, 805.

137. Kise, H.; Shirato, H. *Tetrahedron Lett.* **1985**, *26*, 6081; Ingalls, R. G.; Squires, R. G.; Butler, L. G. *Biotech. Bioeng.* **1975**, *17*, 1627; Reslow, M.; Adlercreutz, P.; Mattiasson, B. *Appl. Microb. Biotech.* **1987**, *26*, 1; Kise, H.; Shirato, H. *Enz. Microb. Tech.* **1988**, *10*, 582.

138. Reinhoudt, D. N.; Eendebak, A. M.; Nijenhuis, W. F.; Verboom, W.; Kloosterman, M.; Schoemaker, H. E. *J. Chem. Soc., Chem. Commun.* **1989**, 399.

139. Inada, Y.; Takahashi, K.; Yoshimoto, T.; Ajima, A.; Matsushima, A.; Saito, Y. *TIBTECH* **1986**, 190.

140. See section 2.1 for subtilisin; Roper, J. M.; Pauer, D. P. *Synthesis* **1983**, 1041.

141. Schutt, H.; Schmidt-Kastner, G.; Arens, A.; Preiss, M. *Biotech. Bioeng.* **1985**, *27*, 420.

142. Chen, S.-T.; Wang, K.-T. *Synthesis* **1987**, 58.

143. Wu, S.-H.; Lo, L.-C.; Chen, S.-T.; Wang, K.-T. *J. Org. Chem.* **1989**, *54*, 4220.

144. Kvittingen, L.; Partali, V.; Braenden, J. U.; Anthonsen, T. *Biotechnol. Lett.* **1991**, *13*, 13.

145. Chen, S.-T.; Wang, K.-T.; Wong, C.-H. *J. Chem. Soc., Chem. Commun.* **1986**, 1514.

146. Chen, S.-T.; Chen, S. Y.; Hsiao, S. C.; Wang, K.-T. *Int. J. Peptide Protein Res.* **1991**, *37*, 347.

147. Imperiali, B.; Prins, T. J.; Fisher, S. L. *J. Org. Chem.* **1993**, *58*, 1613.

148. Mazdiyasni, H.; Konopacki, D. B.; Dickman, D. A.; Zydowsky, T. M. *Tetrahedron Lett.* **1993**, *34*, 435.

149. Holla, E. W.; Sinnwell, V.; Klaffke, W. *Synlett* **1992**, 413.

150. Riva, S.; Chopinean, J.; Kieboom, A. P. G.; Klibanov, A. M. *J. Am. Chem. Soc.* **1988**, *110*, 584.

151. Dordick, J. S. *TIBTECH* **1992**, *10*, 287.

152. Margolin, A. L.; Delinck, D. L.; Whalon, M. R. *J. Am. Chem. Soc.* **1990**, *112*, 2849.

153. Cai, S.; Hakomori, S.; Toyokuni, T. *J. Org. Chem.* **1992**, *57*, 3431.

154. Chinsky, N.; Margolin, A. L.; Klibanov, A. M. *J. Am. Chem. Soc.* **1989**, *111*, 386.

155. Kitaguchi, H.; Fitzpatrick, P. A.; Huber, J. E.; Klibanov, A. M. *J. Am. Chem. Soc.* **1989**, *111*, 3094.

156. Terradas, F.; Teston-Henry, M.; Fitzpatrick, P. A.; Klibanov, A. M. *J. Am. Chem. Soc.* **1993**, *115*, 390.

157. Wescott, C. R.; Klibanov, A. M. *J. Am. Chem. Soc.* **1993**, *115*, 1629.

158. Sakurai, T.; Margolin, A. L.; Russell, A. J.; Klibanov, A. M. *J. Am. Chem. Soc.* **1988**, *110*, 7236.

159. Margolin, A. L.; Fitzpatrick, P. A.; Dubin, P. L.; Klibanov, A. M. *J. Am. Chem. Soc.* **1991**, *113*, 4693.

160. Kim, J. M.; Hennen, W. J.; Sweers, M.; Wong, C.-H. *J. Am. Chem. Soc.* **1988**, *110*, 6481.

161. Liu, J. L.-C.; Shen, G.-J.; Ichikawa, Y.; Rutan, J. F.; Zapata, G.; Vann, W. F.; Wong, C.-H. *J. Am. Chem. Soc.* **1992**, *114*, 3901.

162. Bianchi, D.; Cabri, W.; Cesti, P.; Francalanci, F.; Ricci, M. *J. Org. Chem.* **1988**, *53*, 104.

163. Wong, C.-H.; Ho, M. F.; Wang, K.-T. *J. Org. Chem.* **1978**, *43*, 3604.

164. Drueckhammer, D. G.; Barbas, C. F.; Nozaki, K.; Wong, C.-H. *J. Org. Chem.* **1988**, *53*, 1607.

165. Shin, C.; Seki, M.; Takahashi, N. *Chem. Lett.* **1990**, 2089.

166. Cantacuzene, D.; Pascal, F.; Guerreiro, A. *Tetrahedron* **1987**, *43*, 1823; Chen, S.-T.; Wang, K.-T. *J. Chem. Soc., Chem. Commun.* **1988**, 327.

167. a) Feichter, C.; Faber, K.; Griengl, H. *J. Chem. Soc., Perkin Trans. 1* **1991**, 653. b) Hirose, Y.; Kariya, K.; Sasaki, I.; Kurono, Y.; Achiwa, K. *Tetrahedron Lett.* **1993**, *34*, 3441.

168. Berger, B.; de Raadt, A.; Griengl, H.; Hayden, W.; Hechtberger, P.; Klempier, N.; Faber, K. *Pure Appl. Chem.* **1992**, *64*, 1085.

169. Fulling, G.; Sih, C. J. *J. Am. Chem. Soc.* **1987**, *109*, 2845.

170. Savidge, T. A.; Cole, M. *Method. Enzymol.* **1976**, *43*, 705.

171. Fuganti, C.; Grasselli, P.; Servi, S.; Lazzarini, A.; Casati, P. *Tetrahedron* **1988**, *44*, 2575; Waldmann, H. *Tetrahedron Lett.* **1989**, 3057; Waldmann, H. *Kontakte* **1991**, *2*, 33.

172. Waldmann, H. *Liebigs Ann. Chem.* **1988**, 1175.

173. Kleine, R.; Rothe, U.; Kettmann, U.; Schelle, H. in *Proteases and Their Inhibitors*, Turk, V.; Vitale, L. eds. Pergamon Press, Oxford, 1981, p201.

174. Hermann, P.; Wiss Z. *Univ. Halle* **1987**, *36*, 17; Waldmann, H. *Kontakte* **1991**, *2*, 33.

175. Sussman, J. L.; Harel, M.; Frolow, F.; Oefner, C.; Goldman, A.; Toker, L.; Silman, I. *Science* **1991**, *253*, 872; Maelicke, A. *TIBS* **1991**, 355; Macphee-Quigley, K.; Taylor, P.; Taylor, S. *J. Biol. Chem.* **1985**, *260*, 12185; Brestkin, A. P.; Rozengart, E. V.; Abduvakhabov, A. A.; Sadykuv, A. A. *Russian Chemical Rev.* **1983**, *52*, 931; Jarv, J. *Bioorg. Chem.* **1984**, *12*, 259.

176. Sussman, J. L.; Harel, M.; Frolow, F.; Oefner, C.; Goldman, A.; Toker, L.; Silman, I. *Science* **1991**, *253*, 872.

177. Dropsy, E. P.; Klibanov, A. M. *Biotech. Bioeng.* **1984**, *26*, 911.

178. Kaneko, C.; Sugimoto, A.; Tanaka, S. *Synthesis* **1974**, 876. Deardorff, D. R.; Myles, D. C.; MacFerrin, K. D. *Tetrahedron Lett.* **1985**, *26*, 5615; Backvall, J.-E.; Bystrom, S. E.; Nordberg, R. E. *J. Org. Chem.* **1984**, *49*, 4619.

179. a) Deardorff, D. R.; Matthews, A. J.; McMeekin, D. S.; Craney, C. L. *Tetrahedron Lett.* **1986**, *27*, 1255. b) Griffith, D. A.; Danishefsky, S. J. *J. Am. Chem. Soc.* **1991**, *113*, 5863. An opposite enantiomer was reported: LeGrand, D. M.; Roberts, S. M. *J. Chem. Soc., Perkin Trans. 1* **1992**, 1751. c) Johnson, C. R.; Penning, T. D. J. Am. Chem. Soc. 1986, 108, 5655. d) Johnson, C. R.; Senanayake, C. H. *J. Org. Chem.* **1989**, *54*, 736. e) Pearson, A. J.; Lai, Y.-S.; Lu, W.; Pinkerton, A. A. *J. Org. Chem.* **1989**, *54*, 3882.

180. Greenzaid, P.; Jencks, W. P. *Biochemistry* **1971**, *10*, 1210.

181. Huang, F.-C.; Hsu, L. F.; Mittal, R. S. D.; Ravikumar, P. R.; Chan, J. A.; Sih, C. J.; Caspi, E.; Eck, C. R. *J. Am. Chem. Soc.* **1975**, *97*, 4144.

182. Ohno, M.; Otsuka, M. *Org. React.* **1989**, *37*, 1.

183. a) Tamm, C. *Pure Appl. Chem.* **1992**, *64*, 1187. b) Renold, P.; Tamm, C. *Tetrahedron: Asymmetry* **1993**, *4*, 2295.

184. Toone, E. J.; Werth, M. J.; Jones, J. B. *J. Am. Chem. Soc.* **1990**, *112*, 4946.

185. Lam, L. K. P.; Browne, C. M.; DeJeso, B.; Lym, L.; Toone, E. J.; Jones, J. B. *J. Am. Chem. Soc.* **1988**, *110*, 4409.

186. Mohr, P.; Waespe-Sarcevic, N.; Tamm, C.; Gawronska, K.; Gawronski, J. K. *Helv. Chim. Acta* **1983**, *66*, 2501; Ohno, M. in *Enzymes in Organic Synthesis: Ciba Foundation Symposium 111*; Clark, R.; Porter, S. Eds.; Pitman: London, **1985**, 171; Bjorkling, F.; Boutelje, J.; Gatenbeck, S.; Hult, K.; Norin, T.; Szmulik, P. *Tetrahedron* **1985**, *41*, 1347.

187. a) Jones, M.; Page, M. I. *J. Chem. Soc., Chem. Commun.* **1991**, 316. b) Barton, P.; Page, M. I. *Tetrahedron* **1992**, *48*, 7731. c) Barton, P.; Page, M. I. *J. Chem. Soc., Perkin Trans. 2* **1993**, 2317.

188. a) Freeman, S.; Irwin, W. J.; Mitchell, A. G.; Nicholls, D.; Thomson, W. *J. Chem. Soc., Chem. Commun.* **1991**, 875. b) Page, M. unpublished.

189. Bjorkling, F.; Boutelje, J.; Gatenbeck, S.; Hult, K.; Norin, T. *Tetrahedron Lett.* **1985**, *26*, 4957; Luyten, M.; Muller, S.; Herzog, B.; Keese, R. *Helv. Chim. Acta* **1987**, *70*, 1250; Andrade, M. A. C.; Andrade, F. A. C.; Phillips, R. S. *BioMed. Chem. Lett.* **1991**, *1*, 373.

190. Caron, G.; Kazlauskas, R. J. *J. Org. Chem.* **1991**, *56*, 7251.

191. Cambou, B.; Klibanov, A. M. *J. Am. Chem. Soc.* **1984**, *106*, 2687.

192. Scott, D. L.; White, S. P.; Otwinowski, Z.; Yuan, W.; Gelb, M. H.; Sigler, P. B. *Science* **1990**, *250*, 1541.

193. Ramirez, F.; Jain, M. K. *Proteins* **1991**, *9*, 229.

194. a) Tsai, T.-C.; Hart, J.; Jiang, R.-T.; Bruzik, K.; Tsai, M.-D. *Biochemistry* **1985**, *24*, 3180. b) Jiang, R.-T.; Shyy, Y.-J.; Tsai, M.-D. *Biochemistry* **1984**, *23*, 1661. c) Bruzick, K.; Tsai, M.-D. *Biochemistry* **1984**, *23*, 1656. d) Bruzik, K.; Tsai, M.-D. *J. Am. Chem. Soc.* **1982**, *104*, 863. e) Bruzik, K.; Tsai, M.-D. *Method Enzymol.* **1991**, *197*, 258.

195. Henrickson, H. S.; Dumdei, E. J.; Batchelder, A. G.; Carlson, G. L. *Biochemistry* **1987**, *26*, 3697; Lin, G.; Liu, S. H. *Tetrahedron Lett.* **1993**, *34*, 1959.

196. a) Shuto, S.; Ueda, S.; Imamura, S.; Fukukawa, K.; Matsuda, A.; Ueda, T. *Tetrahedron Lett.* **1987**, *28*, 199. Shuto, S.; Imamura, S.; Fukukawa, K.; Ueda, T. *Chem. Pharm. Bull.* **1988**, *36*, 5020. b) Wang, P.; Schuster, M.; Wang, Y.-F.; Wong, C.-H. *J. Am. Chem. Soc.* **1993**, *115*, 10487.

197. Satouchi, K.; Pinckard, R. N.; Manus, L. M.; Hanahan, D. *J. Biol. Chem.* **1981**, *256*, 4425.

198. Teslet-Lamant, V.; Archaimbault, B.; Durand, J.; Rigaud, M. *Biochim. Biophys. Acta* **1992**, *1123*, 347.

199. Yoneda, K.; Sasakura, K.; Tahara, S.; Iwasa, J.; Baba, N.; Kaneko, T.; Matsuo, M. *Angew. Chem. Int. Ed. Engl.* **1992**, *31*, 1336.

200. Sutton, L. D.; Froelich, S.; Hendrickson, H. S.; Quinn, D. M. *Biochemistry* **1991**, *30*, 5888.

201. Kazlauskas, R. J. *J. Am. Chem. Soc.* **1989**, *111*, 4953; Org. Syn.

202. Liu, Y.-C.; Chen, C.-S. *Tetrahedron Lett.* **1989**, *30*, 1617.

203. Kazlauskas, R. J.; Weissflch, A. N. E.; Rappaport, A. T.; Cuccia, L.A. *J. Org. Chem.* **1991**, *56*, 2656.

204. Desnuelle, P. The Lipases in: *The Enzymes*, Boyer, P.O. ed., Vol. 7, 575, Academic Press, New York **1972**.

205. Chapusand, C.; Semariva, M. *Biochemistry* **1976**, *15*, 4987.

206. Sarda, L.; Desnuelle, P. *Biochim. Biophys. Acta* **1958**, *30*, 513.

207. Pieterson, W. A.; Vidal, J.C.; Volwerk, J. J.; de Haas, G. H. *Biochemistry* **1974**, *13*, 1455; Verger, R.; Mieras, M. C. F.; de Haas, G. H. *J. Biol. Chem.* **1973**, *248*, 4023.

208. Desnuelle, P.; Sarda, L.; Aihard, G. *Biochim. Biophys. Acta* **1960**, *37*, 570.

209. Winkler, F. K.; D'Arey, A.; Hunziker, W. *Nature* **1990**, *343*, 771; van Tilbeurgh, H.; Sarda, L.; Verger, R.; Cambillau, C. *Nature* **1992**, *359*, 159. For site-directed mutagenesis, Lowe, M. E. *J. Biol. Chem.* **1992**, *267*, 17069.

210. Brady, L.; Brzozowski, A. M.; Derewenda, Z. S.; Dodson, E.; Dodson, G.; Tolley, S.; Turkenburg, J. P.; Christiansen, L.; Huge-Jensen, B.; Norskov, L.; Thim, L.; Menge, U. *Nature* **1990**, *343*, 767; Brzozowski, A. M.; Derewenda, U.; Derewenda, Z. S.; Dodson, G. G.; Lawson, D. M.; Turkenburg, J. P.; Bjorkling, F.; Huge-Jensen, B.; Patkar, S. A.; Thim, L. *Nature* **1991**, *351*, 491; Derewenda, U.; Brzozowski, A. M.; Lawson, D. M.; Derewenda, Z. S. *Biochemistry* **1992**, *31*, 1532; Blow, D. *Nature* **1991**, *351*, 444.

211. Dijkstra, B.W. et al. *J. Mol. Biol.* **1981**, *147*, 97; Brunie, S. et al. *J. Biol. Chem.* **1985**, *260*, 9742; Thunnisen, M. M. et al. *Nature* **1990**, *347*, 689; Scott, D. L.; White, S. P.; Otwinowski, Z.; Yuan, W.; Gelb, M.H.; Sigler, P.B. *Science* **1990**, *250*, 1541.

212. Schrag, J. D.; Li, Y.; Wu, S.; Cygler, M. *Nature* **1991**, *351*, 761.

213. Liu, W. H.; Beppa, T.; Arima, K. *Agric. Biol. Chem.* **1977**, *41*, 131.

214. Chen, C.-S.; Fujimoto, Y.; Girdaukas, G.; Sih, C. J. *J. Am. Chem. Soc.* **1982**, *104*, 7294.

215. Chen, C.-S.; Wu, S.-H.; Girdaukas, G.; Sih, C. J. *J. Am. Chem. Soc.* **1987**, *109*, 2812.

216. Kirchner, G.; Scollar, M. P.; Klibanov, A. M. *J. Am. Chem. Soc.* **1985**, *107*, 7072.

217. a) Sweers, H. M.; Wong, C.-H. *J. Am. Chem. Soc.* **1986**, *108*, 6421; b) Degueil-Castaing, M.; Jeso, B. D.; Drouillard, S.; Maillard, B. *Tetrahedron Lett.* **1987**, *28*, 953.

218. a) Wang, Y.-F.; Wong, C.-H. *J. Org. Chem.* **1988**, *53*, 3127. b) Laumen, K.; Brietgoff, D.; Schneider, M. P. *J. Chem. Soc., Chem. Commun.* **1988**, 1459. c) Wang, Y.-F.; Lalonde, J.J.; Homongan, M.; Bergbreiter, D. E.; Wong, C.-H. *J. Am. Chem. Soc.* **1988**, *110*, 7200.

219. Bianchi, D.; Cesti, P.; Battistel, E. *J. Org. Chem.* **1988**, *53*, 5531; Uemura, A.; Nozaki, K.; Yamashita, J.-I.; Yasumoto, M. *Tetrahedron Lett.* **1989**, *30*, 3817.

220. Frykman, H.; Ohrner, N.; Norin, T.; Hult, K. *Tetrahedron Lett.* **1993**, *34*, 1367.

221. Ghogare, A.; Kumar, G. S. *J. Chem. Soc., Chem. Commun.* **1989**, 1533.

222. Gotor, V.; Pulido, R. *J. Chem. Soc., Perkin Trans .1* **1991**, 491.

223. Pozo, M.; Pulido, R.; Gortor, V. *Tetrahedron* **1992**, *48*, 6477.

224. Berger, B.; Faber, K. *J. Chem. Soc., Chem. Commun.* **1991**, 1198.

225. Mischitz, M.; Poschl, U.; Faber, K. *Biotechnol. Lett.* **1991**, *13*, 653.

226. Fernholz, E.; Schloeder, D.; Liu, K. K.-C.; Bradshaw, C. W.; Huang, H.; Janda, K.; Lerner, R. A.; Wong, C.-H. *J. Org. Chem.* **1992**, *57*, 4756.

227. Pulido R.; Ortiz, F. D.; Gotor, V. *J. Chem. Soc., Perkin Trans. 1* **1992**, 2891; Gotor, V.; Moris, F. *Synthesis* **1992**, 626.

228. Moris, F.; Gotor, V. *J. Org. Chem.* **1992**, *57*, 2490.

229. Moris, F.; Gotor, V. *Tetrahedron* **1992**, *48*, 9869.

230. Moris, F.; Gotor, V. *J. Org. Chem.* **1993**, *58*, 653.

231. Pulido, R.; Gotor, V. *J. Chem. Soc., Perkin Trans. 1* **1993**, 589.

232. Cotterill, I. C.; Sutherland, A. G.; Roberts, S. M.; Grobbauer, R.; Spreitz, J.; Faber, K. *J. Chem. Soc., Perkin trans1* **1991**, 1365.

233. a) Kasel, W.; Hultin, P. G.; Jones, J. B. *J. Chem. Soc., Chem. Commun.* **1985**, 1563. b) Laumen, K.; Schneider, M. *Tetrahedron Lett.* **1985**, *26*, 2073.

234. Hemmerle, H.; Gais, H.-J. *Tetrahedron Lett.* **1987**, 3471.

235. Guanti, G.; Banfi, L.; Narisano, E. *Tetrahedron: Asymmetry* **1990**, *1*, 721; Guanti, G.; Narisano, E.; Podgorski, T.; Thea, S.; Williams, A. *Tetrahedron* **1990**, *46*, 7081.

236. Ramos-Tombo, G. M.; Schar, H.-P.; Fernandez, I.; Busquets, X.; Ghisalba, O. *Tetrahedron Lett.* **1986**, 5707.

237. Breitgoff, D.; Laumen, K.; Schneider, M. P. *J. Chem. Soc., Chem. Commun.* **1986**, 1523.

238. Ladner, W. E.; Whitesides, G. M. *J. Am. Chem. Soc.* **1984**, *106*, 7250.

239. Hultin, P. G.; Jones, J. B. *Tetrahedron Lett.* **1992**, *33*, 1399.

240. Guanti, G.; Banfi, L.; Narisano, E. *J. Org. Chem.* **1992**, *57*, 1540.

241. Ehrler, J.; Seebach, D. *Liebigs Ann. Chem.* **1990**, 379.

242. Wimmer, Z. *Tetrahedron* **1992**, *48*, 8431.

243. a) Wang, Y.-F.; Chen, C.-S.; Girdaukas, G.; Sih, C.J. *J. Am. Chem. Soc.* **1984**, *106*, 3695. b) Laumen, K.; Schneider, M. P. *J. Chem. Soc., Chem. Commun.* **1986**, 1298.

244. See ref. 7, also Bianchi, D.; Cabri, W.; Cesti, P.; Francalanci, F.; Rama, F. *Tetrahedron Lett.* **1988**, *29*, 2455.

245. Ramos Tombo, G. M.; Schar, H.-P.; Busquets, X.F.; Ghisalba, O. *Tetrahedron* **1986**, *27*, 5707; Di Dier, E.; Loubinoux, B.; Ramos Tombo, G. M.; Rihs, G. *Tetrahedron*

1991, *47*, 4941. No enantioselective hydrolysis was observed with pure PPL from Sigma. Also see, Bornemann, S.; Crout, D. H. G.; Dalton, H.; Hutchinson, D. W. *Biocatalysis* **1992**, *5*, 297 for a similar result.

246. a) Grandjean, D.; Pale, P.; Chuche, J. *Tetrahedron Lett.* **1991**, *32*, 3043. Improved ee was observed at low temperature and low pH. b) Vanttinen, E.; Kanerva, L. T. *Tetrahedron: Asymmetry* **1992**, *3*, 1529.

247. Ref. 2 and Ader, U.; Breitgoff, D.; Klein, P.; Laumen, K.E.; Schneider, M. P. *Tetrahedron Lett.* **1989**, *30*, 1793.

248. Hemmerle, H.; Gais, H.-J. *Tetrahedron Lett.* **1987**, 28.

249. Naemura, K.; Takahashi, N.; Chikamatsu, H. *Chem. Lett.* **1988**, 1717.

250. Suemune, H.; Tanaka, M.; Obaishi, K.; Sakai, K. *Chem. Pharm. Bull.* **1988**, 15.

251. Fink, A. L.; Hay, G. W. *Can. J. Biochem.* **1969**, *47*, 353.

252. Shaw, J.-F.; Klibanov, A. M. *Biotech. Bioeng.* **1987**, *29*, 648.

253. Hennen, W. J.; Sweers, H. M.; Wang, Y.-F.; Wong, C.-H. *J. Org. Chem.* **1988**, *53*, 4939.

254. Drueckhammer, D. G.; Hennen, W. J.; Pederson, R. L.; Barbas, C. F.; Gautheron, C. M.; Karch, T.; Wong, C.-H. *Synthesis* **1991**, *7*, 499.

255. Therisod, M.; Klibanov, A. M. *J. Am. Chem. Soc.* **1986**, *108*, 5638.

256. Therisod, M.; Klibanov, A. M. *J. Am. Chem. Soc.* **1987**, *109*, 3977.

257. Ciuffreda, P.; Colombo, D.; Ronchetti, F.; Toma, L. *J. Org. Chem.* **1990**, *55*, 4187.

258. Chinn, M. J.; Iacazio, G.; Spackman, D. G.; Turner, N. J.; Roberts, S. M. *J. Chem. Soc., Perkin Trans . 1* **1992**, 661.

259. Ballesteros, A.; Bernabe, M.; Cruzado, C.; Martin-Lomas, M.; Otero, C. *Tetrahedron* **1989**, *45*, 7077.

260. Guibe-Jampel, E.; Rousseau, G.; Salaun, J. *J. Chem. Soc., Chem. Commun.* **1987**, 1080.

261. Cohen, S. G.; Milovanovic, A. *J. Am. Chem. Soc.* **1968**, *90*, 3495.

262. For early observation of amide formation Zaks, A.; Klibanov, A. M. *Proc. Natl. Acad. Sci.* **1985**, *82*, 3192. For peptide synthesis, West, J. B.; Wong, C.-H. *Tetrahedron Lett.* **1987**, *28*, 1629; Margolin, A. L.; Klibanov, A. M. *J. Am. Chem. Soc.* **1987**, *109*, 3802. For reaction with amine or the amine of aminoalcohols, Fernandez, S.; Brieva, R.; Rebolledo, F.; Gotor, V. *J. Chem. Soc., Perkin Trans. 1* **1992**, 2885. For cyclization of amino esters to form 4-8 membered lactams and bislactams, Gutman, A. L.; Meyer, E.; Yue, X.; Abell, C. *Tetrahedron Lett.* **1992**, *33*, 3943.

263. a) Gutman, A. L.; Zuob, K.; Boltanski, A. *Tetrahedron Lett.* **1987**, *26*, 3861. b) Gutman, A. L.; Oren, D.; Boltanski, A.; Bravdo, T. *Tetrahedron Lett.* **1987**, *28*, 5367. c) Gutman, A. L.; Zuob, K.; Bravdo, T. *J. Org. Chem.* **1990**, *55*, 3546. d) Sugai, T.; Ohsawa, S.; Yamada, H.; Ohta, H. *Synthesis* **1990**, 1112; Sugai, T.; Noguchi, H.;

Ohta, H. *Biosci. Biotech. Biochem.* **1992**, *56*, 122. Henkel, B.; Kunath, A.; Schick, H. *Tetrahedron: Asymmetry* **1993**, *4*, 153.

264. a) Margolin, A. L.; Crenne, J.-Y.; Klibanov, A. M. *Tetrahedron Lett.* **1987**, *28*, 1607. b) Wallace, J. S.; Morrow, C. J. *J. Polymer Sci.* **1989**, *27*, 3271. c) Wallace, J. S.; Morrow, C. J. *J. Polymer Sci.* **1989**, *27*, 2553. d) Wallace, J. S.; Reda, K. B.; Williams, M. E.; Morrow, C. J. *J. Org. Chem.* **1990**, *55*, 3544.

265. Howell, J. A. S.; Palin, M. G.; Jaouen, G.; Top, S.; Hafa, H. E.; Cense, J. M. *Tetrahedron: Asymmetry* **1993**, *4*, 1241.

266. Boland, W.; Frobl, C.; Lorenz, M. *Synthesis* **1991**, 1049.

267. Hughes, D. L.; Bergan, J. J.; Amato, J. S.; Bhupathy, M.; Leazer, J. L.; McNamara, J. M.; Sidler, D. R.; Reider, P. J.; Grabowski, E. J. J. *J. Org. Chem.* **1990**, *55*, 6252.

268. a) Breitgoff, D.; Laumen, K.; Schneider, M. P. *J. Chem. Soc., Chem. Commun.* **1988**, 1523. b) Laumen, K.; Breitgoff, D.; Schneider, M. P. *J. Chem. Soc., Chem. Commun.* **1988**, 1459.

269. a) Kazlauskas, R. L.; Weissfloch, A. N. E.; Rapport, A. T.; Cuccia, L. A. *J. Org. Chem.* **1991**, *56*, 2656. b) Johnson, C. R.; Golebiowski, A.; McGill, T. K.; Steensma, D. H. *Tetrahedron Lett.* **1991**, *32*, 2597. c) Ader, U.; Andersch, P.; Berger, M.; Goergens, U.; Seemayer, R.; Schneider, M. *Pure Appl. Chem.* **1992**, *64*, 1165. d) Burgess, K.; Jennings, L. D. *J. Am. Chem. Soc.* **1991**, 113, 6129. e) Kim, M.-J.; Cho, H. *J. Chem. Soc., Chem. Commun.* **1992**, 1411.

270. Bianchi, D.; Cesti, P.; Golini, P. *Tetrahedron* **1989**, 45, 869.

271. Itoch, T.; Takagi, Y.; Nishiyama, S. *J. Org. Chem.* **1991**, *56*, 1521.

272. Burgess, K.; Henderson, I.; Ho, K.-K. *J. Org. Chem.* **1992**, *57*, 1290.

273. Scitimati, A.; Ngooi, T. K.; Sih, C. J. *Tetrahedron Lett.* **1988**, *29*, 4927.

274. Sugai, T.; Ohta, H. *Agric. Biol. Chem.* **1990**, *54*, 3337; Sugai, T.; Ohta, H. *Tetrahedron Lett.* **1991**, *32*, 7063; Sugai, T.; Ritzen, H.; Wong, C.-H. *Tetrahedron: Asymmetry* **1993**, *4*, 1051.

275. Matsuo, N.; Ohno, N. *Tetrahedron Lett.* **1985**, *26*, 5533; Almsick, A. V.; Buddrus, J.; Honicke-Schmidt, P.; Laumen, K.; Schneider, M. P. *J. Chem. Soc., Chem. Commun.* **1989**, 1391.

276. Wang, Y.-F.; Chen, S.-T.; Liu, K.K.-C.; Wong, C.-H. *Tetrahedron Lett.* **1989**, *30*, 1917.

277. Inagaki, M.; Hiratake, J.; Nishioka, T.; Oda, J. *J. Am. Chem. Soc.* **1991**, *113*, 9360.

278. Baba, N.; Takeno, K.; Iwasa, J.; Oda, J. *Agric. Biol. Chem.* **1990**, *54*, 3349.

279. a) Boaz, N. W. *Tetrahedron Lett.* **1989**, *30*, 2061. b) Nakamura, K.; Ishihara, K.; Ohno, A.; Uemura, M.; Nishimura, H.; Hayashi, Y. *Tetrahedron Lett.* **1990**, *31*, 3603. c) Yamazaki, Y.; Hosono, K. *Agric. Biol. Chem.* **1990**, *54*, 3357. d) Uemura, M.;

Nishimura, H.; Yamada, S.; Nakamura, K; Hayashi, Y. *Tetrahedron Lett.* **1993**, *34*, 6581.

280. Bianchi, D.; Cesti, P. *J. Org. Chem.* **1990**, *55*, 5657.

281. a) Gardossi, L.; Bioanchi, D.; Klibanov, A. M. *J. Am. Chem. Soc.* **1991**, *113*, 6328. b) Takaoka, Y.; Kajimoto, T.; Wong, C.-H. *J. Org. Chem.* **1993**, *58*, 4809.

282. a) Holla, E. W. *Angew. Chem. Int. Ed. Engl.* **1989**, *28*, 220. b) Berkowitz, D. B.; Danishefsky, S. J. *Tetrahedron Lett.* **1991**, *32*, 5497.

283. Ciuffreda, P.; Colombo, D.; Ronchetti, F.; Toma, L. *J. Org. Chem.* **1990**, *55*, 4187.

284. Nozaki, K.; Uemura, A.; Yamashita, J.-I.; Yasumoto, M. *Tetrahedron Lett.* **1990**, *31*, 7327; Uemura, A.; Nozaki, K.; Yamashita, J.-I.; Yasumoto, M. *Tetrahedron Lett.* **1989**, *30*, 3819.

285. Guo, Z.W.; Sih, C. J. *J. Am. Chem. Soc.* **1988**, *110*, 1999; Ngooi, T. K.; Scilimati, A.; Guo, Z.-W.; Sih, C. J. *J. Org. Chem.* **1989**, *54*, 911.

286. a) Pederson, R. L.; Liu, K. K.-C.; Rutan, J. F.; Chen, L.; Wong, C.-H. *J. Org. Chem.* **1990**, *55*, 4897. b) Hsu, S.-H.; Wu, S.-S.; Wang, Y.-F.; Wong, C.-H. *Tetrahedron Lett.* **1990**, *31*, 6403. c) Bianchi, D.; Cesti, P.; Battistel, E. *J. Org. Chem.* **1988**, *53*, 5531. d) Nuscimento, M. G.; Rezende, M. C.; Vecchia, R. D.; de Jesus, P.C.; Aguiar, L. M. Z. *Tetrahedron Lett.* **1992**, *33*, 5891. e) Jouglet, B.; Rousseau, G. *Tetrahedron Lett.* **1993**, *34*, 2307. f) Akita, H.; Umezawa, I.; Matsukura, H.; Oishi, T. *Chem. Pharm. Bull.* **1991**, *39*, 1632.

287. a) Irinchijima, S.; Kojima, N. *Agric. Biol. Chem.* **1982**, *46*, 1153. b) Eycken, J. V. D.; Vandewalle, M.; Heinemann, G.; Laumen, K.; Schneider, M. P.; Kredel, J.; Sauer, J. *J. Chem. Soc., Chem. Commun.* **1989**, 306. c) Amici, M. D.; Micheli, C. D.; Carrea, G.; Spezia, S. *J. Org. Chem.* **1989**, *54*, 2646. d) Uemura, A.; Nozaki, K.; Yamashita, J.-I.; Yasumoto, M. *Tetrahedron Lett.* **1989**, *30*, 3817; Uemura, A.; Nozaki, K.; Yamashita, J.-I.; Yasumoto, M. *Tetrahedron Lett.* **1989**, *30*, 3819. e) Xie, Z. F.; Sakai, K. *Chem. Pharm. Bull.* **1989**, *37*, 1650. f) Xie, Z. F.; Suemune, H.; Sakai, K. *J. Chem. Soc., Chem. Commun.* **1989**, 838. g) Laumen, K.; Breitgoff, D.; Seemayer, R.; Schneider, M. P. *J. Chem. Soc., Chem. Commun.* **1989**, 148. h) Yamazaki, T.; Ichikawa, S.; Kitazume, T. *J. Chem. Soc., Chem. Commun.* **1989**, 253. i) Mori, K.; Bernotas, R. *Tetrahedron: Asymmetry* **1990**, *1*, 87. j) Foelsche, E.; Hickel, A.; Honig, H.; Seufer-Wasserthal, P. *J. Org. Chem.* **1990**, *55*, 1749. k) Liang, S.; Paquette, L. A. *Tetrahedron: Asymmetry* **1990**, *1*, 445. l) Grisenti, P.; Ferraboschi, P.; Manzocchi, A.; Santaniello, E. *Tetrahedron* **1992**, *48*, 3827. m) Sparks, M. A.; Panek, J. S. *Tetrahedron Lett.* **1991**, *32*, 4085. n) Goergens, U.; Schneider, M. P. *J. Chem. Soc. Chem. Comun.* **1991**, 1066. o) Ferraboschi, P.; Grisenti, P.; Marzocchi, A.; Santaniello, E. *J. Chem. Soc., Perkin Trans. 1* **1992**, 1159. p) Bianchi, D.; Bosetti, A.; Cesti, P.; Golini, P. *Tetrahedron Lett.* **1992**, *33*, 3231. q) Schneider, M. P.; Goergens,

U. *Tetrahedron: Asymmetry* **1992**, *3*, 525; Goergens, U.; Schneider, M. P. *Tetrahedron: Asymmetry* **1992**, *3*, 1149; Chadha, A.; Goergens, U.; Schneider, M. P. *Tetrahedron: Asymmetry* **1993**, *4*, 1449. r) Boaz, N. W. *J. Org. Chem.* **1992**, 57, 4289. s) Takeshita, M.; Yaguchi, R.; Akutsu, N. *Tetrahedron: Asymmetry* **1992**, *3*, 1369. t) Bevinakatti, H. S.; Banerji, A. A. *J. Org. Chem.* **1992**, *57*, 6003. u) Takabe, K.; Sawada, H.; Satani, T.; Yamada, T.; Katagiri, T.; Yoda, H. *Bioorg. Med. Chem. Lett.* **1993**, *3*, 157. v) Naoshima, Y.; Kamezawa, M.; Tachibana, H.; Munakata, Y.; Fujita, T.; Kihara, K.; Raku, T. *J. Chem. Soc., Perkin Trans. 1* **1993**, 557. w) Momose, T.; Toyooka, N.; Jin, M. *Tetrahedron Lett.* **1992**, *33*, 5389. x) Gotor, V.; Astorga, C.; Rebolledo, F. *Synlett.* **1990**, 387. y) Merlo, V.; Reece, F. J.; Roberts, S. M.; Gregson, M.; Storer, R. *J. Chem. Soc., Perkin Trans. 1* **1993**, 1717. z) Fukazawa, T.; Hashimoto, T. *Tetrahedron: Asymmetry* **1993**, *4*, 2323. aa) Wirz, B.; Barner, R.; Hubscher, J. *J. Org. Chem.* **1993**, *58*, 3980. bb) Takeshita, M; Miura, M.; Unuma, Y. *J. Chem. Soc., Perkin Trans 1* **1993**, 2901. cc) Kanerva, L. T.; Sundholm, O. *J. Chem. Soc., Perkin Trans 1* **1993**, 2407. dd) Itoh, T.; Chika, J.-I.; Takagi, Y.; Nishiyama, S. *J. Org. Chem.* **1993**, *58*, 5717. ee) Allevi, P; Anastasia, M.; Cajone, F.; Ciuffreda, P.; Sanvito, A. M. *J. Org. Chem.* **1993**, *58*, 5000.

288.   Yee, C.; Blythe, T. A.; McNabb, T. J.; Walts, A. E. *J. Org. Chem.* **1992**, *57*, 3525.

289.   Kirchner, G.; Scollar, M. P.; Klibanov, A. M. *J. Am. Chem. Soc.* **1985**, *107*, 7072.

290.   Koshiro, S.; Sonomoto, K.; Tanaka, A.; Fukui, S. *J. Biotech.* **1985**, *2*, 47; Langrand, G.; Baratti, J.; Buono, G.; Triantaphylides, C. *Tetrahedron Lett.* **1986**, *27*, 29.

291.   Oberhauser, T.; Bodenteich, M.; Faber, K.; Penn, G.; Griengl, H. *Tetrahedron* **1987**, *43*, 3931; Saf, R.; Faber, K.; Penn, G.; Grieng, H. *Tetrahedron* **1988**, *44*, 389; Konigsberger, K.; Faber, K.; Marschner, C.; Penn, G.; Baumgartner, P.; Griengl, H. *Tetrahedron* **1989**, *45*, 673; Berger, B.; Rabiller, C. G.; Konigsberger, K.; Faber, K.; Griengl, H. *Tetrahedron: Asymmetry* **1990**, *1*, 541.

292.   Sugai, T.; Kakeya, H.; Ohta, H.; Morooka, M.; Ohba, S. *Tetrahedron* **1989**, *45*, 6135. For other microbial hydrolysis of enol esters, Matsumoto, K.; Tsutsumi, S.; Ihori, T.; Ohta, H. *J. Am. Chem. Soc.* **1990**, *112*, 9614.

293.   Pottie, M.; van der Eycken, J.; Vandervalle, M.; Dewanckele, J. M.; Roper, H. *Tetrahedron Lett.* **1989**, 5319.

294.   Dumortier, A.; van der Eycken, J.; Vandewaller, M. *Tetrahedron Lett.* **1989**, 3201.

295.   Gu, Q.-M.; Chen, C.-S.; Sih, C. J. *Tetrahedron Lett.* **1986**, *27*, 1763.

296.   Riva, S.; Bovara, R.; Ottolina, G.; Secundo, F.; Carrea, G. *J. Org. Chem.* **1989**, *54*, 3161.

297.   Sugai, T.; Kakeya, H.; Ohta, H. *J. Org. Chem.* **1990**, *55*, 4643.

298.   Parida, S.; Dordick, J. S. *J. Org. Chem.* **1993**, *58*, 3238.

299.   Parida, S.; Dordick, J. S. *J. Am. Chem. Soc.* **1991**, *113*, 2253.

300. Oberhauser, T.; Faber, K.; Griengl, H. *Tetrahedron* **1989**, *45*, 1679.

301. Wu, S.-H.; Guo, Z.-W.; Sih, C. J. *J. Am. Chem. Soc.* **1990**, *112*, 1990.

302. Guo, Z.-W.; Sih, C. J. *J. Am. Chem. Soc.* **1989**, *111*, 6836.

303. Wu, S.-H.; Chu, F.-Y.; Wang, K.-T. *Bioorg. Med. Chem. Lett.* **1991**, *1*, 339.

304. Holmberg, E.; Hult, K. *Biocatalysis* **1992**, *5*, 289.

305. Adelhorst, K.; Bjorkling, F.; Godtfredsen, S. E.; Kirk, O. *Synthesis* **1990**, 112.

306. Holla, E.W.; Sinnwell, V.; Klaffke, W. *Synlett* **1992**, 413.

307. Chang, K.-Y.; Wu, S.-H.; Wang, K.-T. *Carbohydrate Res.* **1991**, *222*, 121.

308. Pedrocchi-Fantoni, G.; Servi, S. *J. Chem. Soc., Perkin Trans. 1* **1992**, 1029.

309. Bjorkling, F.; Frykman, H.; Godtfredsem, S. E.; Kirk, O. *Tetrahedron* **1992**, *48*, 4587.

310. Gotor, V.; Menedez, E.; Mouloungui, Z.; Gaset, A. *J. Chem. Soc., Perkin Trans 1* **1993**, 2453.

311. a)Kitazume, T.; Sato, T.; Kobayashi, T.; Lin, J. T. *J. Org. Chem.* **1986**, *51*, 1003. b) Kanerva, L. T.; Kiljunen, E.; Huuhtanen, T. T. *Tetrahedron: Asymmetry* **1993**, *4*, 2355. c) Allenmark, S. G; Anderson, C. *Tetrahedron: Asymmetry* **1993**, *4*, 2371. d) Waagen, V.; Hollingsater, I.; Partli, V.; Thorstad, O; Anthonsen, T. *Tetrahedron: Asymmetry* **1993**, *4*, 2265. e) O'Hagan, D.; Zaidi, N. A. *J. Chem. Soc., Perkin Trans. 1* **1993**, 2389. f) Moris, F.; Gotor, V. *Tetrahedron* **1993**, *49*, 10089. g) Bevinakatti, H.S.; Banerji, A. A.; Newadkar, R. V. *J. Org. Chem.* **1989**, *54*, 2453; Bhalerao, U.T.; Dasaradhi, L.; Neelakantan, P.; Fadnavis, N. W. *J. Chem. Soc., Chem. Commun.* **1991**, 1197; Kanerua, L. T.; Sundholm, O. *J. Chem. Soc., Perkin Trans. 1* **1993**, 1385; Gil, G.; Ferre, E.; Meou, A.; Triantaphyllides, C. *Tetrahedron Lett.* **1987**, *28*, 1647; Takano, S.; Inomata, K.; Ogasawara, K. *J. Chem. Soc., Chem. Commun.* **1989**, 271.

312. Okumura, S.; Iwai, M.; Tominaga, Y. *Agric. Biol. Chem.* **1984**, *48*, 2805. b) Sonnet, P.E. *J. Org. Chem.* **1987**, *52*, 3477.

313. Chan, C.; Cox, P. B.; Roberts, S. M. *J. Chem. Soc., Chem. Commun.* **1988**, 971.

314. MacFarlane, E. L. A.; Roberts, S. M.; Turner, N. J. *J. Chem. Soc., Chem. Commun.* **1990**, 569.

315. Fulling, G.; Sih, C. J. *J. Am. Chem. Soc.* **1987**, *109*, 2845.

316. a) Laumen, K.; Schneider, M. P. *J. Chem. Soc., Chem. Commun.* **1986**, 1298. b) Johnson, C. R.; Penning, T. D. *J. Am. Chem. Soc.* **1986**, *108*, 5655. c) Laumen, K.; Schneider, M. *Tetrahedron Lett.* **1984**, *25*, 5875. d) Wang, Y.-F.; Chen, C.-S.; Girdankas, G.; Sih, C. J. *J. Am. Chem. Soc.* **1984**, *106*, 3695; Laumen, K.; Schneider, M. *Tetrahedron Lett.* **1984**, *25*, 5875.

317. Babiak, K. A.; Ng, J. S.; Dygos, J. H.; Weyker, C. L.; Wang, Y.-F.; Wong, C.-H. *J. Org. Chem.* **1990**, *55*, 3377; Theil, F.; Ballschuh, S.; Schick, H.; Haupt, M.; Hafner, B.; Schwarz, S. *Synthesis* **1988**, 540.

318. Roberts, S.M. *Phil Trans R. Soc. London B* **1989**, *324*, 557.

319. Binns, F.; Roberts, S. M.; Taylor, A.; Williams, C. F. *J. Chem. Soc., Perkin Trans. 1* **1993**, 899.

320. Estermann, H.; Prasad, K.; Shapiro, M. J.; Repic, O.; Hardtmann, G. E.; Bolsterli, J. J.; Walkinshaw, M. D. *Tetrahedron Lett.* **1990**, *31*, 445.

321. a) Braun, P.; Waldmann, H.; Vogt, W.; Kunz, H. *Synlett* **1990**, 105. b) Braun, P.; Waldmann, H.; Vogt, W.; Kunz, H. *Libigs Ann. Chem.* **1991**, 165. c) Waldmann, H. *Kontakte* **1991**, *2*, 33.

322. Gou, D.-M.; Liu, Y.-C.; Chen, C.-S. *J. Org. Chem.* **1993**, *58*, 1287.

323. Honig, H.; Seufer-Wasserthal, P.; Weber, H. *Tetrahedron* **1990**, *46*, 3841.

324. Brieva, R.; Crich, J. Z.; Sih, C. J. *J. Org. Chem.* **1993**, *58*, 1068.

325. Tawaki, S.; Klibanov, A. M. *Biocatalysis* **1993**, *8*, 3.

326. a) Itoh, T.; Kuroda, K.; Tomosada, M.; Takagi, Y. *J. Org. Chem.* **1991**, *56*, 797. b) Akita, H.; Matsuknia, H.; Oishi, T. *Tetrahedron Lett.* **1986**, *27*, 5241. c) Hsiao, K.-F.; Wu, S.-H.; Wang, K.-T. *Bioorg. & Med. Chem.* **1993**, *3*, 2125. d) Khan, R.; Gropen, L.; Konowicz, P. A.; Matulova, M; Paoletti, S. *Tetrahedron Lett.* **1993**, *34*, 7767.

327. a) Gu, R. L.; Lee, I. S.; Sih, C. J. *Tetrahedron Lett.* **1992**, *33*, 1953. b) Crich, J. Z.; Brieva, R.; Marquardt, P.; Gu, R.-L.; Flemming, S.; Sih, C. J. *J. Org. Chem.* **1993**, *58*, 3252.

328. De Jersey, J. *Biochemistry* **1970**, *9*, 1761.

329. Bevinakatti, H. S.; Banerji, A. A.; Newadkar, R. V.; Mokashi, A. A. *Tetrahedron: Asymmetry* **1992**, *3*, 1505.

330. a) Riva, S.; Klibanov, A. *J. Am. Chem. Soc.* **1988**, *110*, 3291. b) Margolin, A. L.; Delinck, D. L.; Whalon, M. R. *J. Am. Chem. Soc.* **1990**, *112*, 2849.

331. Mitsuda, S.; Nabeshima, S.; Hirohara, H. *Appl. Microbiol. Biotechnol.* **1989**, *31*, 334. For active site model, see Umemura, T.; Hirohara, H. in *Biocatalysis in Agricultural Biotechnology*, Whitaker, J. R.; Sonnet, P. E. Eds. American Chemical Society 1989, chapter 26.

332. Yee, P.; Blythe, T.A; McNabb, T. J.; Walts, A. E. *J. Org. Chem.* **1992**, *57*, 3525.

333. Senanayake, C. H.; Bill, T. J.; Larsen, R. D.; Leazer, J.; Reider, P. J. *Tetrahedron Lett.* **1992**, *33*, 5901.

334. Schrag, J. D.; Li, Y.; Wu, S.; Cygler, M. *Nature* **1991**, *351*, 761.

335. Feichter, C.; Faber, K.; Griengl, H. *Tetrahedron Lett.* **1989**, *30*, 551.

336. Kalaritis, P.; Regenye, R. W.; Partridge J. S.; Coffen, D. L. *J. Org. Chem.* **1990**, *55*, 812.

337. Laane, C.; Boeren, S.; Hilhorst, R.; Veeger, C. in *Biocatalysis in Organic Media*, Laane, C.; Tramper, J.; Lilly, M. D. eds., Elsevier, Amsterdam 1987, p65.

338. Secundo, F.; Riva, S.; Carrea, G. *Tetrahedron: Asymmetry* **1992**, *3*, 267.

339. a) Kamal, A.; Damayanthi, Y.; Rao, M. V. *Tetrahedron: Asymmetry* **1992**, *3*, 1361; b) Mischitz, M.; Faber, K. *Tetrahedron Lett.* **1994**, *35*, 81. c) Chenevert, R.; Rhild, R. B.; Letourneau, M.; Gagnon, R.; D'Astous, L. *Tetrahedron: Asymmetry* **1993**, *4*, 1137. d) Suckling, C. J.; Zhu, L.-M. *Bioorg. & Med. Chem.* **1993**, *3*, 531. e) Matthews, D.J.; Wells, J.A. *Science* **1993**, *260*, 1113. f) Lane, J. W.; Hong, K.; Schwabacker, A. W. *J. Am. Chem. Soc.* **1993**, *115*, 2078.

340. Nagasawa, T.; Yamada, H. *Pure Appl. Chem.* **1990**, *62*, 1441.

341. Nagasawa, T.; Yamada, H. *TIBTECH* **1989**, *7*, 153.

342. Bui, K.; Arnaud, A.; Galzy, P. *Enzyme Microb. Technol.* **1982**, *4*, 195.

343. Asano, Y.; Yasuda, T.; Tani, Y.; Yamada, H. *Agric. Biol. Chem.* **1982**, *46*, 1183; Ryuno, K.; Nagasawa, T.; Yamada, H. *Agric. Biol. Chem.* **1988**, *52*, 1813.

344. Sugiura, Y.; Kuwahara, J.; Nagasawa, T.; Yamada, H. *J. Am. Chem. Soc.* **1987**, *109*, 5848.

345. Honicke-Schmidt, P.; Schneider, M.P. *J. Chem. Soc., Chem. Commun.* **1990**, 648-650.

346. de Raadt, A.; Klempier, N.; Faber, K.; Griengl, H. *J. Chem. Soc., Perkin Trans. 1* **1992**, 137.

347. Kakeya, H.; Sakai, N.; Sugai, T.; Ohta, H. *Tetrahedron Lett.* **1991**, *32*, 1343-1346.

348. Kakeya, H.; Sakai, N.; Sano, A.; Yokoyama, M.; Sugai, T.; Ohta, H. *Chem. Lett.* **1991**, 1823-1824.

349. a) Crosby, J. A.; Parratt, J. S.; Turner, N. J. *Tetrahedron: Asymmetry* **1992**, *3*, 1547. b) Beard, T.; Cohen, M. A.; Paratt, J. S.; Turner, N. J. *Tetrahedron: Asymmetry* **1993**, *4*, 1085. c) de Raadt, A.; Griengl, H.; Klempier, N.; Stutz, A. E. *J. Org. Chem.* **1993**, *58*, 3179. d) Geresh, S.; Giron, Y.; Gilboa, Y.; Glaser, R. *Tetrahedron* **1993**, *49*, 10099.

350. Kobayashi, M.; Komeda, H.; Yanaka, N.; Nagasawa, T.; Yamada, H. *J. Biol. Chem.* **1992**, *267*, 20746.

351. Catelani, G.; Mastrorilli, E. *J. Chem. Soc., Perkin Trans. 1* **1983**, 2717.

352. Barili, P. L.; Berti, G.; Catelani, G.; Colonna, F.; Mastrorilli, E. *J. Org. Chem.* **1987**, *52*, 2886.

353. Wistuba, D.; Schurig, V. *Angew. Chem. Int. Ed. Engl.* **1986**, *25*, 1032.

354. a) Bellucci, G.; Berti, G.; Chiappe, C.; Fabri, F.; Marioni, F. *J. Org. Chem.* **1989**, *54*, 968. b) Bellucci, G.; Chiappe, C.; Marioni, F.; Benetti, M. *J. Chem. Soc., Perkin Trans. 1* **1991**, 361. c) Lacourciere, G. M.; Armstrong, R. N. *J. Am. Chem. Soc.* **1993**, *115*, 10466.

355. Blee, E.; Schuber, F. *J. Biol. Chem.* **1992**, *267*, 11881.

356. a) Ohno, M. *Ferment. Ind. Tokyo* **1979**, *37*, 836. b) Sato, H. Jap. Patent 75140684 (1975). c) Allen, R. H.; Jakoby, W. B. *J. Biol. Chem.* **1969**, *244*, 2078.

357. Wong, C.-H.; Whitesides, G. M. *J. Org. Chem.* **1983**, *48*, 3199.

358. a) Herdewijn, P.; Balzarini, J.; De Cercq, E.; Vanderhaeghe, H. *J. Med. Chem.* **1985**, *28*, 1385. b) Bortawick, A. D.; Butt, S.; Biggadike, K.; Exall, A. M.; Roberts, S. M.; Youds, P. M.; Kirk, B. E.; Booth, B. R.; Cameron, J. M.; Cox, S. W.; Marr, C. L. P.; Shill, M. D. *J. Chem. Soc., Chem. Commun.* **1988**, 656.

359. Scollar, M. P.; Sigal, G.; Klibanov, A. M. *Biotech. Bioeng.* **1985**, *27*, 247.

360. a) Pradines, A.; Klaebe, A.; Perie, J.; Paul, F.; Monsan, P. *Tetrahedron* **1988**, *44*, 6373. b) Unverzaght, C.; Kunz, H.; Paulson, J. C. *J. Am. Chem. Soc.* **1990**, *112*, 9308.

361. Pradines, A.; Klaebe, A.; Perie, J.; Paul, F.; Monsan, P. *Enzyme Microb. Technol.* **1991**, *13*, 19.

Chapter 3. Oxidoreductions

Most enzyme-catalyzed redox processes involve the transfer of the equivalent of two electrons by either two one-electron steps or one two-electron step. The one-electron process is a radical process which very often involves the use of cofactors such as flavin, quinoid-coenzymes (vitamins C, E, and K, and coenzyme Q), and transition metals. The two electron process can be a hydride transfer or a proton abstraction followed by two-electron transfer.[1]

1. Nicotinamide Cofactor Dependent Oxidoreductions

Nicotinamide adenine dinucleotide (NAD) and the analogous 2'-phosphate (NADP) (Figure 1) are involved in many 2-electron oxidations catalyzed by dehydrogenases. The nicotinamide ring system is redox active, accepting a hydride or two electrons and a proton to form the 1,4-dihydronicotinamide derivatives (NADH or NADPH).[2] The reversible hydride transfer from a reduced substrate to NAD(P), and that from NAD(P)H to an oxidized substrate, is stereoselective and characteristic of individual enzymes. Each enzyme is able to stereo-specifically transfer one of the diastereotopic methylene hydrogens at C-4 of NAD(P)H to a substrate carbonyl group or an equivalent sp^2 center (C=C or C=N) with high enantiofacial or diastereofacial selectivity.

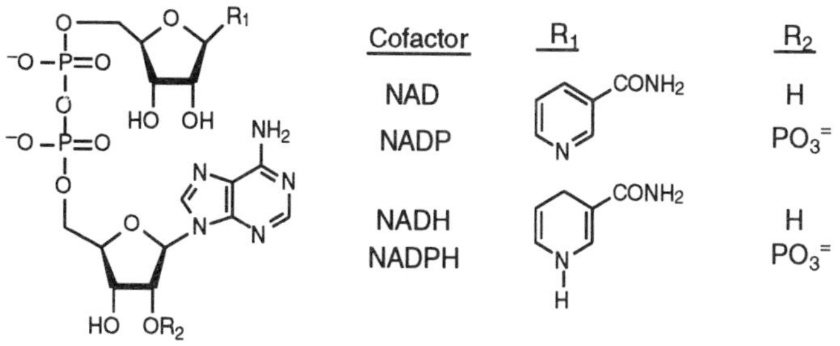

Cofactor	R_1	R_2
NAD	CONH$_2$	H
NADP		PO$_3^=$
NADH	CONH$_2$	H
NADPH		PO$_3^=$

Figure 1

The nicotinamide cofactors are usually not covalently bound to the enzymes and readily dissociate ($K_D = 10^{-3} - 10^{-5}$ M). They are too expensive to be used as stoichiometric reagents in large-scale synthesis: the current prices for them range from $500 per mole for NAD to about $200,000 per mole for NADPH. Recycling of the cofactor is thus required if enzymes requiring nicotinamide cofactors are to be used in preparative synthesis.

A regeneration method must be capable of recycling the cofactor 10^2-10^6 times to be economical. The exact number of turnovers required depends on the initial cost of the cofactor and the value of the product formed by the enzymatic process. High turnover numbers require high selectivity for the formation of enzymatically active cofactor. 1,4-Dihydro-NAD or NADP is the enzymatically active form instead of the 1,2- or 1,6-dihydro-species. If 50% of the original cofactor activity is to remain after 100 turnovers, the regeneration reaction must be 99.3% specific for the formation of the active cofactor. If 10^6 turnovers are required, the recycling step must be 99.99993% efficient. Almost invariably, this requirement for high regioselective recycling necessitates enzymatic catalysis, particularly in the recycling of NADH or NADPH.[3,4]

Cofactor regeneration can accomplish three major objectives in addition to reducing the contribution of cofactor to the cost of synthesis.[3,4] First, it can influence the position of equilibrium. A thermodynamically unfavorable reaction can be driven by coupling with a favorable cofactor regeneration reaction. Second, cofactor regeneration prevents the accumulation of cofactor byproduct which may inhibit the forward process. Third, by eliminating the need for stoichiometric quantities of NAD(P)(H), cofactor regeneration can simplify the reaction work-up. In practice, the oxidized forms of the cofactors—NAD or NADP—are used initially because they are less expensive than the reduced forms.

Any method for regenerating nicotinamide cofactors must be compatible with the synthetic reaction and must be practical, inexpensive, and convenient. Enzymes, reagents, and equipment required should be readily available, inexpensive, and easily manipulated. The reaction should be operated under condition(s) in which enzymes and cofactors are stable enough to achieve high turnover numbers (TN). The TN is defined as the number of moles of product formed per mole of cofactor or enzyme per unit time. This is indicative of the rate or the productivity of the process. The total turnover number (TTN) is the total numbers of moles of product formed per mole of cofactor or enzyme during the course of a complete reaction.

$$\text{TN for cofactor} = \frac{\text{mol product}}{(\text{mol cofactor})\,(\text{time})}$$

$$\text{TN for enzyme} = \frac{\text{mol product}}{(\text{mol enzyme})\,(\text{time})}$$

$$\text{TTN} = \frac{\text{mol product formed}}{\text{mol cofactor or enzyme present in reaction}}$$

The TTN is useful for an estimate of the operational cost of the cofactor or the enzyme.

$$\frac{\text{Operational cost of}}{\text{cofactor and enzyme}} = \frac{\text{cost of cofactor}}{\text{TTN of cofactor}} + \frac{\text{cost of enzyme}}{\text{TTN of enzyme}}$$

For a one-time synthesis, the cost of the cofactor or enzyme is calculated based on its initial cost. For a repeated process, the cost may be calculated as the cost of only the activity lost during the reaction. When estimating the cost of a reaction per unit time, the rate of the reaction must be considered. The regeneration cost, RC, may be defined as the cost of the components (enzyme, reagents, cofactor) required to regenerate one mole of cofactor per unit time.

$$RC = \frac{\text{(cost of components)}}{\text{(mol of cofactor regenerated) (time)}}$$

It is apparent from the this analysis that a reaction with high TTN for NAD(P) and low TTN for enzyme may be more expensive than one with low TTN for NAD(P) and high TTN for enzyme. The maximum TTN can be set by the maximum allowed substrate concentration (determined by substrate solubility, product inhibition, and enzyme stability) divided by the minimum cofactor or enzyme concentration required to effect acceptable reaction rates. It is further limited by the competing rates of productive vs nonproductive reactions (for example, degradation of cofactors and enzymes). Thus, both the initial costs of the cofactor and the enzyme and the efficiency with which they are utilized and regenerated during the synthetic process determine their contribution to the cost of the product.

NAD is presently isolated from yeasts[5] and NADP is normally prepared by phosphorylation of NAD using NAD kinase [EC 2.7.1.23] and ATP.[6] NAD(P)H can be prepared from NAD(P) by chemical[7], enzymatic[8], or microbial reduction.[9] A combined chemical and enzymatic procedure for the synthesis of NAD(P) has been developed[10,11] which may be useful for the preparation of NAD(P) analogs.

1.1   Stability of NAD(P) and NAD(P)H

The reduced and oxidized forms of nicotinamide cofactors are both subject to decomposition in aqueous solution. In general, the reduced forms are stable in base but labile in acid; the oxidized forms are stable in acid but labile in base.[12] Since both oxidized and reduced forms are present in solution during enzymatic reactions, a pH region may be used that represents a comprimise in stability. A pH range of 7-7.5 and 8-8.5 is usually used for NAD-NADH and NADP-NADPH, respectively. The stability and activity of the synthetic enzymes, and the overall equilibrium of the coupled reactions must, however,also be considered. In most cases, reactions are carried out at pH 7-8.[13]

The decomposition of NAD(P)H in pH 5-9 is mainly due to general acid catalysis;[12,13] i.e., a rate-determining protonation of the nicotinamide ring at C-5 followed by a rapid rearrangement to a cyclic ether product.[13] The decomposition of NADPH reflects an additional intramolecular acid catalysis by the 2'-phosphate group.[12] Figure 2 indicates the mechanism and relevant rate equations. The values for k_{HA} depend on the type of acid used and generally follow a Bronsted catalysis law ($\alpha = 0.544$).[12,13] The most effective, general acid catalysts are those with pKa approximately equal to the solution pH. Inorganic phosphate is a better catalyst than expected.[12] The half-lives for NADH and NADPH in 0.1M sodium phosphate at pH 7.0, 25°C are 27 h and 13 h respectively ($k_{HA} = 0.41$ M^{-1}h^{-1}).[12] Some organic buffers such as imidazole, triethanolamine, Hepes and Tris buffers are less effective than expected.[12] The half-lives for NADH and NADPH in 0.1M Tris at pH 7.0 and 25°C are 330 h and 31 h respectively. For large-scale enzymatic reactions, one may use a low concentration of one of these organic buffers (5-10 mM) and adjust the reaction pH by adding sodium hydroxide or hydrochloric acid through a peri-

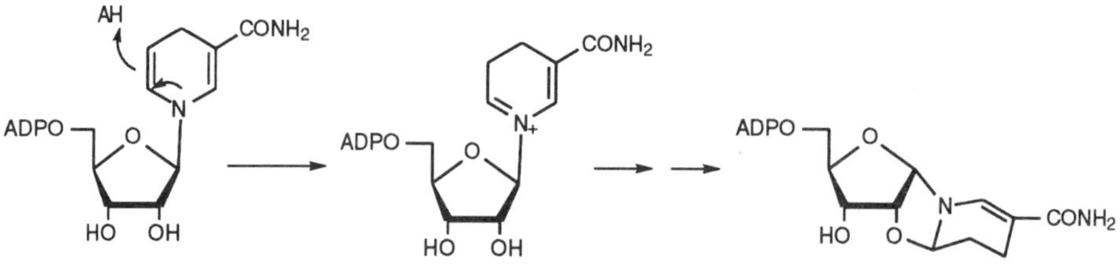

$$-\frac{d\ln[NADH]}{dt} = k^{obs}_{NADH} = k_{H^+}[H^+] + k_{H_2O}[H_2O] + k_{HA}[HA]$$

$$-\frac{d\ln[NADPH]}{dt} = k^{obs}_{NADPH} = k^{obs}_{NADH} + \frac{k_I[H^+]}{k_a+[H^+]} = k^{obs}_{NADH} + \frac{k_I}{1+10^{pH-pKa}}$$

$$k_{H_2O} = 9.4 \times 10^{-3} \text{ M}^{-1}\text{h}^{-1}, \; k_{H^+} = 9.4 \times 10^3 \text{ M}^{-1}\text{h}^{-1}, \; k_I = 0.15 \text{ h}^{-1}$$

Figure 2

staltic pump connected to a pH controller. The ionic strength of the solution may be adjusted by adding inert salts such as NaCl.

The major pathway for the decomposition of NAD(P) under enzymatic reaction is nucleophilic addition, which may be uncatalyzed or catalyzed by the enzyme, at C-4 to form a 1,4-dihydropyridine structure.[14] The nucleophiles could be substrates or other species in solution. For example, in the presence of 0.1M pyruvate at pH 7, NAD, has a half-life of 6.9 h.

1.2   Regeneration Systems for NAD(P)H

A useful regeneration system for NADH and NADPH must be highly regioselective, compatible with the desired enzymatic reduction, and be capable of recycling cofactor 10^2 to $>10^5$ times. At present only enzymatic catalysis provides such high selectivity for the reduction of NAD(P) to NAD(P)H. Other methods based on chemical, electrochemical and photochemical strategies are not selective enough to provide high turnover processes.[3,4] The disadvantages of enzymatic cofactor regeneration are the expense and limited stability of enzymes.

There have been many enzymatic systems developed for the regeneration of NADH and NADPH.[3,4] The most convenient and useful systems for NAD(P)H regeneration are formate/formate dehydrogenase (FDH) from *Candida boidinii* species,[15,16] isopropanol and the alcohol dehydrogenase from *Thermoanaerobium brockii* (for NADP),[17] yeast[17b], or *Pseudomonas sp* (for NAD),[18] and glucose/glucose dehydrogenase (GDH) from *Bacillus* species.[19] The FDH system has the advantages that the reductant is inexpensive and the product isolation is easy. The enzyme has low activity (3 U/mg), however, and is specific for NAD. The advantages of the GDH system include the inexpensive reductant, high specific activity of the enzyme (250 U/mg), high stability of the enzyme, and its ability to regenerate both NADH and NADPH. The byproduct gluconate may, however, complicate product isolation. Furthermore, a high concentration of monovalent cations such as Na^+ or K^+ is required to stabilize the enzyme. Figure 3 illustrates these three regeneration schemes.

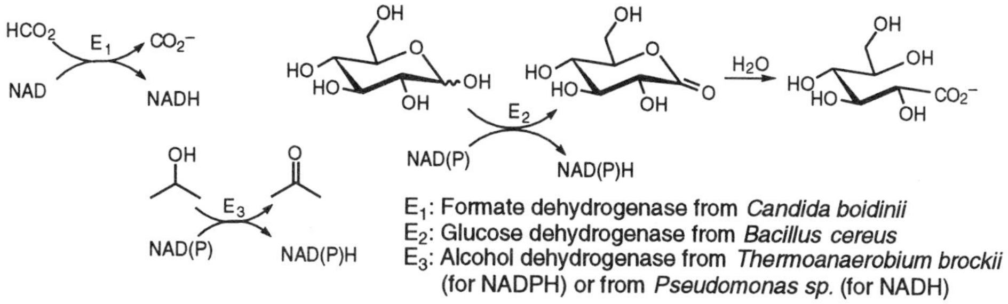

E_1: Formate dehydrogenase from *Candida boidinii*
E_2: Glucose dehydrogenase from *Bacillus cereus*
E_3: Alcohol dehydrogenase from *Thermoanaerobium brockii* (for NADPH) or from *Pseudomonas sp.* (for NADH)

Figure 3

The NADH regeneration based on formate/FDH performed in a membrane-contained reactor has been used for large-scale processes. Up to 600,000 moles of product can be produced per mole of cofactor lost in the process.[16] The cofactor was attached to polyethylene glycol to prevent leakage from the membrane. Today, cofactor regeneration schemes for both NADH and NADPH are successful to the extent that cofactor is no longer the dominant cost in preparative reductions. The cost of enzymes or reagents is usually much greater than the cost of cofactor. Further advances in NAD(P)H dependent enzymatic reactions will come through improved methods of enzyme isolation and stabilization, and of reactor design.

A number of other reactions have been explored for use in reduction of NAD(P) to NAD(P)H, and although they have not so far proved practical in synthesis, they may provide the basis for future processes. Many viologen derivatives can be reduced via one-electron process by catalytic hydrogenation,[20-21] electrochemical reduction,[20,22,23] photochemical reduction[24-26] or by hydrogenase catalyzed reduction.[20,27-31] The reduced viologens have been used as mediators for enzymatic regeneration of NAD(P)H. The enzymes used in this process often contain a prosthetic group that can be reduced by a viologen radical cation. With an appropriate choice of the mediator,[20] the redox potential can be adjusted over a wide range to perform the desired reaction. A typical example is the use of reduced methylviologen for regeneration of NADH catalyzed by diaphorase.[30] When a carbamoylmethylviologen is used with diaphorase, NAD regeneration from NADH becomes favorable (Figures 4 and 5).[32] In the photochemical process, photosensitizers have been used to mediate the reduction of viologen cations to radical cations. Mercaptoethanol,[26] EDTA,[24] and NADH,[25] for example, have been used to reduce $Ru(bpy)_3^{3+}$ to $Ru(bpy)_3^{2+}$ which then reduces V^{2+} to $V^{+\cdot}$ upon illumination (Figure 6).

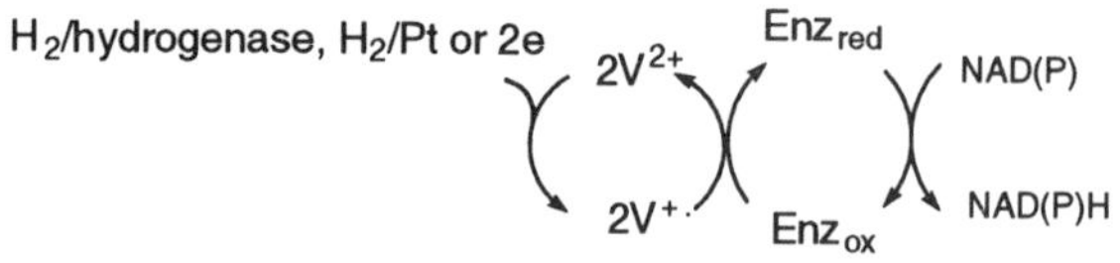

Figure 4

Figure 5

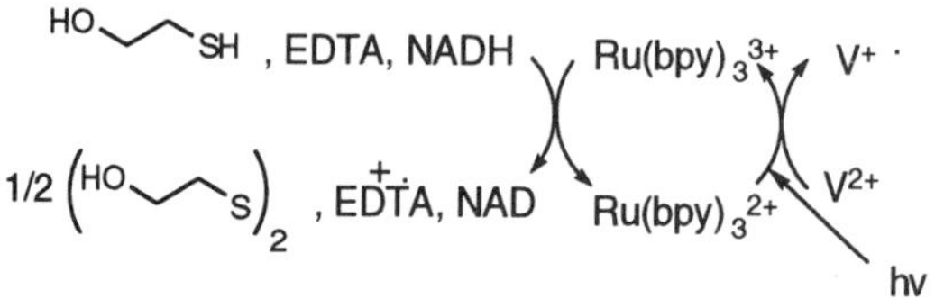

Figure 6

The homogenous Rh complex shown in Figure 7 was reported to be an effective catalyst for reduction of NAD and NADP to NADH and NADPH, respectively, in the presence of formate as hydride donor. The reduced Rh can also be regenerated electrochemically.[33] When the Rh complex is attached to polyethyleneglycol of MW 20,000, the catalytic efficiency is the same

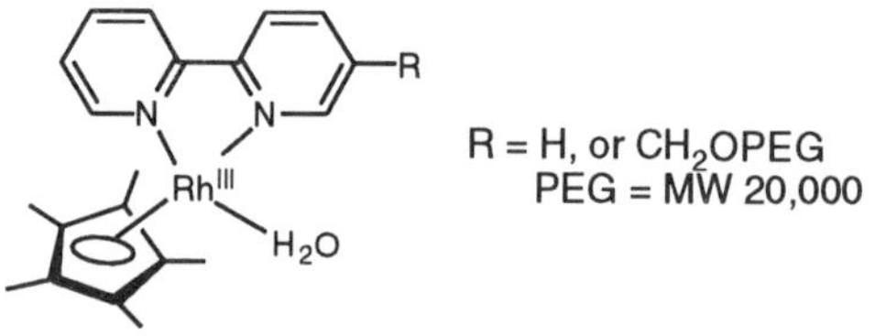

R = H, or CH$_2$OPEG
PEG = MW 20,000

Figure 7

as the unmodified ($\sim$ 26 h^{-1}), and the polymer-bound catalyst can be used in a membrane reactor. A turnover number of 1000 for NADH regeneration has been achieved. A rhodium catalyst compatible with enzymes was also used to reduce pyruvate to DL-lactate by H$_2$. Lactate dehydrogenase was then used to reduce NAD to NADH.[34]

1.3 Regeneration of NAD(P)

The oxidized nicotinamide cofactors NAD(P) are used for the synthesis of ketones from the racemic mixture of the hydroxy compounds. Regeneration of NAD(P) from NAD(P)H is somewhat problematic because of unfavorable thermodynamics and product inhibition. The regioselectivity, however, is not a problem in the oxidation of NAD(P)H. Both enzymatic and non-enzymatic methods can be used for NAD(P) regeneration. Enzymatic methods seem to be preferred because they are simpler and more compatible with biochemical systems.[3,4] Electrochemical,[35] chemical[36,37] and photochemical[38] methods coupled with electron transfer mediators such as methylene blue, phenazine methosulfate, methyl viologen, and ruthenium tris (bipyridine) have been used for small scale regeneration of NAD(P). A direct oxidation of NADH with FMN followed by spontaneous reoxidation of the reduced FMN (FMNH$_2$) by O$_2$ has also been utilized.[39] The reaction is, however, is too slow to be practical. Of several enzymatic methods reported,[3,4] ketoglutarate/glutamate dehydrogenase (GluDH),[40-44] pyruvate/lactate dehydrogenase (LDH)[3,45] and FMN/FMN reductase[46] seem to be the most useful (Figure 8).

The GluDH system (40 U/mg) accepts both NAD and NADP and is thermodynamically favorable. Both ketoglutarate and GluDH are inexpensive, stable and innocuous to enzymes. The disadvantage of this system is that glutamate produced may complicate the workup. The LDH system has the advantage that LDH is stable, inexpensive and has high specific activity (~1000 U/mg). Although pyruvate tends to polymerize in solution and react with NAD in a process catalyzed by LDH, regeneration of NAD based on LDH for enzymatic oxidations of 10-100 mmol of material has been successfully carried out. The disadvantage of this system is that LDH is specific for NAD. The system based on FMN/FMN reductase accepts both NAD and NADP and has the most favorable thermodynamics due to the ultimate oxidation with molecular oxygen. Catalase is often added to destroy hydrogen peroxide. The system is, however, not suitable for enzymes sensitive to molecular oxygen. This problem was also observed in regeneration of NAD(P) based on electron-transfer dyes. Of many dyes evaluated for preparative synthesis,[36] the one based on methylene blue/O$_2$ catalyzed by diaphorase is considered to be the best. Figure 8 illustrates the three commonly used systems for NAD(P) regeneration. Other enzymatic[47] and biological[48] methods were also reported to be useful for nicotinamide cofactor regeneration.

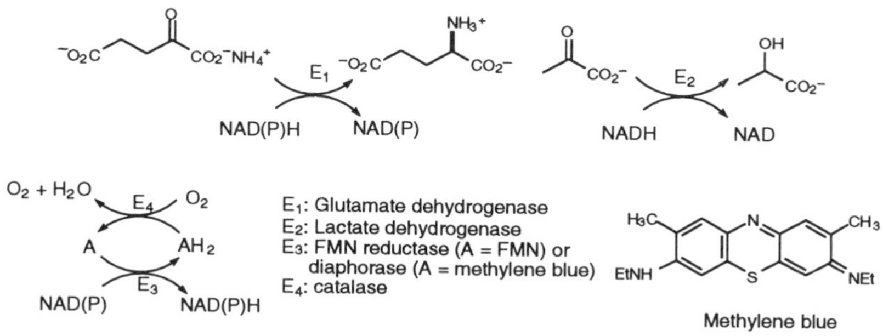

Figure 8

1.4 Reactor Configuration

The reactor configuration most convenient for use in laboratory-scale synthesis is the batch reactor. Typically, reagents and cofactor are present in soluble form, and enzymes are either free in solution or immobilized on polymer supports.[49-53] To have a reasonable reaction rate and high turnover number, the cofactor concentration is usually set at around its K_m value (ca. 0.1mM in most cases) or some what higher and the substrate concentration is in the range of 0.1-0.5M. Batch reactors, however, can have problems of substrate or product inhibition, of separation of enzyme from product , and of substrate solubility. Some of these problems can be solved by the use of systems that are biphasic,[36,54-56] use reverse micelles,[31] or incorporate organic solvents.[17b,56] These systems are particularly suitable for compounds with low solubility in water and for reactions with severe substrate or product inhibition. Problems may, however, still be encountered: for example, slow reaction rate and enzyme inactivation at the interface between organic solvents and water. Immobilization of enzymes may improve their stability. A detailed study of the biphasic system with regard to the relationship between the equilibrium constants for the oxidation or reduction reactions and the Michaelis and product inhibition constants provides useful guidelines for identifying candidates for practical-scale synthesis.[36] When a second enzyme is used for cofactor regeneration in a biphasic system, the regeneration system can be synthetically useful if the overall reaction is favorable and the products are easily separated. Horse liver alcohol dehydrogenase catalyzed oxidation of a meso-diol to a chiral lactone coupled with glutamate dehydrogenase catalyzed reductive amination of α-ketoglutarate to L-α-aminoadipate is a typical example.[55b] In this system, the lactone was extracted into the organic phase while the amino acid product was retained in the aqueous phase; this scheme simplified product separation and minimized product inhibition. Another interesting approach to minimize product inhibition is to deposit the enzyme and the cofactor onto the surface of glass beads or other supports, which are then suspended in a water-immiscible organic solvent containing the substrate.[56]

More economical and practical for large-scale synthesis than for laboratory synthesis are column, hollow fiber, and membrane-based reactors.[47,53] These configurations allow continuous processing of an influent substrate solution while product is removed from the effluent steam. The concentration of cofactor is so low that it can be present in the feed solution. If the enzymatic reaction follows an ordered bi-bi mechanism with the cofactor binding to the enzyme first, the cofactor may remain in the active site during the recycling process. This situation is particularly suitable for a single enzyme system (i.e. one enzyme for synthesis and cofactor regeneration) in a continuous process. Asymmetric reductions catalyzed by both *Thermoanaerobium brockii* alcohol dehydrogenase and *Pseudomonas* alcohol dehydrogenase are typical examples. If cofactors are to be retained in the reactor for a long period of time, they must be attached to some insoluble or high-molecular weight, soluble support. The most successful practice of cofactor immobilization is that reported by Kula and Wandrey where NAD is linked to polyethylene glycol of molecular weight 10^4 (PE-10000) through an ethylene amine linker connected to the amino group of adenine. The PEG-10000-NAD was retained in a membrane reactor for large-scale enzymatic synthesis.[16]

1.5 Stereochemistry and Stereoselectivity of NAD(P)-Dependent Oxidoreductions

Both NADH and NADPH contain two diastereotopic hydrogens (pro-*R* or pro-*S* hydrogen) that can be transferred as a hydride to an oxidized substrate such as aldehyde, ketone or imine. These substrates also contain two diastereotopic or enantiotopic faces (*re* or *si* face) at the sp^2 carbon center to be reduced. In principle, any NAD(P)H dependent oxidoreduction should fall into one of these four types of stereospecificity (Figure 9). In practice, most alcohol dehydrogenases catalyze the transfer of pro-*R* hydrogen to the *re* face of a carbonyl substrate, a process summarized by Prelog's rule[57] (for simplicity, the *re* face here refers to the top face of a carbonyl group with a small substituent on the left and a large substituent on the right side). This rule covers enzymes from yeast, horse liver,[49] and *Thermoanaerobium brockii*.[17] The alcohol dehydrogenase from *Mucor javanicus* is specific for the pro-*S* hydrogen of NADH and *si* face of carbonyl substrates[49] and that from a *Pseudomonas* species is specific for the pro-*R* hydrogen of NADH and *si* face of carbonyl substrates.[18] Some representative (*R*)-alcohols prepared by the *R*-enzyme are shown in Figure 9. A pro-*S*/*re* face specific alcohol dehydrogenase has not been reported yet. To determine whether the enzyme is pro-*R* or pro-*S* specific for the reduced cofactor, ^{1}H-NMR is considered to be the most convenient.[58] The pro-*R* hydrogen has a chemical shift of 2.77 ppm and the pro-*S* hydrogen has a shift of 2.67 ppm. Using deuterated substrate and NAD or NADP in the presence of enzyme, one can examine the chemical shift of the isolated reduced cofactor [4-^{2}H]-NAD(P)H to determine the stereospecificity. One can also use either [4R-^{2}H] or [4S-^{2}H]-NAD(P)H to reduce a carbonyl substrate in the presence of enzyme. The presence or absence of the 4-H ($\delta = 9$ ppm) of the oxidized cofactor recovered will be diagnostic of deuteride or hydride transfer from the labeled reduced cofactor.

A useful application of NAD(P) dependent oxidoreductases is to prepare deuterium or tritium labeled compounds such as alcohols and amino acids from the corresponding keto precursor.[20,59] If a cofactor regeneration system is incorporated into the synthesis, the stereospecificity of hydride transfer from or to the cofactor for the regeneration system may not necessarily be the same as that for the synthesis system. If both synthesis and regeneration have the same stereospecificity for the cofactor, every regeneration step will transfer the labeled hydrogen from the regeneration system through the cofactor to the product. If both systems have a different stereospecificity, the labeled hydrogen will only be transferred to the cofactor, and not to the product in the first regeneration cycle, but will be transferred to the product in the following regeneration cycles. The content of the labeled hydrogen in the product will thus be determined by the regeneration number of the cofactor.[60] For other oxidoreductases which require a mediator (e.g. flavin, lipoic acid, and iron cluster) between NAD(P)H and the keto substrate, the labeled hydrogen may be lost during the transfer as the hydrogens of the reduced mediator become exchangeable with the medium. In this case, the reaction may be carried out in D$_2$O or HTO to prepare the labeled product.[20]

1.6   Other Practical Issues

In addition to the problems associated with cofactor regeneration already described, many NAD(P)-dependent enzymatic reactions suffer from product inhibition; especially in the oxidation of alcohols, the ketone or aldehyde products tend to bind to the enzyme more strongly than do the alcohols. A detailed analysis of product inhibition related to the reaction rate revealed that the ratio

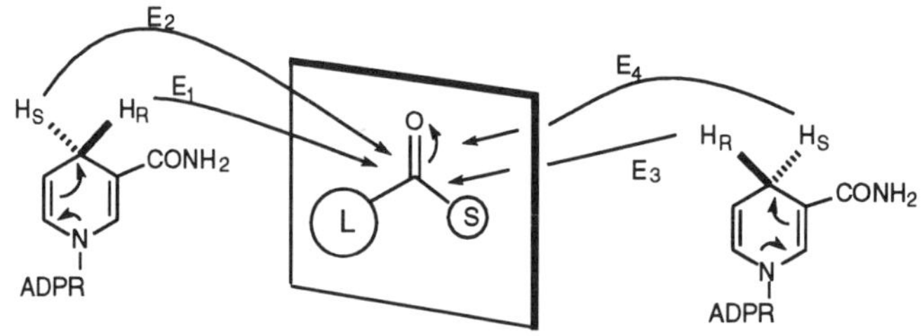

E_1: pro-*R*/si face, Alcohol dehydrogenase from *Pseudomonas sp.* (NAD) or *Lactobacillus kefir* (NADP)
E_2: pro-*S*/si face, from *Mucor javanicus*
E_3: pro-*R*/re face, from yeast, horse liver, and *Thermoanaerobium* species
E_4: pro-*S*/re face, unknown.

Representative reduction products from E_1 (*Pseudomonas sp., NAD dependent*)[18]

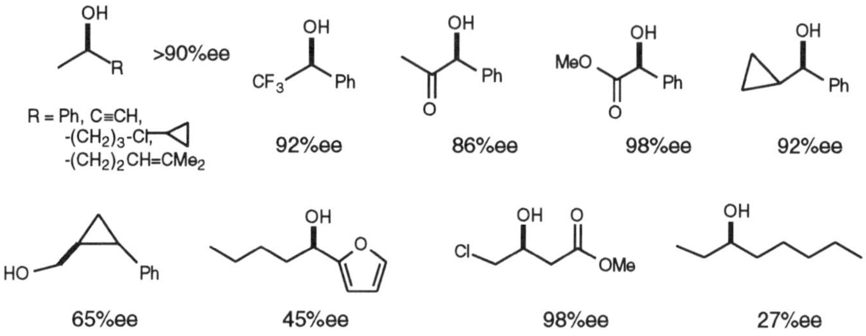

From *Lactobacillus kefir* (NADP dependent)[18b]

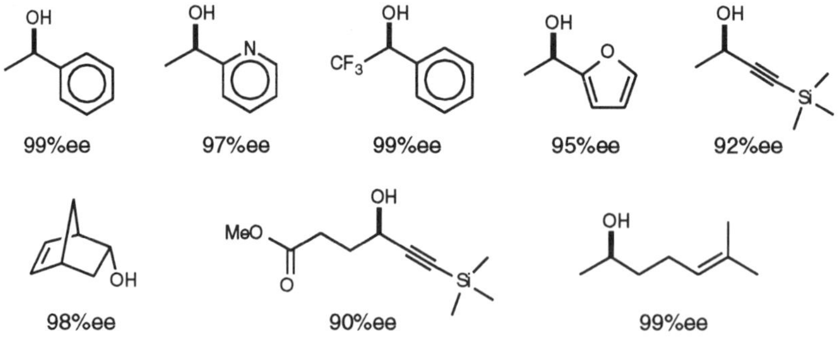

Figure 9

of product inhibition constant to substrate affinity constant (K_i/K_m) will determine the efficacy of the oxidation process.[36,42] If the ratio is greater than 1, the reaction may proceed to acceptable conversions; however, the product concentration may be maintained at a very low level to lessen the inhibition problem. If the ratio is less than 1, the reaction can never proceed efficiently. Of various types of inhibition studied, noncompetitive and mixed inhibitions are the most serious. The only way to overcome this problem seems to be to limit the product concentration. For example, the reaction can be carried out in diluted solution, biphasic system, or with water-miscible organic solvents. For competitive inhibition, one can simply increase reactant concentration to increase the efficiency of the conversion. An alternative solution is to employ a cofactor analog that will influence the K_i/K_m ratio. Many NAD(P) dependent oxidations proceed through a ternary complex (enzyme-ketone-NAD(P)H) from which the ketone product releases first, followed by the release of reduced cofactor. Use of a cofactor analog may affect the binding of ketone or the cofactor to the enzyme, and thereby reduce the degree of product inhibition.[62] Many active NAD(P)H analogs with modification at the amide group of the nicotinamide ring may be useful for this purpose. These analogs can be prepared by NADase catalyzed exchange reactions (also see Chapter 5).[63] The most useful oxidation reaction without significant product

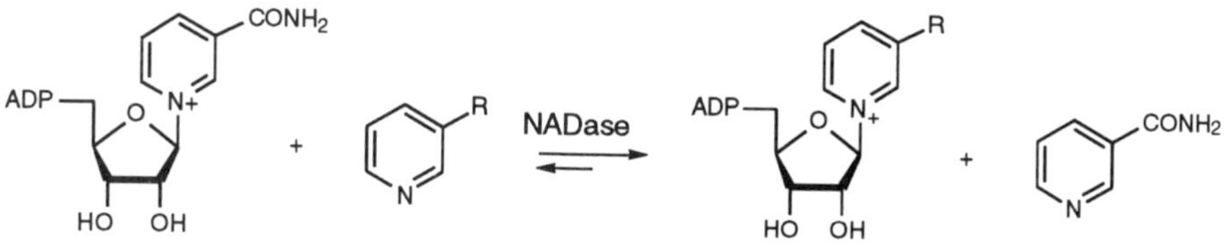

Figure 10

inhibition is perhaps the oxidation of many 1,4-diol systems to chiral lactones catalyzed by horse liver alcohol dehydrogenase.[61]

It is difficult to carry out preparative cofactor-dependent oxidoreductions in a column. Coimmobilization of enzyme and the cofactor is necessary to make the system practical.[64] Either enzyme or cofactor or both, however, may decompose and the immobilized system will have to be replaced. If the K_m value for the cofactor is very low and a single enzyme is used for both synthesis and cofactor regeneration, a column reactor can be carried out practically with a small amount of free cofactor dissolved in the mobile phase.[65] This situation is particularly efficient if the cofactor binds to the enzyme first and releases last. For reaction in a continuous membrane reactor,[16] the cofactor can be attached to polyethylene glycol to increase the molecular weight so that both enzyme and cofactor are retained inside the membrane reactor and the desired product is passed through the membrane. Reverse micelle[31] systems have also been used in the synthesis of compounds such as steroids with low solubility in water. With this methodology, both enzymes and cofactors were entrapped in the micelle, and the substrate and product are retained mostly in the bulk organic phase. Mass transport is often the rate-limiting step in this system.

1.7   Examples

1.7.1 Horse Liver Alcohol Dehydrogenase [EC 1.1.1.1]

Horse liver alcohol dehydrogenase (HLADH) has received more attention than any other alcohol dehydrogenase.[49] The enzyme can differentiate between prochiral groups or faces in symmetrical or meso compounds and make distinctions between enantiotopic groups or faces and geometric isomers. In general, mammalian enzymes exhibit a broader constitutional specificity than microbial enzymes, and HLADH is an excellent example.

HLADH is stable at 25°C for more than a week under normal reaction conditions. Temperatures over 30°C tend to inactivate the enzyme. The range of pH over ehich the enzyme is stable is from 5 to 10, although oxidations are normally run at 8-9 and reductions at approximately 7. HLADH contains bound zinc ions acting as cofactors and thus is susceptible to deactivation by metal chelating reagents.[66] The zinc ion can be replaced with other metals.[67] Inhibition may also be observed with high concentrations of anions. In certain instances, HLADH is subject to substrate and product inhibition.

In order for any enzyme to be a useful catalyst, one must be able to predict whether a particular compound will be a substrate and if so, the stereospecificity of the reaction. In a qualitative manner, one can predict the stereospecificity of the HLADH reaction by using any one of a number of models.[57,68-71] The cubic space section model has proven to be widely applicable[68] and accurate, particularly for cyclic compounds. The development of the model was based upon the x-ray structure of the enzyme[72] combined with kinetic data of known substrates. The enzyme active site is represented as layers of cubes of arbitrary size, depending on the accuracy needed. Certain cubes are considered forbidden and others allowed or limited for occupation by the substrate. The cofactor nicotinamide ring is firmly in place with transfer of the pro-*R* hydride to the *re* face of the carbonyl group and similarly with abstraction of the substrate pro-*R* hydrogen to the coenzyme *re* face during oxidation. The active site zinc atom must also be coordinated to the carbonyl or hydroxyl oxygen atom. Thus, the placement of the active substrate and cofactor moieties during oxidation or reduction is fixed and the remainder of the substrate is rotated to obtain an optimum fit within allowed cubes (Figure 11). In this manner, the stereospecificity of a number of HLADH catalyzed reactions can be rationalized, including different isomers of the same compound. This model has proven to be very versatile, although for compounds with a large number of possible conformations—for example linear acyclic ketones and alcohols—the prediction of stereospecificity is difficult.[68,73]

HLADH can oxidize primary alcohols. In addition the enzyme will reduce a significant number of aldehydes. Each of these reactions is potentially useful since only the pro-*R* hydrogen of the alcohol substrate is removed in the oxidation and hydride transfer occurs to the *re* face of the substrate in the reduction. The aldehyde substrates may be converted to chiral alcohols if tritium or deuterium is transferred from the nicotinamide cofactor. Primary alcohols labelled at either the pro-*R* or pro-*S* position can be obtained easily (Figure 12).[12,49,59] HLADH is unable to oxidize methanol, attributed to a lack of productive binding in the active site. For substrates insoluble in aqueous solution, biphasic systems,[36,56,74] reverse micells,[75] or organic solvents[56] can be used.

1.7.1.1 Oxidation of Substrates

The enantioselective oxidation of primary alcohols catalyzed by HLADH provides a route to obtain chiral aldehydes (Table 1). A series of racemic α-hydroxy and α-amino alcohols were oxidized to the corresponding aldehydes with good yield and high ee.[44] Low enantioselectivity was observed with bulky *R* groups at the β position. Since both D- and L-amino alcohols are readily available from the corresponding amino acids, HLADH-catalyzed oxidation provides useful routes to amino aldehydes.[44,73] The drawbacks of this process are that the reaction is not

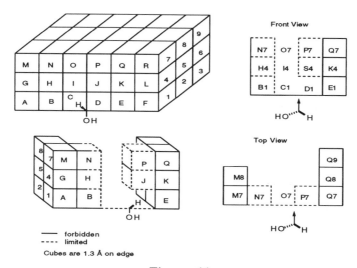

Figure 11

thermodynamically favorable and that the reaction requires NAD (regeneration of NAD is thus required).

Further oxidation of the α-hydroxy aldehydes *in situ* to hydroxy acids with aldehyde dehydrogenase yields several useful compounds. For example, L-α-halolactic acid is a precursor to the synthon L-glycidic acid and 3-amino-2-hydroxypropionic acid has the correct configuration for the preparation of a number of β-adrenoceptor blocking agents.[44]

The oxidation of an enantiotopic group of prochiral meso diols yields lactones with a significant degree of enantioselectivity.[61,76] The reaction presumably proceeds via the hemiacetal after the initial primary alcohol oxidation. The oxidations occur with high pro-*S* selectivity. These reactions provide a method to obtain valuable five- and six-membered chiral lactones, which are useful synthetic intermediates (Table 1).

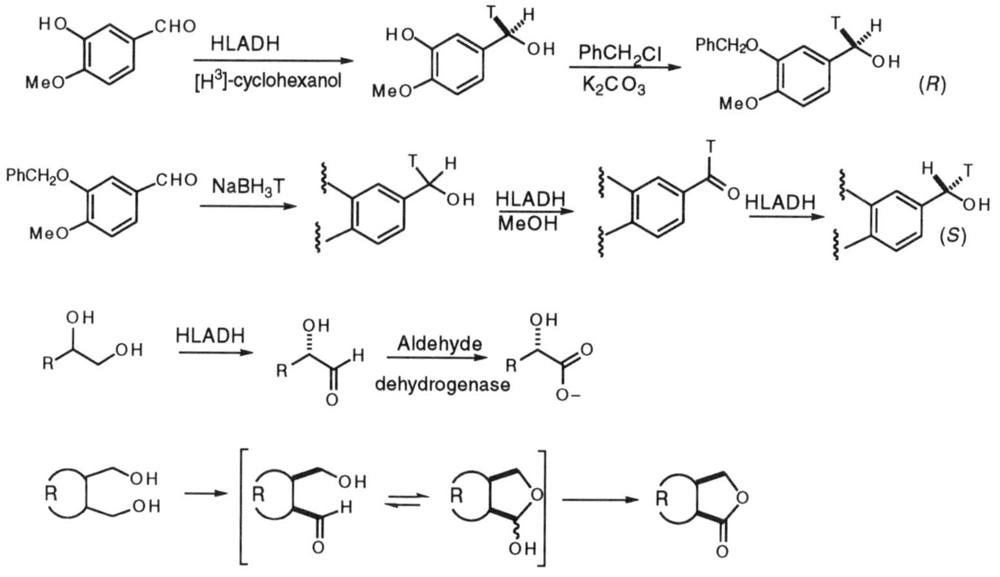

Figure 12

Monocyclic meso-diols can similarly provide bicyclic chiral lactones (Table 2). Oxidation occurs with the same absolute stereospecificity in each case, with the pro-*S* hydroxyl preferentially oxidized with all carbocyclic and the previously mentioned acyclic substrates. The heterocyclic compounds exhibit pro-*R* enantiotopic selection. Oxidation of the hydroxyl group attached to monocyclic systems exhibited low enantioselectivity. Reduction of the corresponding ketones also exhibited low enantioselectivity.

The bicyclic meso diols round out the range of useful substrates by providing chiral lactones available from HLADH oxidations (Table 3). As stated before, the pro-*S* hydroxyl groups of the carbocyclic substrates and pro-*R* hydroxyl groups of the heterocyclic compounds are preferentially oxidized. These chiral lactones can be used as synthetic precursors for many molecules.

Oxidation of bridged systems with a secondary alcohol group showed lower enantioselectivity. Many models have been proposed to explain the stereoselectivity of HLADH catalyzed oxidation of bridged bicyclic substrates. While the cubic space section is an excellent model,[68] some compounds are dealt with optimally via other means. Therefore, the quadrant rule has been proposed to account for the selectivities of bridged bicyclic substrates, both in the oxidative and reductive directions.[77] The carbonyl oxygen atom occupies the center of a quadrant system, as indicated in Figure 13. The substrates are classified as either C2 or C1. C2 groups are those with a two fold symmetry axis passing through the carbonyl bond. C1 ketones belong to the C1 point group with no symmetry element passing through the carbonyl center. The C2 ketones have the larger parts of the substrate occupying the upper left and lower right quadrants. Hydride transfer occurs from the bottom in the direction of the vertical axis. C1 ketones have the most favorable orientation with the larger group on the right of the carbonyl. The orientation with the greatest displacement above the horizontal axis is favored with hydride transfer from below. The quadrant rule is a rule of thumb that can be quickly applied to complex molecules. The rule also applies to the oxidation of bridged bicyclic alcohols, and the majority of the substrates presented here obey the rule.

1.7.1.2 Reduction of substrates

Acyclic ketone substrates are typically poor substrates for HLADH and few have been reported. Cyclic compounds are good substrates for HLADH. The enzyme can also tolerate heterocyclic rings containing oxygen or sulfur without a change in stereospecificity. Nitrogen apparently coordinates the active site zinc atom and cannot be used in the cyclic substrates. For mono and bicyclic ketones, the hydride transfer from NADH is to the *re* face of the carbonyl and primarily in the equatorial direction.

Although acyclic ketones are generally not substrates for HLADH, the thiopyranones provide an alternate method for obtaining such alcohols. Substrates with sulfur heterocycles experience an increase in the rate of reduction. In addition, desulfurization with lithium in diethylamine results in the breakage of one or both sulfur bonds, depending upon reaction conditions, to yield an acyclic chiral alcohol.[78]

HLADH catalyzes the reduction of many different cis and trans decalindiones. These highly symmetrical compounds are potentially useful chiral synthons. The stereospecificity of the enzyme with all of the mono and bicyclic compounds is explained by the cubic active site model.

Horse liver alcohol dehydrogenase also accepts a variety of cage shaped molecules as substrates. These compounds—including meso diketones—are reduced with high enantioselectivity. The aforementioned quadrant rule adequately predicts the observed stereoselectivities for these substrates. Although the specificity of HLADH has been probed in great detail, it is obvious that many new applications of HLADH may yet be discovered. Table 4 summarizes some reductions catalyzed by HLADH.

Table 1. HLADH-catalyzed oxidation of primary alcohol

	% ee	Ref

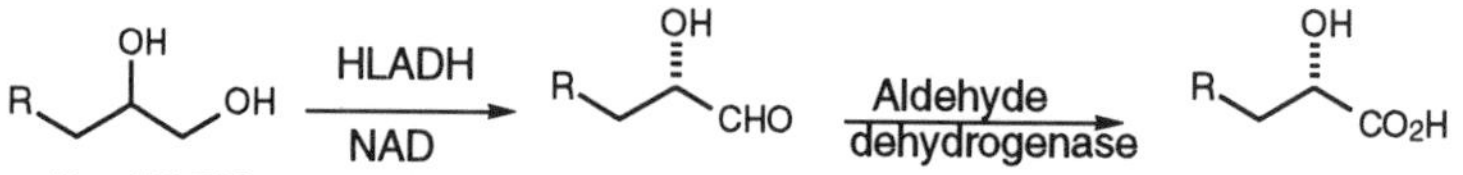

R = HOCH$_2$-, FCH$_2$-, ClCH$_2$-, BrCH$_2$-, CH$_3$-, H$_2$NCH$_2$-, CH$_2$=CH-, CH$_3$CH$_2$-   > 97   [a]

R = (CH$_3$)$_2$-,   < 10   [a]

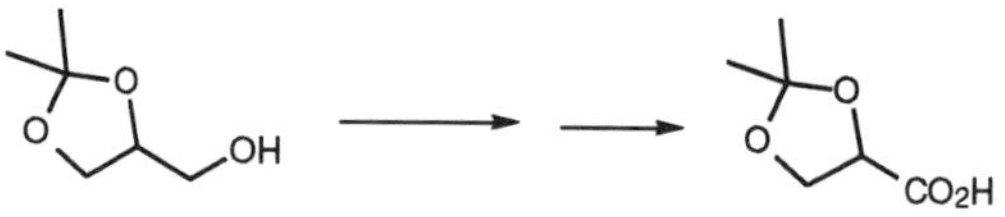

     <10   [a]

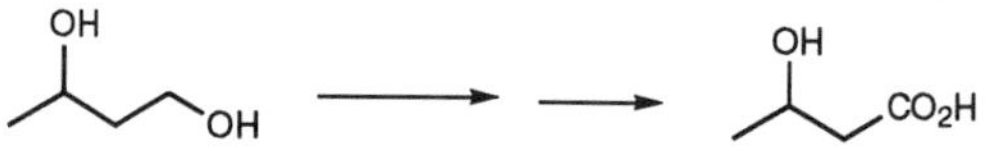

     <10   [a]

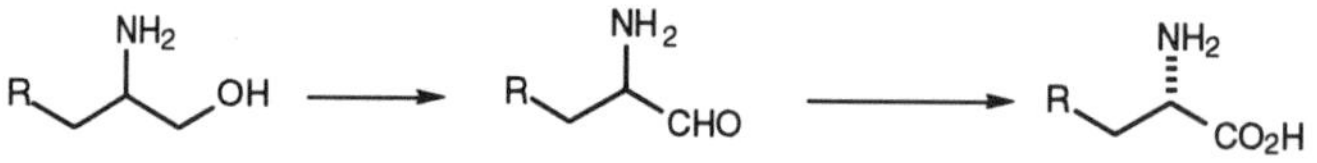

R = HOCH$_2$-   96   [a]

R = CH$_3$-   84

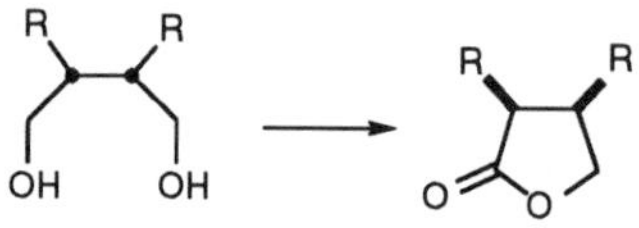

     100   [b]

R = CH$_3$-, CH$_3$CH$_2$-

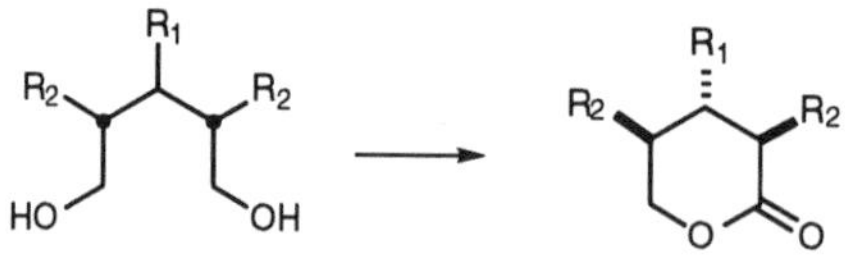

R$_1$	R$_2$		
H	CH$_3$	100	[c]
-CH$_3$	H	90	[c]
-CH$_2$CH$_3$	H	74	[c]
-CH$_2$(CH$_3$)$_2$	H	46	[c]
-CH$_2$Ph	H	20	[c]

[a] Matos, J. R.; Smith, M. B.; Wong, C. -H. *Bioorg. Chem.* **1985**, *13*, 121; Wong, C. -H.; Matos, J. R. *J. Org. Chem.* **1985**, *50*, 1992.
[b] Ng, G. S.-Y.; Yuan, L. - C.; Jakovac, I. J.; Jones, J. B. *Tetrahedron* **1984**, *40*, 1235.
[c] Jones, J. B.; Lok, K. *Can. J. Chem.* **1979**, *57*, 1025.

Table 2. HLADH-catalyzed oxidation of monocyclic alcohols

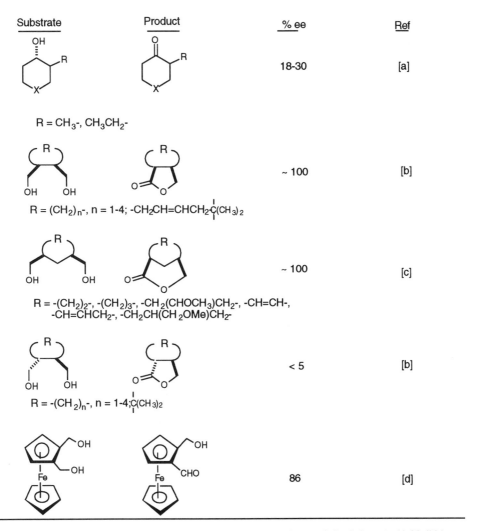

Substrate	Product	% ee	Ref
		18-30	[a]

R = CH₃-, CH₃CH₂-

R = CH₃-, CH₃CH₂-

| | | ~ 100 | [b] |

R = (CH₂)ₙ-, n = 1-4; -CH₂CH=CHCH₂-C(CH₃)₂

| | | ~ 100 | [c] |

R = -(CH₂)₂-, -(CH₂)₃-, -CH₂(CHOCH₃)CH₂-, -CH=CH-,
-CH=CHCH₂-, -CH₂CH(CH₂OMe)CH₂-

| | | < 5 | [b] |

R = -(CH₂)ₙ-, n = 1-4;C(CH₃)₂

| | | 86 | [d] |

[a] Jones, J. B.; Takemura, T.; *Can. J. Chem.* **1984**, *62*, 77; Jones, J. B.; Schwartz, H. M. *ibid* . **1981**, *59*, 1574.
[b] Jakovac, J. J.; Goodbrand, H. B.; Lok, K. P.; Jones, J. B. *J. Am. Chem. Soc.* **1982**, *104*, 4559.
[c] Bridges, A. J.; Raman, P. S.; Ng, G. S. Y.; Jones, J. B. *J. Am. Chem. Soc.* **1984** , *106*, 1461.
[d] Yamazaki, Y.; Hosono, K. *Tetrahedron Lett.* **1988**, *29*, 5769.

Table 3. HLADH-catalyzed oxidation of bridged bicyclic substrates

Substrate	Product	% ee	Ref

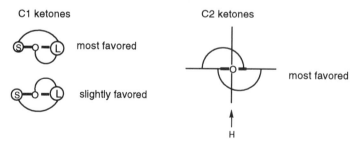

X = O, -CH₂-, -CH₂CH₂-   > 97   [a]

X = O, -CH₂-, -CH₂CH₂-   > 97   [a]

n = 1 - 2   30-60   [b]

X = -CH₂-, Y = -CH₂-

X = -CH₂-, Y = -CH₂CH₂-   73

X = -CH₂CH₂-, Y = -CH₂-   29   [c]

                             73

n = 1   63, 23   [b]

n = 2   100, 76

[a] Jones, J. B.; Francis, C. J.; *Can. J. Chem.* **1984**, *62*, 2578; Lok, K. P.; Jakovac, I. J.; Jones, J. B. *J. Am. Chem. Soc.* **1985**, *107*, 2521.

[b] Nakazaki, M.; Chikamatsu, H.; Naemura, K.; Suzuki, T.; Iwasaki, M.; Sasaki, Y.; Fujii, T. *J. Org. Chem.* **1981**, *46*, 2726.

[c] Nakazaki, M.; Chikamatsu, H.; Fujii, T.; Sasaki, Y.; Ao, S. *J. Org. Chem.* **1983**, *48*, 4337.

C1 ketones

most favored

slightly favored

C2 ketones

most favored

H

Figure 13

Table 4. HLADH catalyzed reductions

Substrate	Product	% ee	Ref
(ketone with OMe, OMe)	OH, OMe, OMe	100	[a]
(ketone with OMe, OMe)	OH, OMe, OMe	89	[a]
(diketone OEt)	OH, O, OEt	84	[a]
Cl (diketone OEt)	Cl, OH, O, OEt	98	[a]
(ketoester OEt)	OH, OEt	98	[b]
(benzaldehyde-D)	OH, D	95	[c]
X = -CH₂-, S; R = -CH₃, CH₃CH₂-	OH, R; X	45-85	[d]
X = -CH₂-, S, O; R = (CH₃)₃C-, CH₃CH₂CH₂-, Ph, (CH₃)₂CH-	OH; X, R	~ 100	[e]
(thiolane ketone) n = 1 - 3	HO''' ; n	10 - 30	[f]
R = CH₃, CH₂=CH-, Ph	'''OH; R	50 - 85	[g]
R = CH₂=CH-, CH₃CH=CH- (cis), CH₃CH₂CH=CH- (cis)	'''OH; '''R	20 - 80	[g]
(norbornanone)	OH	64	[h]
(bicyclic ketone)	OH	83	[h]

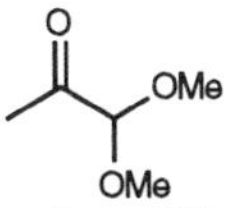

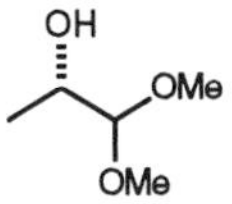

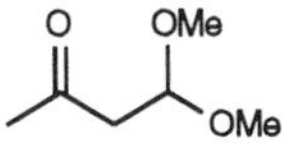

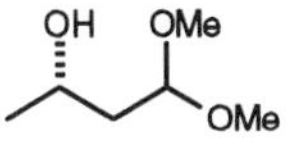

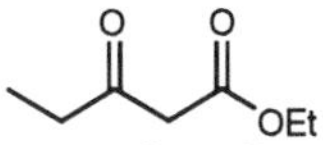

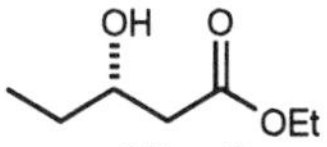

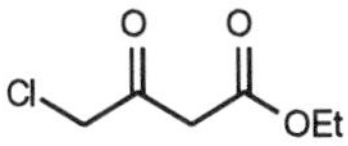

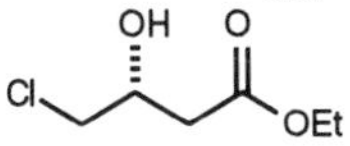

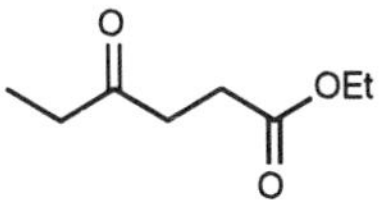

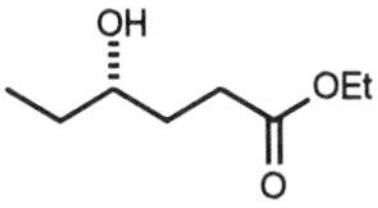

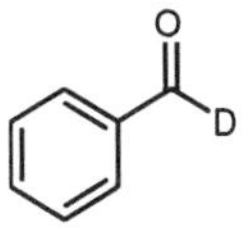

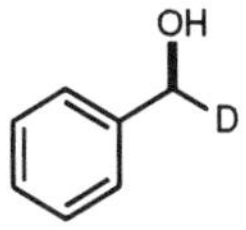

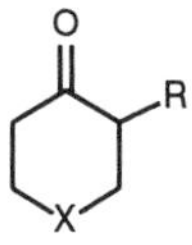

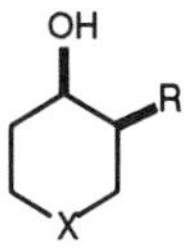

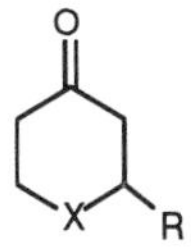

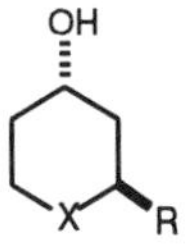

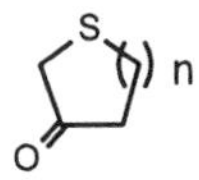

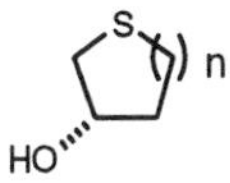

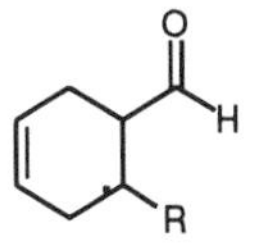

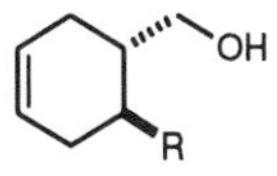

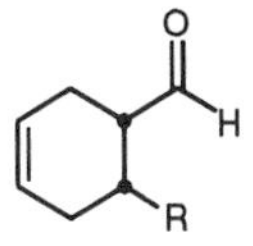

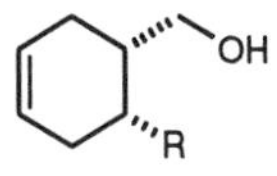

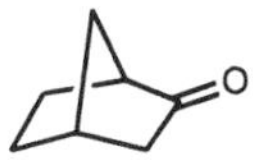

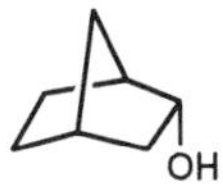

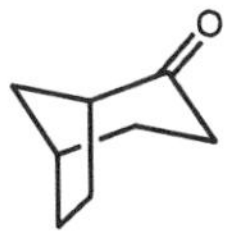

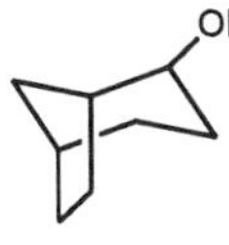

Table 4. continue

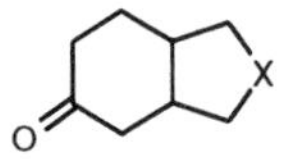

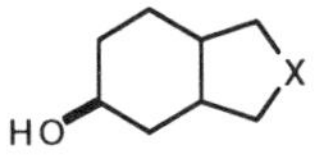

	X = O, S X = CH₂, CO,	> 97 ~ 20 - 60	[i] [j]
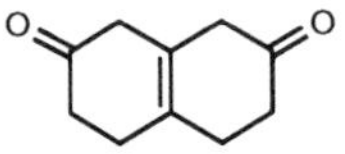	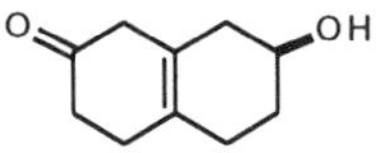	> 98	[k]
R = H, CH₃ CH₂OH	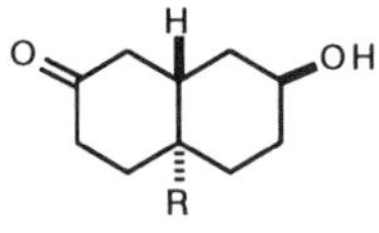	> 98	[k]
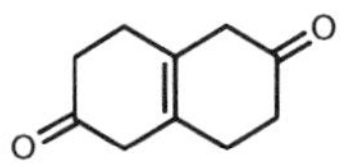	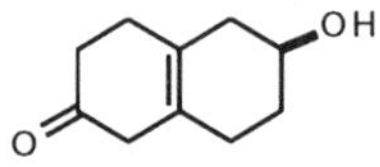	> 98	[k]
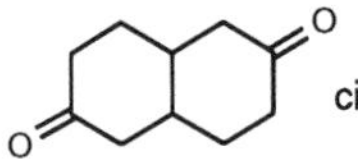 cis or trans	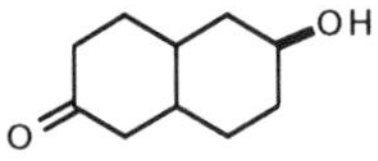	> 98	[k]
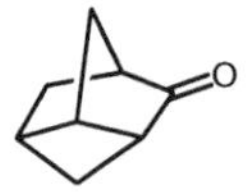	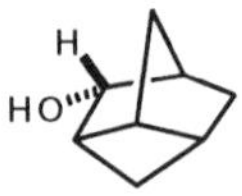	72	[l]
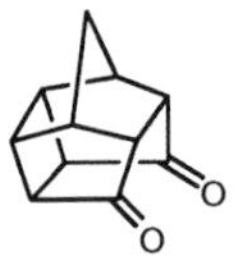	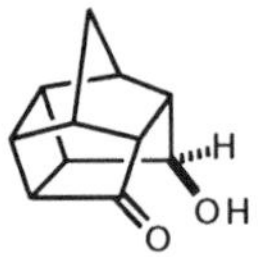	25	[m]
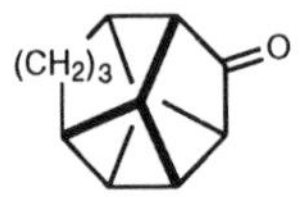 (CH₂)₃	 (CH₂)₃	70	[n]

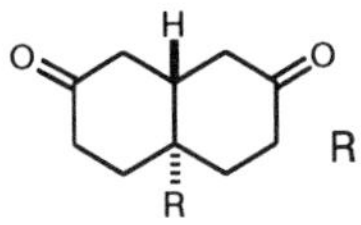

[a] Wong, C.-H.; Drueckhammer, D. G.; Sweers, H. M. *J. Am. Chem. Soc.* **1985**, *107*, 4028.
[b] Drueckhammer, D. G.; sadozai, S. K.; Wong, C.-H.; Roberts, S. M. *Enzyme Microb. Technol.* **1987**, *9*, 564.
[c] Wong, C.-H.; Witesides, G. M. *J. Am. Chem. Soc.* **1981**, *103*, 4890.
[d] Jones, J. B.; Takemura, T. *Can. J. Chem.* **1984**, *62*, 77.
[e] Davies, J.; Jones, J. B. *J. Am. Chem. Soc.* **1979**, *101*, 5405; Haslegrave, J. A.; Jones, J. B. *ibid.* **1982**, *104*, 4666; Sadozai, S. K.; Lepoivre, J. A.; Dommisse, R. A.; Alderweireldt, F. C. *Bull. Soc. Chim. Belg.* **1980**, *89*, 637.
[f] Jones, J. B.; Schwartz, H. M. *Can. J. Chem.* **1981**, *59*, 1574.
[g] Boland, W.; Niedermeyer, U. *Synthesis* **1987**, 27.
[h] Irwin, A. J.; Jones, J. B. *J. Am. Chem. Soc.* **1976**, *98*, 8476.
[i] Lam, L. K. P.; Gair, I. A.; Jones, J. B. *J. Org. Chem.* **1988**, *53*, 1611.
[j] Krawczyk, A. R.; Jones, J. B. *J. Org. Chem.* **1989**, *54*, 1795.
[k] Dodds, D. R.; Jones, J. B. *J. Am. Chem. Soc.* **1988**, *110*, 577.
[l] Nakazaki, M.; Chikamatsu, H.; Fujii, T.; Sasaki, Y.; Ao, Y. *J. Org. Chem.* **1983**, *48*, 4337.
[m] Naemura, K.; Fujii, T.; Chikamatsu, H. *Chem. Lett.* **1986**, 923.
[n] Naemura, K.; Katoh, T.; Chikamatsu, H.; Nakazaki, M. *Chem. Lett.* **1984**, 1371.

1.7.2 *Thermoanaerobium brockii* Alcohol Dehydrogenase [EC 1.1.1.2]

The alcohol dehydrogenase isolated from the ethanologenic thermophile *Thermoanaerobium brockii* (TADH) is an NADP-dependent enzyme, which catalyzes the reduction of ketones and the oxidation of secondary alcohols with high stereospecificity.[17] It is specific for the pro-(*R*) hydrogen of NADPH.[18] Hydride transfer generally occurs to the *re* face of the carbonyl to give the (*S*) alcohol. An interesting reversal of stereochemistry is seen with smaller substrates (Figure 14) where the hydride is delivered to the *si* face to give the (*R*) alcohol.[17] A similar temperature dependent phenomenum was observed with a secondary alcohol dehydrogenase from *Thermoanaerobacter ethanolicus*[79] and was attributed to the relationship between enthalpy and entropy of activation. The enantiomeric excess of the reduced alcohols decreases when temperature increases. The general trend of reactivity for TADH is secondary alcohols >> linear ketones > cyclic ketones > primary alcohols.

The range of ketones reduced includes haloketones and ketoesters (Figure 14). These alcohols have been used as precursors to valuable chiral synthons, obtained with high enantiomeric excess.

Several monocyclic ketones and the bicyclic ketones, 4-norbornanone and bicyclo[3.2.0]hept-2-en-6-one are substrates (Figure 14). The latter compound and derivatives are synthons for optically active natural products. The reduction of bicyclo[3.2.0]hept-2-en-6-one proceeds with high selectivity.

TADH uses NADP(H) as a coenzyme. The reduction of ketones can be coupled with the TADH catalyzed oxidation of 2-propanol to regenerate the cofactor in a one enzyme system for synthetic purposes.[65] TADH is also a useful enzyme for the regeneration of NADP(H) in other systems.

TADH exhibits extraordinary thermostability. The activity is unchanged at 65°C and also shows good stability at 85°C. Normal experimental ranges are from 10-50°C. The optimal pH range is 7.5-8.0. The enzyme has been used equally well as free or immobilized form. The enzyme also shows tolerance to organic solvents and is frequently used in biphasic systems. In conjunction with cofactor regeneration, 2-propanol can be utilized as cosolvent.

1.7.3 Yeast Alcohol Dehydrogenase [EC 1.1.1.1.]

The *in vivo* role of yeast alcohol dehydrogenase (YADH) is to reduce acetaldehyde and oxidize ethanol. The purified enzyme has limited applicability in organic synthesis, but readily reduces a variety of aldehydes and oxidizes acyclic primary alcohols with stoichiometric consumption of NAD or NADH.[49] The purified enzyme is available commercially as either a lyophilized powder or in a buffered suspension. A substantial body of work has accumulated wherein the enzyme is used in a whole cell preparation, either baker's or brewer's yeast, primarily *Saccharomyces cerevisiae*.[48] The properties of these whole cell oxidations and

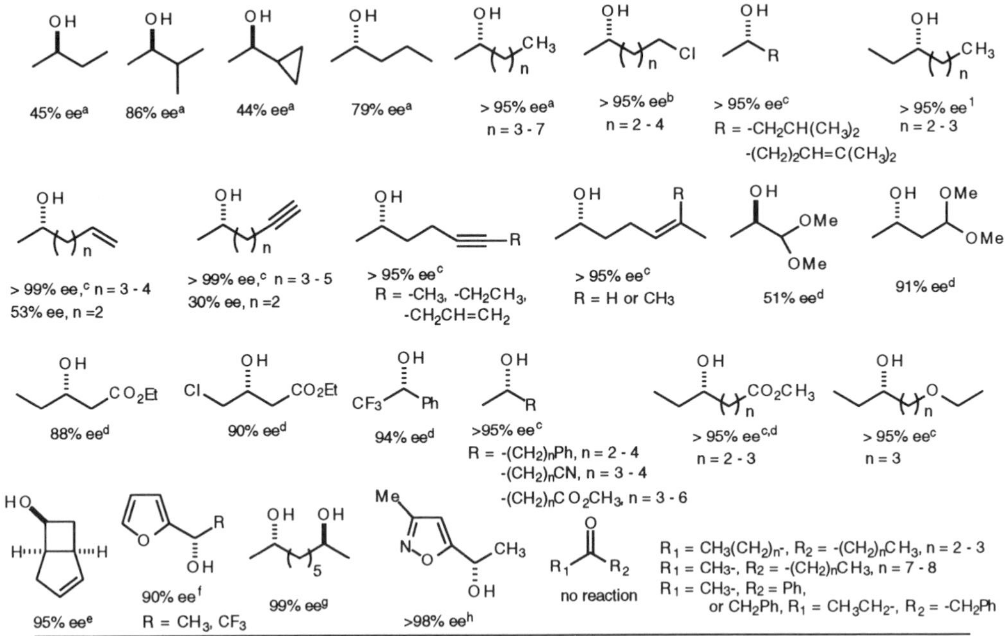

Figure showing various chiral alcohols with ee values:

45% ee[a], 86% ee[a], 44% ee[a], 79% ee[a]

> 95% ee[a] n = 3 - 7

> 95% ee[b] n = 2 - 4

> 95% ee[c] R = -CH₂CH(CH₃)₂ -(CH₂)₂CH=C(CH₃)₂

> 95% ee[1] n = 2 - 3

> 99% ee,[c] n = 3 - 4
53% ee, n =2

> 99% ee,[c] n = 3 - 5
30% ee, n =2

> 95% ee[c] R = -CH₃, -CH₂CH₃, -CH₂CH=CH₂

> 95% ee[c] R = H or CH₃

51% ee[d]

91% ee[d]

88% ee[d], 90% ee[d], 94% ee[d]

>95% ee[c] R = -(CH₂)ₙPh, n = 2 - 4 -(CH₂)ₙCN, n = 3 - 4 -(CH₂)ₙCO₂CH₃, n = 3 - 6

> 95% ee[c,d] n = 2 - 3

> 95% ee[c] n = 3

95% ee[e], 90% ee[f], 99% ee[g]

R = CH₃, CF₃

>98% ee[h]

no reaction

R₁ = CH₃(CH₂)ₙ-, R₂ = -(CH₂)ₙCH₃, n = 2 - 3
R₁ = CH₃-, R₂ = -(CH₂)ₙCH₃, n = 7 - 8
R₁ = CH₃-, R₂ = Ph,
or CH₂Ph, R₁ = CH₃CH₂-, R₂ = -CH₂Ph

(a) Keinan, E.; Jafeli, E. K.; Seth, K. K.; Lamed, R. *J. Am. Chem. Soc.* **1986**, *108*, 162.
(b) Keinan, E.; Seth, K. K.; Lamed, R. *J. Am. Chem. Soc.* **1986**, *108*, 3474.
(c) Keinan, E.; Serth, K. K.; Lamed, R.; Ghirlando, R.; Singh, S. P. *Biocatalysis* **1990**, *3*, 57.
    The 5-OH-ester leads to lactone.
(d) Wong, C. -H.; Drueckhammer, D. G.; Sweers, H. M. *J. Am. Chem. Soc.* **1985**, *107*, 4028.
    The 4-OH-ester leads to lactone.
(e) Butt, S.; Davies, H. G.; Dawson, M. J.; Lawrence, G. C.; Leaver, J.; Roberts, S. M.; Turner, M. K.;
    Wakefield, B. J.; Wall, W. F.; Winders, J. A. *Tetrahedron Lett.* **1985**, *26*, 5077.
(f) Drueckhammer, D. G.; Barbas, C. F.; Nozaki, K.; Wong, C. -H. *J. Org. Chem.* **1988**, *53*, 1607.
(g) Keinan, E.; Sinha, S.C.; Sinha-Bagchi, A. *J. Org. Chem.* **1992**, *57*, 3631.
(h) De Amici, M.; De Micheli, C.; Carrea, G.; Spezia, S. *J. Org. Chem.* **1989**, *54*, 2646.

Figure 14

reductions are very different from the purified enzyme, presumably due to the action of other oxidoreductases, and may be superior for many applications.

Reactions are run at temperatures at or below 30 °C, above which the enzyme is unstable and inactivates quickly. The value of pH for optimum reaction appears to be 8, but the enzyme may be used in the pH 6-8.5 range. Substrate or product inhibition poses little or no problems with YADH. Organic solvents such as 20% acetone, 30% glycerol, and 30% ethylene glycol are tolerated by the enzyme.

The pro-*R* hydride of the nicotinamide ring is transferred to the *re* face of the substrate carbonyl during reduction and vice versa during oxidation.

YADH can oxidize allenic alcohols to allenic aldehydes, which may be difficult to obtain chemically.[80a] The normal decrease in reaction rate with increasing chain length observed for saturated substrates was less severe for ethylenic substrates and even less so for allenic alcohols. The stereospecificity remains to be determined with allenic substrates. Site-directed mutagenesis has been conducted to change the enzyme specificity for NAD to NAD and NADP,[80b] and for short-chain alcohol to long-chain alcohol.[80c]

1.7.4 Other Alcohol Dehydrogenases

The commercial availability of an enzyme may preclude its development as an asymmetric catalyst. An otherwise useful set of alcohol dehydrogenases that has been limited by their availability include pig liver alcohol dehydrogenase (PLADH), *Mucor javanicus* alcohol dehydrogenase (MJADH), and *Curvularia falcata* alcohol dehydrogenase (CFADH).[49] All three enzymes require the phosphorylated nicotinamide cofactor, NADP(H). The specificity of each has been explored and each exhibits synthetic potential in organic synthesis. Theoretically, microbial enzyme sources can be grown in large quantities and the enzyme isolated by standard methods. Thus, a microbial enzyme with broad constitutional specificity and strict stereospecificity could prove very useful or even replace animal sources. CFADH and MJADH could both fulfill this type of role.

PLADH delivers the pro-*S* hydride of the nicotinamide ring to the *re* face of the carbonyl substrate and abstracts the pro-*R* hydrogen from alcohol substrates during oxidation. The enzyme accepts cyclic ketones and alcohols as substrates. The hydride addition to the substrate is typically in the equatorial direction. Purification of this unstable enzyme can be difficult and particular attention must be paid to reaction conditions. PLADH is relatively sensitive to pH and concentration of electrolytes.

MJADH has unusual stereospecificity, delivering the hydride to the *si* face of the carbonyl to give an (*R*) alcohol. Heat denaturation of the enzyme is a problem even in the 16-25°C range. MJADH is stable in the pH range from 4.5 to 8.5 and inactivates quickly outside these ranges. The natural substrate of MJADH is dihydroxy acetone, but the enzyme will accept a variety of substrates.

CFADH is stable at room temperature and in the pH range 5 to 10. The stereospecificity of hydride transfer is to the *re* face of the carbonyl. The hydride is delivered in the axial direction for cyclic ketones. CFADH possesses broad substrate specificity, reducing a variety of cyclic and bicyclic ketones. CFADH may have the potential to become synthetically useful, especially since the microorganism can be cultured in bulk quantities and the enzyme purified by standard methods.

Two recently discovered alcohol dehydrogenases from *Pseudomonas sp.* (available from ATCC 49688) catalyzed the transfer of the pro-*R* hydrogen from NADH to the *si* face of a number of acyclic ketones to give (*R*)-alcohols.[18] Other new alcohol dehydrogenases useful for synthesis have also been discovered.[81] For example, a new NADPH-dependent alcohol dehydrogenase from *Lactobacillus kefir* (ATCC 35411) catalyzes the reduction of acetophenone to (*R*)-phenylethanol.[18b,81ab] The alcohol dehydrogenase from *Drosophila melanogaster* catalyzes the transfer of the pro-*S* hydrogen from NADH to the *si* face of acetaldehyde.[81b] The alcohol dehydrogenase from *Candida parapsilosis* has quite broad substrate specificity, transferring the pro-*R* hydride of NADH to yield (*S*)-alcohols.[81c] Some other alcohol dehydrogenases have also been reported,[81d] but not well utilized in synthesis.

2. Dehydrogenases Which Utilize Ketoacids as Substrates

The ketoacid dehydrogenases are not formally a group of enzymes, but for the purposes of this chapter refer to the enzymes that require a carboxylic acid moiety next to the carbonyl being reduced and the hydroxyl or amino group being oxidized. The substrate specificities are generally not as broad as the alcohol dehydrogenases, but the ketoacid dehydrogenase catalyzed reactions proceed with very high enantioselectivity. Frequently, the enzymes of this group have overlapping specificities and will reduce or oxidize the same substrate, sometimes with the same stereospecificity. This group of enzymes has proven to be useful in the regeneration of the nicotinamide cofactor in addition to the synthesis of chiral hydroxy and natural and unnatural amino acids. Generally, the ketoacid dehydrogenases all utilize the unphosphorylated cofactor.

2.1 L-Lactate Dehydrogenase [EC 1.1.1.27]

L-Lactate dehydrogenase (L-LDH) catalyzes the reduction of pyruvate to L-lactate in the presence of NADH. The commercial sources of this enzyme are numerous: rabbit muscle, porcine heart, bovine heart, chicken liver, spiny dogfish, lobster tail, and *Bacillus stearothermophilus*. The enzyme is inexpensive and fairly stable if immobilized on a solid support and protected against autooxidation. The equilibrium of the reaction lies in favor of the reduction.[82]

L-lactate dehydrogenase accepts a variety of 2-oxo acids as substrates (Figure 15). The (*S*) hydroxyacids produced are valuable chiral synthons. Several of the products have been used as precursors for useful chiral products. Chlorolactic acid was converted to the enantiomerically pure chiral epoxide, epoxyacrylic acid.[82a]

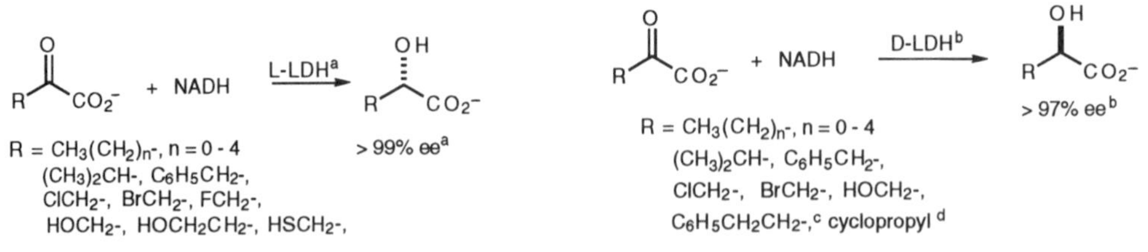

$R = CH_3(CH_2)_n\text{-}, n = 0 - 4$
$(CH_3)_2CH\text{-}, C_6H_5CH_2\text{-},$
$ClCH_2\text{-}, BrCH_2\text{-}, FCH_2\text{-},$
$HOCH_2\text{-}, HOCH_2CH_2\text{-}, HSCH_2\text{-},$

$> 99\% \text{ ee}^a$

$R = CH_3(CH_2)_n\text{-}, n = 0 - 4$
$(CH_3)_2CH\text{-}, C_6H_5CH_2\text{-},$
$ClCH_2\text{-}, BrCH_2\text{-}, HOCH_2\text{-},$
$C_6H_5CH_2CH_2\text{-},^c \text{ cyclopropyl }^d$

$> 97\% \text{ ee}^b$

(a) Kim, M. J.; Whitesides, G. M. *J. Am. Chem. Soc.* **1988**, *110*, 2959.
(b) Simon, E. S.; Plante, R.; Whitesides, G. M. *Appl. Biochem,. Biotechnol.* **1989**, *22*, 169.
(c) Bradshaw, C. W.; Wong, C.-H.; Hummel, W.; Kula, M-R. *Bioorg. Chem.* **1991**, *19*, 29.
(d) Kim, M-J.; Kim, J.Y. *J. Chem. Soc.,Chem. Commun.* **1991**, 326.

Figure 15

Due to the cost of the enzyme and ease of isolation of products, L-lactate dehydrogenase represents a useful synthetic catalyst in addition this enzyme provides an excellent system for the regeneration of the oxidized cofactor. By coupling electrochemical regeneration of NAD and electrochemical reduction of pyruvate, a complete conversion of L-lactate to D-lactate was accomplished with L-LDH.[82b]

L-Lactate dehydrogenase from *Bacillus stearothermophilus* is more specific for short-chain than for long-chain α-ketoacids. The enzyme from testes, however, is better able to accept larger substrates such as α-ketobutyrate. A narrower substrate specificity of the dogfish enzyme was observed.[82c] Comparison of the sequences at the substrate binding site enables modification of the enzyme substrate specificity. For example, the mutations Gln Lys Pro (102 → 105) Met Val Ser and Ala Ala (236 → 237) Gly Gly were made for the enzyme from *Bacillus stearothermophilus* to increase the k_{cat} for α-ketoisocaproate by a factor of 55 and to decrease the k_{cat} for pyruvate by a factor of 14.[83] The mutant enzyme still retains high thermostability. The enzyme was also converted to a specific, highly active malate dehydrogenase.[84] Analysis using molecular graphics indicate that Gln_{102} of L-lactate dehydrogenase from the *Bacillus* species interacts with the substituents adjacent to the α-keto group of the substrate. Replacement of Gln_{102} with the smaller Asn, however, only marginally increases the k_{cat} values for bulky α-keto acids.[85] The $Gln_{102}\rightarrow$Arg, $Arg_{171}\rightarrow$Trp double mutant accepts oxalacetate as a better substrate than does the wild-type enzyme.[82d] No drastic change of substrate specificity (e.g. L to D) has, however, been accomplished by mutagenesis.

2.2   D-Lactate Dehydrogenase [EC 1.1.1.28]

D-Lactate dehydrogenase (D-LDH) catalyzes the reduction of pyruvate to D-lactate *in vivo*. The enzyme from *Leuconostoc mesenteroides* reduces a range of 2-oxo acids to the corresponding (*R*) hydroxy acids with high enantioselectivity.[86] In a sense, D and L lactate dehydrogenase are complementary to each other in that the opposite enantiomer of a number of substrates are obtained. The different stereoselectivity of the two enzymes allowed the synthesis of both enantiomers of the chiral epoxides glycidic acid and 1-butene oxide (Ref. 1, Figure 15). The substrate specificity of D-lactate dehydrogenase is, however, substantially narrower than that of the L-enzyme. The D-enzyme showed less tolerance for side chains longer than three carbons, as these compounds were reduced only slowly. D-LDH from *Staphylococcus epidermidis* reduces α-ketobutyric acid, α-ketopentanoic acid, phenylpyruvate and α-ketocyclopropyl acetic acid to the corresponding D-α-OH acid with very high[86c] enantioselectivity (>99%). As with L-lactate dehydrogenase, this enzyme is inexpensive and exhibits a high specific activity. The reduction is similarly favored and the enzyme must be protected from autooxidation.

2.3   Hydroxyisocaproate Dehydrogenases

The hydroxyisocaproate dehydrogenases (HICAPDH) catalyze oxidoreductions similar to LDH. D-hydroxyisocaproate dehydrogenases from *Lactobacillus casei* reduces 2-oxoacids to the corresponding (*R*)-hydroxyacids and oxidizes solely the (*R*)-hydroxyacids to the ketoacids (Figure 16).[87] Likewise, L-hydroxyisocaproate dehydrogenase from *Lactobacillus confusus*

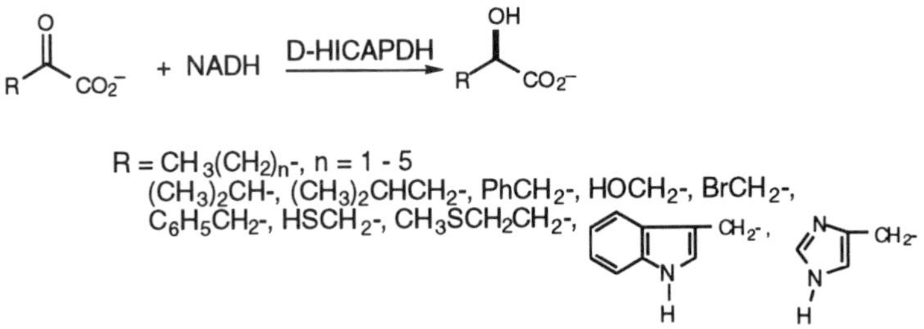

$$R = CH_3(CH_2)_n\text{-}, \ n = 1 - 5$$
$$(CH_3)_2CH\text{-}, \ (CH_3)_2CHCH_2\text{-}, \ PhCH_2\text{-}, \ HOCH_2\text{-}, \ BrCH_2\text{-},$$
$$C_6H_5CH_2\text{-}, \ HSCH_2\text{-}, \ CH_3SCH_2CH_2\text{-},$$

Figure 16

yields (*S*)-α-hydroxyacids as reduction products and oxidizes the (*S*)-α-hydroxyacids preferentially. Both enzymes use NAD(H) as cofactor and show no activity with NADP(H), and also reduce and oxidize a variety of similar substrates. In some instances, the specificity appears to be the same as the lactate dehydrogenases. However, the hydroxyisocaproate dehydrogenases are better able to utilize substrates with longer side chains. The lactate dehydrogenases optimally catalyze the reduction of pyruvate, with slower rates for all other substrates. The hydroxyisocaproate dehydrogenases predominantly catalyze the reduction of ketoisocaproate. Comparing the D and L-hydroxyisocaproate dehydrogenases, the D enzyme appears to have a somewhat broader substrate specificity and a greater synthetic value.

2.4   Glutamate Dehydrogenase [EC 1.4.1.3]

The natural substrate for glutamate dehydrogenase in the oxidative direction is glutamic acid. The oxidation of racemic amino acids destroys a chiral center of the D-enantiomer, and one can prepare the L-enantiomer in this manner. In the reductive amination direction, though 2-oxoglutaric acid is the best substrate, the enzyme accepts other mono and dicarboxylic acids (Figure 17).[88] Glutamate dehydrogenase can therefore be used to synthesize several unnatural amino acids such as L-α-aminoadipic acid[56] and L-β-fluoroglutamic acid.[89] The usefulness of the enzyme as a synthetic catalyst can be coupled with its ability to regenerate NADP. Glutamate dehydrogenase remains useful in coupled systems for cofactor regeneration and has been used in biphasic systems for this purpose. This commercially available enzyme can be easily immobilized on solid support to increase its stability.

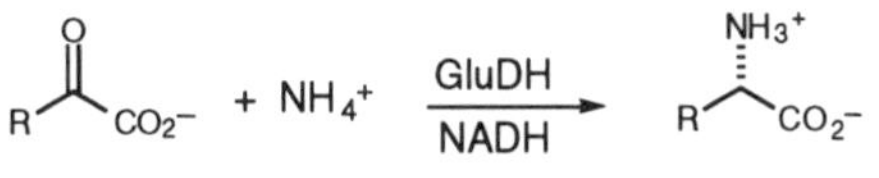

R = HO$_2$C(CH$_2$)$_n$-, n = 1 - 3
CH$_3$(CH$_2$)$_n$-, n = 1 - 2
HO$_2$CCH$_2$CHF-

Figure 17

2.5 Leucine Dehydrogenase [EC 1.4.1.9]

Leucine dehydrogenase (LeuDH) from *Bacillus cereus* exhibits broad substrate specificity as does glutamate dehydrogenase (Figure 18). Although less developed than other ketoacid dehydrogenases, leucine dehydrogenase has the potential to synthesize unnatural amino acids with branched side chains such as *tert*-L-Leucine.[90] The enzyme has been used in a membrane reactor for the continuous production of L-amino acids via reductive amination of 2-oxoacids. As with the other ketoacid dehydrogenases the enzyme utilizes NAD(H) as cofactor. These chiral molecules are precursors to a number of valuable compounds.

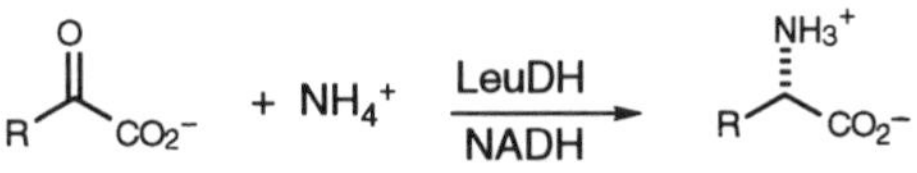

R = CH$_3$(CH$_2$)$_n$-, n = 1 - 3
(CH$_3$)$_2$CH-, (CH$_3$)$_2$CHCH$_2$-
CH$_3$CH$_2$CH(CH$_3$)-, MeS(CH$_2$)$_2$-
(CH$_3$)$_3$C-

Figure 18

2.6 Phenylalanine Dehydrogenase

Based on the early studies of substrate specificity, phenylalanine dehydrogenase (PheDH) may prove to be a very useful synthetic catalyst. The enzyme from *Rhodococcus*[91] accepts a range of aromatic 2-oxoacids as substrates, and produces unnatural amino acids including *p*-halogenated phenylalanine, and homophenylalanine (Figure 19). The *in vivo* role of phenylalanine dehydrogenase is the reductive amination of 2-oxo-3-phenylpropanoic acid (phenylpyruvate) to yield phenylalanine, a compound used in the manufacture of Aspartame. The enzyme isolated from *Rhodococcus spec.* retains high specific activity for the substrate. Phenylalanine dehydrogenase has also been isolated from *Thermoactinomyces intermedius*.[92] Both preparations are suitable for continuous product synthesis in a bioreactor.

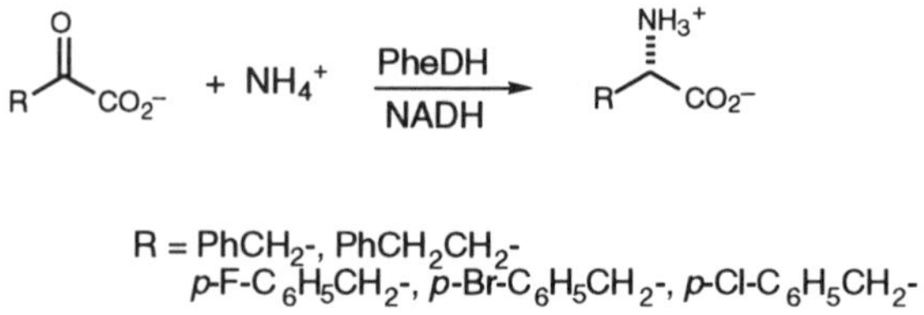

R = PhCH$_2$-, PhCH$_2$CH$_2$-
p-F-C$_6$H$_5$CH$_2$-, *p*-Br-C$_6$H$_5$CH$_2$-, *p*-Cl-C$_6$H$_5$CH$_2$-

Figure 19

Phenylalanine dehydrogenase from *Bacillus sphaericus* SCRC-R79a was also explored for synthesis.[94] Coupled with formate/formate dehydrogenase for NADH regeneration, the enzyme was used for the synthesis of a number of natural and unnatural L-amino acids, including phenylalanine, tyrosine, 4-(fluorophenyl)alanine, 2-amino-4-phenylbutyric acid, and 2-amino-nonanoic acid, with the enzyme enclosed in a dialysis tube. The gene for the enzyme was cloned and expressed in *E. coli*, and used for the synthesis. Acetone-dried cells of the *Bacillus* species also proved effective for the synthesis.

2.7 Other Ketoacid Dehydrogenases

The enzyme 2,3-dihydro-2,3-dihydroxybenzoate dehydrogenase catalyzes the NAD-dependent oxidation of 2,3-dihydro-2,3-dihydroxybenzoate to 2,3-dihydroxybenzoate. It also catalyzes the oxidation of 3-hydroxy cyclohexanecarboxylate analogs to the corresponding 3-keto species. The enzyme recognizes 1*R*, 3*R*-dihydro substrates and is potentially useful for the synthesis of a number of optically active cyclic ketones and hydroxy compounds (Figure 20).[95]

Another recently isolated ketoacid dehydrogenase is NAD-dependent (*R*)-mandelate dehydrogenase from *Streptococcus faccalis*.[96]

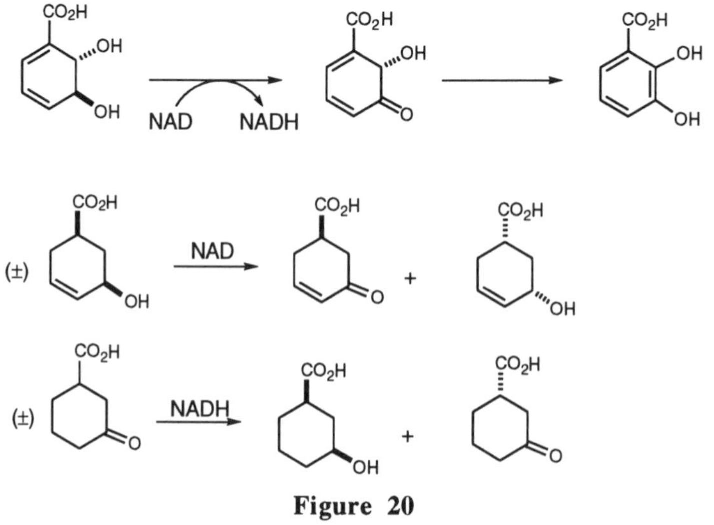

Figure 20

3. Other NAD(P)-Dependent Dehydrogenases

3.1 Hydroxysteroid Dehydrogenases

Hydroxysteroid dehydrogenase (HSDH) catalyzes the regiospecific oxidation of a hydroxy group or reduction of a ketone functionality of a steroid. The natural specificity of each particular enzyme is designated by its name. For example, 7-α HSDH catalyzes the oxidation and reduction of steroids at position-7 (Figure 21). The enzymes selective for positions 3, 7, 12, and 20 have been reported. Due to high regio- and stereoselectivity, this group of enzymes shows great potential in the synthesis of steroids, bile acids and other steroid derivatives. Not only can these enzymes tolerate different side chains on the steroid skeleton, but some can catalyze transformations of unrelated molecules. In particular, 3-α HSDH appears to be especially useful, oxidizing a range of aromatic trans diols with high enantioselectivity. Additionally, 3-α, 20-β HSDH also accepts non-steroid substrates. The coupling of different HSDH such as 3-α HSDH followed by 3-β HSDH or vice versa can invert the hydroxy center selectively at the 3 position. Inversion at C-7 has also been accomplished similarly.[97]

Some steroid derivatives poorly soluble in aqueous solution have been used in biphasic systems in reactions coupled with cofactor regeneration

3.2 Glycerol Dehydrogenase [EC 1.1.1.6]

Glycerol dehydrogenase (GDH) reduces α-hydroxy ketones to chiral (*R*)-1,2 diols. The chiral hydroxy ketones are obtained by oxidation of 1,2-diols or via the kinetic resolution of racemic hydroxy ketones. The *in vivo* role of the NAD(H) requiring GDH is the interconversion of dihydroxyacetone and glycerol. The purified enzyme is commercially available from several sources. The enzyme is stable and can be immobilized but must be used under an inert atmosphere. Many substrates have been found for this enzyme (Figure 22) and can be either acyclic or cyclic.[98] Hydroxy aldehydes are not substrates and hydroxy esters are substrates of one strain, *Geotricum candidum*. The *Geotricum* strain has the additional advantage that it can simultaneously oxidize 2-propanol to regenerate NADH, thus alleviating the need for another dehydrogenase to maintain the cofactor concentration.[99] For synthetic applications, the enzyme from *Cellulomonas sp.* is superior because of its higher specific activity and lower cost.[98]

The substrate specificity of glycerol dehydrogenase is quite different from that of horse liver alcohol dehydrogenase, yeast alcohol dehydrogenase, lactate dehydrogenase and other alcohol dehydrogenases. Some keto acid esters are reduced to (*R*)-α-hydroxy acid esters[99] (Figure 22).

Reactions proceeding in the direction of oxidation usually suffer from slow rates and incomplete conversions due to product inhibition. Like other alcohol dehydrogenase-catalyzed oxidations, product inhibition seems the most serious limitation to the use of these enzymes as catalysts for oxidative organic synthesis.

3.3 Dihydrofolate Reductase [EC 1.5.1.3]

The NADP(H) requiring dihydrofolate reductase (DHFR) catalyzes the *in vivo* interconversion of folate and tetrahydrofolate. The enzyme tolerates changes in the substituents at both the 6 and 7 positions (Figure 23).[100] The methyl group can be substituted at the 7 position. The differences at the other position are much more diverse. Thus, DHFR provides a means to obtain chiral derivatives of dihydrofolate. Preparative syntheses with the enzyme from *E. coli* proceed with high stereoselectivity. The dihydrofolate derivative leucovorin has been used in cancer therapy.

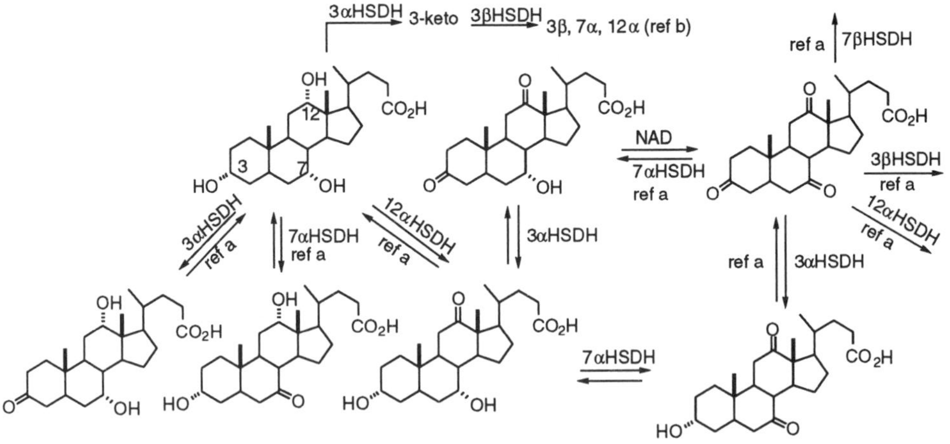

Enzyme source and cofactor: 3α, 7 α, and 3β-HSDH are from Sigma. All require NAD;
7β-HSDH from *Clostridium absonum*, NADP dependent;
12α-HSDH from *Clostridium groupp* P, NADP dependent.

Cofactor regeneration: NAD(P), 2-ketoglutarate/glutamate dehydrogenase
NADH, formate/formate dehydrogenase from yeast
NADPH, glucose/glucose dehydrogenase from *Bacillus*

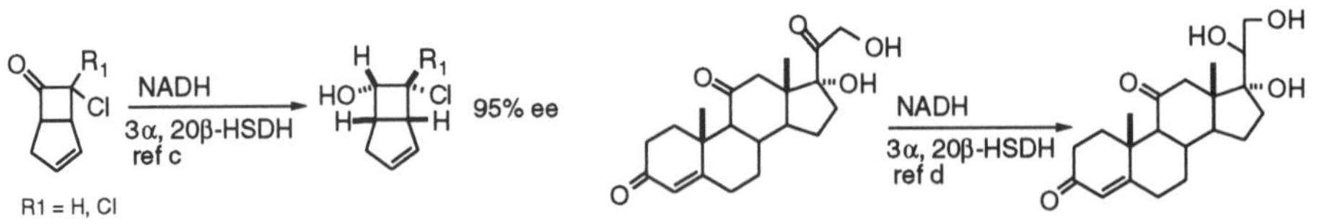

R1 = H, Cl

Ref [a] Riva, S.; Bovara, R.; Pasta, P.; Carrea, G. *J. Org. Chem.* **1986**, *51*, 2902.
Enzymes are immobilized on Sepharose CL-4B

[b] Riva, S.; Bovara, R.; Zetta, L.; Pasta, P.; Ottolina, G.; Carrea, G. *J. Org. Chem.* **1988**, *53*, 88.
Enzymes were immobilized on Eupergit C.

[c] Davies, H. G.; Gartenmann, T. C. C.; Leaver, J.; Roberts, S. M.; Turner, M. K. *Tetrahedron Lett.* **1986**, *27*, 1093.
The enzyme was immobilized on Eupergit C.

[d] Drueckhammer, D. G.; Sadozai, S. K.; Wong, C.-H. *Enz. Microb. Tech.* **1987**, *9*, 564;
Cremonesi, P.; Ferrara, L.; Antonini, E. *Biotech. Bioeng.* **1985**, *17*, 1101.

Figure 21

3.4 Other Oxidoreductases

Many other oxidoreductases have synthetic applicability by virtue of their substrate specificity, but remain to be developed for large scale synthesis. The enantiomeric excess characteristics of these reactions has also not been defined. Among those enzymes are: alanine dehydrogenase (EC 1.4.1.1) from *Bacillus* for reductive amination,[101a] glycerol 3-phosphate dehydrogenase (EC 1.1.1.8) for the oxidation of analogous phosphonate and difluoromethylene phosphonate,[101b] Sepiapterin reductase [EC 1.1.1.153] from *Drosophila*[101] and carbonyl reductase from *Mucor ambiguus*[102b] both for reduction of diketones. The latter reduces cyclic conjugated and cyclic ketones to diols. The former enzyme reduces cyclic substrates in addition to linear diketones to the diols. Dihydroxyacetone reductase[103] transfers the hydride from the nicotinamide cofactor to the *si* face of a variety of acyclic and cyclic subtrates with high enantioselectivity. As with other redox enzymes, these reductases have great potential for further development in organic synthesis.

Selective enzymatic oxidation of one of the several hydroxyl groups of an alditol or aminoalditol has been reported for the synthesis of ketoses. These reactions perhaps involve the class of enzymes called polyoldehydrogenase. Figure 24 illustrates some examples of such microbial and enzymatic transformations. A recently described microbial oxidoreduction of sugar-like α-hydroxy- or α-keto- acids appears to be quite useful.[103b]

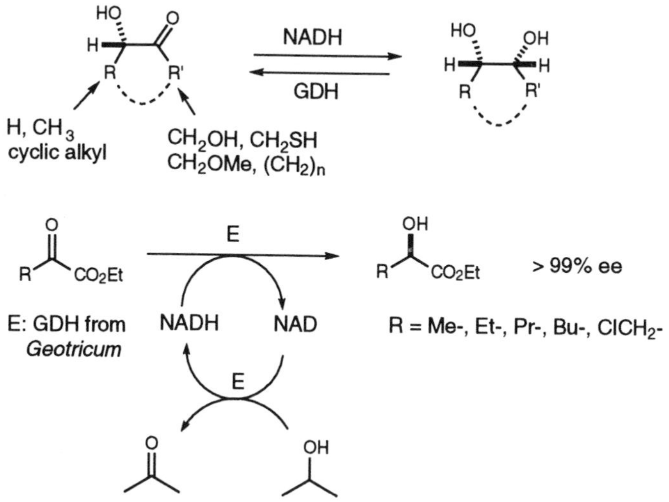

Figure 22

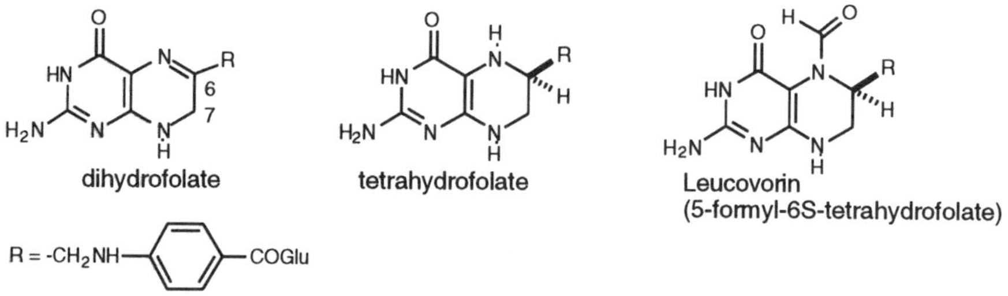

dihydrofolate   tetrahydrofolate   Leucovorin
                                                 (5-formyl-6S-tetrahydrofolate)

R = -CH₂NH—⟨benzene ring⟩—COGlu

Figure 23

4. Oxidoreductases that are Metalloenzymes

4.1   Enolate Reductase [EC 1.3.1.31] and Related Enzymes

Enoate reductase isolated from *Clostridia* species[20] catalyzes stereospecific reduction of a number of α,β-unsaturated carboxylates, aldehydes, and ketones. Mechanistic studies indicate that the enzyme contains an iron-sulfur-flavin cluster that accepts and transfers electrons to substrates. Proton exchange between the reduced cluster and water was observed. The iron-sulfur cluster can be reduced with NADH, with reduced methyl and benzyl viologen generated electrochemically, with hydrogen catalyzed by hydrogenase present in the cell, or with Pd, Pt or Raney nickel modified with fluorine-containing surfactants such as Zonyl-FSC (Figure 25).[104] In practice, whole cells instead of free enzyme are used for reactions, presumably due to the instability of the free enzyme. The productivity numbers of the system are often 10-100 times higher than those found for reduction with yeasts. The system has also been used to prepare isotopically labeled chiral δ-aminolevulinic acid.[105]

In addition to α,β-unsaturated systems, many ketones and 3-oxocarboxylic acid esters can also be reduced stereoselectively with *Clostridia* in the presence of H_2 with very high ee. Some examples are illustrated in Figure 26. Many α-keto mono- or di-carboxylates can also be reduced by *Proteus vulgaris* in the presence of methyl viologen or benzyl viologen to the corresponding (*R*)-hydroxy species with very high ee (Figure 27). The viologen mediator can be regenerated either electrochemically or by hydrogen and a hydrogenase present in the microorganism, or by formate and a formate dehydrogenase, also found in the cell.[20] When carbamoyl methyl viologen (CAV) is used as a mediator ($E_0' = -295$ mV, E_0' MV^{++} = -439 mV),

Figure 24

D-mannitol → (Acetobacter suboxydans[a]) → D-fructose

D-sorbitol → (Acetobacter suboxydans[a]) → L-sorbose

D-ribitol → (Acetobacter suboxydans[a]) → L-erythropentutose

L-mannitol → (Bacterium MD-13[b]) → L-fructose

(Gluconobacter oxydans[c]) X = H, Cbz-

Sorbitol dehydrogenase NADH regeration

R	Ref
D-CH(OH)CH$_3$	d
CH$_2$SC$_6$H$_5$	e
CH(OCH$_2$CH$_3$)$_2$	f
CH=CH$_2$	g
CH$_2$CH=CH$_2$	g
(D)-HOCHCH=CH$_2$	g
(L)-HOCHCH=CH$_2$	g

References

(a) Schnarr, G. W.; Szarek, W. A.; Jones, J. K. N. *Appl. Environ. Microbiol.* **1977**, *33*, 732.
(b) Dhawale, M. R.; Szarek, W. A.; Hay, G. W.; Kropinski, A. M. B. *Carbohydr. Res.* **1986**, *155*, 262.
(c) Kinast, G.; Schedel, M. *Angew. Chem. Int. Ed. Eng.* **1981**, *20*, 805.
(d) Wong, C.-H.; Mazenod, F.P.; Whitesides, G.M. *J. Org. Soc.* **1983**, *48*, 3493.
(e) Heidlas, J.; Schmid, W.; Mathias, J.P.; Whitesides, G.M. *Liebigs Ann. Chem.* **1992**, 95.
(f) Borysenko, C.W.; Spaltenstein, A.; Straub, J.A.; Whitesides, G.M. *J. Am. Chem. Soc.* **1989**, *111*, 9275.
(g) Kobori, Y.; Myles, D.C.; Whitesides, G.M. *J. Org. Chem.* **1992**, *57*, 5899.

Figure 24

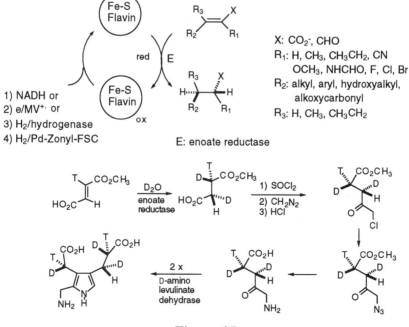

1) NADH or
2) e/MV$^{+\cdot}$ or
3) H$_2$/hydrogenase
4) H$_2$/Pd-Zonyl-FSC

E: enoate reductase

X: CO$_2^-$, CHO
R$_1$: H, CH$_3$, CH$_3$CH$_2$, CN
OCH$_3$, NHCHO, F, Cl, Br
R$_2$: alkyl, aryl, hydroxyalkyl, alkoxycarbonyl
R$_3$: H, CH$_3$, CH$_3$CH$_2$

Figure 25

enantioselective oxidation of racemic α-hydroxy acids can be achieved. Electrochemical oxidation of CAV$^+$ to CAV^{++} can be carried out at a carbon electrode at about 400 mV (vs the standard calomel electrode). Air oxidation can also be used to regenerate CAV^{++}, since the enzyme is stable under air when used as a dehydrogenase.[107] It is, however, sensitive to oxygen when used as a reductase. Similar reduction of α,β-unsaturated systems was carried out with Baker's yeast or fungi.[108]

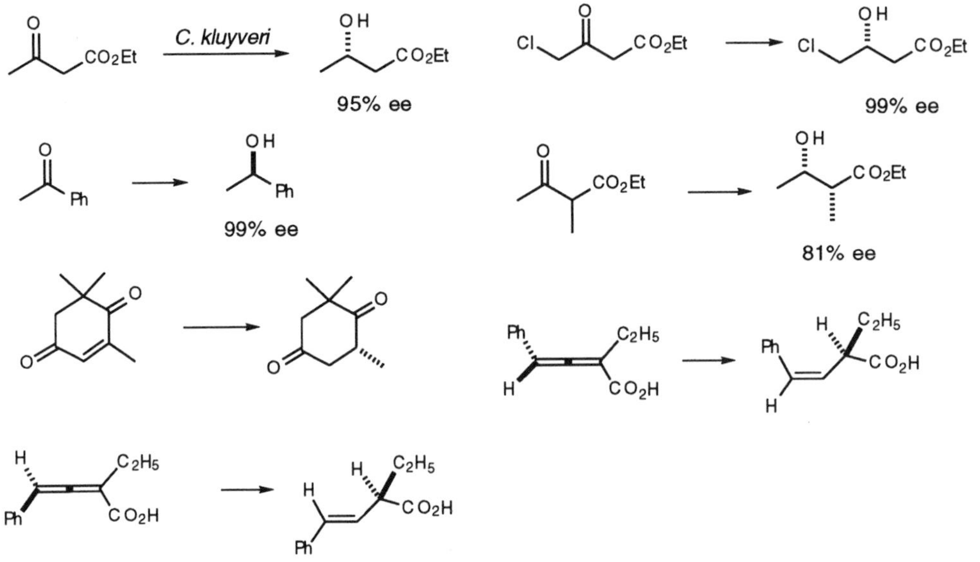

Figure 26

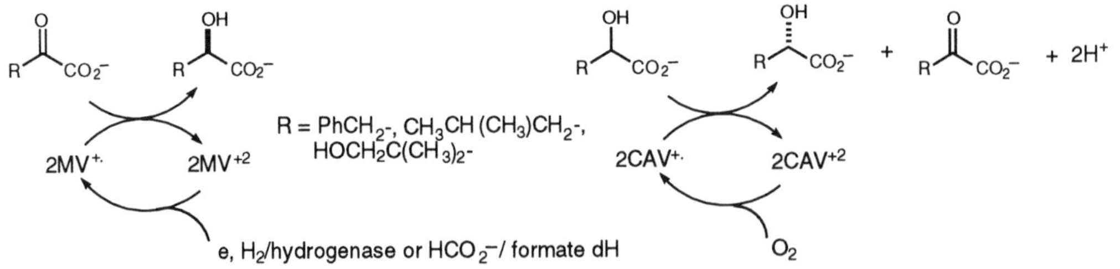

Figure 27

4.2   Galactose Oxidase [EC 1.1.3.9]

Galactose oxidase (GO) from *Dactylium deudroides* contains one atom of Cu(II) per molecule as the cofactor and catalyzes the oxidation of D-galactose at C-6 position in the presence of oxygen to the corresponding aldehyde and hydrogen peroxide.[109] Recent synthetic investigations indicate that the enzyme catalyzes the stereospecific oxidation of glycerol and 3-halo-1,2-propanediols to the corresponding L-aldehydes.[110] Investigation of the substrate specificity of this enzyme using a number of polyols has permitted the development of a model of the enzyme active site.[111] Several unusual L-sugars have been prepared using this enzyme. It appears that a substrate alcohol with configuration similar to that of D-galactose from C-4 to C-6 would be oxidized to aldehyde. The byproduct hydrogen peroxide usually must be destroyed with peroxidase to avoid inactivation of the enzyme. One problem with this enzyme reaction is the product inhibition that results in low yields (10-15%). Recent mechanistic studies indicate that the active form contains Cu(II) associated with a cation radical[112a] which may be considered as an equivalent of a Cu(III) center.[112b] Electron or hydrogen atom abstraction from the hydroxylic substrate by the radical would generate a hydroxyl radical species capable of reducing the copper to Cu(I). The crystal structure of the enzyme[113] indicates that the coordination site of the copper ion contains two His and two Tyr residues, and one of the Tyr residues , probably Tyr_{272}, forms a thioester bond to Cys_{228}. This thioester cofactor exists as a radical cation at the activated state. Oxygen would then be reduced to peroxide and the active enzyme be regenerated.

4.3   Lipooxygenase

Lipoxygenase is a non-heme iron-containing dioxygenase, which catalyzes the oxidation of polyunsaturated fatty acids to a hydroperoxide. In the case of arachidonic acid, the lipoxygenation could occur at carbon 5,8,9,11,12 or 15 depending on the specific lipoxygenase used. Arachidonic acid can be converted to the (S)-5-hydroperoxide[114] on a synthetic scale (15%) using potato lipoxygenase, while a soybean lipoxygenase will convert the fatty acid to the (S)-15-hydroperoxide[115] (Figure 29). Linoleic acid was converted to (S)-13-hydroperoxide.[115] The (S)-5-hydroperoxide is a useful intermediate for the synthesis of leukotrienes.[114] The 5-lipoxygenase contains a high-spin Fe(III) center in the active form which specifically abstracts the $7\text{-}H_S$ hydrogen[116] from the substrate. The reaction was proposed to proceed through a free radical intermediate that reacts directly with oxygen[117] or a γ organoiron intermediate.[118] Based on the results of kinetic isotope effects, stereochemical outcome, and inhibition studies,[118] it seems that the organoiron process is more likely. This conclusion is further supported by the observation of a displaceable water coordinated to the iron(III) center.[119] The detectable radical intermediate could result from the homolysis of C-Fe bond. Recent investigation of the enzyme[120] regarding its application to the synthesis of chiral diols (Figure 29) indicates that a decrease of the hydrophobicity for R and increase of X lead to the increase of the product with the OH group next to R.

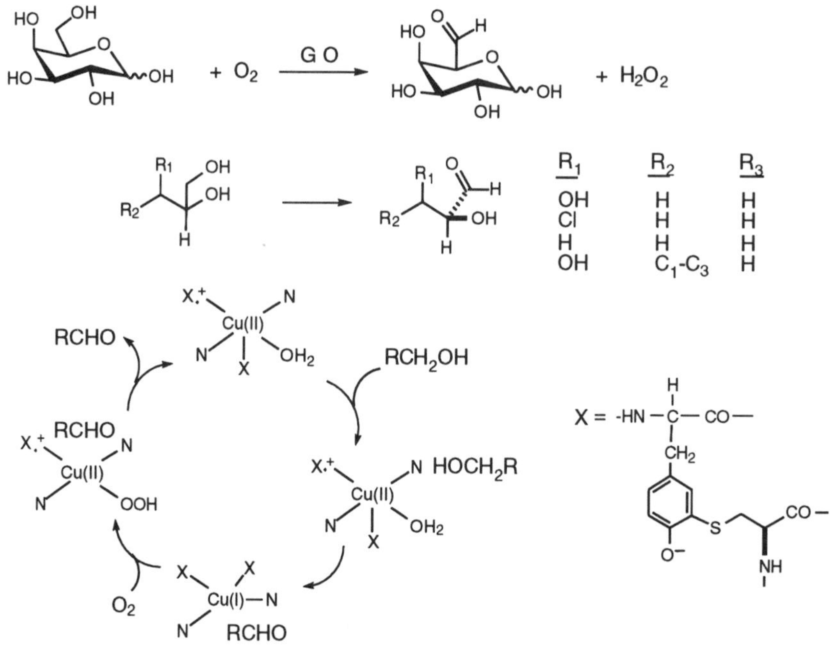

Figure 28

4.4   Arene Dioxygenase

Since 1970 Gibson and coworkers reported the enantioselective oxidation of toluene to
cis-dihydrotoluenediol and benzene to *cis*-cyclohexadienediol by a mutant of *Pseudomonas putida*
(Figure 30),[121] many arenes have been shown to yield diols through microbial techniques.[122] All
reactions were carried out with whole cell processes. *cis*-Dihydrotoluenediol has been used for
synthesis of terpene and prostanoid synthons[123] and cyclohexadienediol has been used in the
synthesis of polybenzene which may have potential in the manufacture of liquid crystals,[124]
fibres, and films. Cyclohexadienediol was also used in the synthesis of carbocyclitols such as
pinitol,[125] conduritols and hydroxyenones (Figure 31).[126] A dioxygenase from *Pseudomonas
putida* also catalyzes the oxidation of styrene, chlorobenzene and derivatives to the corresponding
cyclohexadienes cis-diol, which were used for synthesis of natural products (Figure 32).[127]

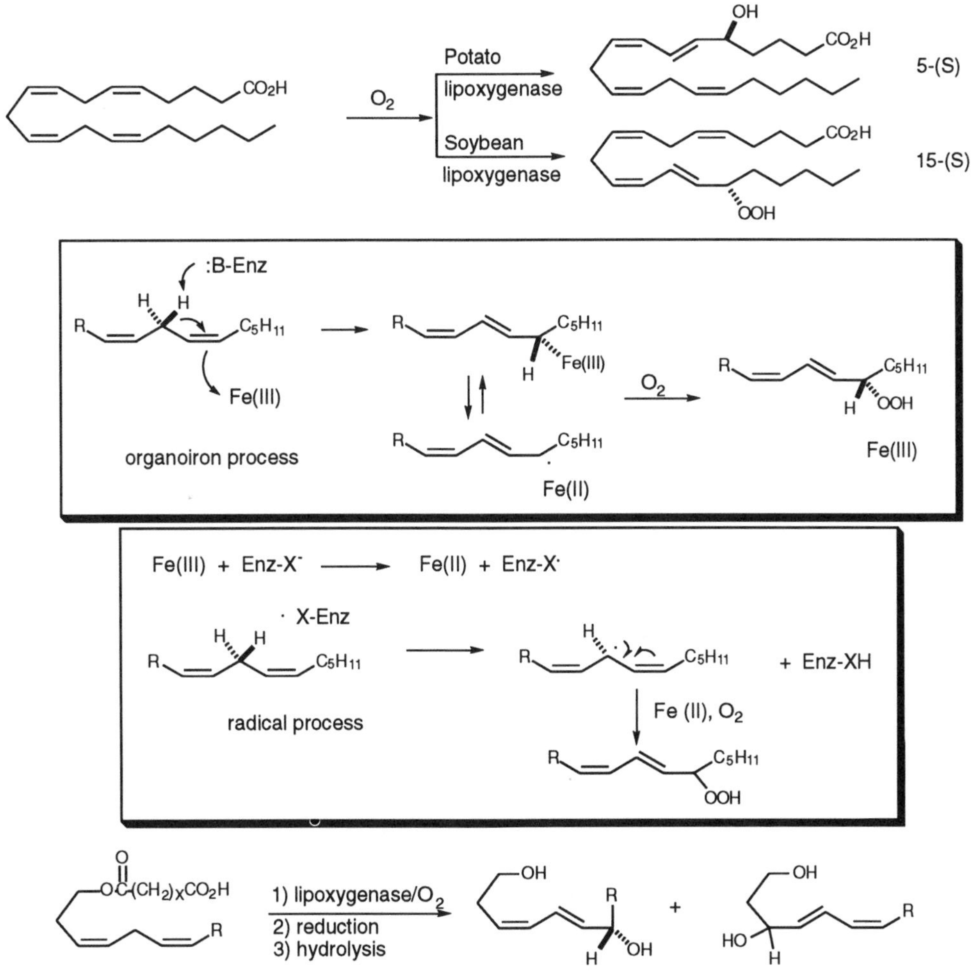

Figure 29

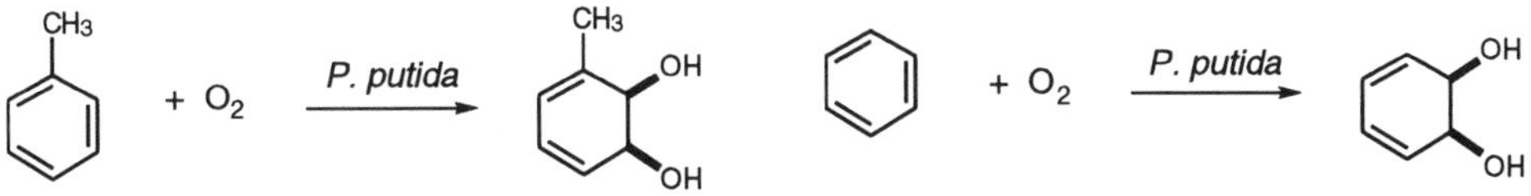

Figure 30

Examples of other dioxygenases which may find use in syntheses are the non-heme iron(III) enzymes catechol 1,2-dioxygenase and protocatechuate 3,4-dioxygenase. Most of the dioxygenases require NADH or NADPH as an indirect electron donor. The prosthetic group can be heme or non-heme iron or copper.[1]

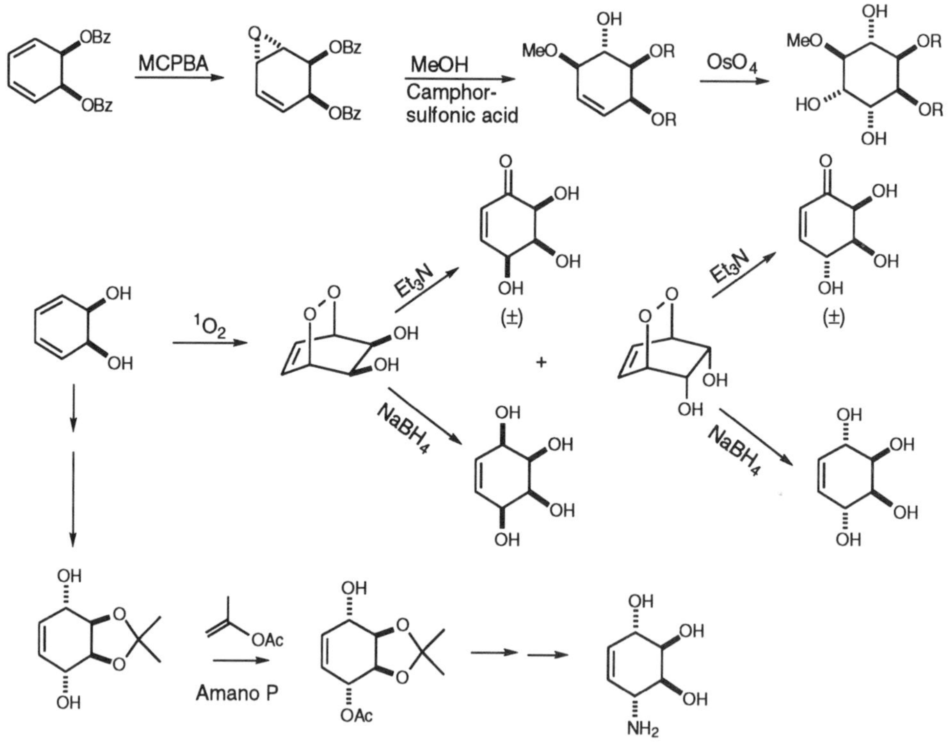

Figure 31

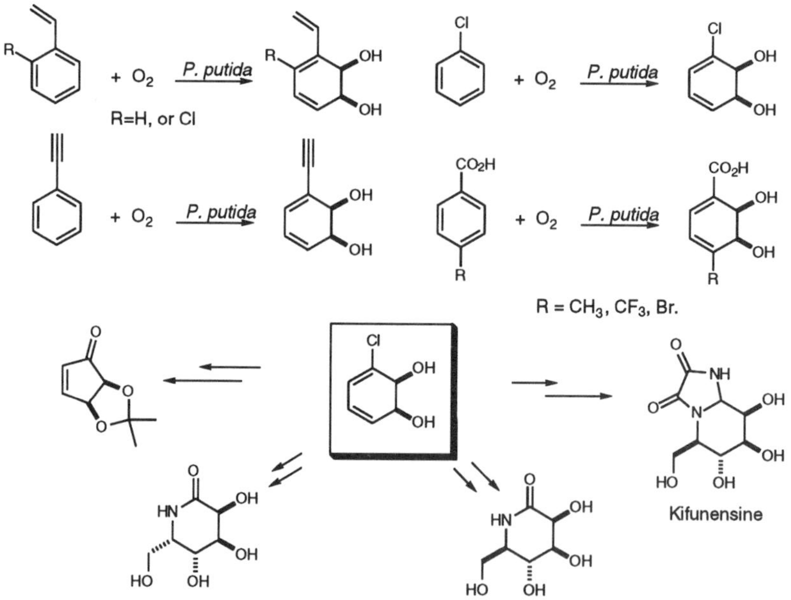

R = CH₃, CF₃, Br.

Kifunensine

Figure 32

4.5 Baeyer-Villiger Oxidations Catalyzed by Monooxygenase

The oxidation of ketones to esters and lactones by the Baeyer-Villiger reaction is a well established synthetic transformation and is typically effected by a peroxide or peracid reagent.[128] The enzymatic Baeyer-Villiger reaction requires flavoenzyme in the presence of NAD(P)H and oxygen. Mechanistic studies[129] of the enzyme cyclohexanone monooxygenase from *Acinetobacter* NCIB 9871[130] indicate that FAD-4a-hydroperoxide intermediate acts as a nucleophile reacting with the carbonyl group (Figure 33). In addition to cyclohexanone, many other cyclic ketones, acyclic ketones, aldehydes, and boronic acids are also substrates, all proceeding with the stereochemistry similar to that of the chemical reaction. Thus, the enzymatic reaction proceeds with $3^o > 2^o > 1^o$ order of group migration with retention of configuration at the migrating group.[131] The enzyme also catalyzes enantioselective transformation of substituted cyclic ketones which may or may not contain enantiotopic substituents,[132] a valuable process

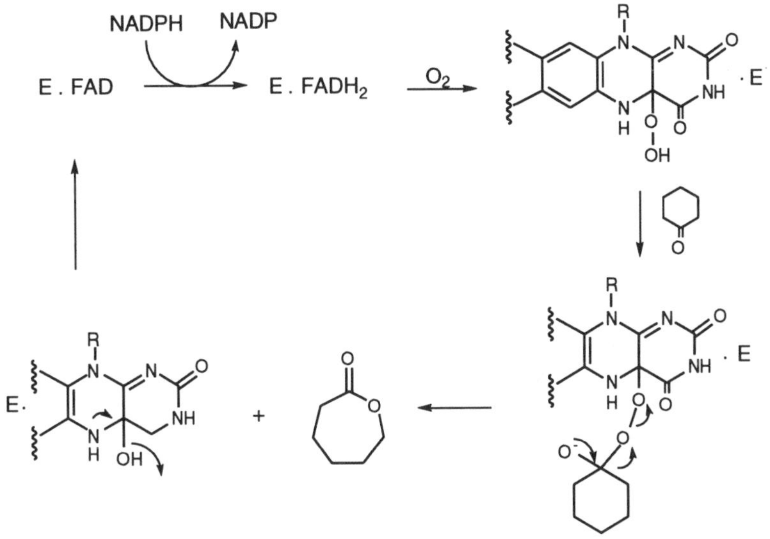

Figure 33

from synthetic point of view (Figure 34). In large-scale reactions, the required NADPH cofactor must be regenerated from NADP. The enzyme, however, is not able to oxidize olefins. Formation of epoxide was not observed. The flavin peroxide also functions as an electrophile, oxidizing heteroatoms such as S, N, Se, and P.[131b] A recent study revealed that *Pseudomonas putida* NCIMB 10007 contains a monooxygenase which catalyzes Baeyer-Villiger oxidation of some bicyclic ketones.[133]

4.6   Other Monooxygenase: ω-Hydroxylase, Methane Monooxygenase, Isopenicillin-N Synthetase and P450

Monooxygenases are metalloenzymes capable of activating molecular oxygen for oxidation of a number of inert organic compounds including alkanes, alkenes, aromatics and

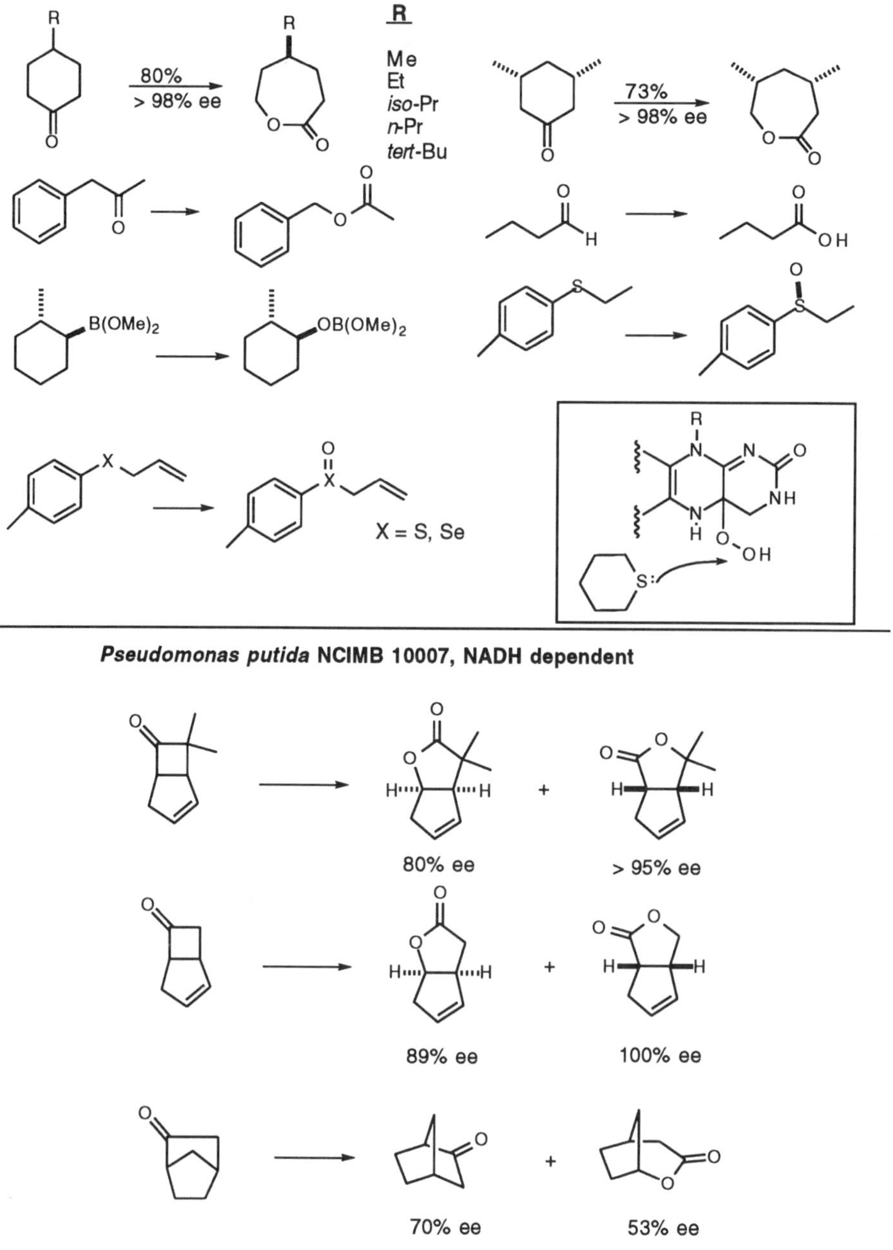

Pseudomonas putida **NCIMB 10007, NADH dependent**

Figure 34: Baeyer-Villiger Oxidations

heteroatoms.[134] The non-heme monooxygenase from *Pseudomonas oleovorans* is an NADH/O_2 dependent enzymatic system that contains three components: reductase, rubredoxin, and hydroxylase.[135] The reductase is a flavoprotein that transfers two electrons from NADH to the iron sulfur center of rubredoxin via its bound FAD. The rubredoxin, a non-heme iron protein, accepts and transfers two electrons from reduced rubredoxin one by one to the hydroxylase, which then accepts a substrate and a molecular oxygen, activates the oxygen, and inserts the oxygen to the substrate. Both the reductase and rubredoxin have been isolated and characterized.[135] The hydroxylase was too unstable to be purified to homogeneity. Mechanistic investigations of the enzyme[136] regarding the activation of oxygen indicates similarity to the heme-containing P450 system[137,138] where the active oxygen species has been established to be a high valent iron oxo species, $Fe^V=O$ or $Fe^{IV}=O$ coordinated with a cation radical ligand (Figure 35).The mechanism of oxygen transfer from the iron oxo species to substrates is still not well established and three possible mechanisms have been proposed (Figure 36). These include

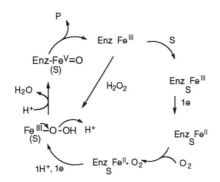

Figure 35

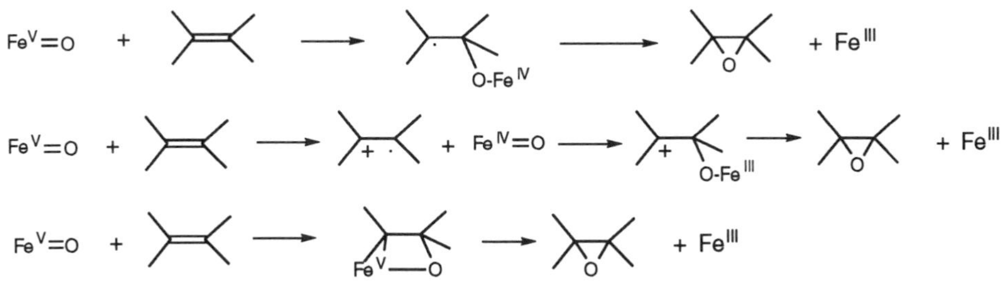

Figure 36

radical, cation radical, and concerted pathways.[139] Recent study on ω-hydroxylase seems to favor the radical mechanism.[136] The iron carbene intermediate was ruled out due to the loss of terminal olefin configuration and the lack of hydrogen exchange, and the cation intermediate was ruled out due to the lack of atom migration.

The enzyme catalyzes hydroxylation of the terminal methyl group of alkanes, epoxidation of terminal alkenes to (R)-epoxides, stereoselective sulfoxidation of methyl thioethers, oxidation of terminal alcohols to aldehydes and oxidative O-demethylation of branched alkyl methyl and vinyl methyl ethers to secondary alcohols and ketones, respectively.[136] For preparative-scale synthesis, NADPH was regenerated with glucose 6-phosphate and glucose 6-phosphate dehydrogenase.[136] Allyl alcohols were shown to be inhibitors instead of substrates for the enzyme. O-Protected allylalcohols, however, can be epoxidized to the corresponding (R)-epoxides (Figure 37).[140-142] Reaction of the enzyme with trans-2-phenyl-1-methylcyclopropane

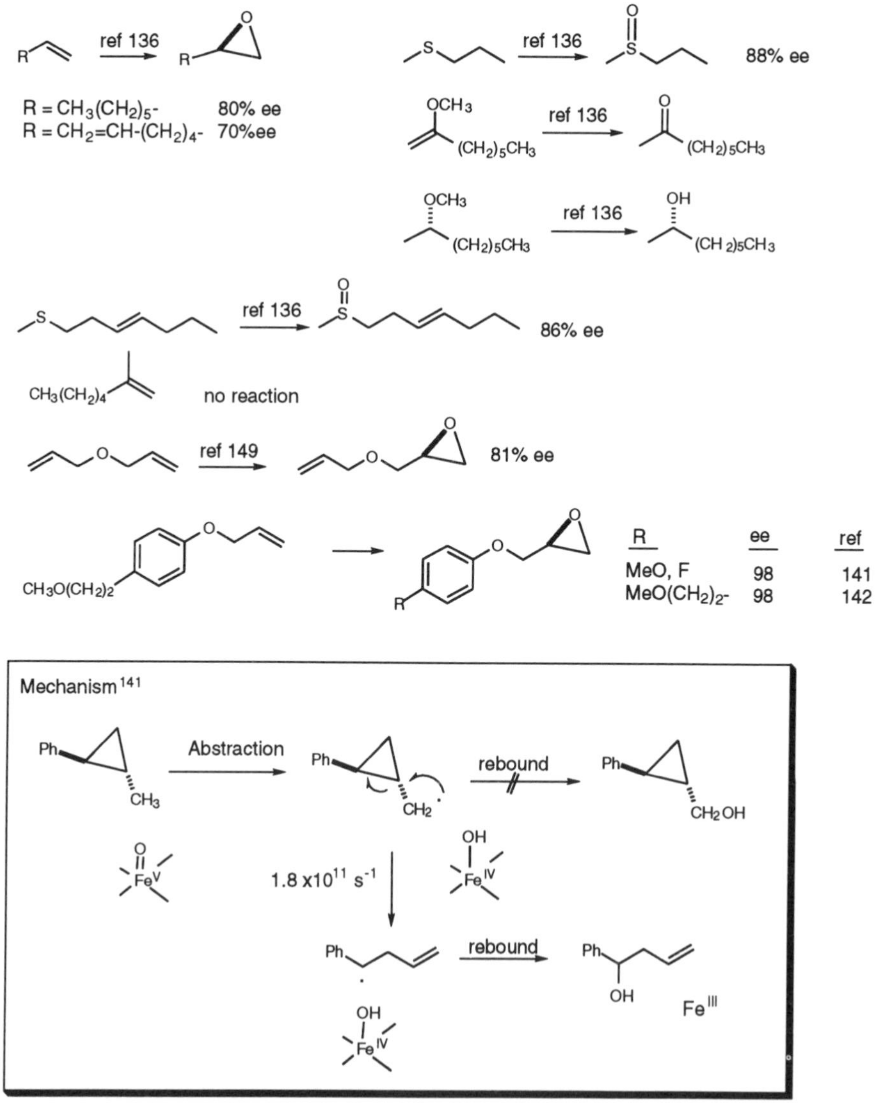

Figure 37

gave 1-phenyl-3-buten-1-ol as a single product,[141] which, together with the observed high kinetic isotope effect (k_H/k_d ca. 8), led to the suggestion of a nonconcerted radical process. This is the first demonstration of the use of a radical probe with the radical ring opening rate exceeding the oxygen rebound rate in enzymatic hydroxylation. A number of oxidation reactions catalyzed by different monooxygenases have been reported based on the whole cell process, and are summarized in Figure 38.

Methane monooxygenase from methanotrophic bacteria catalyzes the oxidation of methane to methanol. It also catalyzes the oxidation of other alkanes containing two to eight carbons to the corresponding primary and secondary alcohols, alkenes to epoxides, thioethers to sulfoxides, and aromatics to the monohydroxy aromatics.[143] The enzymes from *Methylococcus capsulatus* and *methylosinus trichosporium* are complexes of three proteins: Protein A, a non-heme iron component that contains an (μ-oxo) diiron center and acts as oxygenase; Protein C, a 2Fe-2S/FAD protein that accepts electrons from NADH and acts as reductase; and Protein B, a small regulatory protein that regulates the interaction of substrates with protein A.[144] A steady state kinetic analysis indicates that methane binds to the enzyme complex first followed by NADH to form a ternary complex.[145] This complex then binds oxygen to form a secondary ternary complex which gives rise to methanol and water. The isolated protein A can be reduced chemically with methyl viologen or dithionite and catalyzes the oxidation of substrates in the presence of oxygen.[146] Mechanistic study of the enzyme from *M. trichosporium* using the radical probe 1,1-dimethylcyclopropane indicates the reaction goes through a radical as a major pathway and a cation intermediate as a minor pathway.[147a] The enzyme from *M. capsulatus*, however, seems to proceed through a concerted mechanism as no ring-opened product was observed in the hydroxylation of trans-2-phenyl-1-methyl cyclopropane (Figure 39).[147b] The 2.2 Å crystal structure of this enzyme has been determined.[147c]

Isopenicillin-N synthetase is another non-heme monooxygenase that catalyzes the cyclization of tripeptide δ-(L-α-aminoadipyl)-L-cysteinyl-D-valine (ACV) to isopenicillin-N in the presence of oxygen, Fe(II) and ascorbate. This step is a key to the biosynthesis of all penicillin and cephalosporin antibiotics.[148] The enzyme from different species has been purified[149] and the gene from *C. acremonium* has been cloned[150] Isopenicillin-N synthetase from *C. acremonium* has been extensively studied with respect to its substrate specificity,[151] and these studies have led to the elucidation of the mechanism of its action (Figure 40). The ring closure reaction is presumably mediated by the oxo iron species to form the β-lactam ring first, followed by the C-S bond formation to construct the thiazolidine ring. The enzyme possesses a relaxed requirement for variation of the Val residues. When tripeptides with saturated Val analogs were used as substrates, the products from so called "desaturative process"[152] were obtained (Figure 41).

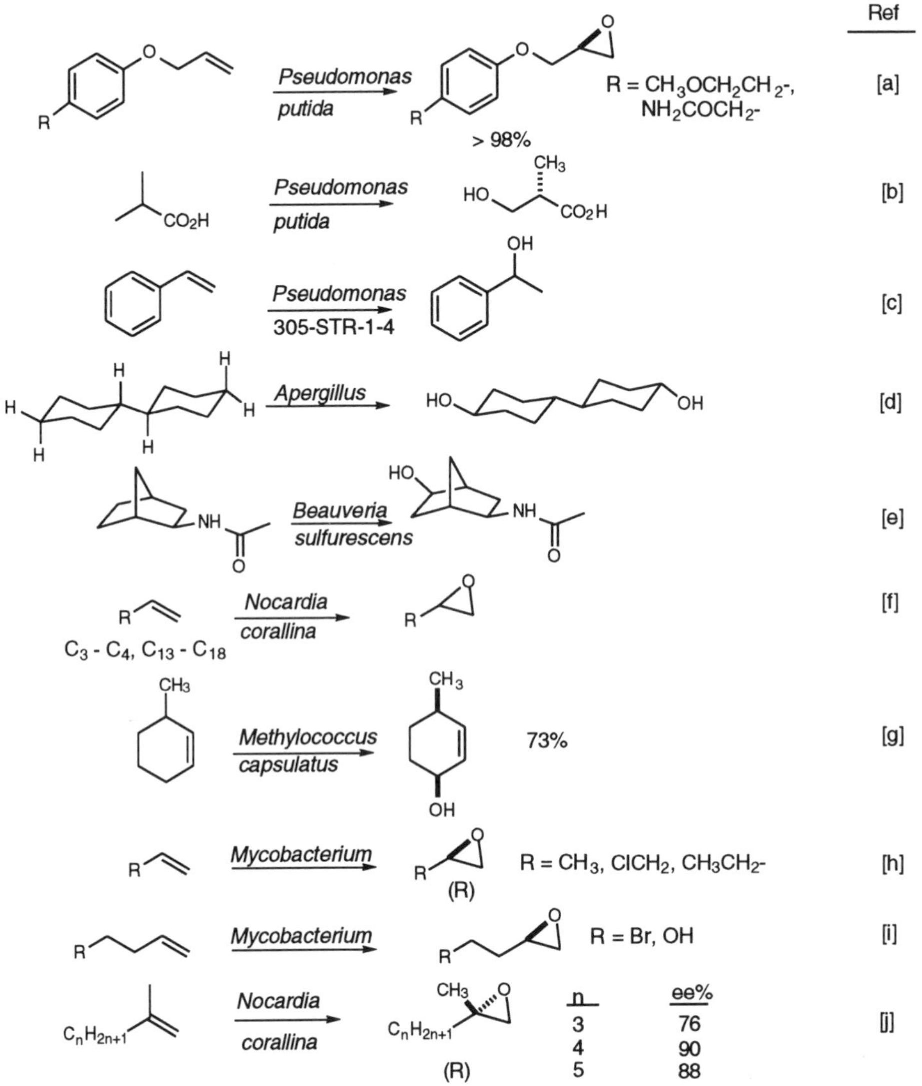

Ref

> 98%   R = CH$_3$OCH$_2$CH$_2$-, NH$_2$COCH$_2$-   [a]

[b]

[c]

[d]

[e]

[f]

73%   [g]

[h]

[i]

n	ee%
3	76
4	90
5	88

[j]

References
[a] Johnstone, S. L.; Philips, G. T.; Robertson, B. W.; Watts, P. D.; Bertola, M. A.; Koger, H. S.; Marx, A. F. in *Biocatalysis in organic media* (Lanne, C.; Tramper, J.; Lilly, M. D. eds), Elsevier, Amsterdam, **1986**, p387.
[b] Goodhue, C. T.; Schaeffer, J. R. *Biotech. Bioeng.* **1971**, *13*, 203.
[c] Shirai, K.; Hisatsuka, K. *Agric. Biol. Chem.* **1979**, *43*, 1399.
[d] Davies, H. G.; Dawson, M. J.; Lawrence, G. C.; Mayall, J.; Noble, D.; Roberts, S. M.; Turner, M. K.; Wall, W. F. *Tetrahedron Lett.* **1986**, *27*, 1089.
[e] Fourneron, J.- D.; Archelas, A.; Vigne, B.; Furstoss, R. *Tetrahedron* **1987**, *43*, 2273.
[f] Furuhashi, K.; Taoka, A.; Uchida, S.; Karube, I.; Suzuki, S. *Eur. J. Appl. Microbiol. Biotechnol.* **1981**, *12*, 39.
[g] Leak, D. J.; Ellison, S. L. R.; Murricane, C.; Baker, P. B. *Biocatalysis* **1988**, *1*, 197.
[h] Habets-Crutzen, A. Q. H.; Carlier, S. J. N.; de Bont, J. A. M.; Wistuba, O.; Schurig, V.; Hartmans, S.; Tramper, J. *Enzyme Microb. Technol.* **1985**, *7*, 17.
[i] Archelas, A.; Hartmans, S.; Tramper, J. *Biocatalysis* **1988**, *1*, 283.
[j] Takahashi, O.; Umezawa, J.; Furuhashi, K.; Takagi, M. *Tetrahedron Lett.* **1989**, *30*, 1583.

Figure 38

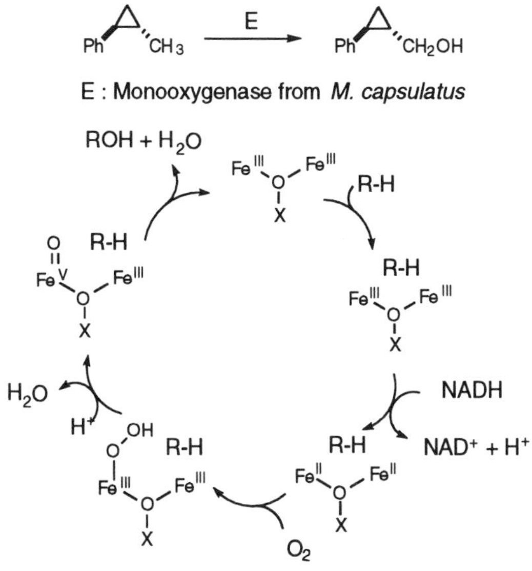

E : Monooxygenase from *M. capsulatus*

Figure 39

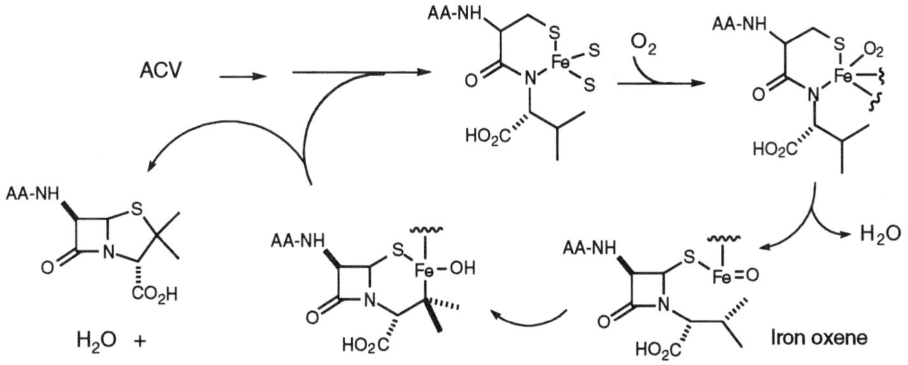

Figure 40: Isopenicillin N Synthetase: Mechanism

When tripeptides containing unsaturated Val analogs were used as substrates, products from both desaturative and hydroxylative pathways were obtained (Figure 42).[152] Expansion of penicillins to cephalosporins is catalyzed by the expansion enzyme from the same species. Substrate specificity studies indicate that the enzyme also has a broad side chain specificity (Figure 43). A new monooxygenase, clavaminate synthetase, has recently been isolated and shown to catalyze the double oxidative cyclization of proclavaminic acid to clavaminic acid in the presence of Fe(II), O_2 and α-ketoglutarate.[153a] This enzyme was used in the synthesis of a new class of β-lactam analogs.[153b]

The cytochrome P450 system is a heme containing monooxygenase that catalyzes the activation of molecular oxygen to form an oxo-iron species for oxidation of alkanes, olefins and heteroatoms.[137] This enzyme system has received substantial attention regarding the scope and

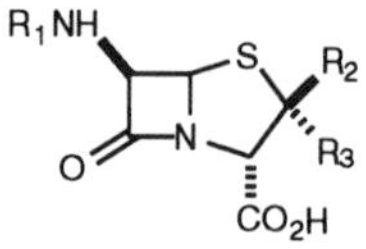

R_1	R_2	R_3	Rel rate	Ref
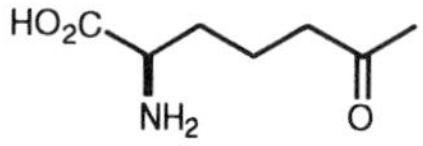	CH_3	CH_3	100	[a],[b]
	CH_3	H	4	[a]
	CH_2CH_3	CH_3	36	[a]
	OCH_3	CH_3	25	[a]
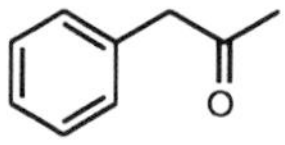	CH_3	CH_3	0.15	[a],[c]
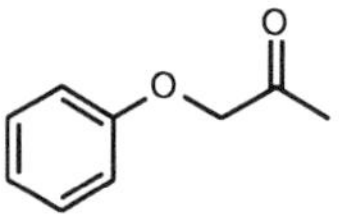	CH_3	CH_3	0.05	[a]
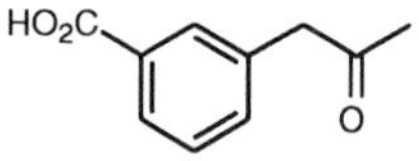	CH_3	CH_3	49	[a]
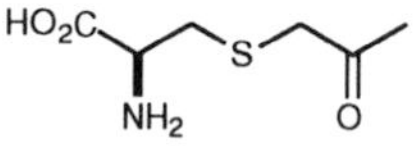	CH_3	CH_3	54	[b]

[a] Baldwin, J. E.; Adlington, R. M.; Crabbe, M. J. C.; Knight, G. C.; Nomoto, T.; Schofield, C. J. *J. Chem. Soc.,Chem Commun.* **1987**, 806
[b] Wolfe, S.; Demain, A. L.; Jensen, S. E.; Westlake, D. W. S. *Science* **1984**, *226*, 1386.
[c] Luengo, J. M.; Alemany, M. T.; Sato, F.; Ramos, F.; Lopez-Nieto, M. J.; Martin, J. F. *Bio/Technology.* **1986**, *4*, 44.

Figure 41

the mechanism of action.[154] The reactions may go through a radical or a cation radical intermediate. Using bicyclo[2.1.0]pentane as a probe for microsome P450,[155] a radical pair was determined to form in the hydroxylation step that collapses with stereochemical specificity at a rate $1.5 \times 10^9 s^{-1}$.[156] Oxidation of alkyl sulfides by microsome P450 indicates that both *S*-oxidation and *S*-dealkylation occur,[157] and *S*-dealkylation takes place more rapidly with substrates bearing a more acidic α-hydrogen, indicating the cation radical mechanism. In any case, chiral sulfoxides can be prepared via monooxygenase reactions. Oxidation at Se also occurs similarly followed by a [2,3] sigmatropic rearrangement (Figure 45).[158] The alcohols obtained were achiral; perhaps racemization of selenoxide is too fast.[158] Cyt P450 also catalyzes oxidative *O*-dealkylation and *N*-dealkylation.[137] The most studied P450 system is that from *Pseudomonas putida*. The amino acid sequence,[159] and crystal structure in the presence[160] and absence[161] of camphor has been determined. The gene has been identified and expressed in *E. coli*,[162] and a mechanism of reaction has been proposed (Figure 35).[137,163]

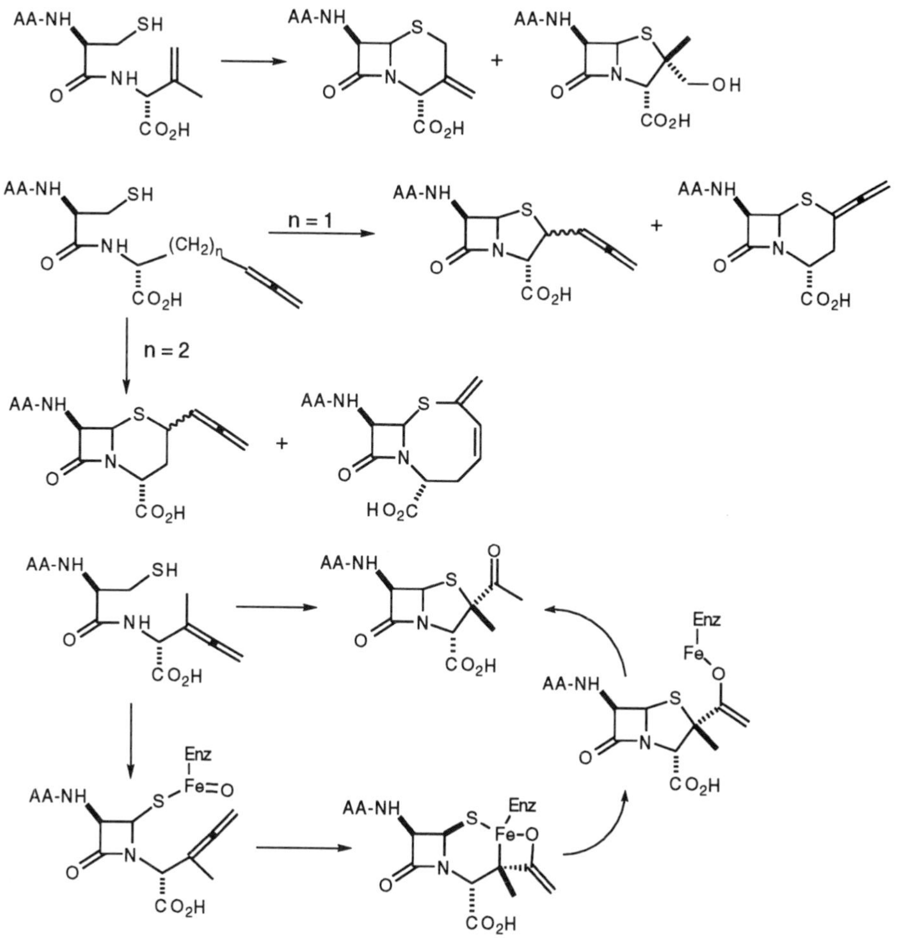

Figure 42

Xanthine oxidase is a molybdenum iron-sulfur flavin hydroxylase that catalyzes the hydroxylation of purine and pyrimidines and the oxidation of benzaldehyde.[1,164] Phenol-, salicylate, melilotate and p-hydroxybenzoate hydroxylase are flavoproteins that catalyze monohydroxylation of aromatic hydroxy compounds utilizing dioxygen and reduced FAD.[165] Enantioselective oxidation of sulfide to sulfoxide can only be achieved with certain substrates and certain microorganisms (Figure 46).[166] Although most monooxygenases catalyze this type of reaction, the enantioselectivity is often low.

4.7 Peroxidases

Chloroperoxidase (EC 1.11.1.10) from *Caldariomyces fumago* has recently been used in the presence of *tert*-butyl and other peroxides for the enantioselective oxidation of aromatic sulfides to (R)-sulfoxides with very high enantioselectivity.[167] Chiral hydroperoxides such as 1-

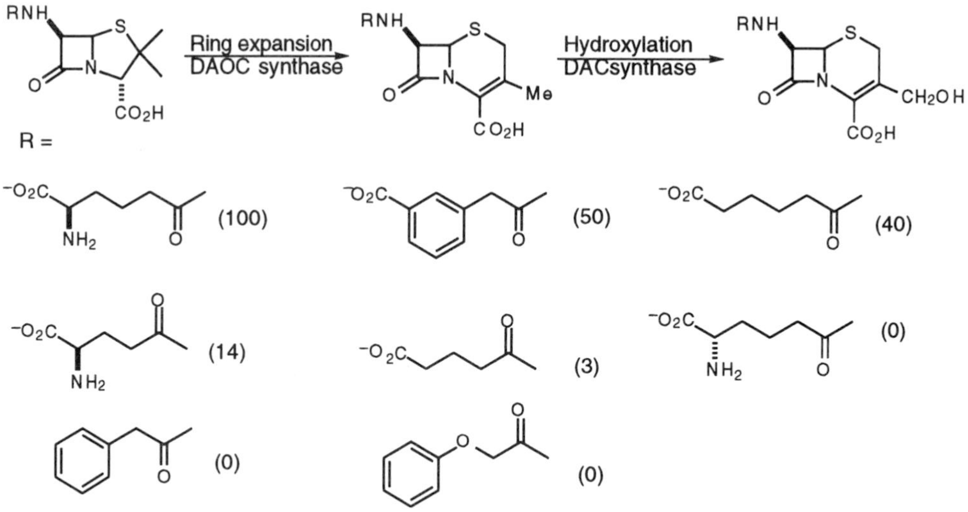

Ref: Baldwin, J. E.; Adlington, R. M.; Coates, J. B.; Crabbe, M. J. C.; Keeping, J. W.; Knight, G. C.; Nomoto, T.; Schofield, C. J.; Ting, H. H. *J. Chem. Soc., Chem Commun.* **1987**, 374.
The number in parenthesis is relative rate.

Both enzymes contain Fe^{++} and α-ketoglutarate. In *C. acremonium* both steps are carried out by a single bifunctional enzyme (DAOC/DAC synthase), while in S. spp. two separate enzymes are involved. For cloning and expression of the enzyme: Ingolia, T. D.; Queener, S. W. *Med. Res. Rev.* **1989**, *9*, 245.

For further specificity and mechanistic studies: Baldwin, J.; Adlington, R. M.; Crouch, N. P.; Pereira,

I. A. C. *J. Chem. Soc., Chem. Commun.* **1992**, 1448; Baldwin, J. E.; Goh, K.-C.; Wood, M. E.; Schofield, C. J. *BioMed. Lett.* **1991**, *1*, 421; Baldwin, J. E.; Adington, R. M.; Crouch, N. P.; Heath, R. J.; Pereira, I. A. C.; Sutherland, J. D. *BioMed. Lett.* **1992**, *2*, 669.

Figure 43

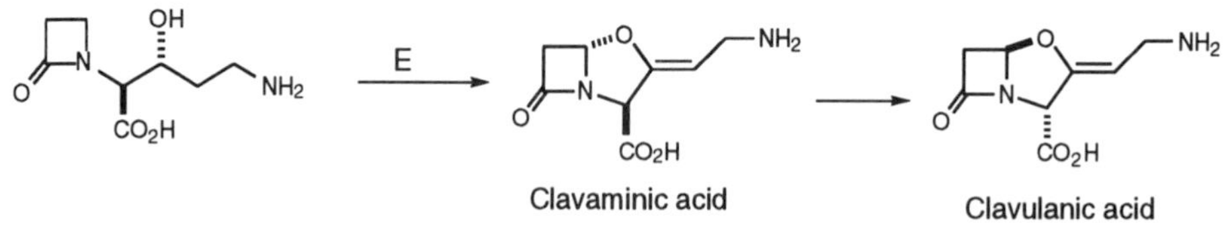

E : clavaminic acid synthase

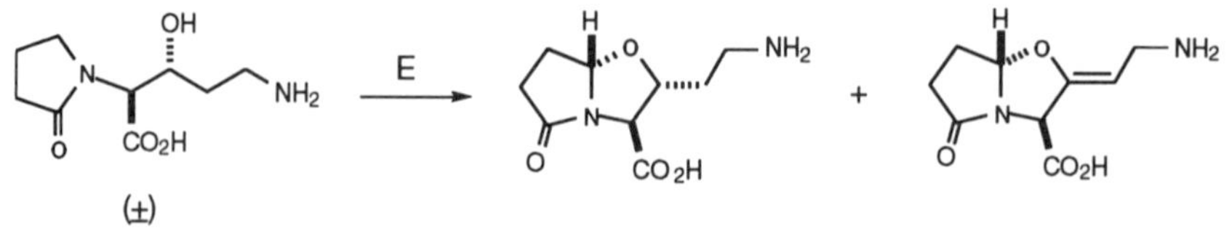

Figure 44

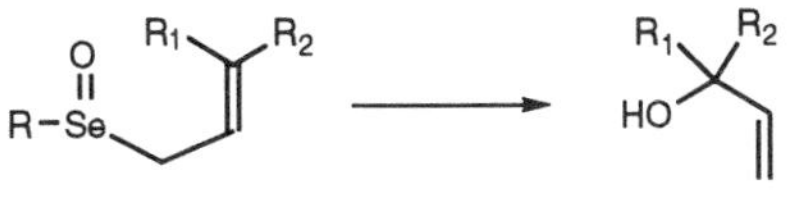

Figure 45

phenylethyl hydroperoxides are also acceptable as oxidants and the (*R*)-enantiomers are selectively accepted as substrates[167b] (Figure 47). The reactions with horseradish peroxidase, however, showed very low enantioselectivity.[167a] The chloroperoxidase from *Caldariomyces*

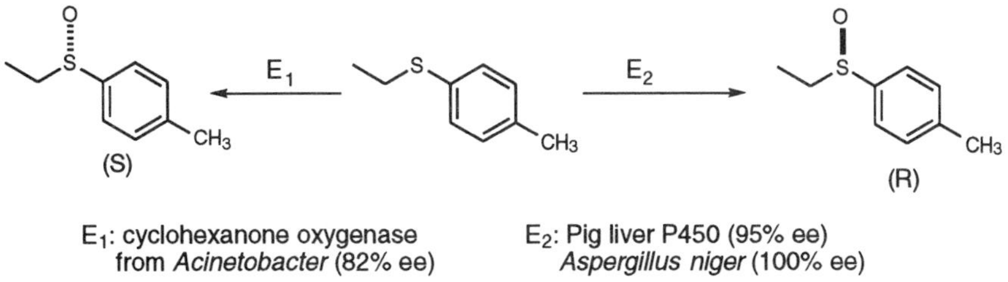

E₁: cyclohexanone oxygenase from *Acinetobacter* (82% ee)

E₂: Pig liver P450 (95% ee) *Aspergillus niger* (100% ee)

Figure 46

fumago catalyzes the oxidation of all halide ions except fluoride,[168] lactoperoxidase oxidizes bromide and iodide,[170] and horseradish peroxidase only oxidizes iodide ion.[171] All reactions require H_2O_2 as natural oxidation reagent. In the presence of an unsaturated acceptor, the products formed by haloperoxidase catalyzed reactions are consistent with the reaction with hypohalous acid.[170,171] When glycals are used as substrates, halohydration occurs regioselectively to give 2-halo-sugars.[167b] The stereoselectivity is, however, quite low with some exceptions. In the presence of high concentrations of a second nucleophile, such as another halide, mixed dihalides were obtained.[172] In addition to monooxygenases, horseradish peroxidase,[173] cytochrome C peroxidase,[174] haloperoxidases[173b,c] and lactoperoxidase[175] all utilize ferryl-oxo ($Fe^{IV}=O$) species for reaction with substrates. This reactive species perhaps forms a bound hypohalite which could undergo regio- and stereo-selective bromohydration of glycals[167b] (Scheme 47). Similar to monooxygenases, the chloroperoxidase from *Caldarimyces fumago* also catalyzes the epoxidation of olefins in the presence of hydrogen peroxide.[176] Horseradish peroxidase and lactoperoxidase were, however, inactive toward olefins.[176a] Of several olefins tested, cis olefins are the best giving epoxides with very high ee.[176b] Since this enzyme is commercially available, and has been cloned,[177] it appears to be a useful "free monooxygenase" for synthetic application. Peroxidase was used in the polymerization of ethylphenol[178a] in reverse micelles.

Other oxidases have also been utilized in synthesis. Tyrosinase was used in the synthesis of Coumestans and benzofuran derivatives,[178b] and coupling of hindered phenols.[178c]

Glycolate oxidase (EC 1.1.3.15) from spinach was used in the oxidation of glycolate to glycoxylate on 0.5 kg scales. The reaction was performed in the presence of catalase from

Aspergillus niger and ethylenediamine. The reaction is known to proceed via a two-electron transfer from glycolate to FMN and reoxidation of the reduced FMN by oxygen to produce hydrogen peroxide which is decomposed with catalase.[179]

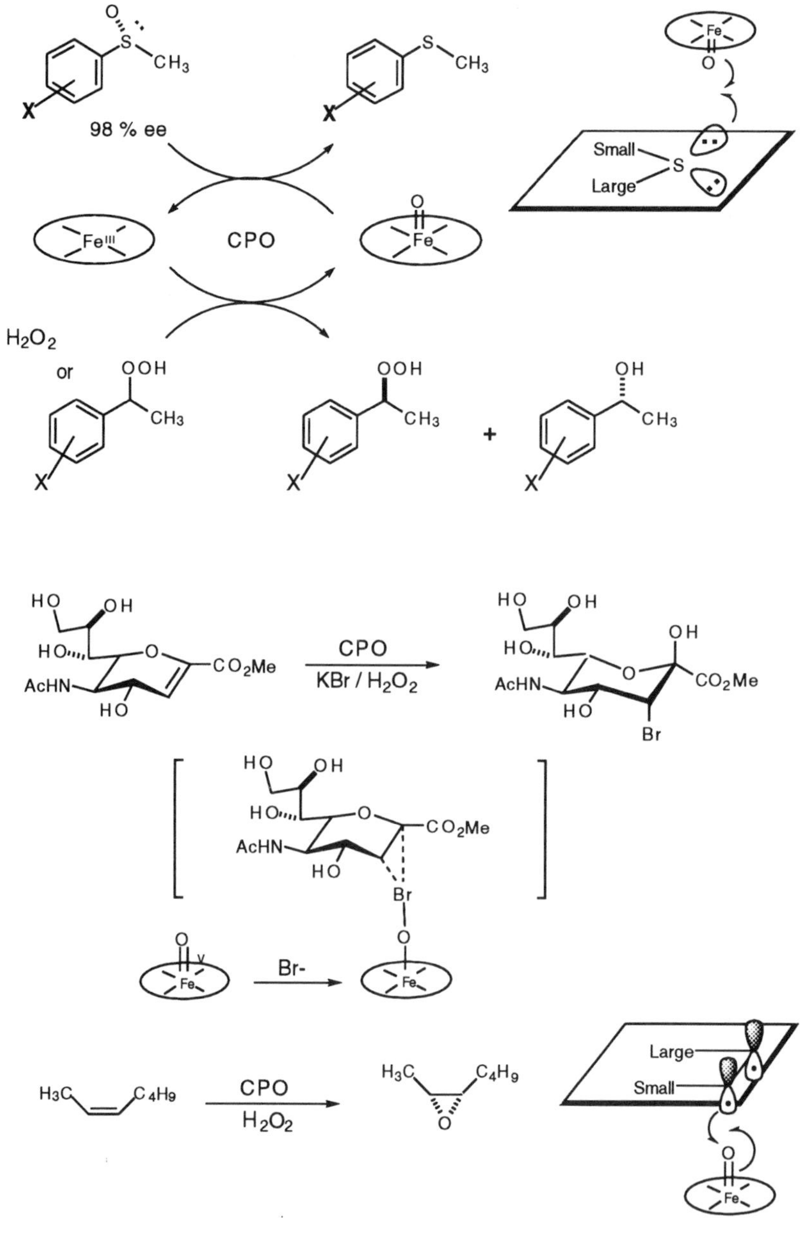

Figure 47

It is worthnoting that many other oxidoreductases exist in nature that utilize different cofactors and transform not only organic but also organometallic compounds.[180] Many NAD-dependent or -independent quinone proteins[181] (e.g. pyrrolquinoline, PQQ, dependent methanol dehydrogenase) or 6-hydroxydopa containing enzymes (e.g. methylamine oxidase) (Figure 48),[182] for example, have recently been discovered.[113] The synthetic value of these enzymes remains to be explored.

PQQ   6-hydroxy DOPA

Figure 48

Table 5.

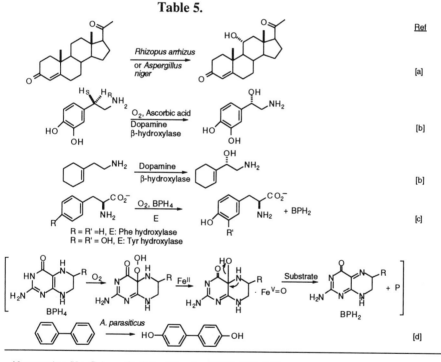

[a]   Ahronowitz, Y.; Cohen, G. *Scientific. Am.* **1981**, *245*, 141.

[b]   Millar, S.; Klinman, J. *Biochemistry* **1985**, *24*, 2114; Ketonization of halophenethylamine, Bossard, M.; Klinman, J. *J. Biol. Chem.* **1986**, *261*, 16421; Phenylaminoalkyl sulfide oxidation, May, S. W.; Phillips, R.; Mueller, P.; Herman, H. *J. Biol. Chem.* **1981**, *256*, 8470; N-dealkylation of benzylic N-substituted analogs, Fitzpatrick, P.; Flory, D.; Villafranca, J. *Biochemistry* **1985**, *24*, 2108; Allylic hydroxylation, Sirimanne, S. R.; May, S. W. *J. Am. Chem. Soc.* **1988**, *110*, 7560.

[c]   Pteridine dependent monooxygenase. For a review, see: Dix, T. A.; Benkovic, S. J. *Acc. Chem. Res.* **1988**, *21*, 101.

[d]   Smith, R. V.; Rosazza, J. P. *Arch. Biochem. Biophys.* **1974**, *161*, 551; Smith, R. V.; Davis, P. J.; Clark, A. M.; Glover-Milton, S. J. *Appl. Bacteriol.* **1980**, *49*, 65; Abramowicz, D. A.; Keese, C. R.; Lockwood, S. H. in *Biocatalysis*; Abramowicz, D. A., ed.; van Nostrand, Reinhold: New York, 1990, p63.

Table 5. Continued.

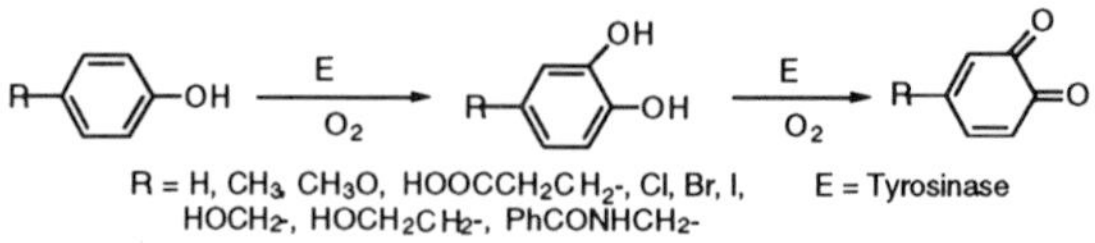

R = H, CH₃, CH₃O, HOOCCH₂CH₂-, Cl, Br, I,   E = Tyrosinase
HOCH₂-, HOCH₂CH₂-, PhCONHCH₂-

[e]

71%

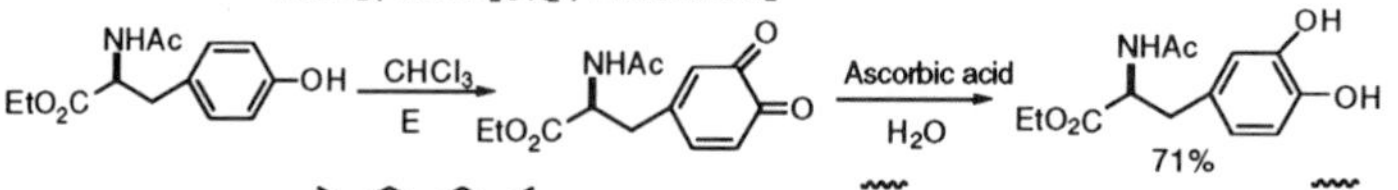

[e]

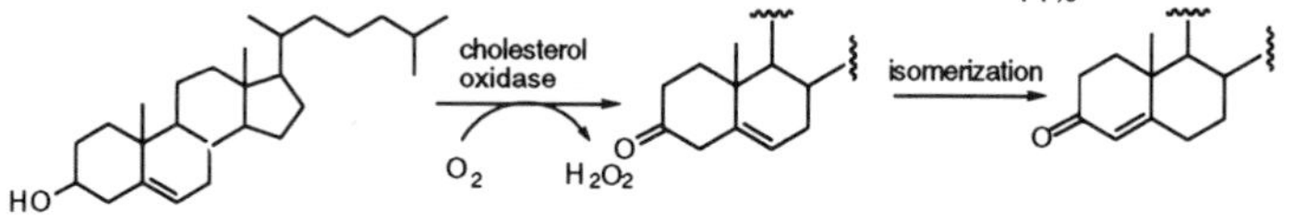

[f]

1) R = -CH₂CH(NH₂)CO₂H(L), X= OH, Y = H.
2) R = -CH(NH₂)CO₂H(L), X= OH, Y = H.
3) R = -CH(OH)CH₂NHCH₃(L), X = H, Y = OH.

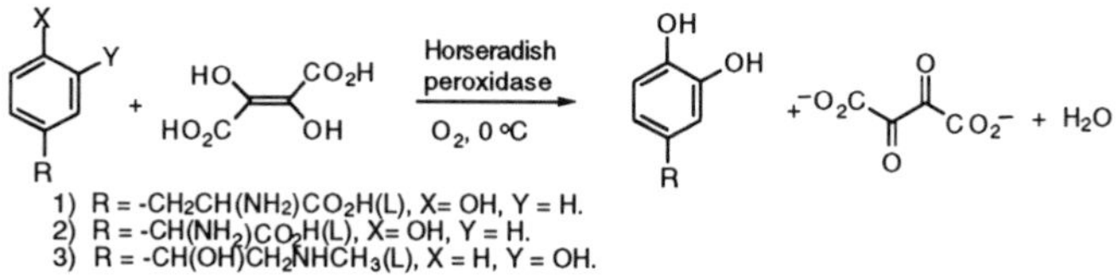

[g]

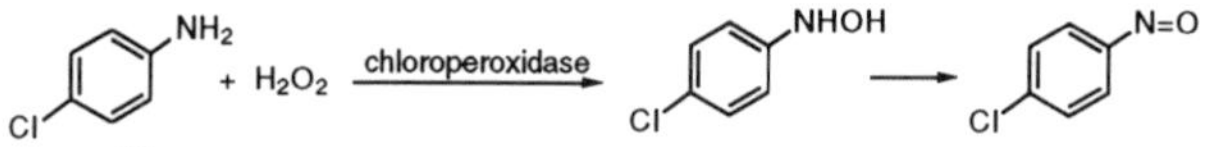

[h]

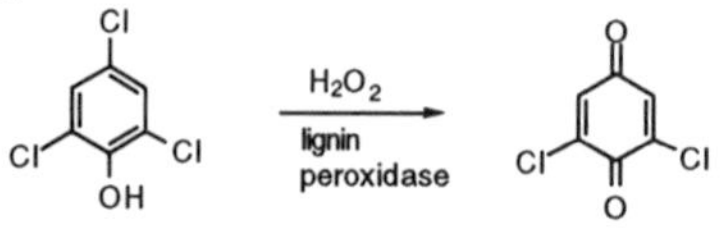

[i]

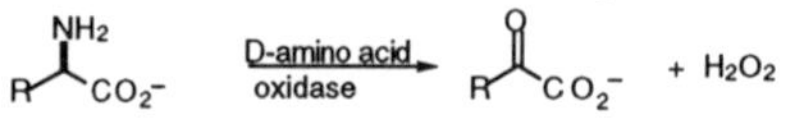

[j]

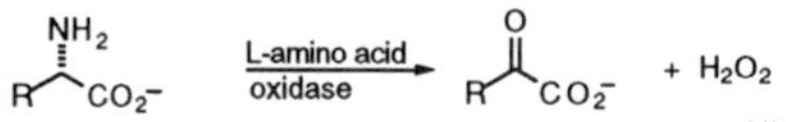

[j]

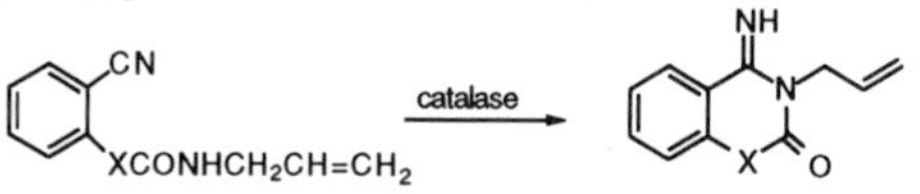

[k]

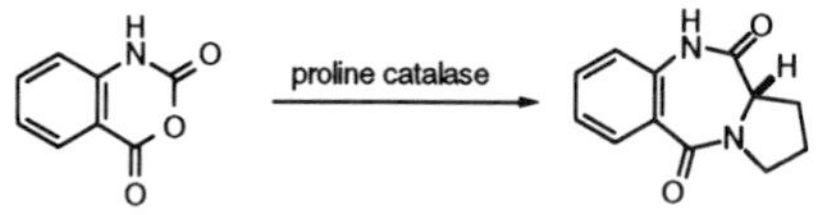

[k]

Aniline   —horse radish peroxidase→   Polyaniline

[l]

---

[e]   Also called polyphenol oxidase (EC 1.14.18.1). Mason, H. S. *Ann. Rev. Biochem.* **1965**, *34*, 594. To avoid polymerization of quinone in water, reactions were carried out in CHCl3. The quinone obtained is stable in the organic solvent and can be reduced nonenzymatically to catechol, Kazandjian, R. Z.; Klibanov, A. M. *J. Am. Chem. Soc.* **1985**, *107*, 5448.

[f   Carried out in microemulsion in the presence of catalase to destroy H2O2 (Lee, K. M.; Biellmann, J. F. *Bioorg. Chem.* **1986**, *14*, 262.) or in supercritical CO2 (Randolph, T. W.; Clark, D. S.; Blanch, H. W.; Prausnitz, J. M. *Science* **1988**, *238*, 387.).

[g]   Buhler, O.; Mason, H. S. *Arch. Biochem. Biophys.* **1961**, *92*, 424. The yield can be improved up to ~ 70% at 0 °C, Klibonov, A. M.; Berman, Z.; Alberti, B. N. *J. Am. Chem. Soc.* **1981**, *103*, 6263.

[h]   Corbett, M. D.; Corbett, B. R. *Experientia* **1983**, *39*, 487.

[i]   Hammel, K.E.; Tardone, P.J. *Biochemistry* **1988**, *27*, 6563.

[j]   Izumi, Y.; Chibata, I.; Itoh, T. *Angew. Chem., Int. Ed. Engl.* **1978**, *17*, 176; *Immobilized Enzymes*, Chibata, I.; Halsted Press, New York, 1978.

[k]   Kamal, A.; Sattur, P. B. *Tetrahedron Lett.* **1989**, *30*, 1133.

[l]   Zemel, H.; American Chemical Soc. National Meeting, Denver, March 1993, *C&E News* **1993**, April 19, p36.

References

1.   Walsh, C. *Enzymatic Reaction Mechanisms*; W.H. Freeman: San Francisco, 1979, p 309.

2.   Bruice, T. C.; Benkovic, S. J. *Bioorganic Mechanisms*; Benjamin: New York, 1965, p 301; Sund, H.; Theorell, M. In *The Enzymes*, 2nd ed.; Boyer, P.D., Lardy, H.; Myrback, K. eds.; Academic: 1963, Vol. VII, p 25.

3.   Chenault, H. K.; Whitesides, G. M. *Appl. Biochem. Biotech.* **1987**, *14*, 147.

4.   Chenault, H. K.; Simon, E. S.; Whitesides, G. M. *Biotech. and Genetic Engineering Rev. 1988*, *6*, 221; Hummel, W.; Kula, M.-R. *Eur. J. Biochem.* **1989**, *184*. 1.

5.   Sakai, T.; Uchida, T.; Chibata, I. *Agric.Biol. Chem.* **1973**, *37*, 1049.

6.   Murata, K.; Kato, J.; Chibata, I. *Biotech. Bioeng.* **1979**, *21*, 887; Hayashi, T.; Tanaka, Y.; Kawashima, K. *Biotech. Bioeng.* **1979**, *21*, 1019.

7.   Lehninger, A. L. *Method. Enzymol.* **1957**, *3*, 885.

8.   Rafter, G. W., Colowick, S. P. *Method. Enzymol.* **1957**, *3*, 887; Suye, S.; Yokoyama, S. Enz. *Microb. Tech.* **1985**, *7*, 418.

9.   Eguchi, S. Y.; Nishio, N.; Nagai, S. *Agric.Biol. Chem.* **1983**, *47*, 2941.

10.   Traub, A.; Kaufman, E.; Teitz, Y. *Anal. Biochem.* **1969**, *28*, 469; Hughes, N. A.; Kenner, G. W.; Todd, A. R. *J. Biol. Chem.* **1950**, *182*, 3733; Kornberg, A. *J. Biol. Chem.* **1950**, *182*, 779.

11.   Walt, D. R.; Findeis, M. A.; Rios-Mercadillo, V. M.; Auge, J.; Whitesides, G. M. *J. Am. Chem. Soc.* **1984**, *106*, 234.

12.   Wong, C.-H.; Whitesides, G. M. *J. Am. Chem. Soc.* **1981**, *103*, 4890.

13.   Oppenheimer, N. J.; Kaplan, N. O. *Biochemistry* **1974**, *13*, 4675.

14.   Johnson, S. L.; Smith, K. W. *Biochemistry* **1976**, *15*, 553; Biellman, J. F.; Lapinte, C.; Haid, E.; Wiemann, G. *Biochemistry* **1979**, *18*, 1212; Everse, J.; Zoll, E. L.; Kahan, L.; Kaplan, N. O. *Bioorg. Chem.* **1971**, *1*, 207.

15.   Shaked, Z.; Whitesides, G. M. *J. Am. Chem. Soc.* **1980**, *102*, 7104.

16.   Wichmann, R.; Wandrey, C.; Buckmann, A. F.; Kula, M.-R. *Biotech. Bioeng.* **1981**, *23*, 2789. For preparation of polyethyleneglycol derivative of NAD(P): Buckmann, A. F. *Biocatalysis* **1987**, *1*, 173; Buckmann, A. F.; Morr, M.; Kula, M. R. *Biotech. Appl. Biochem.* **1987**, *9*, 258; Okuda, K.; Urabe, I.; Okada, H. *Eur. J. Biochem.* **1985**, *151*, 33.

17.   a) Lamed, R.; Keinan, E.; Zeikus, J. G. *Enzyme Microb. Technol.* **1981**, *3*, 144. b) Lemiere, G. L. In *Enzymes as Catalysts in Organic Synthesis*; Schneider, M. P. Ed.; NATO ASI Series; D. Reidel: 1985, p. 19.

18.   a) Shen, G.-J.; Wang, Y.-F.; Wong, C.-H. *J. Chem. Soc., Chem. Commun.* **1990**, *9*, 677; Bradshaw, C. W.; Fu, H.; Shen, G.-J.; Wong, C.-H. *J. Org. Chem.* **1992**, *57*,

1526. b) Bradshaw, C. W.; Hummel, W..; Wong, C.-H. *J. Org. Chem.* **1992**, *57*, 1532.

19.   Wong, C.-H.; Drueckhammer, D. G.; Sweers, H. M. *J. Am. Chem. Soc.* **1985**, *107*, 4028.

20.   Simon, H.; Bader, J.; Gunther, H.; Neumann, S.; Thanos, J. *Angew. Chem. Int. Ed. Engl.* **1985**, *24*, 539.

21.   Chao, S.; Simon, R. A.; Mallouk, T. E.; Wrighton, M. S. *J. Am. Chem. Soc.* **1988**, *110*, 2270; Chao, S.; Wrighton, M. S. *J. Am. Chem. Soc.* **1987**, *109*, 5886.

22.   Gunther, H.; Paxinos, A. S.; Schulz, M.; van Dijk, C.; Simon, H. *Angew. Chem. Int. Ed. Engl.* **1990**, *29*, 1053.

23.   Allen, P. M.; Bowen, W. R. *TIBS* **1985**, *3*, 145.

24.   Mandler, D.; Willner, I. *J. Am. Chem. Soc.* **1984**, *106*, 5352.

25.   Maidan, R.; Willner, I. *J. Am. Chem. Soc.* **1986**, *108*, 1080.

26.   Mandler, D.; Willner, I. *J. Chem. Soc., Chem. Commun.* **1986**, 851.

27.   Klibanov, A. M.; Pugliski, A. V. *Biotech. Lett.* **1980**, *2*, 445.

28.   Danielsson, B.; Winquist, F.; Malpote, J. Y.; Mosbach, K. *Biotech. Lett.* **1982**, *4*, 673.

29.   Payen, B.; Segui, M.; Monsan, P.; Schneider, K.; Friedrich, C. G.; Schlegel, H. G. *Biotech. Lett.* **1983**, *5*, 463.

30.   Wong, C.-H.; Daniels, L.; Orme-Johnson, W. H.; Whitesides, G. M. *J. Am. Chem. Soc.* **1981**, *103*, 6227.

31.   Hilhorst, R.; Laane, C.; Veeger, C. *FEBS* **1983**, *159*, 225; van Berkel-Arts, A.; Dekker, M.; van Dijk, C.; Grande, H. J.; Hagen, W. R.; Hilhorst, R.; Kruse-Wolters, M.; Laane, C.; Veeger, C. *Biochimie* **1986**, *68*, 201.

32.   Gunther, H.; Simon, H. *Appl. Microbiol. Biotechnol.* **1987**, *26*, 9.

33.   Steckhan, E.; Herrmann, S.; Ruppert, R.; Thommes, J.; Wandrey, C. *Angew. Chem. Int. Ed. Engl.* **1990**, *29*, 388.

34.   Abril, O.; Whitesides, G. M. *J. Am. Chem. Soc.* **1982**, *764*, 310.

35.   Kelly, R. M.; Kirwain, D. J. *Biotech. Bioeng.* **1977**, *19*, 1215; Aizawa, M.; Coughlin, R. W.; Charles, M. *Biochim. Biophys. Acta* **1975**, *385*, 362; Jaegfeldt, H.; Torstensson, A. B. C.; Gorton, L. G. C.; Johansson, G. *Anal. Chem.* **1981**, *53*, 1979; Laval, J. M.; Bourdillon, C.; Moiroux, J. *J. Am. Chem. Soc.* **1984**, *106*, 4701; Coughlin, R. W.; Alexander, B. F. *Biotech. Bioeng.* **1975**, *17*, 1379; Jaegfeldt, H. J. *Electroanal. Chem.* **1980**, *110*, 295; Moireux, J.; Elving, P. *J. Anal. Chem.* **1979**, *51*, 346.

36.   Lee, L. G.; Whitesides, G. M. *J. Am. Chem. Soc.* **1985**, *107*, 6999.

37.   Bernofsky, C.; Swan, M. *Anal. Biochem.* **1973**, *53*, 452; Schulman, M. P.; Gupta, N. K.; Omachi, A.; Hoffman, G.; Marshall, W. E. *Anal. Biochem.* **1974**, *60*, 302; Pinder,

S.; Clark, J. B. *Method. Enzymol.* **1971**, *18B*, 20; Legoy, M. D.; Larreta Garde, V.; Le Moullec, J. M.; Ergan, F.; Thomas, D. *Biochimie* **1980**, *62*, 341.

38.   Chambers, R. P.; Ford, J. R.; Allender, J. H.; Baricos, W. H.; Cohen, W. *Enz. Eng.* **1974**, *2*, 195; Julliard, M.; Le Petit, J. *Photochem. Photobiol.* **1982**, *36*, 283.

39.   Jones, J. B.; Taylor, K. E. *Can. J. Chem.* **1976**, *54*, 2969 and 2974.

40.   Wong, C.-H.; McCurry, S. D.; Whitesides, G. M. *J. Am. Chem. Soc.* **1980**, *102*, 7938.

41.   Wong, C.-H.; Pollak, A.; McCurry, S. D.; Sue, J. M.; Knowles, J. R.; Whitesides, G. M. *Methods. Enzymol.* **1982**, *89*, 108.

42.   Lee, L. G.; Whitesides, G. M. *J. Org. Chem.* **1986**, *51*, 25.

43.   Carrea, G.; Bovara, R.; Cremonesi, P.; Lodi, R. *Biotech. Bioeng.* **1984**, *26*, 560; Carrea, G.; Bovara, R.; Longhi, R.; Riva, S. *Enz. Microb. Technol.* **1985**, *1*, 597.

44.   Wong, C.-H.; Matos, J. R. *J. Org. Chem.* **1985**, *50*, 1992.

45.   Wong, C.-H.; Whitesides, G. M. *J. Org. Chem.* **1982**, *47*, 2816.

46.   Drueckhammer, D. G.; Riddle, V. W.; Wong, C.-H. *J. Org. Chem.* **1985**, *50*, 5387.

47.   Chambers, R. P.; Waller, E. M.; Baricos, W. H.; Cohen, W. *Enz. Eng.* **1978**, *3*, 363; Lemiere, G. L.; Lepoivre, J. A.; Alderweireldt, F. C. *Tetrahedron Lett.* **1985**, *26*, 4527; Champbell, J.; Chang, T. M. S. *Enz. Eng.* **1978**, *3*, 371; Kato, T.; Berger, S. J.; Carter, J. A.; Lowry, O. H. *Anal. Biochem.* **1973**, *53*, 86; Lamed, R. J.; Zeikus, J. G. *Biochem. J.* **1981**, *195*, 183; Gwak, S. H.; Ota, Y.; Yagi, O.; Minoda, Y. *J. Ferment. Technol.* **1982**, *60*, 205; Chambers, R. P.; Ford, J. R.; Allender, J. H.; Baricos, W. H.; Cohen, W. *Enz. Eng.* **1974**, *2*, 195; Baricos, W. H.; Chambers, R. P.; Cohen, W. *Anal. Lett.* **1976**, *9*, 257; Hummel, W.; Kula, M.-R. *Eur. J. Biochem.* **1989**, *184*, 1.

48.   Chave, E.; Adamowicz, E.; Burstein, C. *Appl. Biochem. Biotech.* **1982**, *7*, 431; Burstein, C.; Ounsissi, H.; Legoy, M. D.; Gellf, G.; Thomas, D. *Appl. Biochem. Biotech.* **1981**, *6*, 329; Ergan, F.; Thomas, D.; Chang, T. M. S. *Appl. Biochem. Biotech.* **1984**, *10*, 61; Sih, C.; Chen, C.-S. *Angew. Chem. Int. Ed. Engl.* **1984**, *23*, 570; Servi, S. *Synthesis* **1990**, 1; Ward, O. P.; Young, C. S. *Enzym. Microb. Tech.* **1990**, *12*, 482; Csuk, R.; Glanzer, B.I. *Chem. Rev.* **1991**, *91*, 49; Nakamura, K. In *Microbial Reagents in Organic Synthesis*; Servi, S. Ed.; NATO ASI Series, Kluwer Acad.: 1992, p 389.

49.   *Applications of Biochemical Systems in Organic Chemistry*; Jones, J. B.; Sih, C. J.; Perlman, D. eds.; Wiley: New York, N.Y. 1976.

50.   Chibata, I. *Immobilized Enzymes - Research and Development*; Halsted: New York, NY 1978.

51.   Pollak, M.; Blumenfeld, H.; Wax, M.; Baughn, R. L.; Whitesides, G. M. *J. Am. Chem. Soc.* **1980**, *102*, 6324.

52. Klibanov, A. M. *Adv. Appl. Microbiol.* **1983**, *29*, 1; Klibanov, A. M. *Anal. Biochem.* **1978**, *93*, 1.

53. *Method. Enzymol.*; Mosbach, K. Ed.; Academic: New York, NY, 1987, Vols. 135-137.

54. Cremonesi, P.; Carrea, G.; Ferrara, L.; Antonini, E. *Biotech. Bioeng.* **1975**, *17*, 1101.

55. a) Drueckhammer, D. G.; Sadozai, S. K.; Wong, C.-H. *Enz. Microb. Technol.* **1987**, *9*, 564. b) Matos, J. R.; Wong, C.-H. *J. Org. Chem.* **1986**, *51*, 2388.

56. Grunwald, J.; Wirz, B.; Scollar, M. P.; Klibanov, A. M. *J. Am. Chem. Soc.* **1986**, *108*, 6732; Itoh, S.; Terasaka, T.; Matsumiya, M.; Komatsu, M.; Ohshiro, Y. *J. Chem. Soc., Perkin Trans. 1* **1992**, 3253.

57. Prelog, V. *Pure Appl. Chem.* **1968**, *9*, 119.

58. Arnold, L. J.; You, K. S.; Allison, W. S.; Kaplan, N. O. *Biochemistry* **1976**, *15*, 4844.

59. Battersby, A. R. In *Enzymes in Organic Synthesis*; Ciba Foundation Symposium III; Pitman: London, 1985, p 22.

60. Wong, C.-H.; Whitesides, G. M. *J. Am. Chem. Soc.* **1983**, *105*, 5012.

61. Jones, J. B. In *Enzymes in Organic Synthesis*; Ciba Foundation Symposium III; Pitman: London, 1985, p 3.

62. Baici, A.; Luisi, P. L.; Attanasi, O. *J. Mol. Catal.* **1975**, *1*, 223; Shore, J. D.; Theorell, H. *Eur. J. Biochem.* **1967**, *2*, 32; Shore, J. D. *Biochemistry* **1969**, *8*, 1588; Shore, J. D.; Brooks, R. L. *Arch. Biochem. Biophys.* **1971**, *147*, 825; Eklund, H.; Samama, J.-P.; Jones, T. A. *Biochemistry* **1984**, *23*, 5982; Kaplan, N. O.; Ciotti, M. M.; Stolzenbach, F. E. *J. Biol. Chem.* **1956**, *221*, 833; Anderson, B. M.; Kaplan, N. O. *Biol. Chem.* **1959**, *234*, 1226.

63. Schuber, F.; Travo, P.; Pascal, M. *Bioorg. Chem.* **1979**, *8*, 83; Zheng, C.; Phillips, R.S. *J. Chem. Soc., Perkin Trans. 1* **1992**, 1083; Plapp, B.V.; Sogin, D.C.; Dworschack, R.T.; Bohlken, D.P.; Woenckhaus, C.; Jeck, R. *Biochemistry* **1986**, *25*, 5396.

64. Gestrelius, S.; Mansson, M. O.; Mosbach, K. *Eur. J. Biochem.* **1975**, *57*, 529.

65. Keinan, E.; Hafeli, E. K.; Seth, K. K.; Lamed, R. *J. Am. Chem. Soc.* **1986**, *108*, 162.

66. Drum, D. E.; Li, T. K.; Vallee, B. L. *Biochemistry* **1969**, *8*, 3792.

67. Bertini, I.; Gerber, M.; Lanini, G.; Luchinat, C.; Maret, W.; Rawer, S.; Zeppezauer, M. *J. Am. Chem. Soc.* **1984**, *106*, 1826; Maret, W.; Shiemke, A. K.; Wheeler, W. D.; Loehr, T. M.; Sanders-Loehr, J. *J. Am. Chem. Soc.* **1986**, *108*, 6351.

68. Jones, J. B.; Jakovac, I. J. *Can. J. Chem.* **1982**, *160*, 19; Dodds, D. R.; Jones, J. B. *J. Am. Chem. Soc.* **1988**, *110*, 577.

69. Hansch, C.; Bjorkroth, J. P. *J. Org. Chem.* **1986**, *51*, 5461.

70. Lemiere, G. L.; van Osselar, T. A.; Lepoivre, J. A.; Alderweireldt, F. C. *J. Chem. Soc., Perkin Trans. 2* **1982**, 1123.

71. Dutler, H.; Branden, C.-I. *Bioorg. Chem.* **1981**, 101.

72. Eklund, H.; Nordstrom, B.; Zeppezauer, E.; Soderlund, G.; Ohlsson, I.; Boiwe, T.; Soderberg, B. O.; Tapia, O.; Branden, C. I.; Akeson, A. *J. Mol. Biol.* **1976**, 102, 27861; Cedergen-Zeppezauer, E. S.; Andersson, I.; Ottonello, S. *Biochemistry* **1985**, 24, 4000.

73. Matos, J. R.; Smith, M. B.; Wong, C.-H. *Bioorg. Chem.* **1985**, 13, 121.

74. van Elsacker, P. C.; Lemiere, G. L.; Le Poivre, J. A.; Alderweireldt, F. C. *Bioorg. Chem.* **1989**, 17, 28.

75. Larsson, K. M.; Adlercreutz, P.; Mattiasson, B. *Eur. J. Biochem.* **1987**, 166, 157; Lee, K. M.; Biellmann, J. F. *FEBS Lett.* **1987**, 223, 33; Vos, K.; Laane, C.; Van Hoek, A.; Veeger, C.; Visser, A. J. W. G. *Eur. J. Biochem.* **1987**, 169, 275.

76. Jones, J. B. *Tetrahedron* **1986**, 42, 3351.

77. Nakazaki, M.; Chikamatsu, H.; Naemura, K.; Suzuki, T.; Iwasaki, M.; Sasaki, Y.; Fujii, T. *J. Org. Chem.* **1981**, 46, 2726; Nakazaki, M.; Chikamatsu, H.; Naemura, K.; Fujii, T.; Sasaki, Y.; Ao, S. *J. Org. Chem.* **1983**, 48, 4337; Nakazaki, M.; Chikamatsu, H.; Naemura, K.; Sasaki, Y.; Fujii, T. *J. Chem. Soc., Chem. Commun.* **1980**, 626.

78. Jones, J. B.; Schwartz, H. M. *Can. J. Chem.* **1981**, 59, 1574.

79. Pham, V. T.; Phillips, R. S.; Ljungdahl, L. H. *J. Am. Chem. Soc.* **1989**, 111, 1935.

80. a) Ferre, E.; Gil, G.; Barre, M.; Bertrand, M.; Le Petit, J. *Enz. Microb. Tech.* **1986**, 8, 297. b) Fan, F.; Lorenzen, J. A.; Flapp, B. V. *Biochemistry* **1991**, 30, 6397. c) Green, D. W.; Sun, H.-W.; Plapp, B. V. *J. Biol. Chem.* **1993**, 268, 7792.

81. a) Hummel, W. *Appl. Microbiol. Biotech.* **1990**, 34, 15. b) Allemann, R. K.; Hung, R.; Benner, S. A. *J. Am. Chem. Soc.* **1988**, 110, 5555. c) Candida enzyme: Peters, J.; Minuth, T.; Kula, M.-R. *Biocatalysis* **1993**, 8, 31. d) Ammendola, S.; Raia, C. A.; Caruso, C.; Camardella, L.; D'Auria, S.; De Rosa, M.; Rossi, M. *Biochemistry* **1992**, 31, 12514; Persson, B.; Krook, M.; Jornvall, H. *Eur. J. Biochem.* **1991**, 200, 537; Sheehan, M. C.; Bailey, C. J.; Dowds, B. C. A.; McConnell, D. J. *Biochem. J.* **1988**, 252, 661; Gaut, B. S.; Clegg, M. T. *Proc. Natl. Acad. Sci. USA* **1991**, 88, 2060; Hou, C. T.; Patel, R.; Barnabe, N.; Marczak, I. *Eur. J. Biochem.* **1981**, 119, 399.

82. a) Hirschbein, B. L.; Whitesides, G. M. *J. Am. Chem. Soc.* **1982**, 104, 4458. b) Biade, A.E.; Bourdillon, C.; Laval, J.-M.; Mairesse, G.; Moiroux, J. *J. Am. Chem. Soc.* **1992**, 114, 893. c) Hogan, J. K.; Parris, W.; Gold, M.; Jones, J. B. *Bioorg. Chem.* **1992**, 20, 204. d) Kallwass, H. K. W.; Hogan, J. K.; Macfarlane, E. L. A.; Martichonok, V.; Parris, W.; Kay, C.M.; Gold, M.; Jones, J. B. *J. Am. Chem. Soc.* **1992**, 114, 10704.

83. Wilks, H. M.; Halsall, D. J.; Atkinson, T.; Chia, W. N.; Clarke, A. R.; Holbrook, J. J. *Biochemistry* **1990**, 29, 8587.

84. Wilks, H. M.; Hart, K. W.; Freeney, R.; Dunn, C. R.; Muirhead, H.; Chia, W. N.; Barstow, D. A.; Atkinson, T.; Clarke, A. R.; Holbrook, J. J. *Science* **1988**, 242, 1541.

85.   Luyten, M. A.; Bur, D.; Wynn, H.; Parris, W.; Gold, M.; Friesen, J. D.; Jones, J. B. *J. Am. Chem. Soc.* **1989**, *111*, 6800.

86.   a) Simon, E. S.; Plante, R.; Whitesides, G. M. *Appl. Biochem. Biotech.* **1989**, *22*, 169. b) Bradshaw, C. W.; Wong, C.-H.; Hummel, W.; Kula, M.-R. *Bioorg. Chem.* **1991**, *19*, 29. c) Kim, M.-J.; Kim, J.Y. *J. Chem. Soc., Chem. Commun.* **1991**, 326.

87.   a) Hummel, W.; Schutte, H.; Kula, M.-R. *Appl. Microbiol. Biotechnol.* **1984**, *19*, 167 and **1985**, *21*, 7. b) Kallwass, H. K. W. *Enzyme Microb. Technol.* **1992**, *14*, 28.

88.   Struck, J.; Sizer, I. W. *Arch. Biochem. Biophys.* **1960**, *86*, 260.

89.   Vidal-Cros, A.; Gaudry, M.; Marguet, A. *J. Org. Chem.* **1989**, *54*, 498.

90.   Schutte, H.; Hummel, W.; Tsai, H.; Kula, M. R. *Appl. Microbiol. Biotechnol.* **1985**, *22*, 306.

91.   Hummel, W.; Schutte, H.; Schmidt, E.; Wandrey, C.; Kula, M.-R. *Appl. Microbiol. Biotechnol.* **1987**, *26*, 409; Bradshaw, C.; Wong, C.-H.; Hummel, W.; Kula, M.-R. *Bioorg. Chem.* **1991**, *19*, 29.

92.   Takada, H.; Yoshimura, T.; Ohshima, T.; Esaki, N.; Soda, K. *J. Biochem.* **1991**, *109*, 371.

93.   Hummel, W.; Schmidt, E.; Wandrey, C.; Kula, M. R. *Appl. Microbiol. Biotechnol.* **1986**, *25*, 175.

94.   Asano, Y.; Yamada, A.; Kato, Y.; Yamaguchi, K.; Hibino, Y.; Hirai, K.; Kondo, K. *J. Org. Chem.* **1990**, *55*, 5567.

95.   Sakaitani, M.; Rusnak, F.; Quinn, N. R.; Tu, C.; Frigo, T. B.; Berchtold, G. A.; Walsh, C. T. *Biochemistry* **1990**, *29*, 6789.

96.   Yamazaki, Y.; Maeda, H. *Agric. Biol. Chem.* **1986**, *50*, 2621.

97.   Bovara, R.; Canzi, E.; Carrea, G.; Pilotti, A.; Riva, S. *J. Org. Chem.* **1993**, *58*, 499 and references cited.

98.   Lee, L. G.; Whitesides, G. M. *J. Org. Chem.* **1986**, *51*, 25.

99.   Nakamura, K.; Yoneda, T.; Miyai, T.; Ushio, K.; Oka, S.; Ohno, A. *Tetrahedron Lett.* **1988**, *29*, 2453.

100. Rees, L.; Valente, E.; Suckling, C. J.; Wood, H. C. S. *Tetrahedron* **1986**, *42*, 117.

101. a) Kuroda, S.; Tanizawa, K.; Sakamoto, Y.; Tanaka, H.; Soda, K. *Biochemistry* **1990**, *29*, 1009. b) Chambers, R. D.; Jaouhari, R.; O'Hogan, D. *J. Chem. Soc., Chem. Commun.* **1988**, 1169; Adams, P. R.; Harrison, R. *Biochem. J.* **1974**, *141*, 729.

102. Katoh, S.; Sueoka, T. *Biochem. Biophys. Res. Commun.* **1984**, *118*, 859; Sueoka, T.; Katoh, S. *Biochim. Biophys. Acta* **1985**, *843*, 193. Shimizu, S.; Hattori, S.; Hata, H.; Yamada, H. *Eur. J. Biochem.* **1988**, *174*, 37.

103. a) Hochuli, E.; Taylor, K. E.; Dutler, H. *Eur. J. Biochem.* **1977**, *75*, 433; Dutler, H.; VanderBaan, J. L.; Hochuli, E.; Kis, Z.; Taylor, K. E.; Prelog, V. *Eur. J. Biochem.*

1977, *75*, 423. b) Schinschel, C.; Simon, H. *Angew. Chem. Int. Ed. Engl.* **1993**, *32*, 1197.

104. Thanos, I.; Simon, H. *Angew. Chem. Int. Ed. Engl.* **1986**, *25*, 462.

105. Battersby, A. R. *Chem. Ber.* **1984**, *20*, 611.

106. Rambeck, B.; Simon, H. *Angew. Chem. Int. Ed. Engl.* **1974**, *13*, 609.

107. Skopan, H.; Gunther, H.; Simon, H. *Angew. Chem. Int. Ed. Engl.* **1987**, *26*, 128.

108. Leuenberger, H. G. W.; Roguth, W.; Barner, R.; Schmid, M.; Zell, R. *Helv. Chim. Acta* **1979**, *62*, 455.

109. Kwiatkowski, L. D.; Adelman, M.; Pennelly, R.; Kosman, D. J. *J. Inorg. Biochem.* **1981**, *14*, 209.

110. Klibanov, A. M.; Alberti, B. N.; Maletta, M. A. *Biochem. Biophys. Res. Commun.* **1982**, *108*, 804.

111. Root, R. L.; Durrwachter, J. R.; Wong, C.-H. *J. Am. Chem. Soc.* **1985**, *107*, 2997.

112. a) Whittaker, M. M.; Whittaker, J. W. *J. Biol. Chem.* **1988**, *263*, 6074.
    b) Hamilton, G. A.; Adolf, P. K.; de Jersey, J.; DuBois, G. C.; Dyrkacz, G. R.; Libby, R. D. *J. Am. Chem. Soc.* **1978**, *100*, 1899.

113. Ito, N.; Phillips, S. E. V.; Stevens, C.; Ogel, Z. B.; McPherson, M. J.; Keen, J. N.; Yadav, K. D. S.; Knowles, P. F. *Nature* **1991**, *350*, 87; Thomson, A. J. *Nature* **1991**, *350*, 22.

114. Corey, E. J.; Albert, J. O.; Barton, A. E.; Hashimoto, S. I. *J. Am. Chem. Soc.* **1986**, *102*, 1435.

115. Laakso, S. *Lipids* **1982**, *17*, 667; Kwok, P. Y.; Muellner, F. W.; Fried, J. *J. Am. Chem. Soc.* **1987**, *109*, 3692; Lacazio, G.; Langrand, G.; Baratti, J.; Buono, G.; Triantaphylides, C. *J. Org. Chem.* **1990**, *55*, 1690.

116. Corey, E. J.; Lansbury, P. T., Jr. *J. Am. Chem. Soc.* **1983**, *105*, 4093.

117. DeGroot, J. J. M. C.; Veldink, G. A.; Vliegenthart, J. F. G.; Boldingh, J.; Wever, R.; van Gelder, B. F. *Biochim. Biophys. Acta* 1975, 377, 71; Petersson, L.; Slappeudel, S.; Feiters, M. C.; Vliegenthart, J. F. G. *Biochim. Biophys. Acta* **1987**, *913*, 228.

118. Corey, E. J. *Pure Appl. Chem.* **1987**, *59*, 269.

119. Nelson, M. J. *J. Am. Chem. Soc.* **1988**, *110*, 2986.

120. Zhang, P.; Kyler, K. S. *J. Am. Chem. Soc.* **1989**, *111*, 9241; Datcheva, V. K.; Kiss, K.; Solomon, L.; Kyler, K. S. *J. Am. Chem. Soc.* **1991**, *113*, 270.

121. Gibson, D. T.; Hensley, M.; Yoshioka, H.; Mabry, T. J. *Biochemistry* **1970**, *9*, 1626; Gibson, D. T.; Koch, J. R.; Kallio, R. E. *Biochemistry* **1968**, *7*, 2653; Gibson, D. T.; Cardini, G. E.; Maseles, F. C.; Kalio, R. E. *Biochemistry* **1970**, *9*, 1631.

122. Jeffrey, A. M.; Yeh, H. J. C.; Jerina, D. M.; Patel, T. R.; Davey, J. F.; Gibson, D. T. *Biochemistry* **1975**, *14*, 575; McCombie, W. R.; Kwart, L.D.; Gibson, D. T. *J.*

Bacteriol. **1986**, *166*, 1028; Ziffer, H.; Kabuto, K.; Gibson, D. T.; Kobal, M.; Jerina, D. M. *Tetrahedron* **1977**, *33*, 249.

123. Hudlicky, T.; Luna, H.; Price, J. D.; Rulin, F. *Tetrahedron Lett.* **1989**, *30*, 4053; Hudlicky, T.; Luna, H.; Price, J. D.; Rulin, F. *J. Org. Chem.* **1990**, *55*, 4683; Hudlicky, T.; Luna, H.; Barbieri, G.; Kwart, L. D. *J. Am. Chem. Soc.* **1988**, *110*, 4735.

124. Ballard, D. G. H.; Courtis, A.; Shirley, I. M.; Taylor, S. C. *J. Chem. Soc., Chem. Commun.* **1983**, 954.

125. Ley, S. V.; Sternfeld, F.; Taylor, S. *Tetrahedron Lett.* **1987**, *28*, 225; Hudlicky, T.; Price, J. D.; Rulin, F.; Tsunoda, T. *J. Am. Chem. Soc.* **1990**, *112*, 9439; Ley, S. V.; Redgrave, A. J. *Synlett.* **1990**, 393.

126. Carless, H. A. J.; Oak, O. Z. *Tetrahedron Lett.* **1989**, *30*, 1719; Johnson, C. R.; Ple, P. A.; Adams, J. P. *J. Chem. Soc., Chem. Commun.* **1991**, 1006; Johnson, C. R.; Ple, P. A.; Su, L.; Heeg, M. J.; Adams, J. P. *Synlett.* **1992**, 388; Carless, H. A.; Oak, O. Z. *J. Chem. Soc., Chem. Commun.* **1991**, 61.

127. Hudlicky, T.; Luna, H.; Price, J. D.; Rulin, F. *Tetrahedron Lett.* **1989**, *30*, 4053; Hudlicky, T.; Price, J. D. *Synlett.* **1990**, 159; Hudlicky, T.; Price, J. D.; Luna, H.; Andersen, C. M. *Synlett* **1990**, 309; Hudlicky, T.; Luna, H.; Price, J. D.; Rulin, F. *J. Org. Chem.* **1990**, *55*, 4683; Taylor, S. C. Eur. Pat. Appl. EP 76, 606, 1983; Hudlicky, T.; Price, J. D.; Rulin, F.; Tsunoda, T. *J. Am. Chem. Soc.* **1990**, *112*, 9439; Rouden, J.; Hudlicky, T. *J. Chem. Soc., Perkin Trans. 1* **1993**, 1095; Hudlicky, T.; Natchus, M. *J. Org. Chem.* **1992**, *57*, 4740; Hudlicky, T.; Boros, E.E.; Boros, C.H. *Synlett* **1992**, 391.

128. Walsh, C. T.; Chen, Y. C. J. *Angew. Chem. Int. Ed. Engl.* **1988**, *27*, 333.

129. Ryerson, C. C.; Ballou, D. P.; Walsh, C. T. *Biochemistry* **1982**, *21*, 2644.

130. Schwab, J. M.; Li, W. B.; Thomas, L. P. *J. Am. Chem. Soc.* **1983**, *105*, 4800; Donoghue, N. A.; Norris, D. B.; Trudgill, P. W. *Eur. J. Biochem.* **1976**, *63*, 175.

131. a) Latham, J. A.; Walsh, C. T. *J. Chem. Soc., Chem. Commun.* **1986**, 527. b) Branchaud, B. P.; Walsh, C. T. *J. Am. Chem. Soc.* **1985**, *107*, 2153.

132. Taschner, M. J.; Black, D. J. *J. Am. Chem. Soc.* **1988**, *110*, 6892; Taschner, M.J.; Black, D. J.; Chen, Q.-Z. *Tetrahedron: Asymmetry* **1993**, *4*, 1387.

133. Grogan, G.; Roberts, S. M.; Willetts, A. J. *J. Chem. Soc., Chem. Commun.* **1993**, 699.

134. Hayaishi, O. *Molecular Mechanisms of Oxygen Activation*; Academic Press: New York, 1974; Davis, H. G.; Greene, R. H.; Kelly, D. R.; Roberts, S. M. *Biotransformations in Preparative Organic Chemistry*; Academic: 1989, pp 169-219.

135. Ueda, T.; Coon, M. J. *J. Biol. Chem.* **1972**, *247*, 5010; Lode, T.; Coon, M. J. *J. Biol. Chem.* **1971**, *246*, 791; Ruettinger, R. T.; Griffith, G. R.; Coon, M. J. *Arch. Biochim. Biophys.* **1977**, *183*, 528.

136. Colbert, J. E.; Katopodis, A. G.; May, S. W. *J. Am. Chem. Soc.* **1990**, *112*, 3993; Katopodis, A. G.; Wimalasena, K.; Lee, J.; May, S. W. *J. Am. Chem. Soc.* **1984**, *106*, 7928.

137. Ortiz de Montellano, P. R. *Cytochrome P-450*; Plenum: New York, 1986, pp 217-271.

138. Groves, J. T.; McClusky, G. A. *J. Am. Chem. Soc.* **1976**, *98*, 859.

139. Shapiro, S.; Piper, J. U.; Caspi, E. *J. Am. Chem. Soc.* **1982**, *104*, 2301; Collman, J. P.; Brauman, J. I.; Meunier, B.; Raybuck, S. A.; Kodadek, T. *Proc. Natl. Acad. Sci. USA* **1984**, *81*, 3245; Hanzlik, R. P.; Ling, K.-H. J. *J. Org. Chem.* **1990**, *55*, 3992; Traylor, T. G.; Xu, F. *J. Am. Chem. Soc.* **1988**, *110*, 1953; Valentine, J. S.; Burstyn, J. N.; Margerum, L. D. In *Oxygen Activation by Transition Metals* ; Sawyer, D. T., Ed.; Plenum: New York, 1989, pp 175-187.

140. Fu, H.; Shen, G.-J.; Wong, C.-H. *Recueil*, in press.

141. Fu, H.; Newcomb, M.; Wong, C.-H. *J. Am. Chem. Soc.* **1991**, *113*, 5878.

142. Johnstone, S. L.; Phillips, G. T.; Robertson, B. W.; Watts, P. D.; Bertola, M. A.; Koger, H. S.; Marx, A. F. In *Biocatalysis in Organic Media*; Laane, C.; Tramper, J.; Lilly, M. D. eds.; Elsevier: The Netherlands, 1987, p 387.

143. Dalton, H. *Adv. Appl. Microbiol.* **1980**, *26*, 71.

144. Woodland, M. P.; Dalton, H. *J. Biol. Chem.* **1984**, *259*, 53.

145. Green, J.; Dalton, H. *Biochem. J.* **1986**, *236*, 155; Lund, J.; Woodland, M. P.; Dalton, H. *Eur. J. Biochem.* **1985**, *147*, 297.

146. Fox, B. G.; Borneman, J. G.; Wackeh, L. P.; Lipscomb, J. D. *Biochemistry* **1990**, *29*, 6419.

147. a) Ruzicka, F.; Huang, D.-S.; Donnelly, M. I.; Frey, P. A. *Biochemistry* **1990**, *29*, 1696. b) Liu, K. E.; Johnson, C. C.; Newcomb, M.; Lippard, S. J. *J. Am. Chem. Soc.* **1993**, *115*, 939. c) Rosenzweig, A. C.; Frederick, C. A.; Lippard, S. J.; Nordlund, P. *Nature* **1993**, *366*, 537.

148. Abraham, E. P.; Huddlestone, J. A.; Jayatilake, G. S.; Osullivan, J.; White, R. L. in *Recent Advances in the Chemistry of β-Lactam Antibiotics: 2nd International Symposium;*; Gregory, G. I., Ed.; The Royal Society of Chemistry: London, 1981, p. 125.

149. Hollander, I. J.; Shen, Y. Q.; Heim, J.; Demain, A. L.; Wolfe, S. *Science* **1984**, *224*, 610; Peng, C. P.; Chakravarti, B.; TAyatilake, G. S.; Ting, H.-H.; White, R. L.; Baldwin, J. E.; Abraham, E. P. *Biochem. J.* **1984**, *222*, 789; Jensen, S. E.; Leskiw, B. K.; Vining, L. C.; Aharonowitz, Y.; Westlake, W. S. *Can. J. Microbiol.* **1986**, *32*, 953;

Castro, J. M.; Liras, P.; Laiz, L.; Cortes, J.; Martin, J. *J. Microbiol.* **1988**, *134*, 133; Baldwin, J. E.; Gagnon, J.; Ting, H.-H. *FEBS Lett.* **1985**, *188*, 253.

150. Baldwin, J. E.; Killin, S. J.; Pratt, A. J.; Sutherland, J. D.; Turner, N. J.; Crabbe, M. J. C.; Abraham, E. P.; Willis, A. C. *J. Antibiotics* **1987**, *XL(5)*, 652; Ramon, D.; Carramolino, L.; Patino, C.; Sanchez, F.; Penalva, M. A. *Gene* **1987**, *57*, 171; Leskiw, B. K.; Aharonowitz, Y.; Mevarech, M.; Wolfe, S.; Vining, L. C.; Westlake, D. W. S.; Jensen, S. E. *Gene* **1988**, *62*, 187; Cohen, G.; Shiffman, D.; Mevarech, M.; Aharonowitz, Y. *TIBTECH* **1990**, *8*, 105.

151. a) Baldwin, J. E.; Abraham, E. *Nat. Prod. Rep.* **1988**, *5*, 129. b) Huffman, G. W.; Gesellchen, P. D.; Turner, J. R.; Rothenberger, R. B.; Osborne, H. E.; Miller, F. D.; Chapman, J. L.; Queener, S. W. *J. Med. Chem.* **1992**, *35*, 1897. c) Cooper, R. D. G. *Bioorg. Med. Chem.* **1993**, *1*, 1. d) Baldwin, J. E.; Lynch, G. P.; Schofield, C. J. *J. Chem. Soc., Chem. Commun.* **1991**, 736. e) Baldwin, J. E.; Adlington, R. M.; Marquess, D. G.; Pitt, A. R.; Russell, A. T. *J. Chem. Soc., Chem. Commun.* **1991**, 856.

152. Baldwin, J. E.; Adlington, R. M.; King, L. G.; Paris, M. F.; Sobey, W. J.; Sutherland, J. D.; Ting, H. H. *J. Chem. Soc., Chem. Commun.* **1988**, 1635; Baldwin, J. E.; Blackburn, J. M.; Schofield, C. *J. Biocatalysis* **1990**, *3*, 129.

153. a) Krol, W. J.; Basak, A.; Salowe, S. P.; Townsend, C. A. *J. Am. Chem. Soc.* **1989**, *111*, 7625; Salowe, S. P.; Marsh, E. N.; Townsend, C. A. *Biochemistry* **1990**, *29*, 6499. b) Baldwin, J. E.; Adlington, R. M.; Bryans, J. S.; Lloyd, M. D.; Sewell, T. J.; Schofield, C. J.; Baggaley, K. H.; Cassels, R. *J. Chem. Soc.* **1992**, 877.

154. Mansuy, D. *Pure Appl. Chem.* **1987**, *59*, 759; Groves, J. T. *Ann. N.Y. Acad. Sci.* **1986**, *471*, 99; Dolphin, D. *Phil. Trans. R. Soc. London B* **1985**, *311*, 579.

155. Ortiz de Montellano, P. R.; Stearns, R. A. *J. Am. Chem. Soc.* **1987**, *109*, 3415.

156. Newcomb, M.; Manek, M. B.; Glenn, A. G. *J. Am. Chem. Soc.* **1991**, *113*, 949.

157. Holland, H. L. *Chem. Rev.* **1988**, *88*, 473

158. Davis, F. A.; Stringer, O. D.; McCauley, J. P. *Tetrahedron* **1985**, *41*, 4747.

159. Haniu, M.; Armes, L. G.; Tanaka, M.; Yasunobu, K. T. *Biochem. Biophys. Res. Commun.* **1982**, *105*, 889.

160. Poulos, T. L.; Finzel, B. C.; Gunsalus, I. C.; Wagner, G. C.; Kraut, J. *J. Biol. Chem.* **1985**, *260*, 16122.

161. Poulos, T. L.; Finxel, B. C.; Howard, A. J. *Biochemistry* **1986**, *25*, 5314.

162. Unger, B. P.; Gunsalus, I. C.; Sligar, S. G. *J. Biol. Chem.* **1986**, *261*, 1158.

163. Guengerich, F. P.; Macdonald, T. L. *Acc. Chem. Res.* **1984**, *17*, 9.

164. Angelino, S. A. G. F.; Buurman, D. J.; van der Plas, H. C.; Muller, F. *Recueil* **1982**, *101*, 342; Pelsy, G.; Klibanov, A. M. *Biochim. Biophys. Acta* **1983**, *742*, 352.

165. Schreuder, H. A.; Hol, W. G. J.; Drenth, J. *Biochemistry* **1990**, *29*, 3101.

166. Light, D. R.; Maxman, D. J.; Walsh, C. *Biochemistry* **1981**, *21*, 2490; Latham, J. A. Branchaud, B. P.; Chen, Y. C. J.; Walsh, C. T. *J. Chem. Soc., Chem. Commun.* **1986**, 528.

167. a) Colonna, S.; Gaggero, N; Manfredi, A.; Casella, L.; Gullotti, M.; Carrea, G.; Pasta, P. *Biochemistry* **1990**, *29*, 10465; Colonna, S.; Gaggero, N.; Casella, L.; Carrea, G.; Pasta, P. *Tetrahedron: Asymmetry* **1992**, *3*, 95. b) Fu, H.; Kondo, H.; Ichikawa, Y.; Look, G.; Wong, C.-H. *J. Org. Chem.* **1992**, *57*, 7265.

168. Thomas, J. A.; Morris, D. R.; Hager, L. P. *J. Biol. Chem.* **1970**, *245*, 3135.

169. Morrison, M.; Sconbaum, G. R. *Ann. Rev. Biochem.* **1976**, *45*, 861.

170. Neidleman, S. L.; Amon, W. F.; Geigert, J. U.S. Patent 4, 247, 641 91981); Geigert, J.; Neidleman, S. L.; Dalietos, D. J. *J. Biol. Chem.* **1983**, *258*, 2273. Neidleman, S. L.; Diassi, P. A.; Junta, B.; Palmere, R. M.; Pan, S. C. *Tetrahedron Lett.* **1966**, *44*, 5337.

171. Ramakrishnan, K.; Oppenhuizen, M. E.; Saunders, S.; Fisher, J. *Biochemistry* **1983**, *22*, 3271.

172. Neidleman, S. L.; Geigert, J. *Trends Biotech.* **1983**, *1*, 21.

173. a) Oertling, W. A.; Babcock, G. T. *Biochemistry* **1988**, *27*, 3331. b) Ortiz de Montellano, P.R.; Choe, Y.S.; DePillis, G.D.; Catalano, C.E. *J. Biol. Chem.* **1987**, *262*, 11641. c) Libby, R.A.; Thomas, J.A.; Kaiser, L.W.; Hager, L.P. *J. Biol. Chem.* **1982**, *257*, 5030.

174. Varotsis, C.; Babcock, G. T. *Biochemistry* **1990**, *29*, 7357; Spiro, T. G.; Smulevich, G.; Su, C. *Biochemistry* **1990**, *29*, 4497.

175. Doerge, D.R.; Cooray, N.M.; Brewster, M.E. *Biochemistry* **1991**, *30*, 8960.

176. a) Geigert, J.; Lee, T.D.; Dalietos, D.J.; Hirano, D.S.; Neidleman, S.L. *Biochem. Biophys. Res. Commun.* **1986**, *136*, 778. b) Allain, E.J.; Hager, L.P.; Deng, L.; Jacobsen, E.N. *J. Am. Chem. Soc.* **1993**, *115*, 4415. c) Colonna, S.; Gaggero, N.; Casella, L.; Carrea, G.; Pasta, P. *Tetrahedron: Asymmetry* **1993**, *4*, 1325.

177. Fang, G.-H.; Kenigsberg, P.; Axley, M.J.; Nuell, M.; Hager, L.P. *Nucleic Acid Res.* **1986**, *14*, 8061.

178. a) Rao, A.M. *Biotechnol. Bioeng.* **1993**, *41*, 531. b) Pandey, G.; Muralikrishna, C.; Bhalerao, U.T. *Tetrahedron* **1989**, *45*, 6867. c) Pandey, G.; Muralikrishna, C.; Bhalerao, U.T. *Tetrahedron Lett.* **1990**, *31*, 3771.

179. Seip, J.E.; Fager, S.K.; Gavagan, J.E.; Gosser, L.W.; Anton, D.L.; DiCosimo, R. *J. Org. Chem.* **1993**, *58*, 2253.

180. Ryabov, A.D. *Angew. Chem. Int. Ed. Engl.* **1991**, *30*, 931; Bergbreiter, D.E.; Momongan, M. *Appl. Biochem. Biotech.* **1992**, *32*, 55.

181.   Quine, J. A.; Frank, J.; Jongejan, J. A. *Adv. Enzymol.* **1987**, *59*, 169; Geerlof, A.; Van Tol, J.B.A.; Jongejan, J.A.; Duine, J.A. In *Microbial Reagents in Organic Synthesis*; Servi, S., Ed.; NATO ASI Series: 1992, p 411.

182.   Janes, S. M.; Wu, D.; Wemmer, D.; Smith, A. J.; Kaur, S.; Haltby, D.; Durlingame, A. L.; Klinman, J. P. *Science* **1990**, *248*, 981.

Chapter 4. C-C Bond Formation

1. Aldol Condensation

One of the most useful set of chemical methods for C-C bond formation is the catalytic aldol condensation.[1,2] Complimentary to these methods are the use of enzymes called aldolases which catalyze aldol addition reaction. Aldolases, of which over twenty have been identified so far,[3-6] usually catalyze the stereospecific condensation of an aldehyde with a ketone donor. The catalytic activity of these enzymes depends on one of two mechanisms.[7] In type I aldolases, the lysine residue in the active site forms a Schiff base with the donor, which in turn adds stereospecifically to the acceptor. Type I aldolases, which are primarily found in animals and higher plants, do not require a metal cofactor. In contrast, type II aldolases, found predominately in microorganisms, use a Zn^{2+} cofactor which acts as a Lewis acid in the active site. In both cases, the aldolases accept a variety of unnatural substrates, and in many cases, the stereoselectivity of the reaction is highly predictable. The following section describes the catalytic properties and synthetic application of some of the known aldolases.

1.1 FDP (Fructose 1,6-Diphosphate) Aldolase (EC 4.1.2.13)

FDP aldolase catalyzes the reversible aldol addition reaction of dihydroxyacetone phosphate (DHAP) and D-glyceraldehyde 3-phosphate (G3P) to form D-fructose 1,6-diphosphate (FDP) (Figure 1). The equilibrium constant, $K_{eq} \approx 10^4$ M^{-1}, favors FDP formation. Both type I

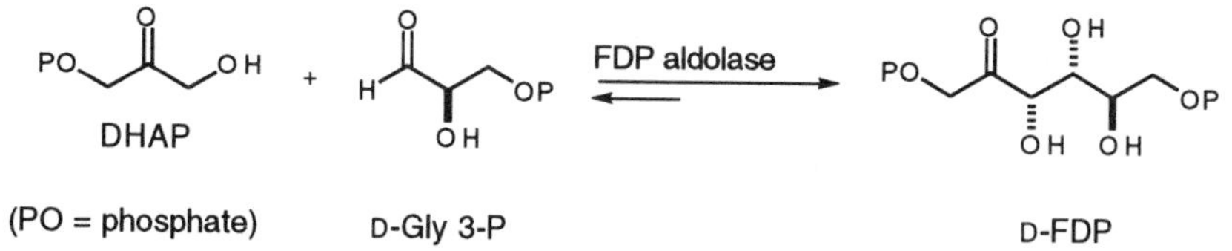

Figure 1. Aldol addition reaction catalyzed *in vivo* by FDP aldolase.

and II enzymes have been isolated from a variety of eukaryotic and prokaryotic sources, which include rabbit muscle,[8,9] human muscle,[10] *E. coli*,[11] *Bacillus stearothermophilus*,[12] *Corynebacterium glutamicum*,[13] *Drosophila melanogaster*,[14] liver,[15] *S. cerevisiae*,[16] spinach,[17] alga,[18] insect,[19] Antartic fishes,[20] other microorganisms,[21] and other animal tissues.[22] Most of the mechanistic studies have been carried out on FDP aldolases from rabbit muscle[23] or yeast,[24] and the X-ray structures of the enzymes from rabbit muscle (2.7 Å resolution),[25] human muscle (3 Å resolution),[26a] and *Drosophila melanogaster*[26b] have been determined and shown to be similar. In rabbit muscle aldolase (RAMA), Lys-229 is recognized as being responsible for

Schiff-base formation with DHAP.[27] The distance between the Schiff-base nitrogen and the phosphate on the G3P in the active site of RAMA has been calculated to be 8.3 Å.[28] Another lysine residue in the active site (Lys_{107}), ca. 8.9 Å from the site of Schiff-base formation,[25] is thought to stabilize the negatively charged phosphate of G3P (Figure 2). The enzyme RAMA has been cloned and overexpressed and site-directed mutagenisis studies have identified Asp-33 as another critical residue in catalysis.[9b]

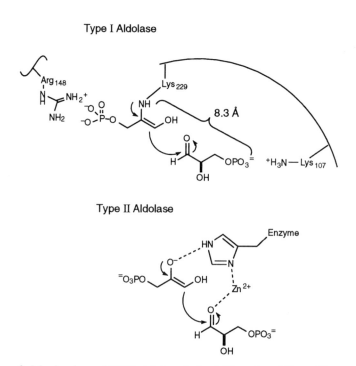

Figure 2. Mechanism of FDP aldolase in both Type I and Type II enzymes.

Generally, the type I FDP aldolases exist as tetramers (M.W. ca. 160 KDa), while the type II enzymes are dimers (M.W. ca. 80 KDa). The sequences of the type I enzymes have a high degree of homology (>50%) with the active site sequence being conserved throughout evolution.[12,25] Significant differences in the *C*-terminal regions have, however, been identified that may be important in mediating substrate specificity.[25] Some of the type I aldolases are commercially available and have useful specific activity (ca. 60 U/mg). These enzymes are not particularly air-sensitive, although there is an active site thiol group, which may oxidize. The half-life of the free enzyme, which is ca. 2 days in aqueous solution at pH 7.0,[29,30] can be lengthened by immobilization or by enclosure in a dialysis membrane. The type II aldolase from yeast is commercially available. Others have recently been cloned and overexpressed from several microbial sources.[13,30–32] The enzyme from *E. coli* has no thiol group in the active site, and has a half life of ca. 60 days in 0.3 mM Zn^{2+} at pH 7.0.[30] Despite the small degree of homology in primary sequence between the enzymes from *E. coli* (type II) and rabbit muscle (type I), studies have shown they possess almost the same substrate specificity.

To date, FDP aldolase, especially RAMA, is the most widely-used aldolase in organic synthesis (Table 1). This enzyme accepts a wide range of aldehyde acceptor substrates with DHAP as the donor to generate $(3S,4R)$ vicinal diols, stereospecifically, with D-*threo* stereochemistry.[28–30,33–59,62,63] Suitable acceptors include unhindered aliphatic aldehydes, α-heteroatom substituted aldehydes,[29] and monosaccharides and their derivatives.[43] Phosphorylated aldehydes react more rapidly than do their unphosphorylated analogs,[43] but aromatic aldehydes, sterically hindered aliphatic and α,β-unsaturated aldehydes are generally not substrates.[29] The specificity for the donor substrate is much more stringent. Only three DHAP analogs have been found to be substrates for RAMA and are all poor (ca. 10% of the activity of DHAP) (Figure 3).[29,44]

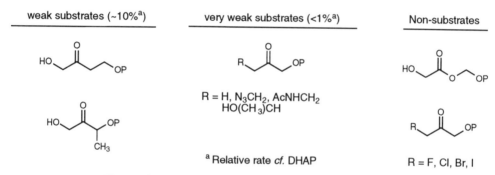

| weak substrates (~10%ᵃ) | very weak substrates (<1%ᵃ) | Non-substrates |

R = H, N₃CH₂, AcNHCH₂
HO(CH₃)CH

ᵃ Relative rate *cf.* DHAP

R = F, Cl, Br, I

Figure 3. Donor substrate specificity of FDP aldolase.

The diastereoselectivity exhibited by FDP aldolase was found to be dependent on the reaction conditions. In a kinetically controlled experiment, the D-enantiomer of G3P was accepted by the enzyme with a 20:1 preference over the L-enantiomer.[29] A lower enantioselectivity was observed for the unphosphorylated aldehyde. Racemic mixtures of unnatural aldehyde acceptors were also partially resolved in a kinetically controlled reaction. In RAMA, the kinetic selectivity is explained by the secondary binding of a charged group with the lysine residue that normally interacts with the phosphate on G3P.[28] When the substrate is constrained by two point binding, nucleophilic *si* face attack onto the *R* aldehyde is faster since the *S* aldehyde has a bulky group on the same face where the nucleophile approaches. A study with a series of carboxylated substrates supports this model (Figure 4).[28] When racemic α-hydroxy aldehyde carboxylates (n = 1-3, 5) were treated with DHAP and RAMA, the substrates that had carboxylates that could best overlap with the second lysine residue gave the highest enantiomeric excess.

Table 1. Products prepared from FDP aldolase-catalyzed reactions with DHAP.

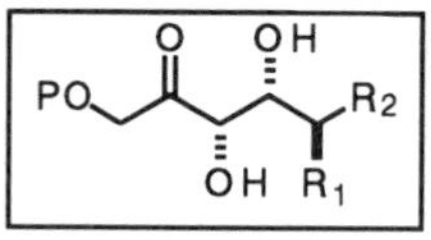

R_1	R_2	Ref.	R_1	R_2	Ref.
HO	$HOCH_2$	[a]			
HO	N_3CH_2	[b,c]			
HO	CH_3	[d,e]	HO	(structure)	[n–q]
N_3	CH_3	[f]			
N_3	CH_2NHAc	[g]			
F	CH_2N_3	[h]	HO	(structure)	[l]
OEt	CH_2N_3	[h]			
HO	(structure)	[i]	H_2N	(structure)	[n]
HO	(structure)	[j]	H	(structure)	[n]
HO	(structure)	[k]			
HO	(structure)	[l]	HO	(structure)	[n]
HO	(structure)	[l,m]	HO	(structure)	[n]

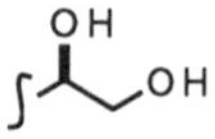

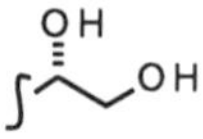

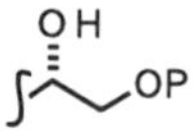

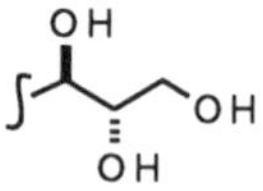

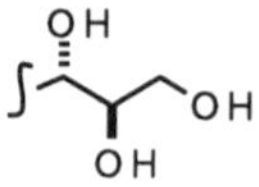

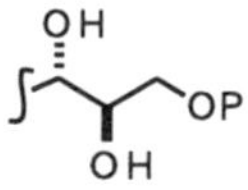

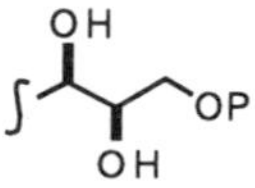

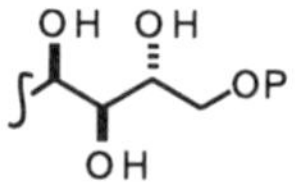

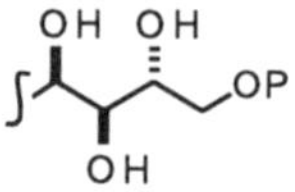

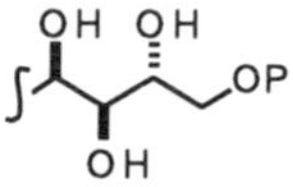

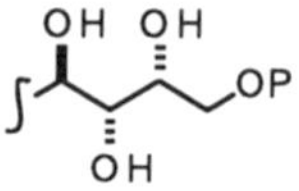

R_1	R_2	Ref.	R_1	R_2	Ref.
HO	N_3CH_2	[b,c]			
HO	CH_3	[e]	HO	(structure)	[n–q]
$CH_2=CHCH_2$	$HOCH_2$	[d]			
CH_3O	$HOCH_2$	[r]			
N_3	CH_3	[f]	HO	(structure)	[l]
N_3	CH_2NHAc	[q]			
HO	(structure)	[s]			
HO	(structure)	[t]	HO	(structure)	[n]

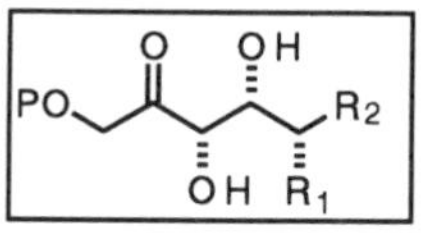

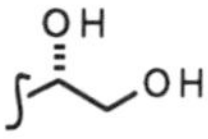

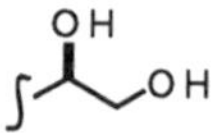

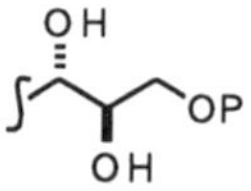

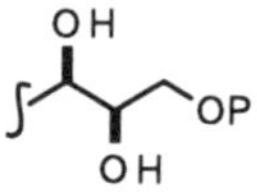

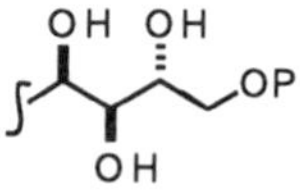

Table 1: continue

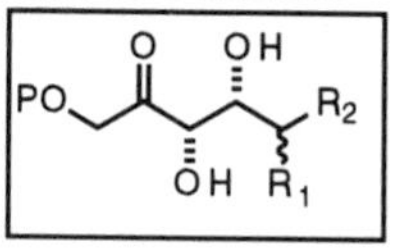

R_1	R_2	Ref.	R_1	R_2	Ref.
H	THPO	[r]	HO	FCH_2	[w]
H	BzO	[r]	HO	CH_3	[y,z]
H	CH_3CH_2	[r,u]	HO	$HOCH_2$	[a,w]
H	CH_3	[r,v]	HO	CH_3CH_2	[e]
H	$(EtO)_2PO$	[u]	HO	N_3CH_2	[b,c,aa]
H	Ph	[u]	BzlO	CH_3	[r]
H	$OHC(CH_2)_3$	[u]	Ph	CH_3	[r]
H	$HOCH_2$	[w]	HO	$MeOCH_2$	[w]
H	CbzNH	[d]	HO	$ClCH_2$	[bb]
H	$CbzNHCH_2$	[d]	H	N_3CH_2	[cc]
H	$CH_3CH(OH)$	[d]	H	$ClCH_2$	[dd]
H	$CF_3CONHCH_2$	[d]	H	$PhSCH_2$	[ee]
H	$(CH_3)_2C(OH)$	[d]	H	$HSCH_2$	[ff]
H	CH_3OCH_2	[v]			
H	$MeO_2CCH(NHAc)$	[x]			

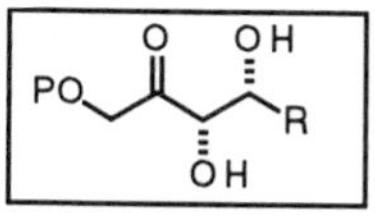

R	Ref.	R	Ref.
CHO	[u]	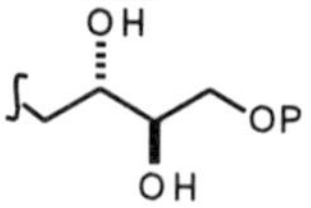	[n]
C(O)OH	[u]	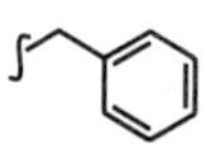	
$HCO(CH_2)_2CH_2$	[u]	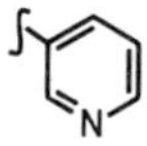	
H	[gg]	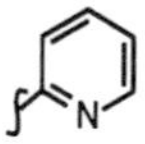	
H_3C	[hh]	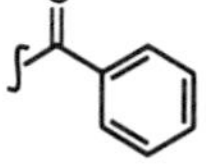	[u]
$POCH_2$	[i]		
	[ii]		[u]
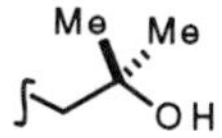	[d]		
	[ii]		[u]
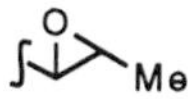	[u]		[u]

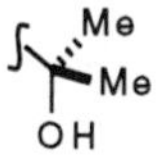

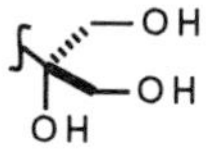

Other Examples

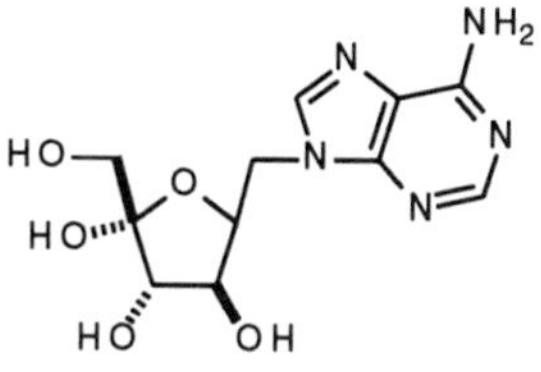

[jj]

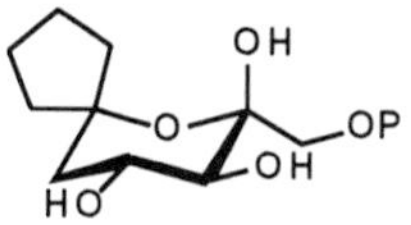

[ii]

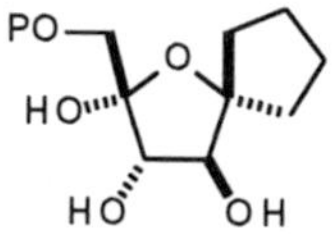

[ii]

Table 1: References

a.　Wong, C.-H.; Whitesides, G. M. *J. Org. Chem.* **1983**, *48*, 3199.

b.　Ziegler, T.; Straub, A.; Effenberger, F. *Angew. Chem. Int. Ed. Engl.* **1988**, *27*, 716.

c.　von der Osten, C. H.; Sinskey, A. J.; Barbas, C. F., III; Pederson, R. L.; Wang, Y.-F.; Wong, C.-H. *J. Am. Chem. Soc.* **1989**, *111*, 3924.

d.　Durrwachter, J. R.; Wong, C.-H. *J. Org. Chem.* **1988**, *53*, 4175.

e.　Wong, C.-H.; Mazenod, F. P.; Whitesides, G. M. *J. Org. Chem.* **1983**, *48*, 3493.

f.　Wang, Y.-F.; Dumas, D. P.; Wong, C.-H. *Tetrahedron. Lett.* **1993**, *34*, 403.

g.　Takaoka, Y.; Kajimoto, T.; Wong, C.-H. *J. Org. Chem.* **1993**, *58*, 4890.

h.　Kajimoto, T.; Liu, K. K.-C.; Pederson, R. L.; Zhong, Z.; Ichikawa, Y.; Porco, J. A., Jr.; Wong; C.-H. *J. Am. Chem. Soc.* **1991**, *113*, 6178.

i.　Jones, J. K. N.; Matheson, N. K. *Can. J. Chem.* **1959**, *37*, 1754.

j.　Horecker, B. L.; Smyrniotis, P. Z. *J. Am. Chem. Soc.* **1952**, *74*, 2123.

k.　Ballou, C. E.; Fischer, H. O. L.; MacDonald, D. L. *J. Am. Chem. Soc.* **1955**, *77*, 5967; Lardy, H. A.; Wiebelhaus, V. D.; Mann, K. M. *J. Biol. Chem.* **1950**, *187*, 325.

l.　Jones, J. K. N.; Sephton, H. H. *Can. J. Chem.* **1960**, *38*, 753.

m.　Haustveit, G. *Carbohydr. Res.* **1976**, *47*, 164.

n.　Bednarski, M. D.; Waldmann, H. J.; Whitesides, G. M. *Tetrahedron Lett.* **1986**, *27*, 5807.

o.　Franke, F. P.; Kapuscinski, M.; MacLeod, J. K.; Williams, J. F. *Carbohydr. Res.* **1984**, *125*, 177.

p.　Mehler, A. H.; Cusic, M. E., Jr. *Science* **1967**, *155*, 1101.

q.　Kapuscinski, M.; Franke, F. P.; Flanigan, I.; MacLeod, J. K.; Williams, J. F. *Carbohydr. Res.* **1985**, *140*, 65.

r.　Bednarski, M.D.; Simon, E. S.; Bischofberger, N.; Fessner, W.-D.; Kim, M. J.; Lees, W.; Saito, T.; Waldmann, H. J.; Whitesides, G. M. *J. Am. Chem. Soc.* **1989**, *111*, 627.

s.　Gorin, P. A. J.; Jones, J. K. N. *J. Chem. Soc.* **1953**, 1537.

t.　Jones, J. K. N.; Kelly, R. B. *Can. J. Chem.* **1956**, *34*, 95.

u.　Effenberger, F.; Straub, A. *Tetrahedron Lett.* **1987**, *28*, 1641.

v.　Lehninger, A. L.; Sice, J. *J. Am. Chem. Soc.* **1955**, *77*, 5343.

w.　Durrwachter, J. R.; Drueckhammer, D. G.; Nozaki, K.; Sweers, H. M.; Wong, C.-H. *J. Am. Chem. Soc.* **1986**, *108*, 7812.

x.　Turner, N. J.; Whitesides, G. M. *J. Am. Chem. Soc.* **1989**, *111*, 624.

y.　Hough, L.; Jones, J. K. N. *J. Chem. Soc.* **1952**, 4052.

z.　Huang, P.; Miller, O. N. *J. Biol. Chem.* **1958**, *330*, 805.

aa.　Pederson, R. L.; Kim, M.-J.; Wong, C.-H. *Tetrahedron Lett.* **1988**, *29*, 4645.

bb.　Liu, K. K.-C.; Pederson, R. L.; Wong, C.-H. *J. Chem. Soc., Perkins Trans. 1* **1991**, 2669.

cc.　Hung, R. R.; Straub, J. A.; Whitesides, G. M. *J. Org. Chem.* **1991**, *56*, 3849.

dd.　Maliakel, B. P.; Schmid, W. *Tetrahedron Lett.* **1992**, *33*, 3297.

ee.　Schmid, W.; Heidlas, J.; Mathais, J. P.; Whitesides, G. M. *Liebigs Ann. Chem.* **1992**, 95.

ff.　Effenberger, F.; Straub, A.; Volker, N. *Liebigs Ann. Chem.* **1992**, 1297.

gg.　Charalampous, F. E. *J. Biol. Chem.* **1954**, *211*, 249.

hh.　Corin, P. A. J.; Hough, L.; Jones, J. K. N. *J. Chem. Soc.* **1953**, 2140.

ii.　Fessner, W.-D.; Walter, C. *Angew. Chem., Int. Ed. Engl.* **1992**, *31*, 614.

jj.　Liu, K .K.-C.; Wong, C.-H. *J. Org. Chem.* **1992**, *57*, 4798.

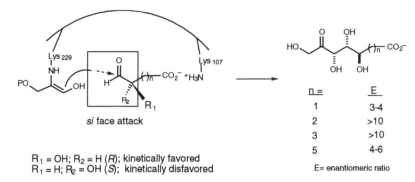

Figure 4. Mechanism of kinetic resolution for FDP aldolase.

In the cases where the product can cyclize to form a six-membered ring hemiketal, racemic aldehydes are resolved if the enzymatic reaction is under thermodynamic control. Since the reaction is reversible, the product with the fewest 1,3-diaxial interactions will predominate after equilibration. For example, with racemic β-hydroxybutyraldehyde[29,36] or 2-hydroxymethyl-4-pentenal[36] as substrates, only a single diastereomer was obtained with the methyl or allyl group in the equatorial position (Figure 5).

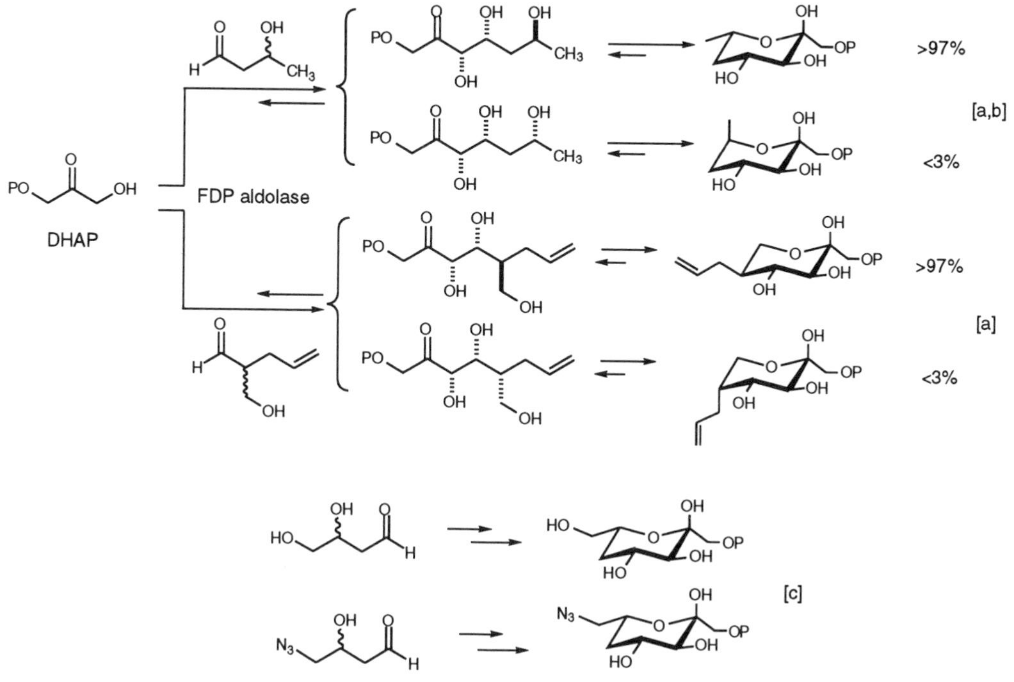

a) Durrwachter, J. R.; Wong, C.-H. *J. Org. Chem.* **1988**, *53*, 4175.
b) Bednarski, M. D.; Simon, E. S.; Bischofberger, N.; Fessner, W.-D.; Kim, M.-J.; Lees, W.; Saito, T.; Waldmann, H.; Whitesides, G. M. *J. Am. Chem. Soc.* **1989**, *111*, 627.
c) Liu, K. K.-C.; Pederson, R. L.; Wong, C. H. *J. Chem. Soc., Perkin Trans 1,* **1991**, 2669.

Figure 5. Thermodynamically-controlled resolution of racemic aldehydes using FDP aldolase.

1.2   Synthesis of Dihydroxyacetone Phosphate (DHAP)

FDP aldolase and three other aldolases mentioned in this chapter require DHAP as the donor substrate. DHAP is available commercially, but due to its cost (>$1000/g), other methods for generating DHAP are used so that enzymatic synthesis can be done on a preparative scale. Most conveniently, DHAP can be generated enzymatically *in situ* from FDP using FDP aldolase and triosephosphate isomerase (TPI).[33,34] FDP aldolase catalyzes the retro-aldol reaction of FDP to give G3P and DHAP. G3P is rapidly isomerized to DHAP with TPI as the catalyst. One drawback to this method is that the thermodynamics of the reaction may favor the formation of FDP rather than the desired product. The overall reaction may not go to completion, and FDP may also interfere with product isolation. Another enzymatic method for DHAP generation employs glycerol kinase to catalyze the phosphorylation of dihydroxyacetone (DHA) using ATP with *in situ* regeneration of ATP.[33,34] Improved procedures for the synthesis of acetylphosphate and phosphoenolpyruvate (PEP) have been developed[45] in which DHAP is synthesized in 83% yield of ca. 87% purity with ATP regeneration. DHAP can also be synthesized non-enzymatically via phosphorylation of dihydroxyacetone dimer with $(PhCH_2O)_2PNEt_2$ followed by subsequent oxidation to the phosphate with H_2O_2 (Figure 6),[46] or with either $POCl_3$.[47a] or $(PhO)_2POCl$.[47b]

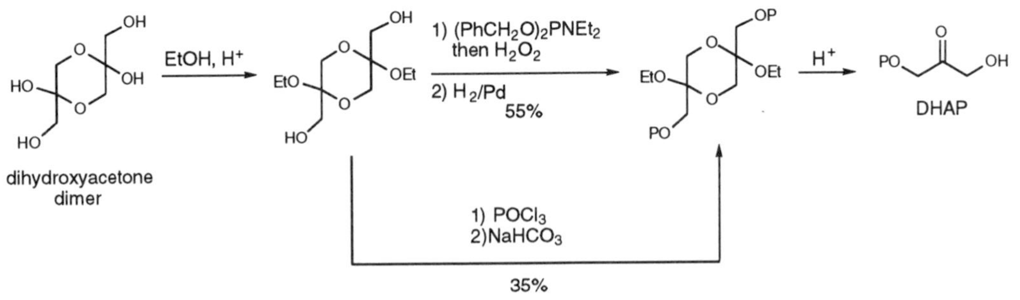

Figure 6. Chemical synthesis of DHAP.

Reductive cleavage of the aromatic rings, or hydrolysis of the chlorophosphate (for the third example), yields a stable dimer precursor of DHAP that can easily be converted into DHAP by acid catalyzed hydrolysis. The overall procedure involving phosphitylation/oxidation gives ca. 55% overall yield of DHAP, compared to ca. 35% using $POCl_3$. A multienzyme system which converts sucrose to DHAP has also been utilized in aldolase reactions.[48]

An alternative to using DHAP employs a mixture of DHA and inorganic arsenate (Figure 7).[35,49] When DHA reversibly reacts with inorganic arsenate (k $\approx 2.4 \times 10^{-3}$ $M^{-1}s^{-1}$) dihydroxyacetone arsenate is formed; this compound is an analog of DHAP and is also a donor substrate for aldolases. In the presence of RAMA, TPI and inorganic arsenate, DHA was converted to D-fructose in almost quantitative yield (Figure 8).[49] The arsenate ester of DHA must also act as a substrate for the isomerase which results in the formation of D-glyceraldehyde. D-Fructose is then generated when the remaining DHA arsenate reacts with D-glyceraldehyde in the aldol reaction. Generally, however, when DHA and inorganic arsenate replace DHAP in aldol additions, precaution should be taken as arsenate is toxic. In addition, reaction rates are very slow and low yields of aldol products are often obtained.

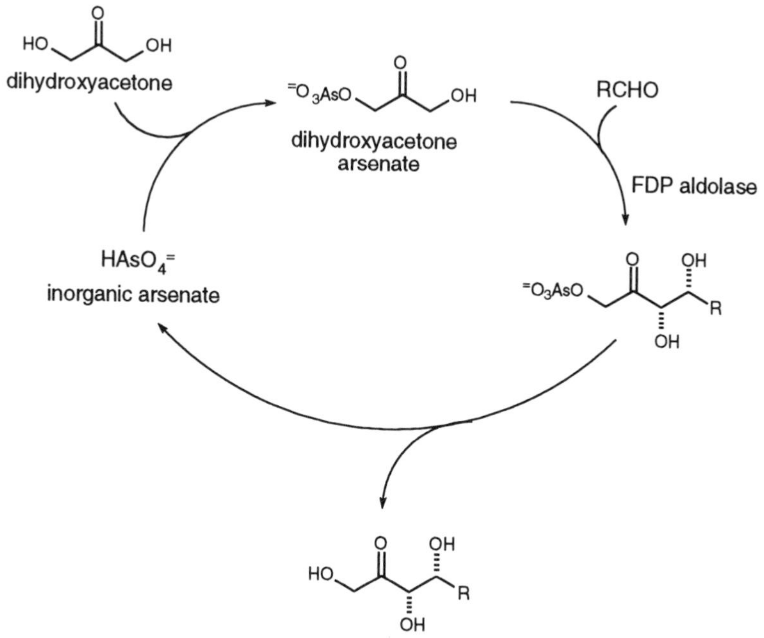

Figure 7. Dihydroxyacetone arsenate as a donor substrate for FDP aldolase.

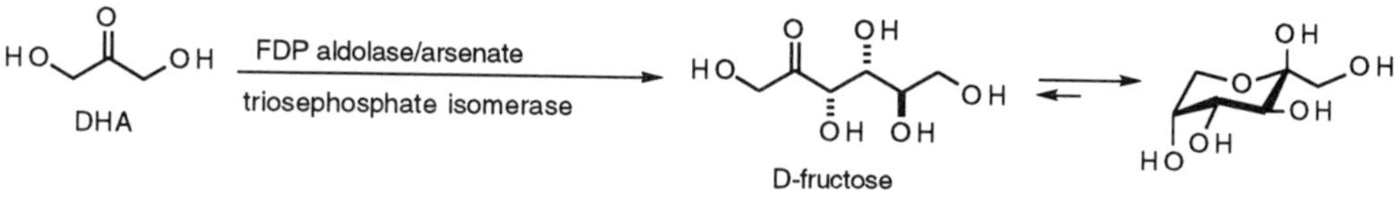

Figure 8. Synthesis of D-fructose from DHA in the presence of RAMA, triose phosphate isomerase and inorganic arsenate.

Inorganic vanadate also reacts with DHA to form the corresponding vanadate ester. This compound cannot be used as a DHAP analog, however, since inorganic vanadate also oxidizes DHA. In other organic phosphate-requiring enzymatic reactions where such oxidation does not occur, inorganic vanadate has been used to generate useful substrates analogs.[49]

1.3   Examples using FDP Aldolase

Numerous examples using FDP aldolase in organic synthesis have appeared recently. These include the production of ^{13}C-labeled sugars,[33,50,51] nitrogen-containing sugars,[30,37–39] deoxysugars,[34,35,36] fluorosugars,[35,52] and 7-, 8- and 9-carbon sugars.[33,34,36,43] The aldehydes necessary for FDP aldolase are used in racemic form, but many non-natural aldehydes are prepared as one enantiomer. In general, two groups are used as "masked" aldehydes: a terminal olefin that can be cleaved by ozonolysis,[44] or an acetal which is deprotected through acid hydrolysis.[30] The α-chiral substituent can be prepared by ring opening by nucleophiles of readily-available (*R*)- and (*S*)-glycidaldehyde acetal, or the corresponding thiirane and aziridine.[53] Both enantiomers of glycidaldehyde acetal are prepared by lipase-catalyzed resolution of 3-chloro-2-hydroxypropanal diethyl acetal.[30,53] Following resolution, the epoxide is simply generated by treating the chloro alcohol with base. Enzymatic synthesis with some of these non-natural aldehydes with FDP aldolase are represented by the examples discussed below.

Aza sugars have become increasingly important targets since they have been found to be inhibitors in glycoprocessing.[54] Two potent glycosidase inhibitors, deoxynojirimycin and deoxymannojirimycin, were readily prepared in three steps with RAMA being used in the key C-C bond forming step.[37,38] Using racemic 3-azido-2-hydroxypropanal and DHAP, diastereomeric 6-azido ketoses was formed. Following the acid phosphatase-catalyzed removal of phosphate and subsequent reductive amination (Figure 9), the products were isolated in a 4:1

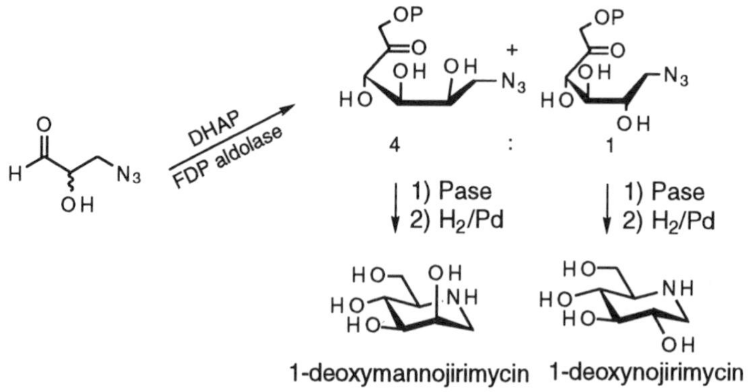

Figure 9. Synthesis of deoxynojirimycin and deoxymannojirimycin from racemic 3-azido-2-hydroxypropanal using FDP aldolase.

ratio favoring the manno-derivative. This observation indicates that the D-aldehyde is a better substrate than L-aldehyde for the enzyme. A similar result was obtained with FDP aldolase from *E. coli* .[30] Deoxynojirimycin and deoxymannojirimycin can also be prepared exclusively by utilizing the respective optically pure azidoaldehydes.[30] Both (*R*)- and (*S*)-3-azido-2-hydroxypropanal were obtained in >98% e.e. by resolution of the racemic acetal precursor catalyzed by LP-80 (Figure 10).[30,53] In a similar strategy several other 3-substituted-2-hydroxy propanal acetals were made that were precursors to a variety of 6-substituted ketoses.[53]

RAMA has also been used in the synthesis of aza sugars corresponding to *N*-acetylglucosamine and *N*-acetylmannosamine, based on the analogous RAMA-catalyzed aldol reaction/reductive amination procedure from (*S*)- and (*R*)-3-azido-2-acetamidopropanal, respectively (Figure 11).[55] Synthesis of the precursor aldehydes necessary for the aldolase reaction started from lipase-resolved (*R*)- and (*S*)-3-azido-2-hydroxypropanal diethyl acetal.[53]

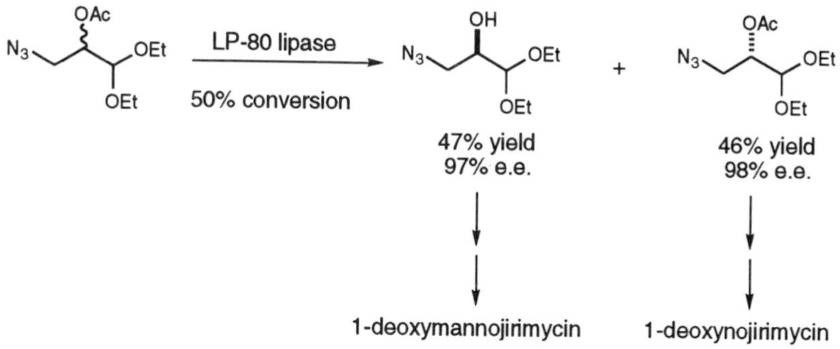

Figure 10. Lipase-catalyzed resolution of 3-azido-2-hydroxypropanal diethyl acetal.

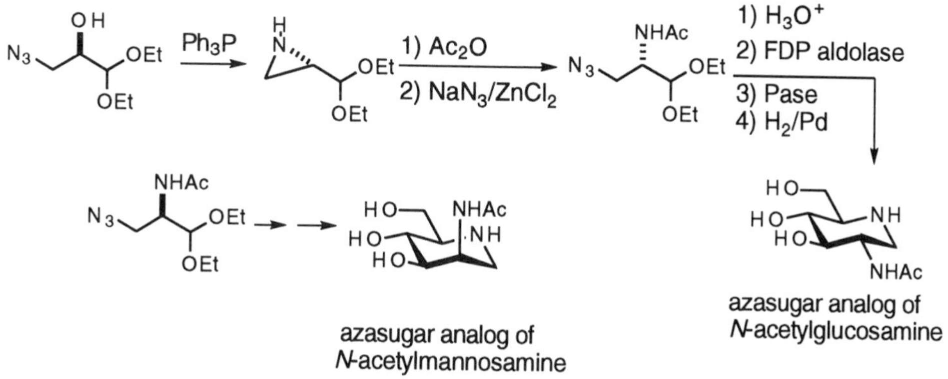

Figure 11. Synthesis of azasugar analogs to *N*-acetylglucosamine and *N*-acetylmannosamine using RAMA.

Reaction with triphenylphosphine formed an aziridine, which was acetylated and regioselectively opened with sodium azide. Hydrolysis of the diethylacetal provided the two chiral aldehydes in high yield and high optical purity.

When 2-azidoaldehydes are used as substrates for the RAMA-catalyzed aldol reaction/reductive amination, a number of polyhydroxylated pyrrolidines were synthesized (Figure 12).[56,57] 1,4-Dideoxy-1,4-imino-D-arabinitol was prepared from Cbz-protected α-aminoacetaldehyde[37] or azidoacetaldehyde,[56] and both (2*R*,5*R*)- and (2*S*,5*R*)-bis(hydroxymethyl)-(3*R*,4*R*)-dihydroxypyrrolidine were synthesized from racemic 2-azido-3-hydroxypropanal, respectively. In the latter case, the kinetic product of the aldol addition was converted to the (2*R*,5*R*) pyrrolidine,[56] while the thermodynamic product gave the (2*S*,5*R*) stereoisomer.[57]

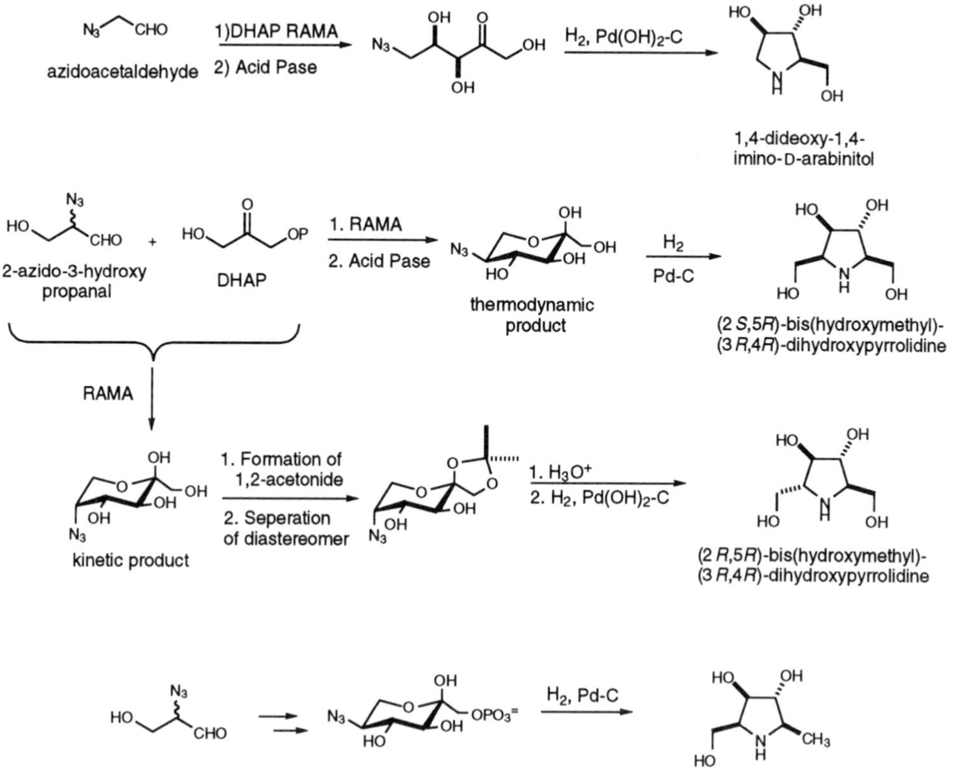

Figure 12. Synthesis of polyhydroxylated pyrrolidines using RAMA.

6-Deoxy aza sugars and their analogs can also be prepared by direct reductive amination of the aldol products prior to removal of the phosphate group (Figure 13).[55] Reduction is thought to proceed through the imine 6-phosphate intermediate instead of reduction of the aza sugar-6-phosphate; glucose 6-phosphate did not react under the same reaction conditions (Figure 13).

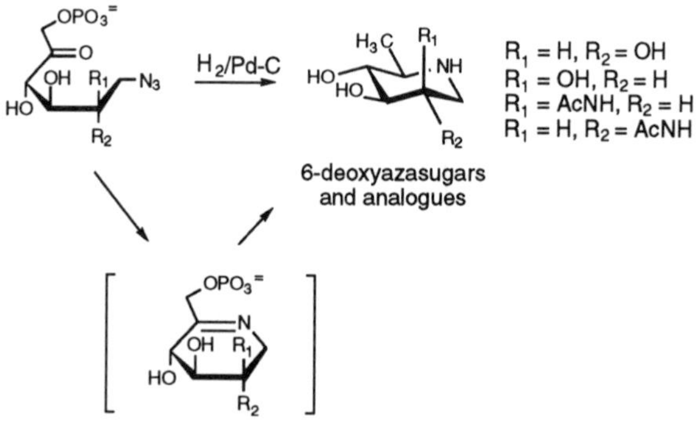

Figure 13. Synthesis of 6-deoxy azasugars and their analogs by direct reductive amination of phosphorylated aldol addition products.

Other aldolases besides FDP aldolase have been used to synthesize a wide range of other polyhydroxylated piperidines and pyrrolidines (*vide infra*).[57,58] Because these aza sugars are so readily prepared, they should be considered as general chiral synthons for the synthesis of other interesting nitrogen heterocycles.

Besides aza sugars, RAMA has been used in the synthesis of a variety of oxygen heterocycles. For example, 3-deoxy-D-arabinoheptulosonic acid (DAHP) is an important intermediate in the shikimate pathway for the biosynthesis of aromatic amino acids in plants and has been synthesized using RAMA(Figure 14).[40] Starting from *N*-acetylaspartate β-semialdehyde

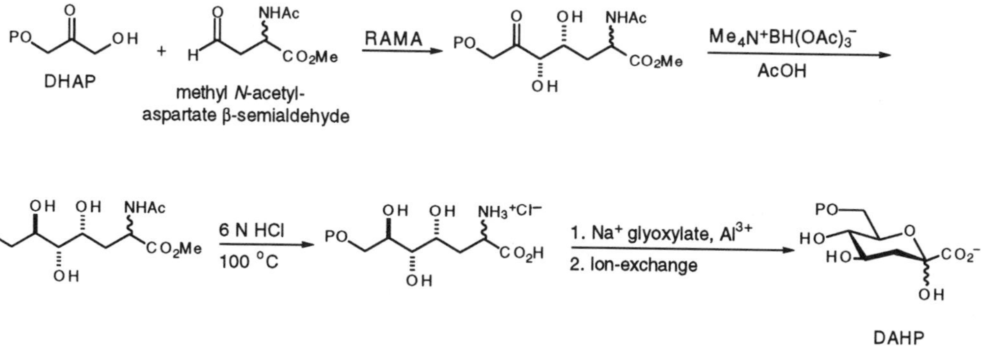

Figure 14. Synthesis of DAHP using RAMA.

as a substrate for RAMA, DAHP was synthesized in four steps in 13% overall yield. The enzymatic step produced the desired D-*threo* stereochemistry, and chemical reduction of the ketone gave the desired (6R)-stereoisomer in 60% diastereomeric excess. Other analogs of DAHP are also potentially available by this route since RAMA utilizes a number of aldehyde acceptors.

Products from the RAMA-catalyzed aldol reaction have also been used to synthesize carbocycles.[42] The protected chloro sugar, generated from the aldolase reaction with chloroacetaldehyde, undergoes nucleophilic attack at the carbonyl carbon with allylmagnesium bromide. Under radical cyclization conditions, a cyclitol is produced(Figure 15). When the Grignard reagent is added in tetrahydrofuran, displacement of the chloride occurs, and gives the *threo*-pentulose-*C*-allylglycoside as the major product (Figure 15). Thus, these two new synthetic routes provide efficient approaches to either cyclitols or *C*-glycosides. In principle, this technique can also be applied to the products of other aldolase-catalyzed reactions to provide stereoisomeric derivatives.

The discovery that RAMA also accepts pentose and hexose phosphates as viable substrates provided a new route to novel high-carbon sugars. Several of these compounds have been synthesized, including analogs of sialic acid and KDO (Figure 16).[43]

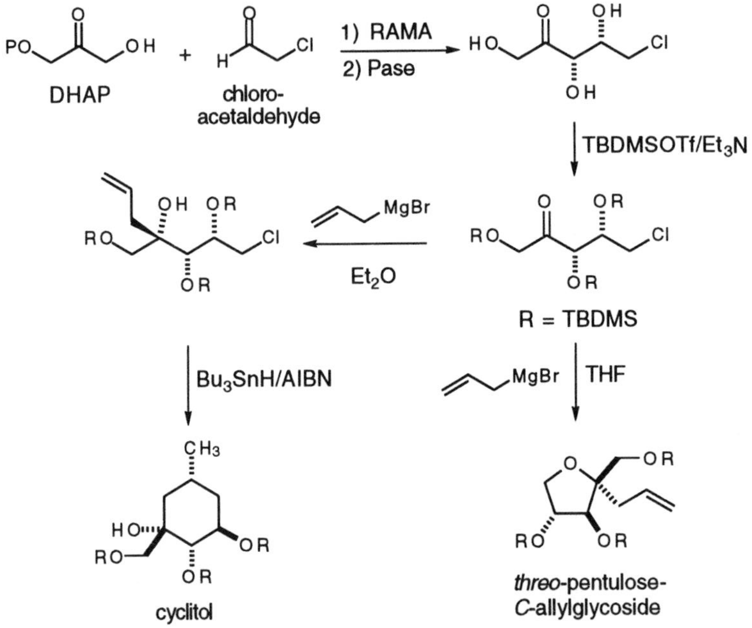

Figure 15. Synthesis of a cyclitol and a C-glycoside using RAMA.

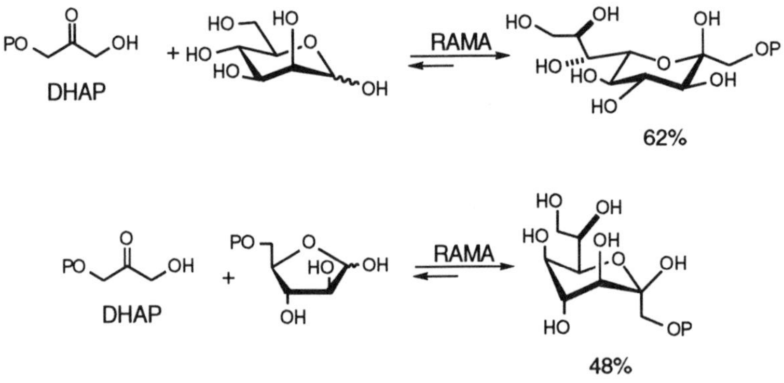

Figure 16. Synthesis of high-carbon sugars using RAMA.

Synthesis with RAMA is not limited to carbohydrate derivatives. In fact, during the synthesis of the naturally occurring beetle pheromone (+)-*exo*-brevicomin the aldolase reaction was used to establish the two chiral centers in the target molecule (Figure 17).[59]

As shown in the many examples above, FDP aldolase generates several types of ketose monosaccharides. Most of the important naturally-occurring carbohydrates are, however, aldoses. One way to generate aldoses from FDP-aldolase products is by using glucose isomerase (GI, or xylose isomerase (EC 5.3.1.5)).[60] This enzyme catalyzes the isomerization of fructose to

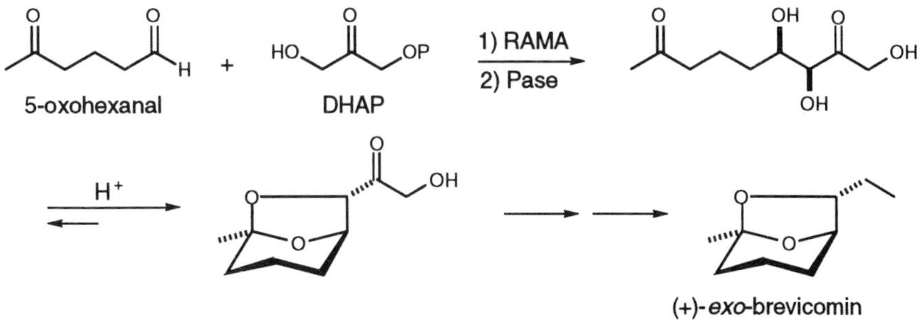

Figure 17. Synthesis of (+)-*exo*-brevicomin using RAMA.

glucose and is used in the food industry for the production of high fructose corn syrup. GI also accepts fructose analogs that are modified at the 3,5 and 6 position (Figure 18).[35] Various FDP aldolase products can be isomerized to a mixture of the ketose and aldose, and subsequently, the two isomers can be separated using Ca^{2+}- or Ba^{2+}-treated cation exchange resins. Aldose analogs including 6-deoxy, 6-fluoro, 6-*O*-methyl and 6-azidoglucose have been synthesized using this FDP aldolase/GI methodology. Not all products from FDP aldolase-catalyzed reaction give the desired aldose when reacted with GI. In some cases, e.g. 5-deoxy-D-fructose, the equilibrium lies completely in favor of the ketose, while in other cases the products from the aldolase reaction are simply not substrates.

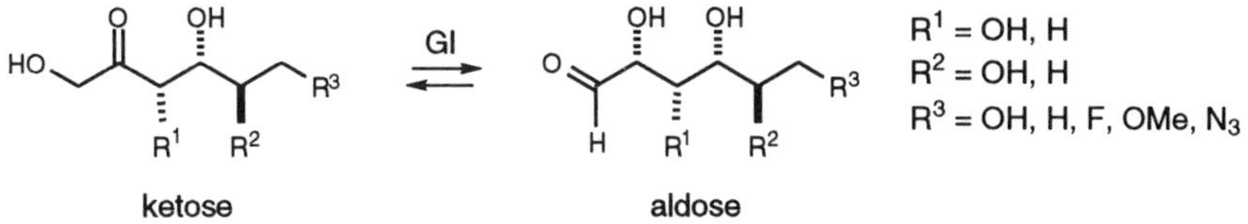

Figure 18. Substrate specificity of GI.

An alternative method for aldose generation from the FDP aldolase products is through the so-called "inversion strategy".[41] Here, monoprotected dialdehydes are used as substrates for FDP aldolase, generating protected aldehyde ketoses. The carbonyl group is then stereoselectively reduced, either chemically or enzymatically with iditol dehydrogenase, and the aldehyde is subsequently deprotected to afford the aldose (Figure 19). The NADH-dependent iditol dehydrogenase (IDH) from *Candida utilis* (also known as sorbitol or polyol dehydrogenase (EC 1.1.14)) was used previously to reduce the carbonyl group of ketoses to give the (*S*)-alcohol.[34,61] The corresponding (*R*)-alcohol was obtained by non-stereoselective reduction of the ketone with $NaBH(OAc)_3$ from which the (*S*)-epimer was selectively removed by oxidation catalyzed by IDH.[41] The synthesis of L-xylose and 2-deoxy-D-*arabino*-hexose (Figure 19) illustrate the two different reduction methods in the "inversion strategy".[41]

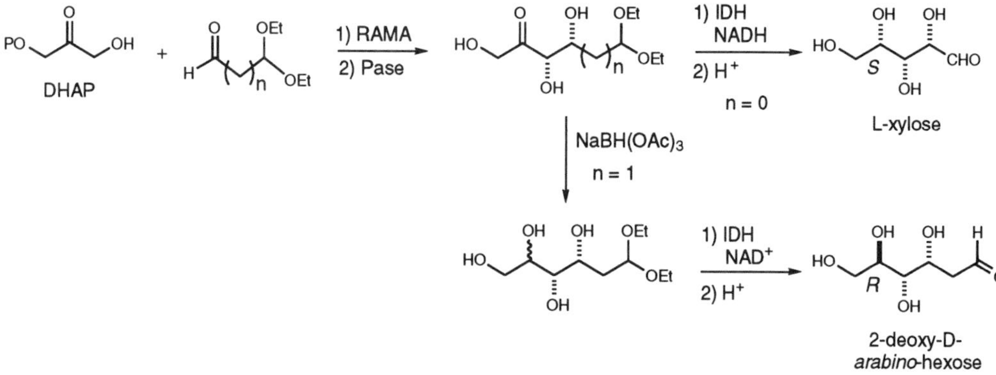

Figure 19. Use of the "inversion strategy" to synthesize L-xylose and
2-deoxy-D-*arabino*-hexose.

While the use of GI and the "inversion strategy" have been developed for the production of aldoses from the products of FDP aldolase-catalyzed reactions, the strategies are also amenable to the products of the other aldolases. Recently, aldose/ketose isomerases with different substrate specificity have been cloned and overexpressed.[62]

While most of the FDP aldolase reactions use the commercially available aldolase from rabbit muscle, there is still considerable interest in determining the utility of FDP aldolase from other sources. Preliminary studies indicate that there are no significant differences in either substrate specificity or stereoselectivity between the various FDP aldolases.[63] Conclusive evidence still does not exist for the type II aldolases which operate by a different mechanism. In fact, the type II aldolase from *E. coli*[30] and yeast,[31] which have been subcloned and overexpressed, have the potential to be useful catalysts for synthesis.

1.4   Fuculose 1-Phosphate (Fuc 1-P) Aldolase (EC 4.1.2.17), Rhamnulose 1-Phosphate (Rha 1-P) Aldolase (EC 4.1.2.19) and Tagatose 1,6-diphosphate (TDP) Aldolase

Besides FDP aldolase, there are three other known aldolases that use DHAP as the donor in the aldol reaction; Fuc 1-P aldolase, Rha 1-P aldolase and TDP aldolase. Fuc 1-P aldolase catalyzes the reversible condensation of DHAP and L-lactaldehyde to give L-Fuc 1-P , and with the same substrates, Rha 1-P aldolase produces L-Rha 1-P (Figure 20). Both of these enzymes are type II aldolases and are found in several microorganisms.[64] Fuc 1-P aldolase[65,66] and Rha 1-P aldolase[66] have been cloned and overexpressed and subsequently purified. Tagatose 1,6-diphosphate (TDP) aldolase, a type I aldolase involved in the galactose metabolism of *cocci*, catalyzes the reversible condensation of G3P with DHAP to give D-TDP (Figure 20).

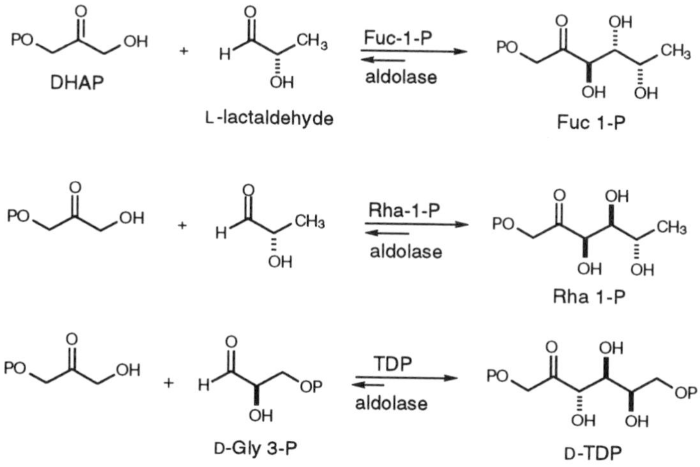

Figure 20. Aldol addition reaction catalyzed *in vivo* by Fuc 1-P aldolase, Rha 1-P aldolase and TDP aldolase.

Both Fuc 1-P and Rha 1-P aldolase accept a variety of aldehydes, and generate vicinal diols with *D-erythro* and *L-threo* configurations, respectively (Figure 21).[64–67] While the enzymes yield products with the (3*R*)-configuration, stereospecifically, the stereoselectivity at C-4 is somewhat diminished for a few substrates.[66] Sterically unhindered 2-hydroxyaldehydes normally give very high diastereoselectivities.

These two aldolases also show significant kinetic preference for the L-enantiomer of 2-hydroxyaldehydes (>95:5), so they facilitate the kinetic resolution of a racemic mixture of these compounds (Figure 22).[62] Both enzymes have been used in the synthesis of rare ketose 1-phosphates[62] and several aza and deoxy aza sugars (Figure 23).[57,68]

Fuc 1-P and Rha 1-P aldolases have also been utilized in whole cell systems with DHA and catalytic inorganic arsenate.[49] With L-lactaldehyde and DHA arsenate as the substrates in the Rha 1-P aldolase reaction, the aldol product L-rhamnulose was subsequently isomerized to L-rhamnose using rhamnose isomerase present in the cell (Figure 24). No such isomerization was observed with L-xylulose when the corresponding aldol product using glycoaldehyde was the substrate. Recent studies have since shown that both rhamnose and fucose isomerase only accept substrates with a (2*R*,3*R*) and (2*S*,3*R*) stereochemistry, respectively.[62]

An alternative method for preparing the same ketose 1-phosphates available from the Rha 1-P and Fuc 1-P aldolases is to use the rhamnose and fucose isomerases (Rha I and Fuc I, Figure 25).[62] The isomerase only partially converts the aldose to the ketose, so a second step which phosphorylated the ketose drives the reaction to completion. Several ketose 1-phosphates were prepared using rhamnulose kinase in the second step to give overall yields of 72–90%.

Aldehyde Substrate	Product	Aldolase	Diastereomeric Ratio *trans* : *cis*	Relative Rate (%)
glycolaldehyde	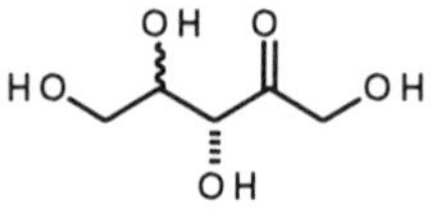	Fuc 1-P	<3 : 97	38
		Rha 1-P	>97 : 3	43
L-lactaldehyde	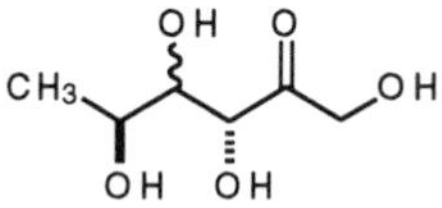	Fuc 1-P	<3 : >97	100
		Rha 1-P	>97 : <3	100
D-Gly	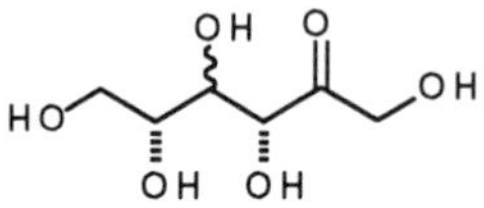	Fuc 1-P	<3 : >97	28
		Rha 1-P	>97 : <3	42
L-Gly	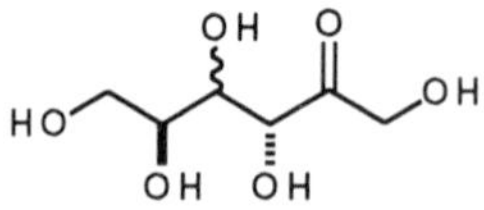	Fuc 1-P	<3 : >97	17
		Rha 1-P	>97 : <3	41
3-hydroxy-propionaldehyde	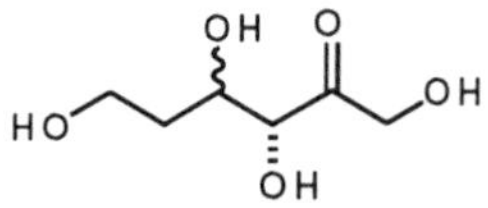	Fuc 1-P	<3 : >97	11
		Rha 1-P	>97 : <3	29
formaldehyde	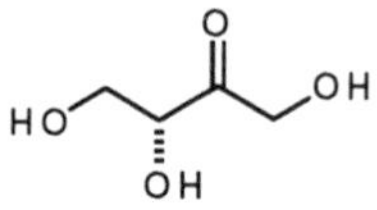	Fuc 1-P	-	44
		Rha 1-P	-	22
acetaldehyde	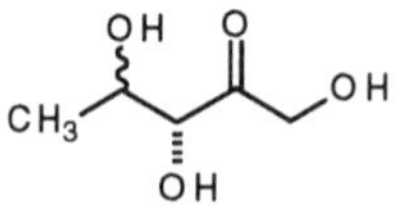	Fuc 1-P	5 : 95	14
		Rha 1-P	69 : 31	32
isobutyraldehyde	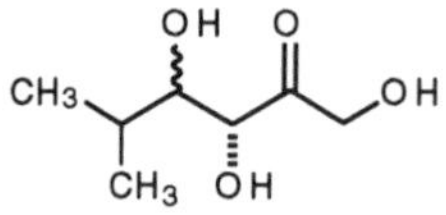	Fuc 1-P	30 : 70	20
		Rha 1-P	97 : 3	22

Figure 21. Acceptor substrate specificity and diastereoselectivity for Fuc 1-P and Rha 1-P aldolases.

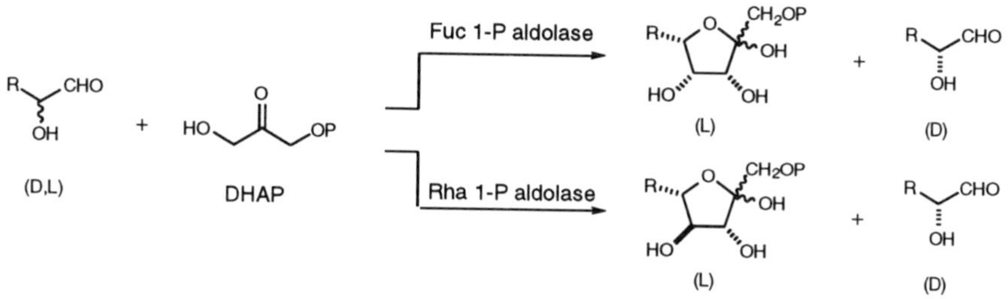

Figure 22. Kinetic resolution of 2-hydroxyaldehydes using Fuc 1-P and Rha 1-P aldolases.

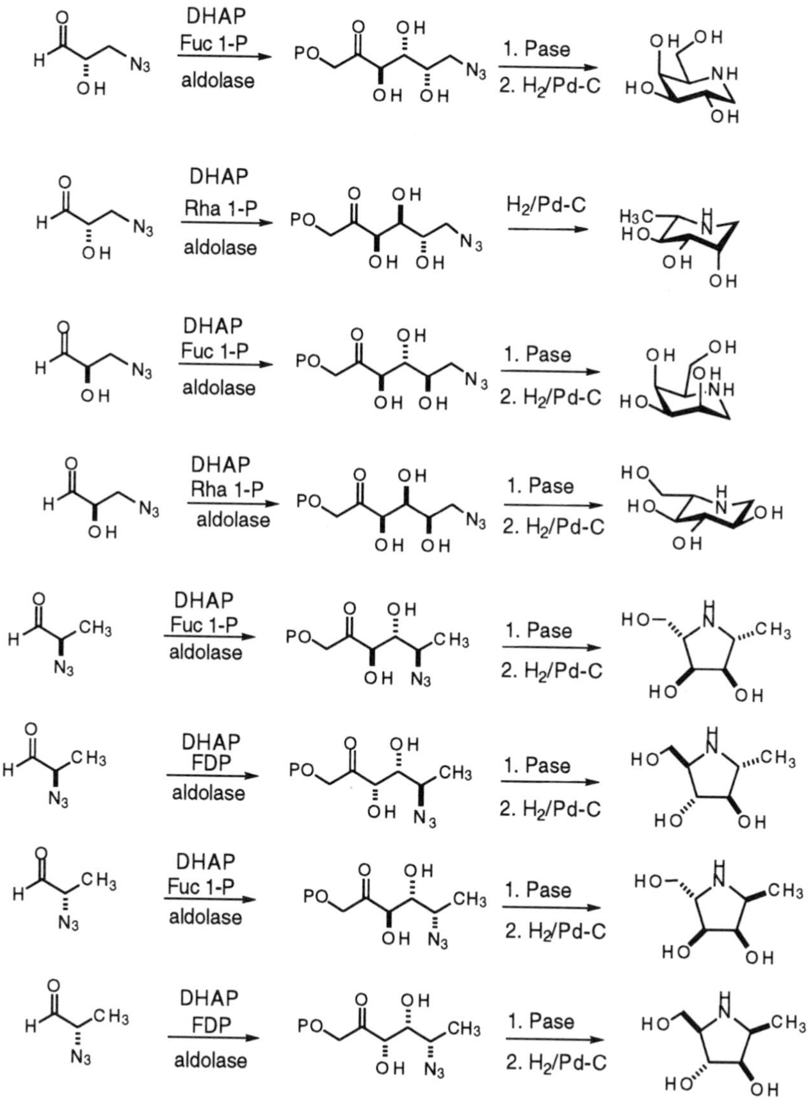

Figure 23. Examples of azasugar synthesis using Fuc 1-P and Rha 1-P aldolases.

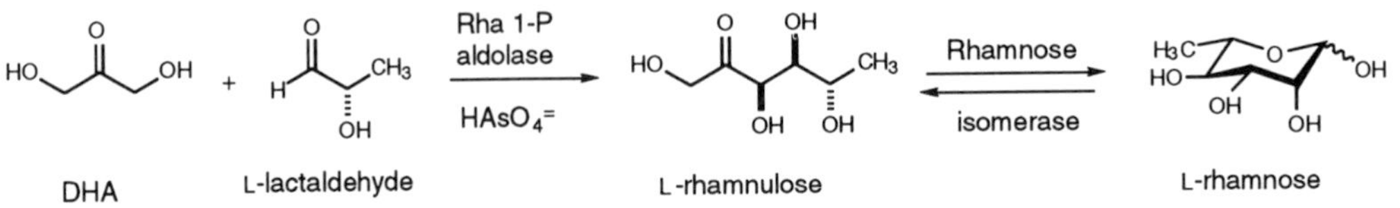

Figure 24. Synthesis of L-rhamnose from L-lactaldehyde and DHA in the presence of inorganic arsenate, Rha 1-P aldolase and rhamnose isomerase.

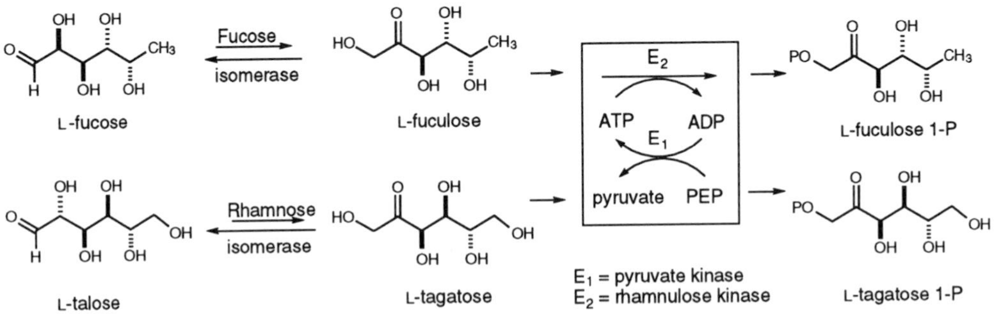

Figure 25. Isomerase/kinase catalyzed synthesis of ketose 1-phosphates.

TDP aldolase has been isolated from several sources.[69] In particular, the enzyme from *E. coli* has a narrow pH profile with an optimum at pH 7.5, but it still displays acceptable activity within pH 6.5-7. Like the other DHAP aldolases, TDP aldolase accepts a variety of acceptor substrates for the aldol reaction, including glycoaldehyde, D- and L-glyceraldehyde, acetaldehyde and isobutyraldehyde.[70] In all cases investigated so far, a diastereomeric mixture of products is formed. Also, instead of exhibiting the expected D-tagatose-like *erythro* configuration, >90% of each product possessed the *threo* configuration similar to D-fructose (Figure 26). Only with the

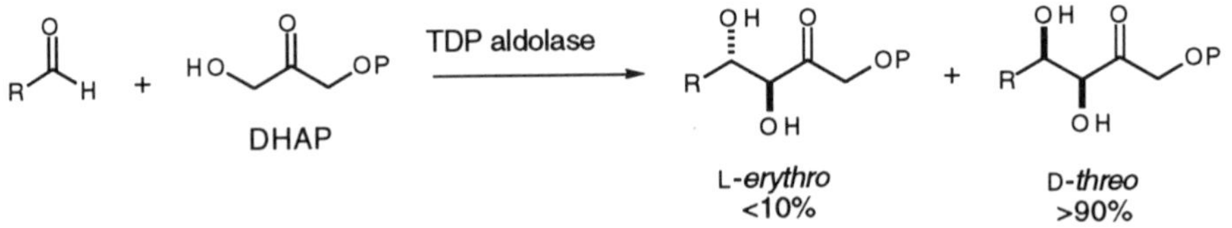

Figure 26. Diastereoselectivity of TDP aldolase.

natural substrate D-glyceraldehyde does the major product (D-TDP) have the tagatose configuration. Due to this lack of stereoselectivity, TDP aldolase is not yet synthetically useful; however, with suitable protein engineering, this situation may change in the future.

From a synthetic point of view, each one of the DHAP dependent aldolases yields a product whose stereochemistry at C3 and C4 is complementary to the other products; i.e., all four stereochemical permutations are provided by these aldolases (Figure 27). FDP, Rha 1-P and Fru 1-P aldolases provide three of the four diastereomers from a variety of non-natural aldehydes. When the stereoselectivity for TDP aldolase is finally improved for other substrates, any one of the C3/C4 stereoisomeric ketoses will be obtainable using these enzymes.

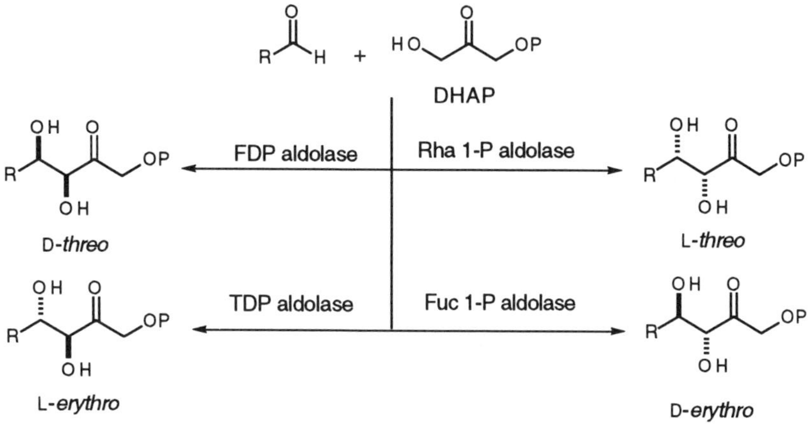

Figure 27. Product stereochemistries generated by the four complementary DHAP aldolases.

1.5 *N*-Acetylneuraminate (NeuAc) Aldolase (EC 4.1.3.3) and NeuAc Synthetase (EC 4.1.3.19)

The enzyme NeuAc aldolase, also known as sialic acid aldolase, catalyzes the reversible condensation of pyruvate with D-*N*-acetylmannosamine (ManNAc) to form *N*-acetyl-5-amino-3,5-dideoxy-D-*glycero*-D-*galacto*-2-nonulosonic acid (NeuAc or sialic acid) (Figure 28).[71,72] Although the β-anomer predominates in solution, the α-anomer of NeuAc is the substrate for the enzyme, and the initial products of aldol cleavage are α-D-ManNAc and pyruvate.[73] The enzyme has a catabolic function, *in vivo*, with an equilibrium constant for the retro-aldol reaction of 12.7 M^{-1}. For synthetic purposes, however, an equilibrium favoring the aldol product can be achieved by using excess pyruvate.[72] NeuAc aldolase has been isolated from both bacteria and animals, and in both cases, it is a Schiff base-forming type I aldolase. The optimum pH for activity is 7.5, but it is still active between pH 6 and 9 and is stable under oxygen.[72,74] The enzymes from *Clostridia* and *E. coli* are now commercially available (Toyobo), and the enzyme from *E. coli* has

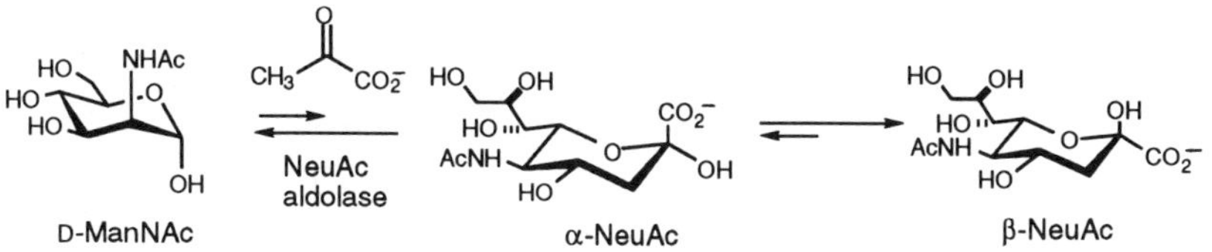

Figure 28. Aldol addition reaction catalyzed *in vivo* by NeuAc aldolase.

been cloned and overexpressed.[75] NeuAc aldolase has been used in both free and immobilized form ,[76–79] and in some instances, it has been enclosed in a dialysis membrane.[80]

Synthetic studies have shown that a high conversion (ca. 90%) of ManNAc to NeuAc was achieved using the isolated enzyme, although several equivalents of pyruvate were required. To eliminate the need for excess pyruvate and to aid in the isolation of products, the NeuAc synthesis can be coupled to a more thermodynamically stable product. One example using ManNAc as the starting material involved the coupling of the NeuAc aldolase reaction with a sialyltransferase reaction to produce sialylsaccharides.[81] Another variant on this process uses a mixture of ManNAc and GlcNAc whereby the GlcNAc was epimerized to ManNAc chemically[82a] or enzymatically[82b] with N-acetylglucosamine 2-epimerase.

Extensive substrate specificity studies were also carried out on this enzyme. Only pyruvate is acceptable as the donor: 3-fluoropyruvate, acetylphosphonate, 3-hydroxybutanoate, 2-oxobutyrate and 3-bromopyruvate are not substrates.[79] In the case of the acceptor substrate the enzyme is more flexible. Substitutions at C-2 -4 and -6 of ManNAc are allowed with only a slight preference that the absolute stereochemistry at C-4, 5 or 6 be the same as the stereochemistry in ManNAc.[79,83–87] Some pentoses and their analogs are also substrates, but two and three carbon molecules are not accepted.

Recently, there has been significant interest in the synthesis of NeuAc analogs due to their role in cellular biology. NeuAc and derivatives are found at the termini of mammalian glycoconjugates and play an important role in biochemical recognition.[88] Polysialic acids are also found in bacteria and mammalian tissues and may be involved in cell adhesion and cell-cell communication.[89] Because NeuAc aldolase accepts a wide variety of substrates and is readily available, many sialic acid derivatives have been prepared that may be used in biological studies (see Figure 29).[51, 79,85,90–95]

Of the many NeuAc derivatives synthesized so far, most give products with *S*-configuration at C-4. Recent observations indicate however, that under thermodynamically-controlled reactions with certain sugars, the stereochemistry at C-4 can become reversed. For example, in the NeuAc aldolase catalyzed synthesis of KDO, a mixture of (*S*)-C-4 and (*R*)-C-4 products were isolated when D-arabinose was the substrate (Figure 30).[86] Also, NMR studies with several other sugars (e.g. L-mannose, D-gulose, 2-azido-2-deoxymannose) showed that over time, products with a C-4 equatorial group predominated in some cases. These examples violate the normal stereochemical preference of the enzyme.[96] Apparently, pyruvate attacks the acceptor sugar to give the thermodynamically more stable product, and the facial selectivity is merely a consequence for the preference to form an C-4 equatorial product. Several biologically interesting L-sugars were synthesized using this method including L-NeuAc, L-KDO and L-KDN (Figure 31). The excess pyruvate can be decomposed with pyruvate decarboxylase to simplify the product isolation.[96b]

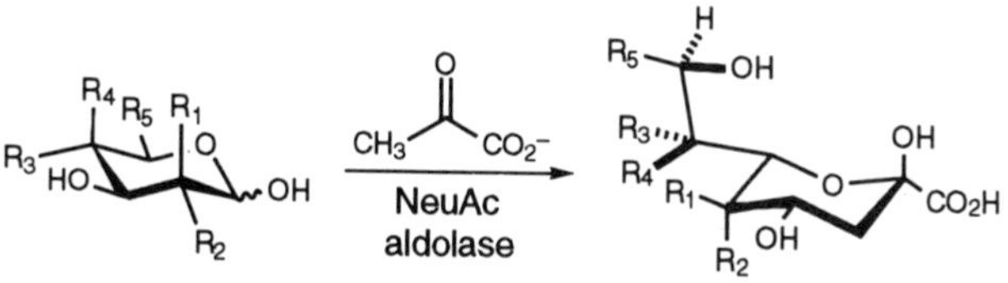

R_1	R_2	R_3	R_4	R_5	Rel rate [c,d]	Ref
AcNH	H	OH	H	CH_2OH	1	[a,b,c]
AcNH	H	OH	H	CH_2OAc	0.2	[a,c]
AcNH	H	OH	H	CH_2N_3	0.6	[d,e]
AcNH	H	OH	H	CH_2F	0.6	[d]
OH	H	OH	H	CH_2OH	2	[f]
OH	H	H	H	CH_2OH	-	[g]
OH	H	H	F	CH_2F	-	[d]
OH	H	OH	H	H	0.1	[h]
H	H	OH	H	CH_2OH	1.3	[h]
H	OH	OH	H	CH_2OH	0.07	[g]
Ph	H	OH	H	CH_2OH	-	[h]
AcNH	H	OH	H	CH_2OCH_3	-	[i]
AcNH	H	OH	H	$CH_2OCOCHOHCH_3$	-	[i]
N_3	H	OH	H	CH_2OH	-	[i]
AcNH	H	OH	H	$CH_2OP(O)Me_2$	-	[d]

a. Augé, C.; David, S.; Gautheron, C. *Tetrahedron Lett.* **1984**, *25*, 4663.
b. Bednarski, M.D.; Chenault, H.K.; Simon, E.S.; Whitesides, G.M. *J. Am. Chem. Soc.* **1987**, *109*, 1283.
c. Kim, M.J.; Hennen, W.J.; Sweers, H.M.; Wong, C.-H. *J. Am. Chem. Soc.* **1988**, *110*, 6481.
d. Liu, J. L.-C.; Shen, G.-J.; Ichikawa, Y.; Rutan, J. R.; Zapata, C.; Vann, W. F.; Wong, C.-H.
  J. Am. Chem. Soc. **1992**, *114*, 3901.
e. Brossmer, R.; Rose, U.; Kasper, D.; Smith, T.L.; Grasmuk, H.; Unger, F.M. *Biochem. Biophys. Res.*
  Comm. **1980**, *96*, 1282.
f. Augé, C.; Gautheron, C. *J. Chem. Soc., Chem. Commun.* **1987**, 859.
g. Augé, C.; Bouxom, B.; Cavaye, B.; Gautheron, C. *Tetrahedron Lett.* **1989**, *30*, 2217; Schrell, A.;
  Whitesides, G.M. *Liebigs Ann. Chem.* **1990**, 111.
h. Augé, C.; Gautheron, C.; David, S. *Tetrahedron* **1990**, *46*, 201.
i. Augé, C.; David, S.; Gautheron, C.; Malleron, A.; Cavaye, B. *New. J. Chem.* **1988**, *12*, 733.

Figure 29. NeuAc analogs synthesized using NeuAc aldolase.

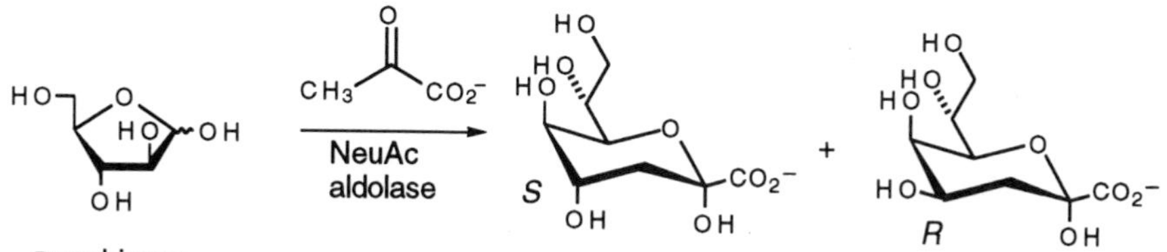

Figure 30. Unusual thermodynamic control of stereoselectivity in aldol addition
reactions catalyzed by NeuAc aldolase.

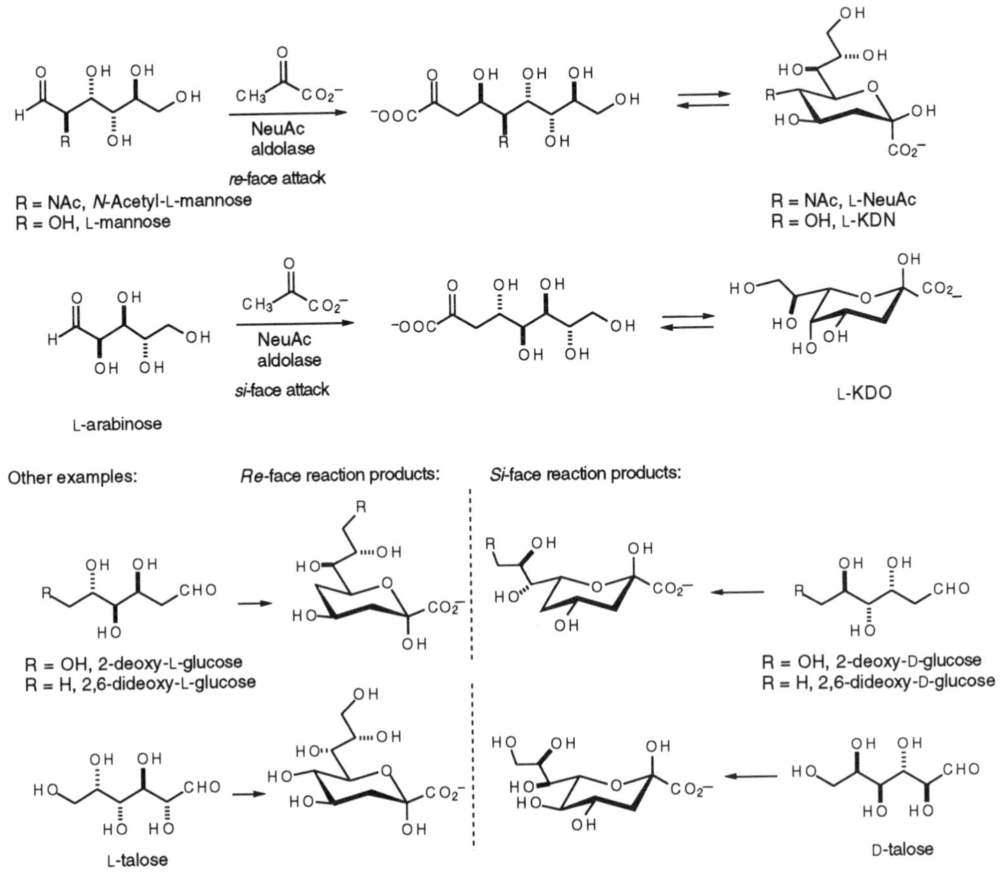
NeuAc
aldolase
re-face attack
R = NAc, N-Acetyl-L-mannose
R = OH, L-mannose
R = NAc, L-NeuAc
R = OH, L-KDN
NeuAc
aldolase
si-face attack
L-arabinose
L-KDO
Other examples:
Re-face reaction products:
Si-face reaction products:
R = OH, 2-deoxy-L-glucose
R = H, 2,6-dideoxy-L-glucose
R = OH, 2-deoxy-D-glucose
R = H, 2,6-dideoxy-D-glucose
L-talose
D-talose

Figure 31. Synthesis of L-NeuAc, L-KDN, and L-KDO using NeuAc aldolase.

A facile synthesis of 9-*O*-acetylNeuAc[77] was improved by regioselective irreversible acetylation of ManNAc catalyzed by subtilisin followed by NeuAc aldolase-catalyzed condensation of the resulting 6-*O*-acetylManNAc with pyruvate (Figure 32).[79] This two-step enzymatic synthesis provided 9-*O*-acetyl-NeuAc in ca. 80% yield. Some other 9-*O*-acylated NeuAc derivatives may also be prepared in this fashion.[79]

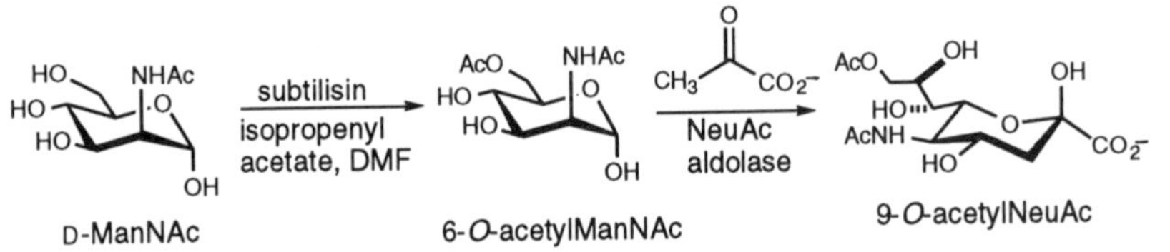
subtilisin
isopropenyl
acetate, DMF
NeuAc
aldolase
D-ManNAc
6-O-acetylManNAc
9-O-acetylNeuAc

Figure 32. Synthesis of 9-*O*-acetyl-NeuAc by combined use of subtilisin and NeuAc aldolase.

Some other noteworthy examples include the synthesis of the α-methyl ketoside of an *N*-unprotected NeuAc for use in preparation of *N*-substituted NeuAc.[85] Also, polyacrylamides bearing pendant α-sialoside[93] groups and polymers of *C*-linked sialosides[94] were prepared which strongly inhibit agglutination of erythrocytes by influenza virus. Finally, a NeuAc aldolase-catalyzed addition of pyruvate to *N*-Cbz-D-mannosamine, followed by a reductive amination of the aldolase product gives an aza sugar that is a precursor to 3-(hydroxymethyl)-6-epicastanospermine (Figure 33).[97a] Another five-membered ring aza sugar has recently been prepared based on sialic acid aldolase (Figure 33).[97b]

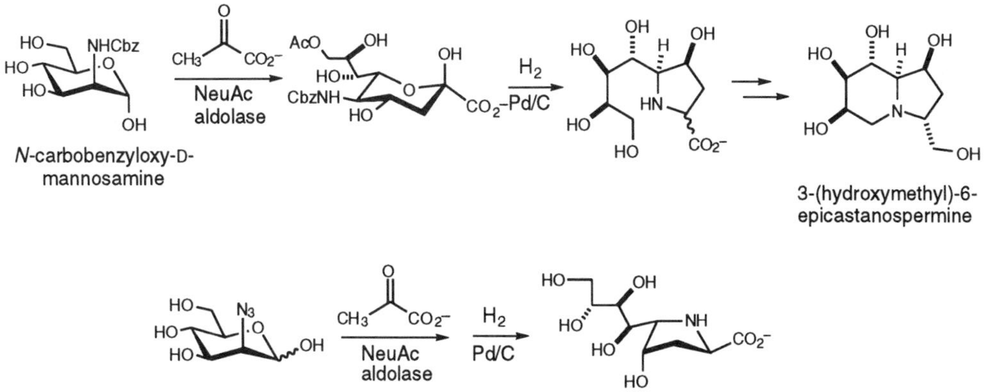

Figure 33. Synthesis of azasugars using NeuAc aldolase.

The synthesis of NeuAc *in vivo* is accomplished using NeuAc synthetase,[74] which catalyzes the irreversible condensation of PEP and *N*-acetylmannosamine. Although this enzyme has not yet been isolated and characterized, it may prove synthetically useful in the future since the forward reaction is favored kinetically.

1.6   3-Deoxy-D-*manno*-2-octulosonate Aldolase (EC 4.1.2.23) and 3-Deoxy-D-*manno*-2-octulosonate 8-Phosphate Synthetase (EC 4.1.2.16)

3-Deoxy-D-*manno*-2-octulosonate aldolase, also known as 2-keto-3-deoxyoctanoate (KDO) aldolase, catalyzes the reversible condensation of pyruvate with D-arabinose to form KDO with an equilibrium constant in the cleavage direction of 0.77 M^{-1} (Figure 34). KDO and its activated form CMP-KDO are key intermediates in the synthesis of the outer membrane lipopolysaccharide (LPS) of gram-negative bacteria[98] so analogs of KDO may inhibit LPS biosynthesis or LPS binding protein.[99–104] KDO aldolase has been isolated and purified from *E. coli*[105] and *Aerobacter cloacae*.[106] Preliminary investigations of this enzyme showed high

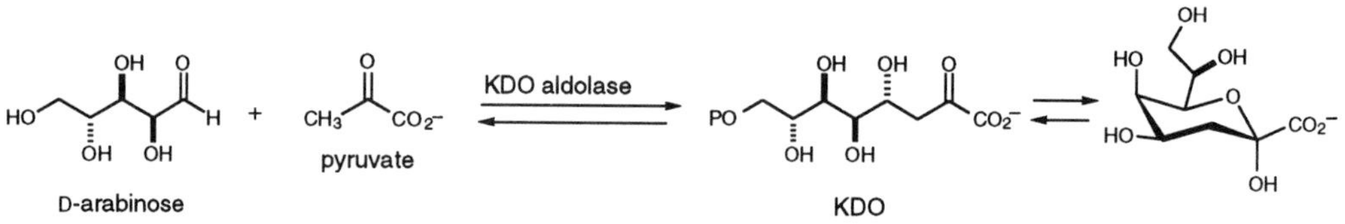

Figure 34. Aldol addition reaction catalyzed *in vivo* by KDO aldolase.

specificity for KDO in the direction of cleavage, whereas the condensation reaction proceeded with some flexibility; several unnatural substrates, including D-ribose, D-xylose, D-lyxose, L-arabinose, D-arabinose 5-phosphate and *N*-acetylmannosamine were reported to be weak substrates (relative rate <5% *cf.* D-arabinose).[105] More recent studies on the substrate specificity showed that the KDO aldolase from *Aureobacterium barkerei*, strain KDO-37-2, accepted an even wider variety of substrates, which included trioses, tetroses, pentoses and hexoses as substrates.[107] The enzyme is specific for substrates having (*R*)-configuration at C-3, but the stereochemical requirements at C-2 are less stringent. Under kinetic control, the C-2 *S* configuration is favored while the C-2 *R* configuration is thermodynamically favored (Figure 35).

R₁ = OH; R₂ = H (*S*); kinetically favored
R₁ = H; R₂ = OH (*R*); thermodynamically favored

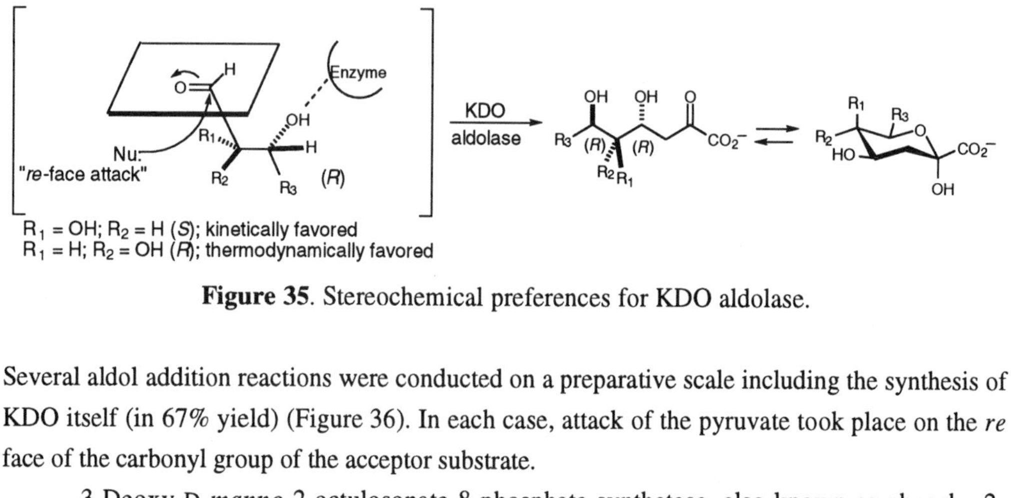

Figure 35. Stereochemical preferences for KDO aldolase.

Several aldol addition reactions were conducted on a preparative scale including the synthesis of KDO itself (in 67% yield) (Figure 36). In each case, attack of the pyruvate took place on the *re* face of the carbonyl group of the acceptor substrate.

3-Deoxy-D-*manno*-2-octulosonate 8-phosphate synthetase, also known as phospho-2-keto-3-deoxyoctanoate (KDO 8-P) synthetase, catalyzes the irreversible aldol reaction of PEP and D-arabinose 5-phosphate to give KDO 8-P (Figure 37).[108] The enzyme has been isolated from *E. coli* B[109] and *Pseudomonas aeruginosa*,[110] and the *E. coli* enzyme has been cloned and overexpressed in *E. coli* and *Salmonella typhimurium*.[111] KDO 8-P itself has been synthesized using KDO 8-P synthetase[109] where the starting material, D-arabinose 5-phosphate, was generated either by hexokinase-catalyzed phosphorylation of arabinose[105] or by isomerase-catalyzed reaction of D-ribose 5-phosphate.[110] To date, the substrate specificity of KDO 8-P

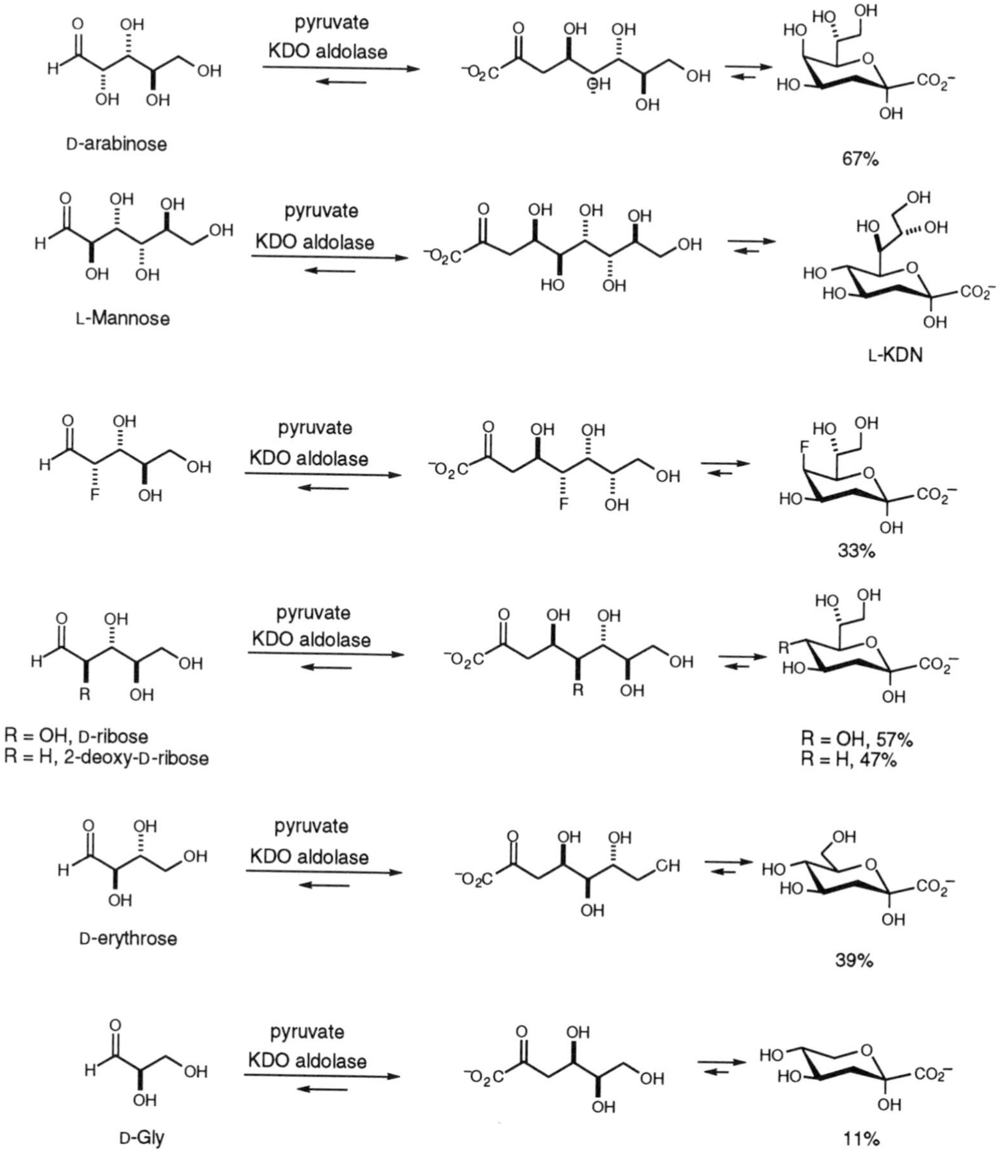

Figure 36. KDO aldolase-catalyzed synthesis of carbohydrates.

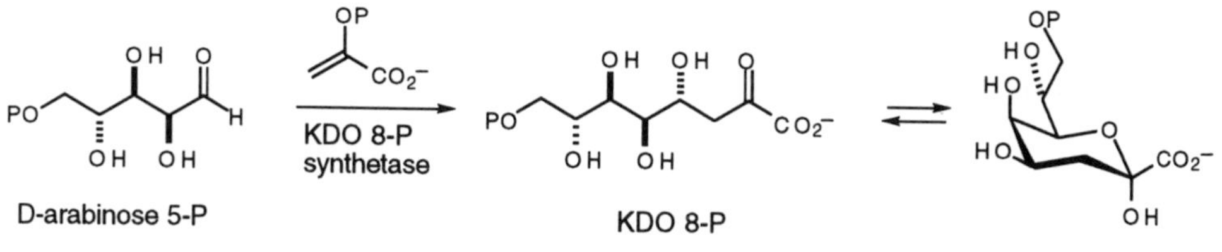

Figure 37. Aldol addition reaction catalyzed *in vivo* by KDO 8-P synthetase.

synthetase has not been well investigated, and initial studies indicate this enzyme is very specific for its natural substrates.

1.7   3-Deoxy-D-*arabino*-2-heptulosonic   Acid   7-Phosphate   (DAHP) Synthetase   (EC   4.1.2.15)

DAHP, also known as phospho-2-keto-3-deoxyheptanoate, is a key intermediate in the shikimate pathway for the biosynthesis of aromatic amino acids in plants.[112] *In vivo*, DAHP synthetase catalyzes the synthesis of DAHP from PEP and D-erythrose 4-phosphate.[113] The enzyme has been cloned[114] and used to synthesize DAHP (Figure 38).[115] The D-erythrose 4-phosphate necessary for the DAHP synthesis was generated *in situ* from Fru 6-P catalyzed by transketolase in the presence of D-ribose 5-phosphate. The Fru 6-P needed for this reaction was generated from D-fructose and ATP catalyzed by hexokinase in the presence of an ATP regeneration system. Further study on this particular enzyme system indicated that it is more efficient and economical to use whole cells containing a DAHP synthetase plasmid.[116] Such a system also provides the other necessary enzymes for the synthesis of DAHP. To date, no study has been conducted to determine the substrate specificity of this enzyme.

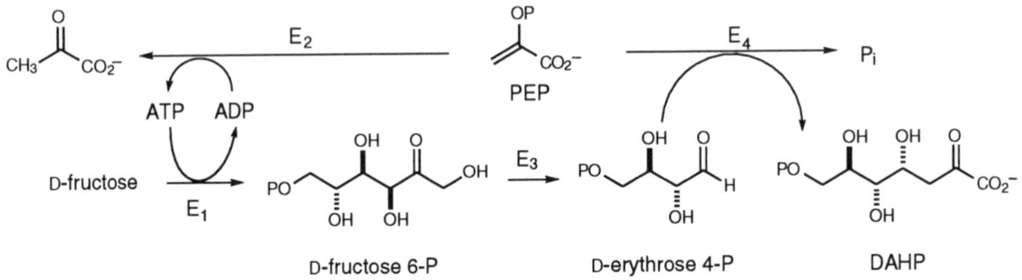

E_1) Hexokinase; E_2) Pyruvate kinase; E_3) Transketolase + D-ribose 5-P; E_4) DAHP synthetase

Figure 38. Multi-enzyme synthesis of DAHP.

1.8   2-Keto-4-hydroxyglutarate (KHG) Aldolase (EC 4.1.2.31)

KHG aldolase, which participates in the terminal step of mammalian catabolism of L-hydroxyproline,[117] catalyzes the reversible condensation of pyruvate and glyoxylate to form KHG (Figure 39).[117,118] The enzyme, isolated and purified from bovine liver and *E. coli* , are both type I aldolases. Studies of substrate specificity on KHG aldolase from bovine liver indicate that it accepts both enantiomers of KHG equally well, and also cleaves 2-keto-3-deoxyglucarate, 2-keto-4,5-dihydroxyvalerate and oxalacetate.[118] In the condensation direction, this enzyme is relatively specific for glyoxylate, but some pyruvate derivatives are accepted(Figure 39). The enzyme from *E. coli* prefers the natural substrate (KHG with (*S*)-configuration) but also cleaves

2-keto-4-hydroxybutyrate and oxalacetate,[119] so racemic KHG can be decomposed selectively to obtain the unreacted (R)-enantiomer.[119b] In the condensation reaction, glyoxylate can be replaced with glyoxaldehyde, formaldehyde, acetaldehyde and formic acid, while pyruvate can be substituted by α-ketobutyrate and bromopyruvate. One anomalous result in these preliminary studies is that some of the cleaving substrates cannot be formed in the aldol reaction. More work is necessary to explain this phenomenon as well as to determine the full synthetic utility of KHG aldolase in organic synthesis.

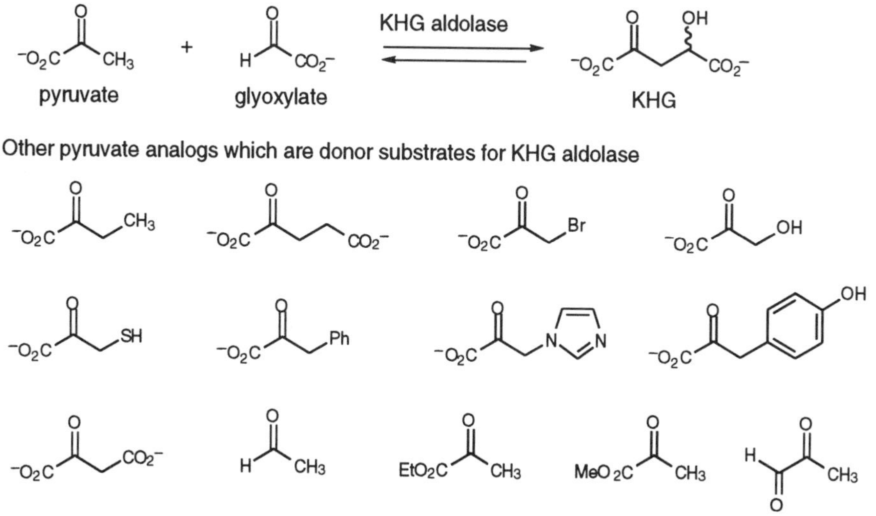

Other pyruvate analogs which are donor substrates for KHG aldolase

Figure 39. Aldol addition reaction catalyzed *in vivo* by KHG aldolase and the donor substrate specificity of this enzyme.

1.9 2-Keto-3-deoxy-6-phosphogluconate (KDPG) Aldolase (EC 4.1.2.14)

KDPG aldolase from *Pseudomonas putida*, whose X-ray structure has been solved,[120] catalyzes the reversible condensation of pyruvate with G3P to form KDPG (Figure 40) which, *in vivo*, favors the condensation product ($K \approx 10^3$ M^{-1}). Studies on the specificity of KDPG aldolase from *P. fluorescens* show that a number of unnatural aldehydes are accepted albeit at rates much lower than the natural substrate (Figure 40).[121a] Unlike several of the other aldolases mentioned above, simple aliphatic aldehydes are not substrates. As long as there is a polar functionality at C-2 or C-3, however, there appears to be no other structural requirement for the acceptor aldehyde. These studies also demonstrate that KDPG aldolase stereospecifically generates a new stereocenter at C-4 with (*S*)-configuration. The enzyme from *E. coli* has been cloned.[121b] Other related enzymes including 2-keto-3-deoxy-6-phospho-D-galactonate aldolase

(E. C. 4.1.2.21), 2-keto-3-deoxy-D-xylonate aldolase (EC 4.1.2.28) and 2-keto-3-deoxy-L-arabonate aldolase (E. C. 4.1.2.18) also showed narrow substrate specificity (see Table 2).

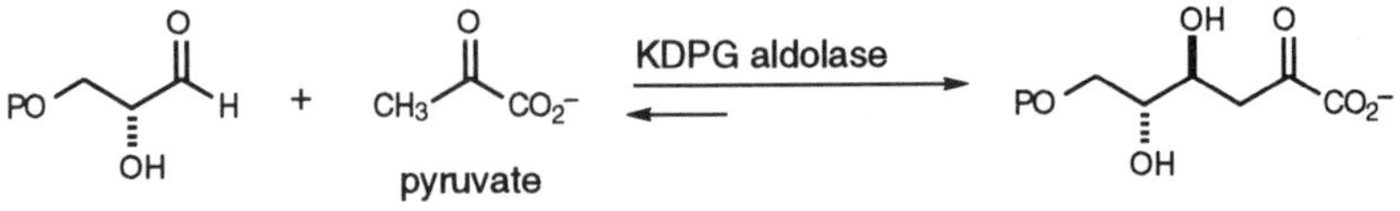

Other acceptor substrates of KDPG aldolase	V_{rel}
nitropropanal	200
chloroacetaldehyde	120
D-glyceraldehyde	100
D-lactaldehyde	27
ribose 5-P	5
erythrose	1.5
glycoaldehyde	1.5
benzaldehyde	0
butyraldehyde	0
ribose	0

Figure 40. Aldol addition reaction catalyzed *in vivo* by KDPG aldolase and the acceptor substrate specificity of this enzyme.

1.10 2-Keto-3-deoxy-D-glucarate (KDG) Aldolase (EC 4.1.2.20)

KDG aldolase catalyzes, *in vivo*, the reversible reaction of pyruvate and tartronic acid semialdehyde to form KDG (Figure 41). Various bacteria have been the source of this enzyme, and it has been isolated and purified from *E. coli*.[122] Several non-natural aldehyde acceptors act as substrates for KDG aldolase, including glycoaldehyde, glyoxylate and D- and L-glyceraldehyde.

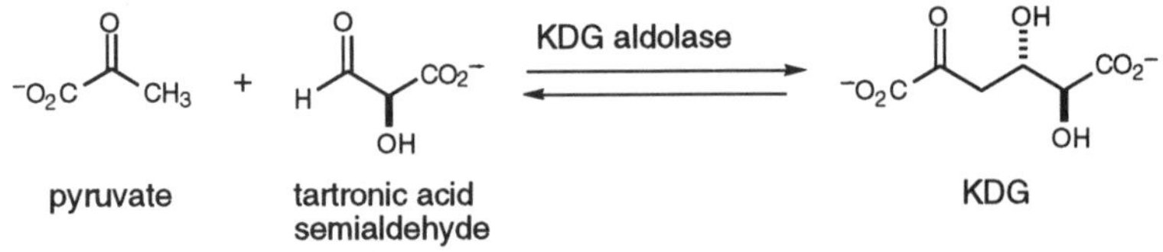

Figure 41. Aldol addition reaction catalyzed *in vivo* by KDG aldolase.

1.11 2-Deoxyribose 5-phosphate aldolase (DERA) (EC 4.1.2.4)

The enzyme DERA[123] is unique among the aldolases in that the donor of the aldol reaction is an aldehyde. *In vivo*, this enzyme catalyzes the reversible condensation of acetaldehyde and G3P to form D-2-deoxyribose 5-phosphate with an equilibrium constant for deoxyribose phosphate of 2×10^{-4} M (Figure 42). DERA is a type I aldolase and has been isolated from animal tissues[124] and several microorganisms.[125] The *E. coli* gene encoding the enzyme has been sequenced,[126] subcloned, and subsequently overexpressed in *E. coli* .[127] At 25 °C and pH 7.5, this enzyme is fairly stable and retains 70% of its activity after 10 days.

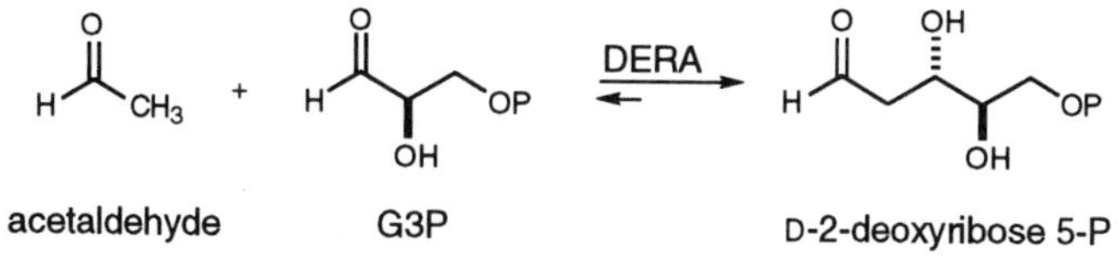

Figure 42. Aldol addition reaction catalyzed *in vivo* by DERA.

A number of unnatural substrates are accepted by DERA, and the newly generated chiral center always has the (*S*)-configuration. Specificity studies have been conducted on DERA from *Lactobacillus plantarum* (specific activity of ca. 6000 U/mg for condensation)[128] and *E. coli* (V_{max} = 21 U/mg for aldol cleavage).[127] The enzyme from *L. plantarum* accepts various acceptor substrates including L-G3P, D-erythrose 4-phosphate, glycoaldehyde phosphate, D-ribose 5-phosphate, D,L-glyceraldehyde, D-erythrose and D-threose, but not D-ribose or glycoaldehyde.[129] Only propionaldehyde could weakly replace acetaldehyde as the donor. The *E. coli* enzyme, on the other hand, is able to utilize several other donor substrates including acetaldehyde, propionaldehyde, acetone and fluoroacetone.[127] A number of aliphatic aldehydes, sugars and sugar phosphates are acceptor substrates. In some cases, however, the rates of the aldol reactions are very slow (0.4 - 1% *cf.* the natural substrates). Among the several compounds synthesized from the *E. coli* enzyme are an fluoro ketose and an trideoxy aza sugar (Figure 43). The aldolases described above, together with many of the other known aldolases of which the substrate specificity remains to be examined, are summarized in Table 2.

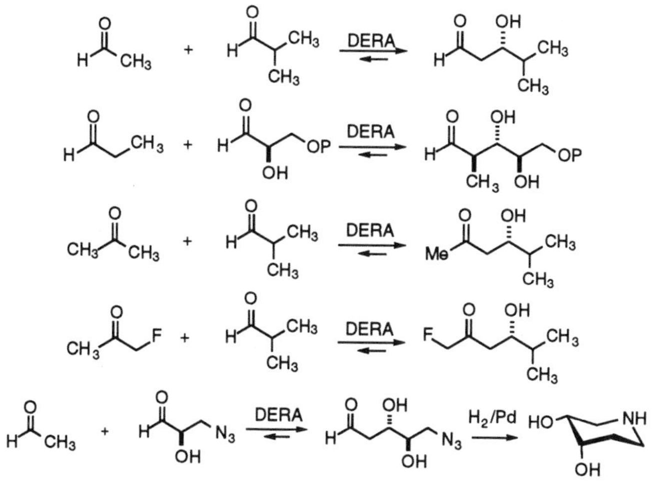

Figure 43. Syntheses of carbohydrate analogs using DERA from *E. coli*.

Carbon-Carbon Bond Formation

Table 2. Aldolases and the reactions they catalyze *in vivo*.

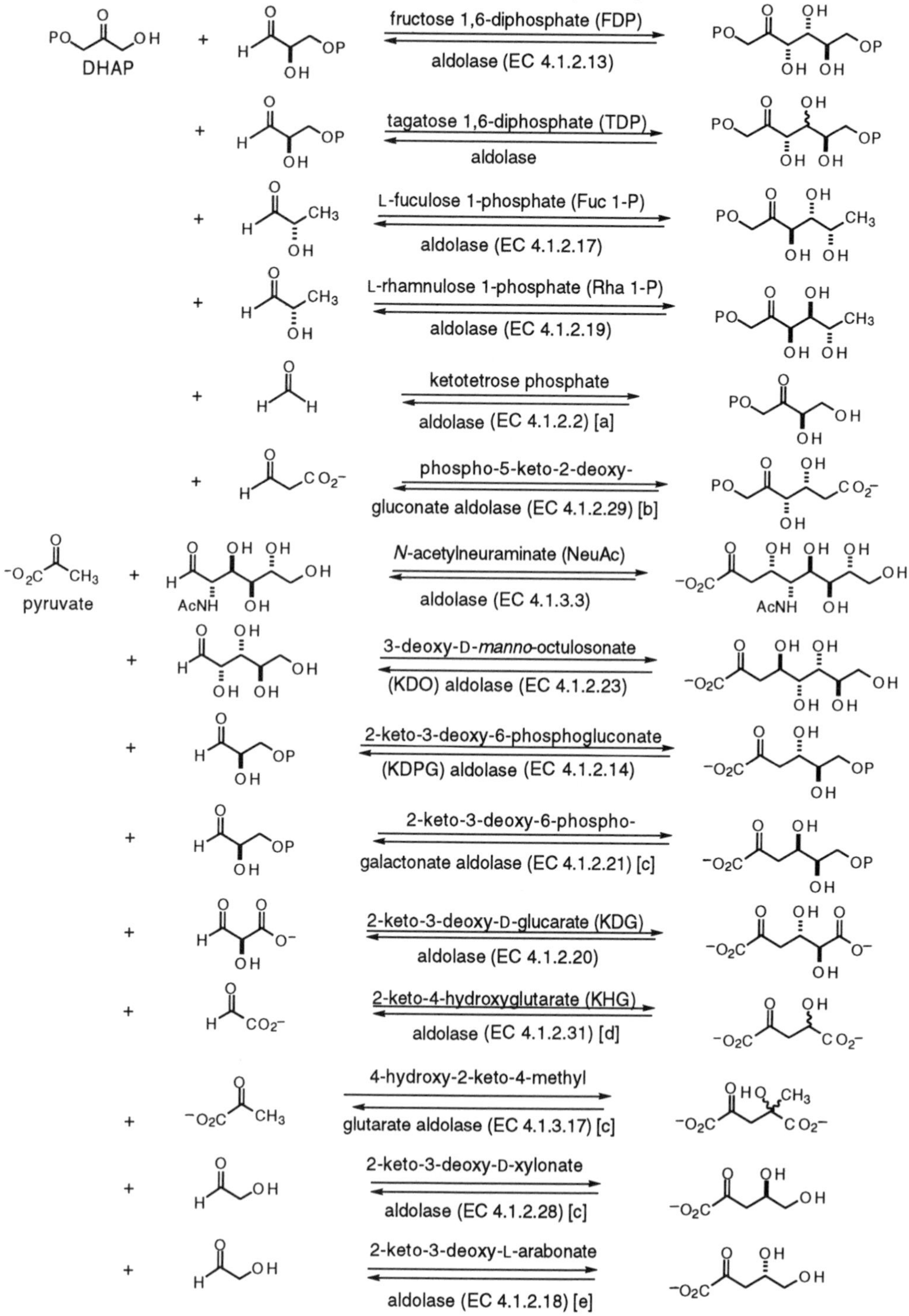

Donor	Acceptor	Enzyme	Product
DHAP		fructose 1,6-diphosphate (FDP) aldolase (EC 4.1.2.13)	
		tagatose 1,6-diphosphate (TDP) aldolase	
		L-fuculose 1-phosphate (Fuc 1-P) aldolase (EC 4.1.2.17)	
		L-rhamnulose 1-phosphate (Rha 1-P) aldolase (EC 4.1.2.19)	
		ketotetrose phosphate aldolase (EC 4.1.2.2) [a]	
		phospho-5-keto-2-deoxy-gluconate aldolase (EC 4.1.2.29) [b]	
pyruvate		N-acetylneuraminate (NeuAc) aldolase (EC 4.1.3.3)	
		3-deoxy-D-*manno*-octulosonate (KDO) aldolase (EC 4.1.2.23)	
		2-keto-3-deoxy-6-phosphogluconate (KDPG) aldolase (EC 4.1.2.14)	
		2-keto-3-deoxy-6-phospho-galactonate aldolase (EC 4.1.2.21) [c]	
		2-keto-3-deoxy-D-glucarate (KDG) aldolase (EC 4.1.2.20)	
		2-keto-4-hydroxyglutarate (KHG) aldolase (EC 4.1.2.31) [d]	
		4-hydroxy-2-keto-4-methyl glutarate aldolase (EC 4.1.3.17) [c]	
		2-keto-3-deoxy-D-xylonate aldolase (EC 4.1.2.28) [c]	
		2-keto-3-deoxy-L-arabonate aldolase (EC 4.1.2.18) [e]	

Table 2: continue

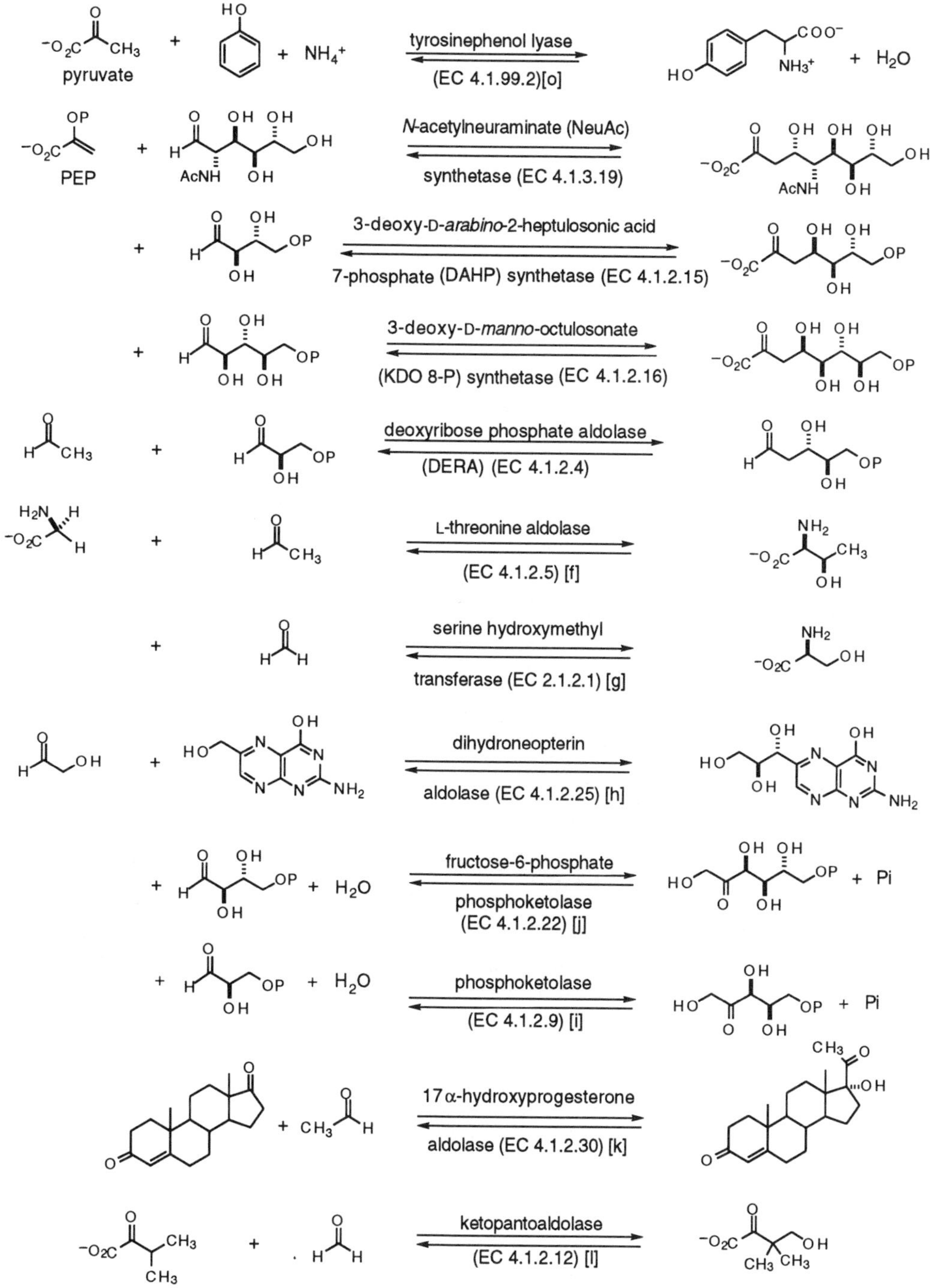

Reactants	Enzyme	Products
pyruvate + phenol + NH₄⁺	tyrosinephenol lyase (EC 4.1.99.2)[o]	tyrosine + H₂O
PEP + AcNH sugar	N-acetylneuraminate (NeuAc) synthetase (EC 4.1.3.19)	NeuAc product
+ phosphosugar	3-deoxy-D-*arabino*-2-heptulosonic acid 7-phosphate (DAHP) synthetase (EC 4.1.2.15)	DAHP
+ phosphosugar	3-deoxy-D-*manno*-octulosonate (KDO 8-P) synthetase (EC 4.1.2.16)	KDO 8-P
acetaldehyde + glyceraldehyde phosphate	deoxyribose phosphate aldolase (DERA) (EC 4.1.2.4)	deoxyribose phosphate
glycine + acetaldehyde	L-threonine aldolase (EC 4.1.2.5) [f]	threonine
+ formaldehyde	serine hydroxymethyl transferase (EC 2.1.2.1) [g]	serine
glycolaldehyde + pterin	dihydroneopterin aldolase (EC 4.1.2.25) [h]	neopterin
+ phosphosugar + H₂O	fructose-6-phosphate phosphoketolase (EC 4.1.2.22) [j]	+ Pi
+ phosphosugar + H₂O	phosphoketolase (EC 4.1.2.9) [i]	+ Pi
steroid + CH₃CHO	17α-hydroxyprogesterone aldolase (EC 4.1.2.30) [k]	17α-hydroxyprogesterone
ketoisovalerate + formaldehyde	ketopantoaldolase (EC 4.1.2.12) [l]	ketopantoate

Table 2: continue

Table 2: References

a.   Isolated from rat liver, see: Charalampous, F. C. *Method Enzymol.* **1962**, *5*, 283; Acetaldehyde, glycoaldehyde or glyceraldehyde cannot replace formaldehyde.

b.   Andeson, W. A.; Magasanik, B. *J. Biol. Chem.* **1971**, *246*, 5662.

c.   Wood, W.A. in *The Enzymes*; Boyer, P.D., Ed; Academic Press: New York, 1970; Vol VII, p. 281; Floyd, N. C.; Liebster, M. H.; Turner, N. T. *J. Chem. Soc., Perkin Trans 1* **1992**, 1085.

d.   This enzyme also catalyzes the aldol addition of pyruvate with formaldehyde to give 4-hydroxy-2-oxobutyrate, originally thought to be catalyzed by hydroxyoxobutyrate aldolase (EC 4.2.1.1). Phenylpyruvate is also a donor substrate, while acetaldehyde, benzaldehyde and crotonaldehyde are not acceptor substrates, see: Hift, H.; Mahler, H. R. *J. Biol. Chem.* **1952**, *198*, 901.

e.   Isolated from *Pseudomonas*, see: Dahm, A. S.; Anderson, R. L. *J. Biol. Chem.* **1972**, *247*, 2238. D-Lactaldehyde can replace glycoaldehyde.

f.   Hetenyi, B.,Jr.; Anderson, P. J.; Kinson, G. A. *Biochem. J.* **1984**, *224*, 355; Dainty, R. H. *Biochem. J.* **1967**, *104*, 46. The reaction is in favor of degradation.

g.   Schirch, L. *Adv. Enzymol.* **1982**, *53*, 83. Stover, P.; Zamora, M. ;Shostak, K.; Gautam-Basak, M.; Schirch, V. *J. Biol. Chem.* **1992**, *267*, 17679. A multicopy plasmid containing the *E. coli* serinehydroxymethyl transferase was introduced to Klebsiella aerogenes for overexpression of the enzyme. The enzyme requires tetrahydrofolate (THF) and pyridoxal phosphate. THF first reacts nonenzymatically with formaldehyde to form $N5,N10$-methylene THF which is then accepted by the enzyme to form serine, see: Hamilton, B. K.; Hsiao, H. Y.; Swanm, W. E.; Anderson, D. M.; Delente, J. J. *Trends in Biotechnology* **1985**, *3*, 64. This enzyme also catalyzes the reversible aldol reaction of glycine with acetaldehyde to give L-allothreonine, originally thought to be catalyzed by L-allothreonine aldolase (EC 4.1.2.6); substrate specificity studies: Lotz, B. T.; Casparski, C. M.; Peterson, K.; Miller, M. J. *J. Chem. Soc., Chem. Commun.* **1990**, 1107; opposite stereoselectivity observed: Herbert, R. B.; Wilkinson, B.; Ellames, G. J.; Kanes, E. K. *J. Chem. Soc., Chem. Commun.* **1993**, 205.

h.   Mathis, J. B.; Brown, G. M. *J. Biol. Chem.* **1970**, *245*, 3015. The reaction requires thiamine pyrophosphate and favors cleavage.

i.   Heath, E. C.; Hurwitz, J.; Horecker, B. L.; Ginsberg, A. *J. Biol. Chem.* **1958**, *231*, 1009. The reaction favors the cleavage of D-xylulose 5-phosphate. The enzyme from *Leuconostoc mesenteroides* also accepts fructose 6-phosphate, hydroxypyruvate and glycoaldehyde as substrates.

j.   Racker, E. *Method Enzymol.* **1962**, *5*, 276. The reaction favors degradation.

k   Johnston, D. E.; Chiao, Y.-B.; Gavaler, J. S.; van Thiel, D. H. *Biochem. Pharm.* **1981**, *30*, 1827.

l.   Maas, W. K.; Vogel, H. J. *J. Bacteriol.* **1953**, *65*, 388; McIntosh, E. N.; Purko, M.; Wood, W. A. *J. Biol. Chem.* **1957**, *228*, 499.

m.   Beisswenger, R.; Kula, M.-R. *Appl. Microbiol. Biotechnol.* **1991**, *34*, 604. Beisswenger, R.; Snatzke, G.; Thiem, J.; Kula, M.-R. *Tetrahedron Lett.* **1991**, *32*, 3159.

n.   Phillips, R. S.; Dua, R. K. *J. Am. Chem. Soc.* **1991**, *113*, 7385.

o.   Faleev, S. B.; Ruvinov, S. B.;Demidkina, T. V.; Myagkikh, I. V.; Gololobov, M. Y.; Bakhmutov, V. I.; Belikov, V. M. *Eur. J. Biochem.* **1988**, *177*, 395.

2. Ketol and Aldol Transfer Reactions

2.1 Transketolase (TK) (EC 2.2.1.1)

One of the enzymes in the pentose phosphate pathway is transketolase (TK)[130] which reversibly transfers the C1-C2 ketol unit from D-xylulose 5-phosphate to D-ribose 5-phosphate to generate D-sedoheptulose 7-phosphate and G3P. The enzyme relies on two cofactors for activity: thiamine pyrophosphate (TPP) and Mg^{2+}. TK from baker's yeast is commercially available,[131] and the enzyme has also been isolated from spinach.[132,133] The enzyme isolated from yeast shows a slightly higher diastereoselectivity (ca. 100%)[133] than the enzyme from spinach (ca. 95%), with the newly-formed hydroxymethine chiral center always possessing (*S*)-configuration. Besides D-xylulose 5-phosphate, D-Erythrose 4-phosphate also acts as a ketol donor which produces Fru 6-P and G3P (Figure 44). β-Hydroxypyruvic acid (HPA) also acts as a ketol

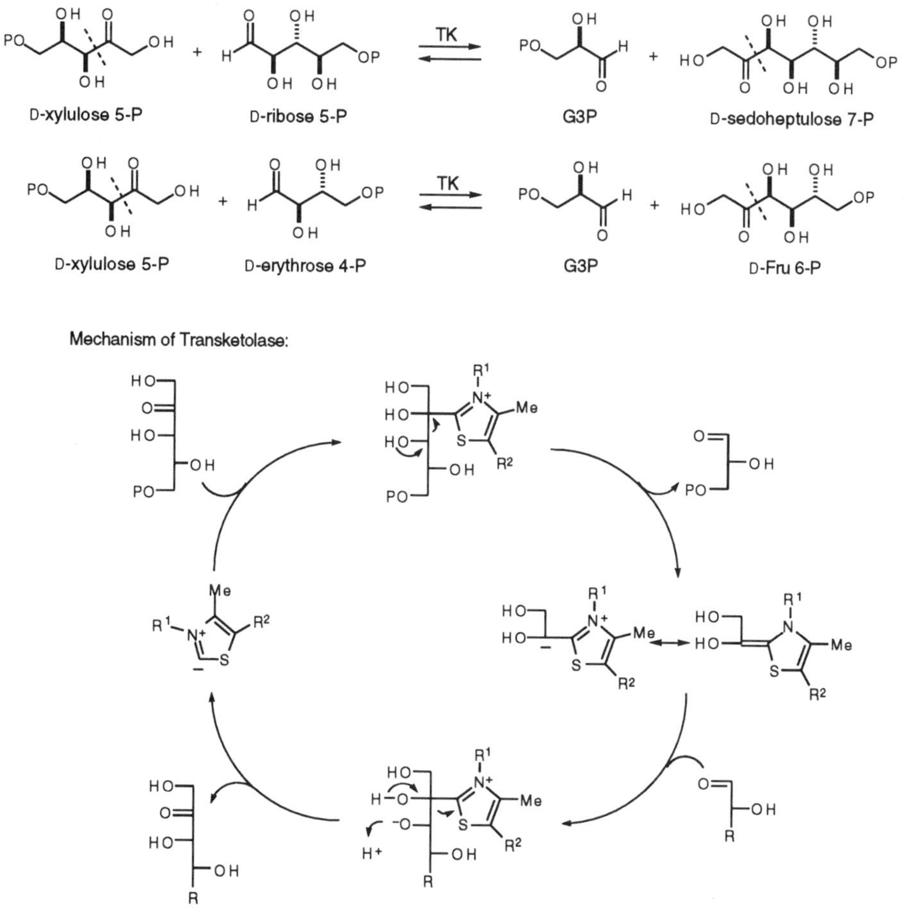

Mechanism of Transketolase:

Figure 44. Ketol transfer reactions in the oxidative pentose phosphate pathway catalyzed by TK.

donor[134] which is transferred to an aldose acceptor with an activity of 4% compared to D-xylulose 5-phosphate.[132] One valuable feature of HPA is that once the ketol unit is transferred by TK, carbon dioxide is lost rendering the overall reaction irreversible. Other analogs of HPA, e.g. 2,3-dioxopropionic acid, 2-oxo-3-hydroxybutyric acid, 2-oxomalonic acid and 2-ketogluconic acid, are not substrates,[135] but a wide range of aldehydes are ketol acceptors including aliphatic, α,β-unsaturated, aromatic and heterocyclic aldehydes (Figure 45).[135,136] The presence of a hydroxy or an oxygen atom at C-2 and/or C-3 has a positive effect on the rate, while steric hindrance near the aldehyde decreases the rate. β-Hydroxy aldehydes epimeric at C-2 can be efficiently resolved by TK since only the D-enantiomers are substrates and give D-*threo* products.[135,137] Beyond the C-2 configuration, the enzyme appears to have no preference for the stereochemistry.

A few synthetic examples that employ TK are illustrated in the synthesis of (+)-*exo*-brevicomin[138] and the aza sugar 1,4-dideoxy-1,4-imino-D-arabinitol.[38] Both syntheses involve the condensation of HPA with racemic 2-hydroxyaldehydes whereby the ketol unit is diastereoselectively transferred to only the D-enantiomer of the aldehyde.

2.2   Transaldolase (TA) (EC 2.2.1.2)

Like transketolase, TA is an enzyme in the oxidative pentose phosphate pathway.[130] TA, which operates through a Schiff base intermediate, catalyzes the transfer of the C1-C3 aldol unit from D-sedoheptulose 7-phosphate to G3P to produce D-Fru 6-P and D-erythrose 4-phosphate (Figure 46). Although it is commercially available, TA has rarely been used in organic synthesis, and no detailed substrate specificity study has yet been performed. In one application, TA was used in the synthesis D-fructose from starch (Figure 47).[139] The aldol moiety was transferred from Fru 6-P to D-glycerol in the final step of this multi-enzyme synthesis of D-fructose.

3.   Addition of HCN to Aldehydes

Cyanohydrin synthetases or oxynitrilases are the class of enzymes that catalyze the addition of HCN to aldehydes to form cyanohydrins.[140–145] The substrate specificity of these enzymes is relatively broad as illustrated by the several different aromatic aldehydes used as acceptors for the (*R*)-oxynitrilase (or mandelonitrile lyase) (EC 4.1.2.10). *R*-Cyanohydrins are the predominant products in theses reactions; in aqueous solutions, however, cyanohydrins racemize and competing non-enzymatic cyanide addition results in products with low optical purity. To circumvent this problem, the enzyme has been immobilized on cellulose or the reaction

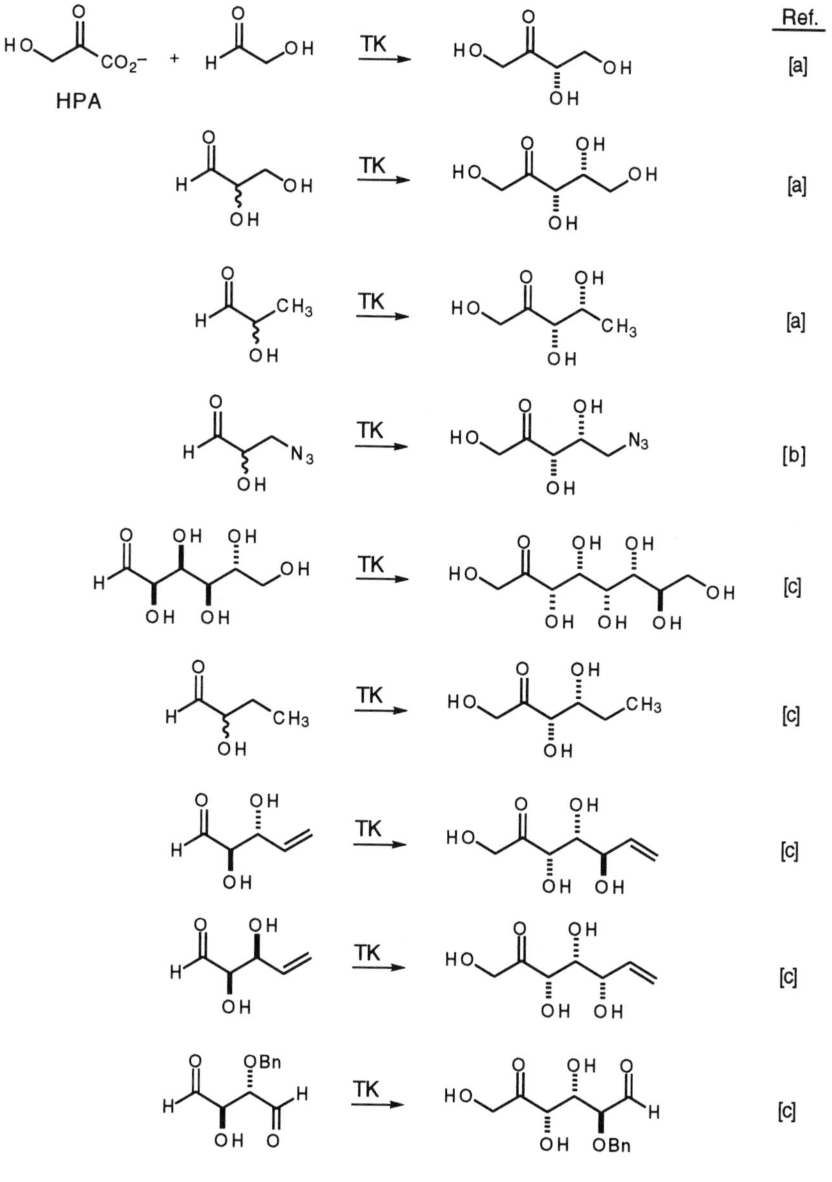

a) Bolte, J.; Demuynck, C.; Samaki, H. *Tetrahedron Lett.* **1987**, *28*, 5525. Hobbs, G. R.; Lilly, M. D.; Turner, N. J.; Ward, J. M.; Willets, A. J.; Woodley, J. M. *J. Chem. Soc., Perkin Trans. I* **1993**, 165.
b) Ziegler, T.; Straub, A.; Effenberger, F. *Angew. Chem., Int. Ed. Engl.* **1988**, *27*, 716.
c.Kobori, Y.; Myles, D. C.; Whitesides, G. M. *J. Org. Chem.* **1992**, *57*, 5899.

Figure 45. Acceptor substrate specificity of TK.

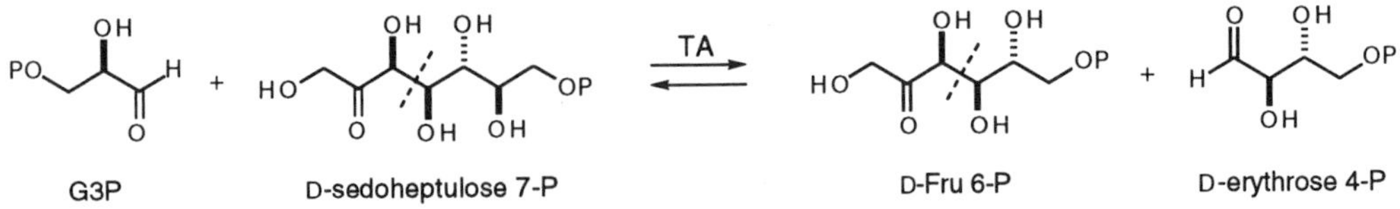

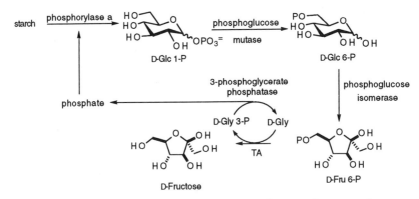

Figure 46. Aldol transfer reaction in the oxidative pentose phosphate pathway catalyzed by TA.

Figure 47. Multi-enzyme synthesis of D-fructose from starch.

has been carried out in ethyl acetate. The result is products with higher yields (77-99%) and improved optical purity (73-99% ee) (Figure 48).[140,141] Another method for generating cyanohydrins uses acetone cyanohydrin as the –CN donor in the enzymatic synthesis of several aliphatic cyanohydrins avoiding the use of the highly toxic HCN gas.[142] In addition to aromatic cyanohydrins, aliphatic (R)-cyanohydrins were prepared by this method (Figure 49).[143]

The (S)-oxynitrilase (EC 4.1.2.11), first studied in the 1960's, was recently used in the synthesis of several S-cyanohydrins derivatives (Figure 48).[144,145] The enzyme from *Sorghum bicolor* is active in ethyl acetate at a pH of 3-4 where racemization and non-enzymatic cyanohydrin formation can be minimized.

4. Acyloin Condensation

Acyloin condensations catalyzed by yeast were first observed by Neuberg and later by Becvarova.[146] Since that time, several other yeast-catalyzed acyloin condensations between acetaldehyde and aromatic aldehydes have been observed, all reportedly having the (R)-configuration (Figure 50).[147,148] In one example, the acyloin formed from benzaldehyde itself is used in the industrial manufacture of (-)-ephedrine.[149] In some cases, the α-hydroxyl ketone is not isolated but a (1R,2S) diol is formed. Presumably, the yeast alcohol dehydrogenase present in

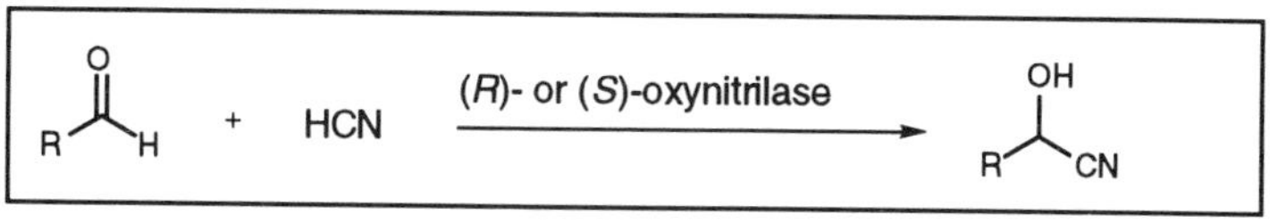

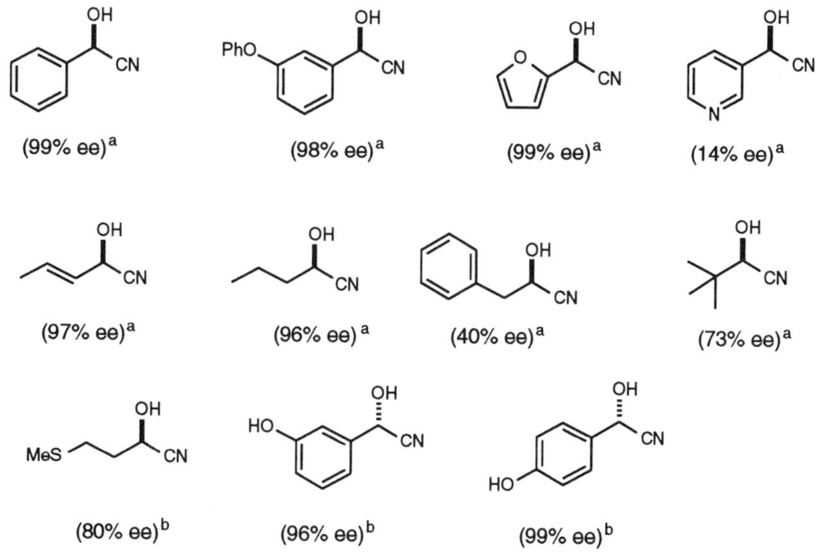

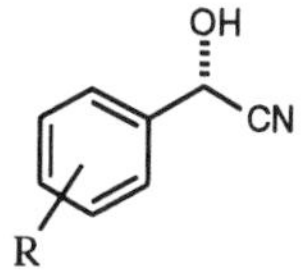

Figure 48. Stereoselective cyanohydrin formation using oxynitrilases.

The scheme shows:

$$R\text{-CHO} + HCN \xrightarrow{(R)\text{- or }(S)\text{-oxynitrilase}} R\text{-CH(OH)-CN}$$

Products with their ee values:

- Phenyl cyanohydrin (99% ee)[a]
- 3-PhO-phenyl cyanohydrin (98% ee)[a]
- 2-Furyl cyanohydrin (99% ee)[a]
- 3-Pyridyl cyanohydrin (14% ee)[a]
- Propenyl cyanohydrin (97% ee)[a]
- Butyl cyanohydrin (96% ee)[a]
- Phenethyl cyanohydrin (40% ee)[a]
- tert-Butyl cyanohydrin (73% ee)[a]
- MeS-ethyl cyanohydrin (80% ee)[b]
- 3-HO-phenyl cyanohydrin (96% ee)[b]
- 4-HO-phenyl cyanohydrin (99% ee)[b]

R	% ee[c]
H	98
4-CH₃	78
4-Cl	54
3-Cl	91
3-Br	90
3-OCH₃	90
3-OPh	96

3-OH or 4-OH   Decomposed

a. Effenberger, F.; Ziegler, T.; Forster, S. *Angew. Chem. Int. Ed. Engl.* **1987**, *26*, 458
b. Niedermeyer, U.; Kula, M-R. *ibid* **1990**, *29*, 386.
c. Effenberger, F.; Horsch, B.; Forster, S.; Ziegler, T. *Tetrahedron Lett.* **1990**, *31*, 1249.

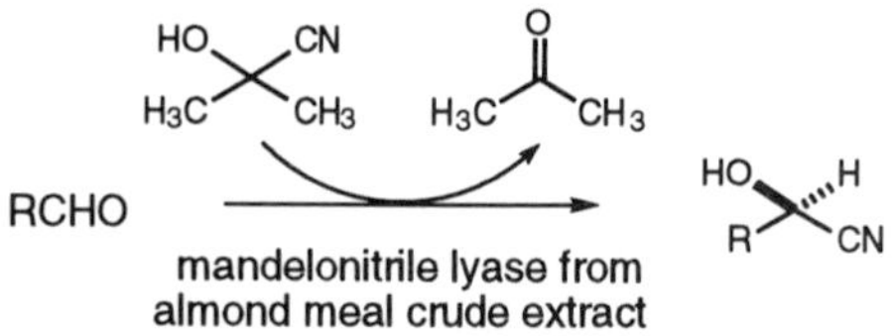

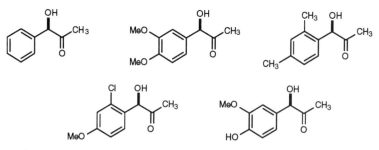

Figure 49. Enzymatic synthesis of cyanohydrins using acetone cyanohydrin.

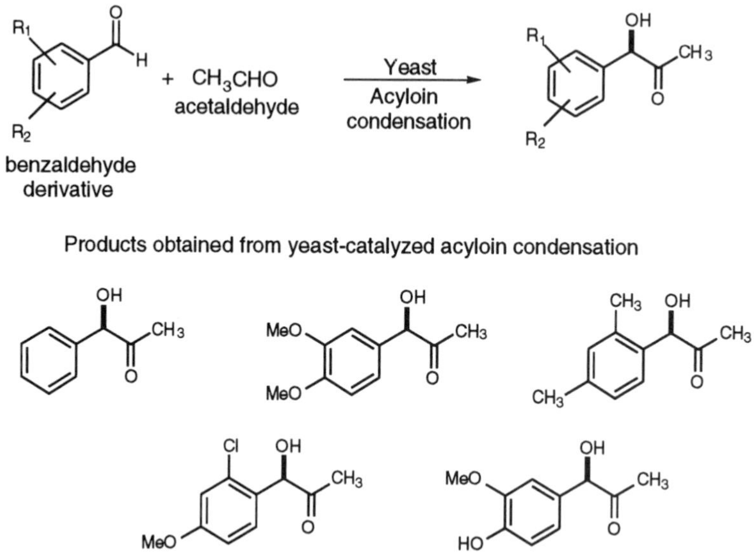

R =	%Conv., %ee
$CH_3CH_2CH_2$	100, 95
$(CH_3)_2CH$	99, 83
$CH_3(CH_2)_2CH_2$	100, 97
$CH_3CH_2(CH_3)CH$	100, 94
$CH_3(CH_2)_3CH_2$	100, 94
$CH_3(CH_2)_5CH_2$	98, 87
$CH_3(CH_2)_7CH_2$	94, 63

benzaldehyde derivative + CH_3CHO acetaldehyde → Yeast / Acyloin condensation

Products obtained from yeast-catalyzed acyloin condensation

Figure 50. Acyloin condensations between acetaldehyde and benzaldehyde derivatives catalyzed by yeast.

the whole cell acts to stereoselectively reduce the ketone of the acyloin (Figure 51).[150] The enzyme responsible for the acyloin condensation was recently isolated from yeast and was identified as pyruvate decarboxylase.[151] In fact, this highly purified decarboxylase yields *R*-hydroxy ketones in greater optical purity than those obtained from the whole cells.[152]

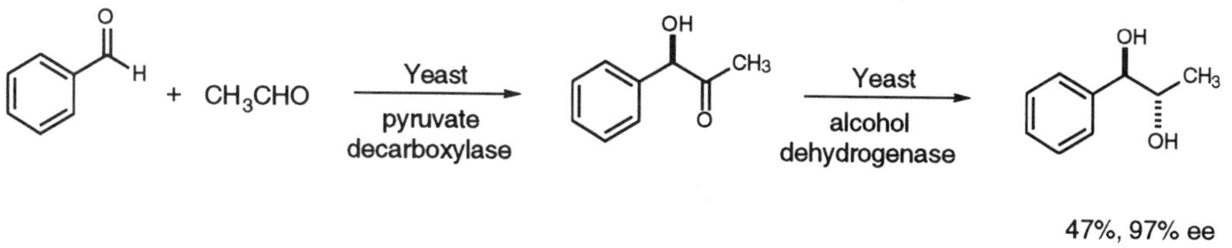

Figure 51. Yeast catalyzed acyloin condensation and reduction.

5. C-C Bond Forming Reactions Involving AcetylCoA

There are a number of enzymatic reactions that utilize coenzyme A (CoA) thioesters as substrates.[153–155] These are involved in the biosynthesis of steroids, terpenoids, macrolides, fatty acids and other natural products. Due to the high cost of CoA, however, these enzymes can only be used practically in organic synthesis if the CoA thioester is recycled (Figure 52).[153,154] Many examples of schemes for CoA recycling have been demonstrated using the synthesis of citric acid from oxaloacetate and acetyl CoA (Figure 52) as an example. In one scheme, phosphotransacetylase (EC 2.3.1.8) catalyzes the transfer of the acetyl group to CoA from acetylphosphate.[154,155] A variant using dextran-bound CoA was also developed, although this system resulted in low yields and low turnover rates due to the poor affinity of the enzymes for the bound CoA.[154] Stability of acetylphosphate has always been a problem in this reaction, so alternative substrates and enzymes have also been used in the recycling step.[153] Propanoylphosphate, butanoylphosphate, propenoylphosphate, chloroacetylphosphate and fluoroacetylphosphate are all accepted by phosphotransacetylase with rates of 93%, 2%, 7%, and 17% with respect to acetylphosphate (Figure 53).

AcetylCoA can also be recycled with carnitine acetyltransferase (EC 2.3.1.7) and acetylcarnitine (Figure 52).[153] The process requires longer reaction times, but the enzyme and acetylcarnitine are relatively stable. Propanoylcarnitine, butanoylcarnitine, crotonylcarnitine and pentenoylcarnitine were also prepared as substrates for carnitine acyltransferase and were accepted at relative rates of 78%, 50%, 5% and 2%, respectively (*cf.* acetylphosphate).(Figure 53). Another enzymatic approach to regenerate acetylCoA uses acetylCoA synthetase (EC 6.2.1.1) (turnover number of 1000 for CoA) (Figure 52)[154] which requires ATP.

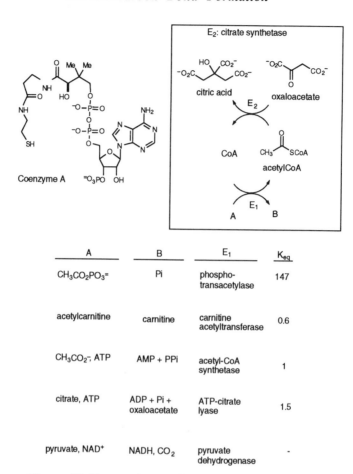

Figure 52. Enzymatic recycling systems of acetylCoA.

A	B	E_1	K_{eq}
$CH_3CO_2PO_3^=$	Pi	phospho-transacetylase	147
acetylcarnitine	carnitine	carnitine acetyltransferase	0.6
$CH_3CO_2^-$, ATP	AMP + PPi	acetyl-CoA synthetase	1
citrate, ATP	ADP + Pi + oxaloacetate	ATP-citrate lyase	1.5
pyruvate, NAD$^+$	NADH, CO_2	pyruvate dehydrogenase	-

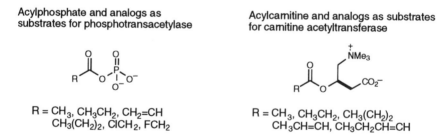

Acylphosphate and analogs as substrates for phosphotransacetylase

R = CH$_3$, CH$_3$CH$_2$, CH$_2$=CH
CH$_3$(CH$_2$)$_2$, ClCH$_2$, FCH$_2$

Acylcarnitine and analogs as substrates for carnitine acetyltransferase

R = CH$_3$, CH$_3$CH$_2$, CH$_3$(CH$_2$)$_2$
CH$_3$CH=CH, CH$_3$CH$_2$CH=CH

Figure 53. Substrate specificity of phosphotransacetylase and carnitine acetyltransferase.

An interesting non-enzymatic regeneration process of acetylCoA utilizes a phase transfer catalyst in a biphasic aqueous–organic system (Figure 54).[156] 4-Dimethylaminopyridine is used as the acetylation catalyst for CoA to form acetylCoA, which in turn can be used to synthesize citric acid (turnover number of 10^3 for CoA). This method can also be used to prepare many different acylCoA derivatives for use as substrates for CoA-dependent enzymes. A new chemoenzymatic approach to CoA and its analogs has also recently been reported.[157]

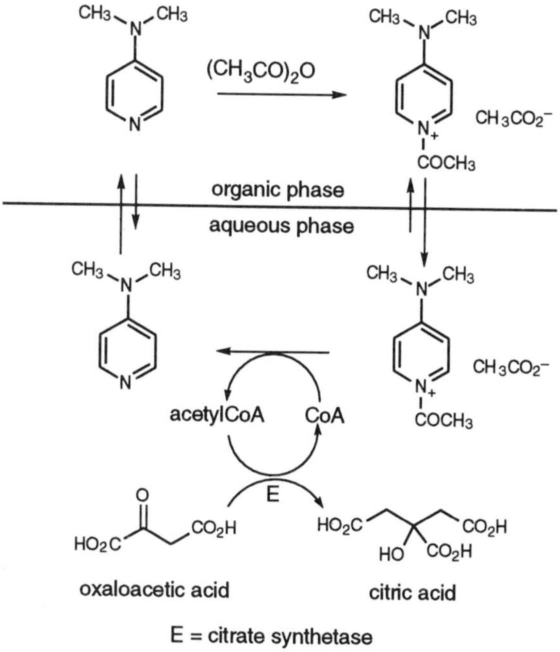

Figure 54. Chemical regeneration of acetylCoA using a phase transfer catalyst.

Another biosynthetic sequence that utilizes acetylCoA is poly-β-hydroxybutyrate synthesis (Figure 55). Many whole cell systems have been used to synthesize this polymer and related materials (x = 1–7).[158] Copolymers consisting of (R)-3-hydroxybutyl and (R)-3-

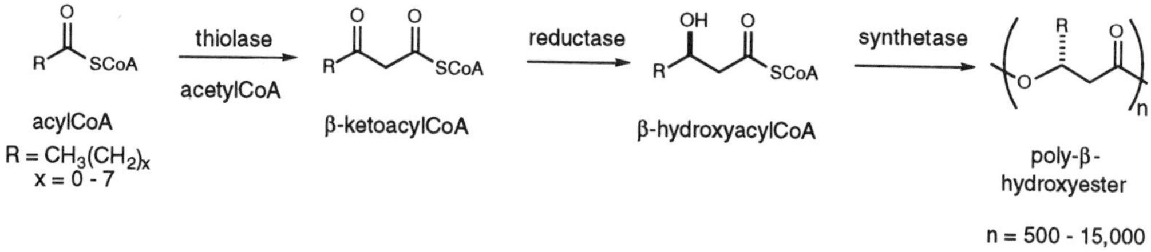

Figure 55. Enzyme-catalyzed reactions involved in the whole-cell synthesis of poly-β-hydroxyesters.

hydroxyvaleryl units (Figure 55, x = 0 and 1, respectively) were prepared by feeding propionate to whole cells of *A. eutrophus* .[158] The first enzyme in the synthesis, acetoacetylCoA thiolase

(EC 2.3.1.9), catalyzes the carbon-carbon bond forming step. The active site of acetoacetylCoA thiolase contains a cysteine, which attacks acetylCoA to form a sulfur-acetyl enzyme intermediate. This species then reacts with the enolate derived from enzymatic deprotonation of the other acetylCoA. Mechanistic studies that were performed on this enzyme from *Zooglea ramigera*[159] showed that the thiolase forms an acyl enzyme intermediates with a number of acylCoA derivatives, but only accepts acetylCoA as the nucleophile. After reduction of the ketone by acetoacetylCoA reductase (EC 1.1.1.36), the resulting β-hydroxy thioester is polymerized by polyhydroxybutyrate synthetase.[159] These polymers are synthetically useful as a source of (*R*)-β-hydroxy acids.[159b] A streptomyces host-vector system has recently been developed for effecient expression of recombinant polyketide synthases. Using this expression system, several novel polyketides have been synthesized *in vivo* in high quantities.[160a] Enzymatic synthesis of novel polyketides may be possible in the future.[160b]

6. Isoprenoid and Steroid Synthesis

Some of the enzymes involved in the biosynthesis of steroids have recently been used in organic synthesis. 2,3-Oxidosqualene lanosterol cyclase catalyzed the synthesis of a number of lanosterol analogs (Figure 56).[161–163] When using an enzyme suspension from Baker's yeast containing this cyclase, catalysis was promoted by ultrasonic irradiation.[162,163] Lanosterol cyclase was also utilized to synthesize a C-30 functionalized lanosterol; in this reaction, the enzyme formed the steroid ring system, while leaving the normally labile vinyl group at C-30 intact (Figure 57).[162] This product was subsequently converted to (+)-30-hydroxylanosterol and the corresponding aldehyde. These compounds are natural receptor-mediated feedback inhibitors of.HMG-CoA reductase, and, therefore, are of interest in the development of hypocholesteremic drugs.[164]

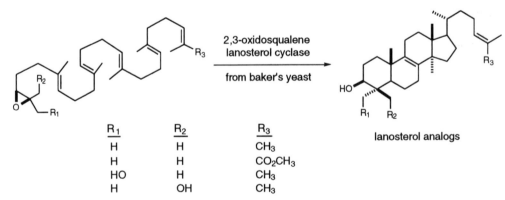

R_1	R_2	R_3
H	H	CH_3
H	H	CO_2CH_3
HO	H	CH_3
H	OH	CH_3

lanosterol analogs

Figure 56. Synthesis of lanosterol analogs using 2,3-oxidosqualene lanosterol cyclase.

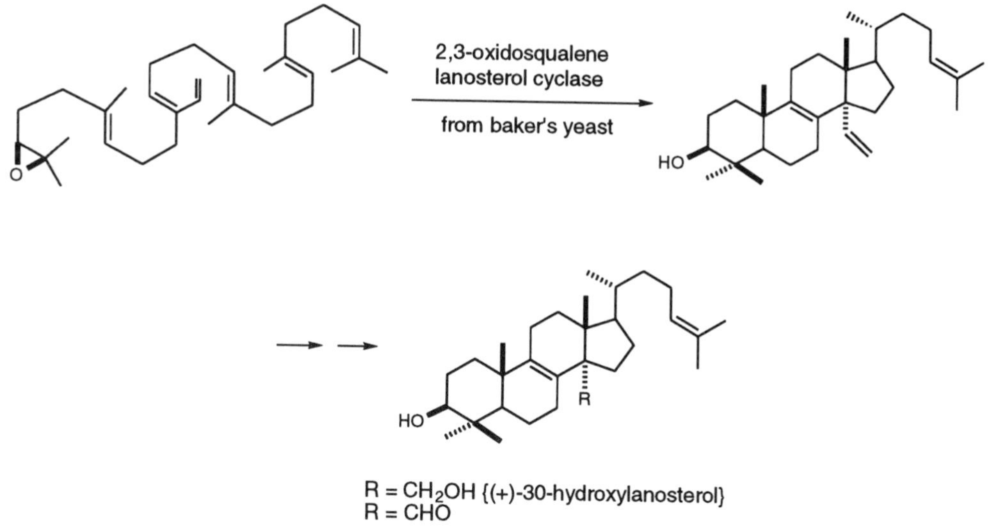

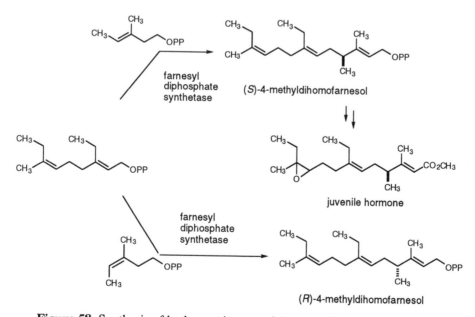

R = CH₂OH {(+)-30-hydroxylanosterol}
R = CHO

Figure 57. Synthesis of C-30 functionalized lanosterols using 2,3-oxidosqualene lanosterol cyclase.

Using farnesyl diphosphate synthetase (EC 2.5.1.10) from pig liver, both enantiomers of 4-methyldihomofarnesol were synthesized, where the (*S*)-enantiomer was used as a precursor for the synthesis of a juvenile hormone (Figure 58).[165]

Figure 58. Synthesis of both enantiomers of 4-methyldihomofarnesol using farnesyl diphosphate synthetase.

7. β-Replacement of Chloroalanine

New methods have recently been developed for the synthesis of novel amino acids using pyridoxal phosphate-dependent enzymes.[166] These enzymes usually catalyze transaminations, α,β-eliminations, α,γ-eliminations and decarboxylations of amino acids. Using β-chloroalanine as a substrate, however, β-replacement can be carried out to produce unusual amino acids. For example, tryptophan synthase (EC 4.2.1.20) from *E. coli* catalyzed the formation of tryptophan and analogs, while tyrosine, DOPA and other analogs were synthesized using tyrosine phenol lyase (EC 4.1.99.2) (Figure 59).

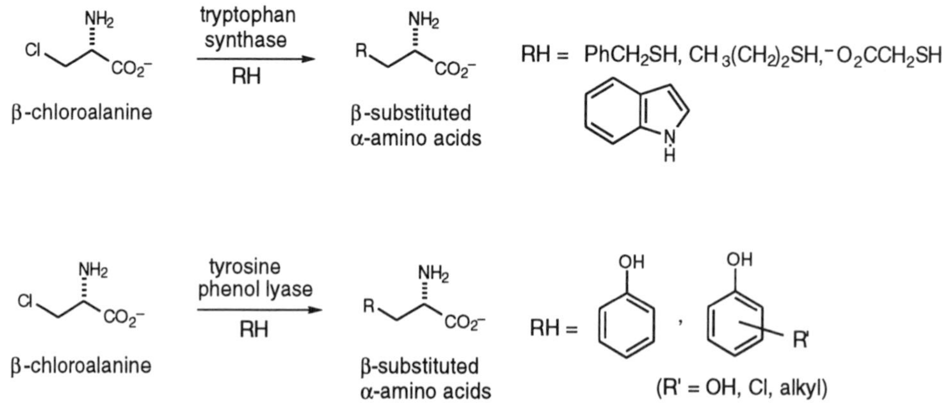

Figure 59. Synthesis of β-substituted α-amino acids from β-chloroalanine using tryptophan synthase and tyrosine phenol lyase.

8. C-C Bond Formation Catalyzed by Vitamin B$_{12}$

Although vitamin B$_{12}$ is not an enzyme, its complex structure and other related Co complexes have been used to catalyze a number of C-C bond forming reactions under reductive conditions. Co^I in B$_{12}$ is a very good nucleophile and reacts rapidly with alkyl halides in an oxidative addition reaction to form an organo-Co derivative. Homolytic cleavage of the C-C bond, either by electrochemical or photochemical methods, gives a carbon radical which undergoes inter- or intramolecular addition to an olefin (Figure 60).[167]

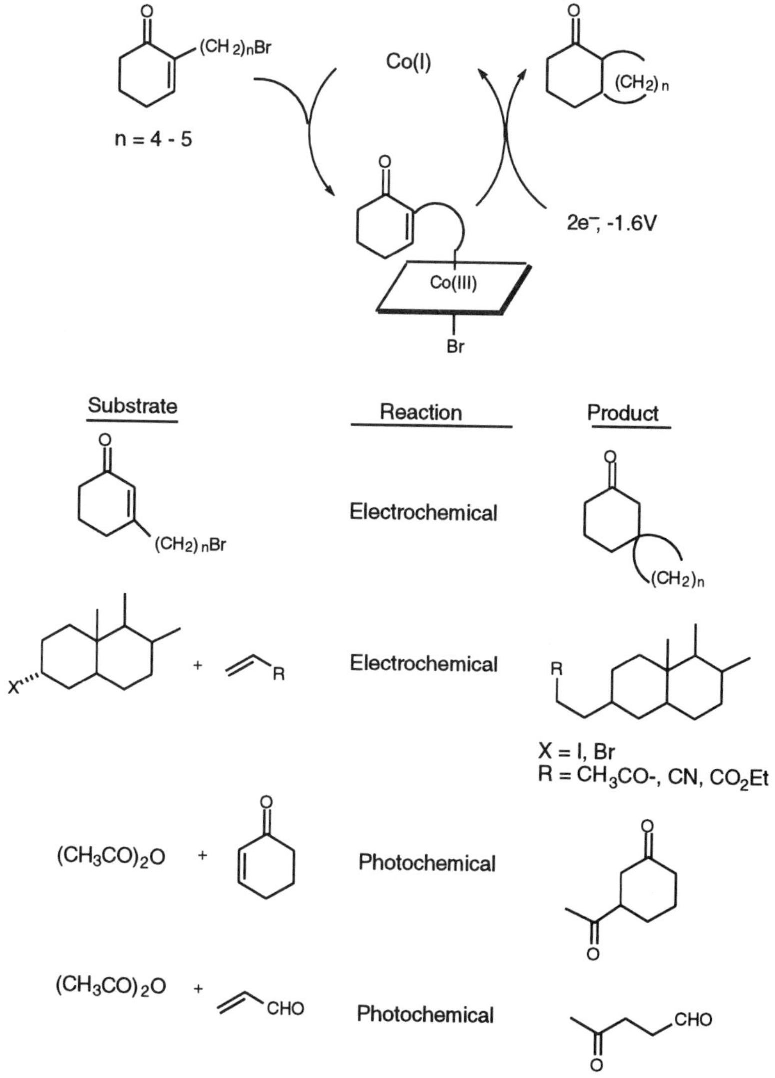

Figure 60. C–C bond formation with vitamin B$_{12}$.

References

1.   Slough, G. A.; Bergman, R. G.; Heathcock, C. H. *J. Am. Chem. Soc.* **1989**, *111*, 938; Ito, Y.; Sawamura, M.; Hayashi, T. *J. Am. Chem. Soc.* **1986**, *108*, 6405.

2.   Examples of catalytic asymmetric aldol reactions: Ho, Y.; Sawamura, M.; Hayashi, T. *J. Am. Chem. Soc.* **1986**, *108*, 6405.; Slough, G. A.; Bergman, R. G.; Heathcock, C. H. *J. Am. Chem. Soc.* **1989**, *111*, 938; Kobayashi, S.; Fujishita, Y.; Mukaiyama, T. *Chem. Lett.* **1990**, 1455; Furuta, K.; Maruyama, T.; Yamamoto, H. *J. Am. Chem. Soc.* **1991**, *113*, 1041.

3.   Whitesides, G. M.; Wong, C.-H. *Angew. Chem. Int. Ed. Engl.* **1985**, *24*, 617.

4.   Wong, C.-H. in *Enzymes as Catalysts in Organic Synthesis* ; Schneider, M. P., Ed.; D. Reidel: Dordrecht, 1986; p. 199.

5.   Wong, C.-H. in *Biocatalysis* ; Abramowics, D. A., Ed.; von Nostrand Reinhold: New York, 1990; p. 319.

6.   Toone, E. J.; Simon, E. S.; Bednarski, M. D.; Whitesides, G. M. *Tetrahedron* **1989**, *45*, 5365.

7.   Horecker, B. L.; Tsolas, O.; Lai, C.-Y. in *The Enzymes* ; Boyer, P. D., Ed.; Academic Press: New York, 1975; Vol. VII, p. 213; Rutter, W. J. *Fed. Proc.* **1964**, *23*, 1248.

8.   Grazi, E; Cheng,. T.; Horecker, B. L.*Biochem. Biophys. Res. Commun.* **1962**, 7, 250.

9.   a) Tolan, D. R.; Amsden, A. B.; Putney, S. D.; Urdea, M. S.; Penhoet, E. E. *J. Biol. Chem.* **1984**, *259*, 1127. b) Morris, A. J.; Tolan, D. R.; *J. Biol. Chem.* **1993**, *268*, 1095.

10.   Scopes, R. K. *Biochem. J.* **1977**, *161*, 253; Izzo, P.; Costanzo, P.; Lupo, A.; Rippa, E.; Paolella, G.; Salvatore, F. *Eur. J. Biochem.* **1988**, *174*, 569; Rottmann, W. H.; Tolan, D. R.; Penhoet, E. E. *Proc. Natl. Acad. Sci.* **1984**, *81*, 2738.

11.   Stribling, D.; Perham, R. *Biochem. J.* **1973**, *131*, 833; Baldwin, S. A.; Perham, R. N.;Stribling, D. *Biochem. J.* **1978**, *169*, 633; Baldwin, S. A.; Perham, R. N. *Biochem. J.* **1978**, *169*, 643; Ruffler, D.; Bock, A. *J. Bacteriol.* **1973**, *116*, 1054; Atherly, A. G.; Russell, P. *Can. J. Microbiol.* **1979**, *25*, 937; Schreyer, R.; Bock, A. *J. Bacteriol.* **1973**, *115*, 268; Carnosus, S.; Brockamp, H. P.; Kula, M.-R. *Tetrahedron Lett.*, **1990**, *31*, 7123.

12.   Sugimoto, S.; Nosoh, Y. *Biochim. Biophys. Acta* **1971**, *235*, 210; Hill, H. A. O.; Lobb, R. R.; Sharp, S. L.; Stokes, A. M.; Harris, J. I.; Jack R. S. *Biochem. J.* **1976**, *153*, 551.

13. von der Osten, C. H.; Barbas, C. F., III; Wong, C.-H.; Sinskey, A. J. *Mol. Microbiol.* **1989**, *3*, 1625.

14. Brenner-Holzach, O. *Arch. Biochem. Biophys.* **1979**, *194*, 321; Malek, A. A.; Hy, M.; Honegger, A.; Rose, K.; Brenner-Holzach, O. *Arch. Biochem. Biophys.* **1988**, *266*, 10.

15. Marquardt, R. *Can. J. Biochem.* **1971**, *49*, 658; Blostein, R.; Rutter, W. J. *J. Biol. Chem.* **1963**, *238*, 3280; Penhoet, E. E.; Kochman, M.; Rutter, W. J. *Biochemistry* **1969**, *8*, 4391; Gracy, R. W.; Lacko, A. G.; Horecker, B. L. *J. Biol. Chem.* **1969**, *244*, 3913; Peanasky, R. J.; Lardy, H. A. *J. Biol. Chem.* **1958**, *233*, 365; Matsushima, T.; Kawabe, S.; Sugimura, T. *J. Biochem. Jap.* **1968**, *63*, 555.

16. Richards, O. C.; Rutter, W. J. *J. Biol. Chem.* **1961**, *236*, 3177.

17. Kruger, I.; Schnarrenberger, C. *Eur. J. Biochem.* **1983**, *136*, 101. Fluri, R.; Ramasarma, T.; Horecker, B. L. *Eur. J. Biochem.* **1967**, *1*, 1127. Valentin, M.-L.; Bolte, J. *Tetrahedron Lett.* **1993**, *34*, 8103.

18. Willard, J. M.; Gibbs, M. *Biochim. Biophys. Acta* **1968**, *151*, 438; Russell, G. K.; Gibbs, M. *Biochim. Biophys. Acta* **1967**, *132*, 145.

19. Fernandez-Sousa, J. M.; Gavilanes, F. G.; Gavilanes, J. G.; Paredes, J. A. *Arch. Biochem. Biophys.* **1978**, *188*, 456.

20. Komatsu, S. K.; Feeney, R. E. *Biochim. Biophys. Acta* **1970**, *206*, 305.

21. Lebherz, H. G.; Rutter, W. J. *J. Biol. Chem.* **1973**, *248*, 1650; Ingram, J. M.; Hochster, R. M. *Can. J. Biochem.* **1967**, *45*, 929; London, J. *J. Biol. Chem.* **1974**, *249*, 7977; Barnes, E. M. Akagi, J. M.; Himes, R. H. *Biochim. Biophys. Acta* **1971**, *227*, 199; Freeze, H.; Brock, T. D. *J. Bacteriol.* **1970**, *101*, 541; Lebherz, H. G.; Bradshaw, R. A.; Rutter, W. J. *J. Biol. Chem.* **1973**, *248*, 1660; Bai, N. J.; Pai, M. R.; Murthy, P. S.; Venkitasubramanian, T. A. *Arch. Biochem. Biophys.* **1975**, *168*, 230.

22. Lebherz, H. G. *Biochemistry* **1972**, *11*, 2243; Kochman, M.; Kwiatkowska, D. *Arch. Biochem. Biophys.* **1972**, *152*, 856; Ting, S.-M.; Sia, C. L.; Lai, C. Y.; Horecker, B. L. *Arch. Biochem. Biophys.* **1971**, *144*, 485.

23. Kuo, D. J.; Rose, I. A. *Biochemistry* **1985**, *24*, 3947; Rose, I. A.; Warms, J. V. B. *Biochemistry* **1985**, *24*, 3952; Wurster, B.; Hess, B. *Biochem. Biophys. Res. Commun.* **1973**, *55*, 985; Ray, B. D.; Harper, E. T.; Fife, W. K. *J. Am. Chem. Soc.* **1983**, *105*, 3731; Grazi, E.; Trombetta, G. *Arch. Biochem. Biophys.* **1984**, *233*, 595; Sygusch, J.; Beaudry, D. *J. Biol. Chem.* **1984**, *259*, 10222; Periana, R. A.; Motiu-Degrood, R.; Chiang, Y.; Hupe, D. J. *J. Am. Chem. Soc.* **1980**, *102*, 3923; Grazi, E.

Biochem. Biophys. Res. Commun. **1974**, *59*, 450; Patthy, L. *Eur. J. Biochem.* **1978**, *88*, 191; Rose, I. A.; Warms, J. V. B.; Kuo, D. J. *J. Biol. Chem.* **1987**, *262*, 692.

24. Kadonaga, J. T.; Knowles, J. R. *Biochemistry* **1983**, *22*, 130; Belasco, J. G.; Knowles, J. R. *Biochemistry* **1983**, *22*, 122; Kobes, R. D.; Simpson, R. T.; Vallee, B. L.; Rutter, W. J. *Biochemistry* **1969**, *8*, 585; Midelfort, C. F.; Gupta, R. K.; Rose, I. A. *Biochemistry* **1976**, *15*, 2178; Smith, G. M.; Mildvan, A. S. *Biochemistry* **1981**, *20*, 4340; Hartman, F. C. *Biochemistry* **1970**, *9*, 1776; Galdes, A.; Hill, H. A. O. *Biochem. J.* **1978**, *171*, 539; Magnien, A.; Le Clef, B.; Biellmann, J.-F. *Biochemistry* **1984**, *23*, 6858. Morris, A. J.; Tolan, D. R. *J. Biol. Chem.* **1993**, *268*, 1095.

25. Sygusch, J.; Beaudy, D.; Allaire, M. *Proc. Natl. Acad. Sci.* **1987**, *84*, 7846.

26. a) Gamblin, S. J.; Cooper, B.; Millar, J. R.; Davies, G. J.; Littlechild, J. A.; Watson, H. C. *FEBS Lett.* **1990**, *262*, 282. b) Hester, G.; Brenner-Holzhach, O.; Rossi, F. A.; Stuck-Donatz, M.; Winterhalter, K. H.; Smit, J. D. G.; Piontek, K. *FEBS Lett.* **1991**, *292*, 237.

27. Lubini, D. G. E.; Christen, P.; *Proc. Natl. Acad. Sci. USA* **1979**, *76*, 2527; Lai, C. Y.; Nakai, N. Chang, D. *Science* **1974**, *183*, 1204.

28. Lees, W. J.; Whitesides, G. M. *J. Org. Chem.* **1993**, *58*, 1887.

29. Bednarski, M. D.; Simon, E. S.; Bischofberger, N.; Fessner, W.-D.; Kim, M.-J.; Lees, W.; Saito, T.; Waldmann, H.; Whitesides, G. M. *J. Am. Chem. Soc.* **1989**, *111*, 627.

30. von der Osten, C. H.; Sinskey, A. J.; Barbas, C. F., III; Pederson, R. L.; Wang, Y.-F.; Wong, C.-H. *J. Am. Chem. Soc.* **1989**, *111*, 3924.

31. Alefounder, P. R.; Baldwin, S. A.; Perham, R. N.; Short, N. J. *Biochem. J.* **1989**, *257*, 529.

32. Schwelberger, H. G.; Kohlwein, S. D.; Paltauf, F. *Eur. J. Biochem.* **1989**, *180*, 301.

33. Wong, C.-H.; Whitesides, G. M. *J. Org. Chem.* **1983**, *48*, 3493.

34. Wong, C.-H.; Mazenod, F. P.; Whitesides, G. M. *J. Org. Chem.* **1983**, *48*, 3493.

35. Durrwachter, J. R.; Drueckhammer, D. G.; Nozaki, K.; Sweers, H. M.; Wong, C.-H. *J. Am. Chem. Soc.* **1986**, *108*, 7812.

36. Durrwachter, J. R.; Wong, C.-H. *J. Org. Chem.* **1988**, *53*, 4175.

37. Pederson, R. L.; Kim, M.-J.; Wong, C.-H. *Tetrahedron Lett.* **1988**, *29*, 4645.

38. Ziegler, T.; Straub, A.; Effenberger, F. *Angew. Chem., Int. Ed. Engl.* **1988**, *27*, 716.

39. Straub, A.; Effenberger, F.; Fischer, P. *J. Org. Chem.* **1990**, *55*, 3926.

40. Turner, N. J.; Whitesides, G. M. *J. Am. Chem. Soc.* **1989**, *111*, 624.

41. Borysenko, C. W.; Spaltenstein, A.; Straub, J. A.; Whitesides, G. M. *J. Am. Chem. Soc.* **1989**, *111*, 9275.

42. Schmid, W.; Whitesides, G. M. *J. Am. Chem. Soc.* **1990**, *112*, 9670.

43. Bednarski, M. D.; Waldmann, H. J.; Whitesides, G. M. *Tetrahedron Lett.* **1986**, *27*, 5807. Fructose and sorbose diphosphates are better substrates than the corresponding 1-phosphates: Richards, O. C.; Rutter, W. J. *J. Biol. Chem.* **1961**, *236*, 3185.

44. Bischofberger, N.; Waldmann, H.; Saito, T.; Simon, E. S.; Lees, W.; Bednarski, M. D.; Whitesides, G. M. *J. Org. Chem.* **1988**, *53*, 3457.

45. Crans, D. C.; Whitesides, G. M. *J. Am. Chem. Soc.* **1985**, *107*, 7019; Crans, D. C.; Kazlauskas, R. J.; Hirschbein, B. L.; Wong, C.-H.; Abril, O.; Whitesides, G. M. *Methods Enzymol.* **1987**, *136*, 263. For phosphoenolpyruvate synthesis: Hirschbein, B. L.; Mazenod, F. P.; Whitesides, G. M. *J. Org. Chem.* **1982**, *47*, 3765.

46. Pederson, R. L.; Esker, J.; Wong, C.-H. *Tetrahedron* **1991**, *47*, 2643.

47. a)Effenberger, F.; Straub, A. *Tetrahedron Lett.* **1987**, *28*, 1641; b) Colbran, R. L.; Jones, J. K. N.; Matheson, N. K.; Rozema, I. *Carbohydr. Res.* **1967**, *4*, 355.

48. Fessner, W.-D.; Walter, C. *Angew. Chem., Int. Ed. Engl.* **1992**, *31*, 614.

49. Drueckhammer, D. G.; Durrwachter, J. R.; Pederson, R. L.; Crans, D. C.; Daniels, L.; Wong, C.-H. *J. Org. Chem.* **1989**, *54*, 70; Crans, D. C.; Sudhakar, K.; Zamborelli, T. J. *Biochemistry* **1992**, *31*, 6812.

50. Serianni, A. S.; Cadman, E.; Pierce, J.; Hayes, M. L.; Barker, R. *Method. Enzymol.* **1982**, *89*, 83.

51. Wong, C.-H.; Shen, G. -J.; Pederson, R. L.; Wang, Y.-F.; Hennen, W. J. *Method. Enzymol.* **1991**, *202*, 591.

52. Wong, C.-H.; Druekhammer, D. G.; Sweers, H. M. in *Fluorinated Carbohydrates: Chemical and Biochemical Aspects*; Taylor, N. F., Ed.; ACS Symposium Series 374; American Chemical Society: Washington, D.C., 1988, pp 29–42.

53. Pederson, R. L.; Liu, K. K.-C.; Rutan, J. F.; Chen, L.; Wong, C.-H. *J. Org. Chem.* **1990**, *55*, 4897.

54. Elbein, A. D. *Ann. Rev. Biochem.* **1987**, *56*, 497; Look, G. C.; Fotsch, C. H.; Wong, C.-H. *Acc. Chem. Res.* **1993**, *26*, 182.

55. Kajimoto, T.; Liu, K. K.-C.; Pederson, R. L.; Zhong, Z.; Ichikawa, Y.; Porco, J. A., Jr.; Wong, C.-H. *J. Am. Chem. Soc.*, **1991**, *113*, 6187.

56. Hung, R. R.; Straub, J. A.; Whitesides, G. M. *J. Org. Chem.* **1991**, *56*, 3849.

57. a) Liu, K. K.-C.; Kajimoto, T.; Chen, L.; Zhong, Z.; Ichikawa, Y.; Wong, C.-H. *J. Org. Chem.* **1991**, *56*, 6280; b) Kajimoto, T.; Chen, L.; Liu, K. K.-C.; Wong, C.-H. *J. Am. Chem. Soc.* **1991**, *113*, 6678.

58. Wang, Y.-F.; Dumas, D. P.; Wong, C.-H. *Tetrahedron Lett.* **1993**, *34*, 403.

59. Schultz, M.; Waldmann, H.; Vogt, W.; Kunz, H. *Tetrahedron Lett.* **1990**, *31*, 867.

60. Hochster, R. M.; Watson, R. W. *Arch. Biochem. Biophys.* **1954**, *48*, 120; Boguslawski, G. *J. App. Biochem.* **1983**, *5*, 186; Makkee, M.; Kieboom, H. van Bekkum, A. P. G. *Recl. Trav. Chim. Pays-Bas* **1984**, *103*, 361; Bock, K.; Meldal, M.; Meyer, B.; Wiebe, L. *Acta Chem. Scand. B* **1983**, *37*, 101; Callens, M.; Kersters-Hilderson, H.; Vangrysperre, W.; De Bruyne, C. K. *Enzym. Microbiol. Technol.* **1988**, *10*, 695; For a revised mechanism involving ring opening followed by a 1,2-hydride shift, see: Collyer, C. A.; Henrick, K.; Blow, D. M. *J. Mol. Biol.* **1990**, *212*, 211; Whitlow, M.; Howard, A. J.; Finzel, B. C.; Poulos, T. L.; Winborne, E.; Gilliland, G. L. *Proteins* **1991**, *9*, 153. For site-directed mutagenisis on the enzyme see: Lambeir, A.-M.; Lauwereys, M.; Stanssens, P.; Mrabet, N. T.; Snauwaert, J.; van Tilbeurgh, H.; Mattyssens, G.; Lasters, I.; De Maeyer, M.; Wodak, S. J.; Jenkins, J.; Chiadmi, M.; Janin, J. *Biochemistry* **1992**, *31*, 5459; Meng, M.; Bagdasarian, M.; Zeikus, J. G. *Bio/Technology* **1993**, *11*, 1157.

61. Christensen, U.; Tuchsen, E.; Andersen, B. *Acta Chem. Scand. B* **1975**, *29*, 81.

62. Fessner, W.-D.; Badia, J.; Eyrisch, O.; Schneider, A.; Sinerius, G. *Tetrahedron Lett.* **1992**, *33*, 5231.

63. Liu, K. K.-C.; Pederson, R. L.; Wong, C.-H. *J. Chem. Soc., Perkin Trans. 1* **1991**, 2669.

64. Chiu, T.-H.; Feingold, D. S. *Biochemistry* **1969**, *8*, 98; Takagi, Y. *Method. Enzymol.* **1966**, *9*, 542; Chiu, T.-H.; Otto, R.; Power, J.; Feingold, D. S. *Biochim. Biophys. Acta* **1966**, *127*, 249.

65. Ozaki, A.; Toone, E. J.; von der Osten, C. H.; Sinskey, A. J.; Whitesides, G. M. *J. Am. Chem. Soc.* **1990**, *112*, 4970.

66. Fessner, W.-D.; Sinerius, G.; Schneider, A.; Dreyer, M.; Schulz, G. E.; Badia, J.; Aguilar, J. *Angew. Chem., Int. Ed. Engl.* **1991**, *30*, 555.

67. Chiu, T.-H.; Feingold, D. S. *Biochem. Biophys. Res. Commun.* **1965**, *19*, 511. Zhou, P.; Mohd. Salleh, H.; Chan, P. C. M.; Lajoie, G.; Hovek, J. F.; Chandra Nambiar, P. T.; Ward, O. P. *Carbohydrate Res.* **1993**, *239*, 155.

68. Lees, W. J.; Whitesides, G. M. *Bioorg. Chem.* **1992**, *20*, 173.

69. Crow, V. L.; Thomas, T. D. *J. Bacteriol.* **1982**, *151*, 600; Bissett, D. L.; Anderson, R. L. *J. Biol. Chem.* **1980**, *255*, 8750. van Rooijen, R. J.; van Schalkwijk, S.; de Vos, W. M. *J. Biol. Chem.* **1991**, *266*, 7176.

70. Fessner, W.-D.; Eyrisch, O. *Angew. Chem., Int. Ed. Engl.* **1992**, *31*, 56; Eyrisch, O.; Sinerius, G.; Fessner, W.-D. *Carbohydr. Res.* **1993**, *238*, 287.

71.   Comb, D. G.; Roseman, S. *J. Biol. Chem.* **1960**, *235*, 2529; Brunett, P.; Jourdian, G. W.; Roseman, S. *J. Biol. Chem.* **1962**, *237*, 2447; Devries, G. H.; Binkley, S. B. *Arch. Biochem. Biophys.* **1972**, *151*, 234.

72.   Uchida, Y.; Tsukada, Y.; Sugimori, T. *J. Biochem.* **1984**, *96*, 507; Uchida, Y.; Tsukada, Y.; Sugimori, T. *Agric. Biol. Chem.* **1985**, *49*, 181.

73.   Deijl,C. M.; Vliegenthart, J. F. G. *Biochem. Biophys. Res. Commun.* **1983**, *111*, 668.

74.   Blacklow, R. S.; Warren, L. *J. Biol. Chem.* **1962**, *237*, 3520; Merker, R. I.; Troy, F. A. *Glycobiology* **1990**, *1*, 93; Warren, L.; Blacklow, R. S. *Biochem. Biophys. Res. Comm.* **1962**, *7*, 433.

75.   Aisaka, K.; Tamura, S.; Arai, Y.; Uwajima, T. *Biotechnol. Lett.* **1987**, *9*, 633; for the DNA sequence of the NeuAc aldolase gene see: Kawakami, B.; Kudo, T.; Narahashi, Y.; Horikoshi, K. *Agric. Biol. Chem.* **1986**, *50*, 2155.

76.   Augé, C.; David, S.; Gautheron, C. *Tetrahedron Lett.* **1984**, *25*, 4663.

77.   Augé, C.; David, S.; Gautheron, C.; Veyrieres, A. *Tetrahedron Lett.* **1985**, *26*, 2439.

78.   Augé, C.; Gautheron, C. *J. Chem. Soc., Chem. Commun.* **1987**, 859.

79.   Kim, W. M.-J.; Hennen, J.; Sweers, H. M.; Wong, C.-H. *J. Am. Chem. Soc.* **1988**, *110*, 6481. Liu, J. L.-C.; Shen, G.-J.; Ichikawa, Y.; Rutan, J. R.; Zapata, C.; Vann, W. F.; Wong, C.-H. *J. Am. Chem. Soc.* **1992**, *114*, 3901.

80.   Bednarski, M. D.; Chenault, H. K.; Simon, E. S.; Whitesides, G. M. *J. Am. Chem. Soc.* **1987**, *109*, 1283.

81.   Ichikawa, Y.; Liu, J. L.-C.; Shen, G.-J. Wong, C.-H. *J. Am. Chem. Soc.* **1991**, *113*, 6300.

82.   a)Simon, E. S.; Bednarski, M. D.; Whitesides, G. M. *J. Am. Chem. Soc.* **1988**, *110*, 7159; b) Kragl, U.; Gygax, D.; Ghisalba, O.; Wandrey, C. *Angew. Chem., Int. Ed. Engl.* **1991**, *30*, 827.

83.   Shukla, A. K.; Schauer, R. *Anal. Biochem.* **1986**, *158*, 158.

84.   Augé, C.; David, S.; Gautheron, C.; Malleron, A.; Cavaye, B. *New J. Chem.* **1988**, *12*, 733.

85.   Schrell, A.; Whitesides, G. M. *Liebigs Ann. Chem.* **1990**, 1111.

86.   Augé, C.; Gautheron, C.; David, S. *Tetrahedron* **1990**, *46*, 201.

87.   Brossmer, R.; Rose, U.; Kasper, D.; Smith, T. L; Grasmuk, H.; Unger, F. M. *Biochem. Biophys. Res. Commun.* **1980**, *96*, 1282.

88.   Schauer, R. *TIBS* **1985**, *10*, 357; Corfield, A. P.; Schauer, R. in *Sialic Acids*; R. Schauer, Ed.; Springer-Verlag: New York, 1982; p. 195; Paulson, J. C. *TIBS* **1989**, *14*, 272.

89.   Finne, J. *TIBS* **1985**, 129; Livingston, B. D.; De Roberts, E. M.; Paulson, J. C. *Glycobiology* **1990**, *1*, 39; Pozsgay, V.; Brisson, J.-R.; Jennings, H. J.; Allen, S.; Paulson, J. C. *J. Org. Chem.* **1991**, *56*, 3377.

90.   Drueckhammer, D. G.; Hennen, W. J.; Pederson, R. L.; Barbas, C. F., III; Gautheron, C. M.; Krach, T.; Wong, C.-H. *Synthesis* **1991**, 499.

91.   Schrell, A.; Whitesides, G. M. *Liebigs Ann. Chem.* **1990**, 1111.

92.   David, S.; Augé, C.; Gautheron, C. *Adv. Carbohydr. Chem. Biochem.* **1991**, *49*, 175; Schwark, J -R.; Halcomb, R. L.; Lin, C.-H.; Ichikawa, Y.; Wong, C.-H. in preparation.

93.   Spaltenstein, A.; Whitesides, G. M. *J. Am. Chem. Soc.* **1991**, *113*, 686.

94.   Nagy, J. O.; Bednarski, M. D. *Tetrahedron Lett.* **1991**, *32*, 3953; Spevak, W.; Nagy, J. O.; Charych, D. H.; Schaefer, M. E.; Gilbert, J. H.; Bednarski, M. D. *J. Am. Chem. Soc.* **1993**, *115*, 1146.

95.   Augé, C.; Bouxom, B.; Cavaye, B.; Gautheron, C. *Tetrahedron Lett.* **1989**, *30*, 2217.

96.   a)Gautheron-Le Narvor, C.; Ichikawa, Y.; Wong, C.-H. *J. Am. Chem. Soc.* **1991**, *113*, 7816; b) Lin, C.-H.; Sugai, T.; Halcomb, R. L.; Ichikawa, Y.; Wong, C.-H. *J. Am. Chem. Soc.* **1992**, *114*, 10138.

97.   a)Zhou, P.; Mohd. Salleh, H.; Honek, J. F. *J. Org. Chem.* **1993**, *58*, 264; b) Halcomb, R. L.; Wong, C.-H. unpublished work.

98.   Raetz, C. R. H.; Dowhan, W. *J. Biol. Chem.* **1990**, *265*, 1235.

99.   Schumann, R. R.; Leong, S. R.; Flaggs, G. W.; Gray, P. W.; Wright, S. D.; Mathison, J. C.; Tobias, P. S.; Ulevitch, R. J. *Science* **1990**, *249*, 1429.

100.   Hammond, S. M.; Claesson, A.; Jansson, A. M.; Larsson, L.-G.; Pring, B. G.; Town, C. M.; Ekstrom, B. *Nature* **1987**, *327*, 730; Luthman, K.; Orbe, M.; Waglund, T.; Claesson, A. *J. Org. Chem.* **1987**, *52*, 3777.

101.   Andersson, F. O.; Classon, B.; Samuelsson, B. *J. Org. Chem.* **1990**, *55*, 4699.

102.   Norbeck, D. W.; Kramer, J. B. *Tetrahedron Lett.* **1987**, *28*, 773.

103.   Ray, P. H.; Kelsey, J. E.; Biogham, E. C.; Benedict, C. D.; Miller, T. A. in *Bacterial Lipopolysaccharides*; ACS Symposium Series 231, American Chemical Society: Washington, D.C., 1983; p. 141.

104.   Danishefsky, S. J.; Deninno, M. P. *Angew. Chem., Int. Ed. Engl.* **1987**, *26*, 15.

105.   Ghalambor, M. A.; Heath, E. C. *J. Biol. Chem.* **1966**, *241*, 3222.

106.   Ghalambor, M. A.; Heath, E. C. *Method. Enzymol.* **1966**, *9*, 534.

107.   Sugai, T.; Shen, G.-J., Ichikawa, Y.; Wong, C.-H. *J. Am. Chem. Soc.* **1993**, *115*, 413.

108.   Ray, P. H. *J. Bacteriol.* **1980**, *141*, 635; Ray, P. H. *Method. Enzymol.* **1982**, *83*, 525.

109. Bednarski, M. D.; Crans, D. C.; DiCosimo, R.; Simon, E. S.; Stein, P. D.; Whitesides, G. M. *Tetrahedron Lett.* **1988**, *29*, 427.

110. Levin, D. H.; Racker, E. *J. Biol. Chem.* **1959**, *234*, 2532.

111. Woisetschlager, M.; Hogenauer, G. *J. Bacteriol.* **1986**, *168*, 437.

112. Haslam, E. in *The Shikimate Pathway*; Halsted: New York, 1974.

113. Sprinson, D. B.; Srinivasan, P. R.; Katagiri, M. *Method. Enzymol.* **1962**, *5*, 394.

114. Ogino, T.; Garner, C.; Markley, J. L., Herrmann, K. M. *Proc. Natl. Acad. Sci.* **1982**, *79*, 5828.

115. Reimer, L. M.; Conley, D. L.; Pompliano, D. L.; Frost, J. W. *J. Am. Chem. Soc.* **1986**, *108*, 8010.

116. Draths; K .M.; Frost, J. W. *J. Am. Chem. Soc.* **1990**, *112*, 1657. Draths; K .M.; Frost; J. W. *ibid.* **1990**, *112*, 9630. Draths, K. M.; Pompliano, D. L.; Conley, D. L.; Frost, J. W.; Berry, A.; Disbrow, G. L.; Staversky, R. J.; Lievense, J. C. *ibid.* **1992**, *114*, 3956.

117. Dekker, E. E. *Bioorg. Chem.* **1977**, *1*, 59; Dekker, E. E.; Kobes, R. D.; Grady, S. R. *Method. Enzymol.* **1975**, *42*, 280.

118. Scholtz, J. M.; Schuster, S. M. *Bioorg. Chem.* **1984**, *12*, 229; Rosso, R. G.; Adams, E. *J. Biol. Chem.* **1967**, *242*, 5524.

119. a)Nishihara, H.; Dekker, E. E. *J. Biol. Chem.* **1972**, *247*, 5079; b) Floyd, N. C.; Liebster, M. H.; Turner, N. J. *J. Chem. Soc., Perkin Trans. 1* **1992**, *1*, 1085.

120. Mavrisdis, L. M.; Hatada, M. H.; Tulinsky, A.; Leboida, L. *J. Mol. Biol.* **1982**, *162*, 419.

121. a) Allen, S. T.; Heintzelman, G. R.; Toone, E. J. *J. Org. Chem.* **1992**, *57*, 426; b) Conway, T.; Yi, K. C.; Egan, S. E.; Wolf, R. E.; Rowley, D. L. *J. Bacteriol.* **1991**, *173*, 5247.

122. Blumenthal, H. J.; Jepson, T. *Bacteriol. Proc.* **1965**, 82; Fish, D. C.; Blumenthal, H. J. *Method. Enzymol.* **1966**, *9*, 529.

123. Racker, E. *J. Biol. Chem.* **1952**, *196*, 347.

124. McGeown, M. G.; Malpress, F. H. *Nature* **1952**, *170*, 575.

125. Hoffee, P. A. *Arch. Biochem. Biophys.* **1968**, *126*, 795.

126. Valentin-Hansen, P.; Boetius, F.; Hammer-Jespersen, K.; Svendsen, I. *Eur. J. Biochem.* **1982**, *125*, 561.

127. Barbas, C. F., III; Wang, Y.-F.; Wong, C.-H. *J. Am. Chem. Soc.* **1990**, *112*, 2013. Chen, L.; Dumas, D. P.; Wong, C.-H. *ibid.* **1992**, *114*, 741.

128. Pricer, W. E., Jr.; Horecker, B. L. *J. Biol. Chem.* **1960**, *235*, 1292.

129. Rosen, O. M.; Hoffee, P.; Horecker, B. L. *J. Biol. Chem.* **1965**, *240*, 1517.

130. Racker, E. in *The Enzymes*; Boyer, P. D.; Lardy, H.; Myrback, K., Eds.; Academic Press: New York, 1961; Vol. V, p 397; Mocali, A.; Aldinucci, D.; Paoletti, F. *Carbohydr. Res.* **1985**, *143*, 288.

131. The DNA sequence of the yeast transketolase gene has recently b...n determined: Fletcher, T. S.; Kwee, I. L.; Nakada, T.; Largman, C.; Martin, B. M. *Biochemistry* **1992**, *31*, 1892.

132. Bolte, J.; Demuynck, C.; Samaki, H. *Tetrahedron Lett.* **1987**, *28*, 5525. Hobbs, G. R.; Lilly, M. D.; Turner, N. J.; Ward, J. M.; Willets, A. J.; Woodley, J. M. *J. Chem. Soc., Perkin Trans. 1* **1993**, 165.

133. Demuynck, C.; Fisson, F.; Bennani-Baiti, I.; Samaki, H.; Mani, J.-C. *Agric. Biol. Chem.* **1990**, *54*, 3073.

134. Srere, P.; Cooper, J. R.; Tabachnick, M.; Racker, E. *Arch. Biochim. Biophys.* **1958**, *74*, 295.

135. Kobori, Y.; Myles, D. C.; Whitesides, G. M. *J. Org. Chem.* **1992**, *57*, 5899.

136. Demuynck, C.; Bolte, J.; Hecquet, L.; Dalmas, V. *Tetrahedron Lett.* **1991**, *32*, 5085.

137. Effenberger, F.; Null, V.; Ziegler, T. *Tetrahedron Lett.* **1992**, *33*, 5157.

138. Myles, D. C.; Andrulis, P. J., III; Whitesides, G. M. *Tetrahedron Lett.* **1991**, *32*, 4835.

139. Moradian, A.; Benner, S. A. *J. Am. Chem. Soc.* **1992**, *114*, 6980.

140. Effenberger, F.; Ziegler, T.; Forster, S. *Angew. Chem., Int. Ed. Engl.* **1987**, *26*, 458.

141. Effenberger, F.; Hörsch, B.; Weingart, F.; Ziegler, T.; Kühner, S. *Tetrahedron Lett.* **1991**, *32*, 2605.

142. Ognyanov, V. I.; Datcheva, V. K.; Kyler, K. S. *J. Am. Chem. Soc.* **1991**, *113*, 6992.

143. Huuhtanen, T. T.; Kanerva, L. T. *Tetrahedron Asymmetry.* **1992**, *3*, 1223.

144. Effenberger, F.; Hörsch, B.; Ziegler, T. *Tetrahedron Lett.* **1990**, *31*, 1249.

145. Niedermeyer, U.; Kula, M.-R. *Angew. Chem., Int. Ed. Engl.* **1990**, *29*, 386.

146. Neuberg, C.; Hirsch, J. *Biochem. Z.* **1921**, *115*, 282; Beevarova, H.; Hane, O.; Mauk, K. *Folia Microbiol.* **1963**, *8*, 165.

147. Behrens, M.; Iwaneff, N. N. *Biochem. Z.* **1921**, *121*, 311; Fuganti, C.; Grasselli, P. *Chem. Ind.* **1977**, 983.

148. For a review see: Servi, S. *Synthesis* **1990**, 1.

149. Rose, A. H. in *Industrial Microbiology*; Butterworths, Washington, **1961**, p. 264; Braun, M.; Wild, H. *Angew. Chem., Int. Ed. Engl.* **1984**, *23*, 723.

150. Ohta, H.; Ozaki, K.; Konishi, J.; Tsuchihashi, G. *Agric. Biol. Chem.* **1986**, *50*, 1261.

151. Kren, V.; Crout, D. H. G.; Dalton, H.; Hutchinson, D. W.; König, W.; Turner, M. M.; Dean, G.; Thomson, N. *J. Chem. Soc., Chem. Commun.* **1993**, 341.

152. Grue-Sorensen, G.; Spenser, I. D. *J. Am. Chem. Soc.* **1988**, *110*, 3714. Bringer-Meyer, S.; Sahm, H. *Biocatalysis* **1988**, *1*, 321. Crout, D. H. G.; Dalton, H.; Hutchinson, D. W.; Miyagoshi, M. *J. Chem. Soc., Perkin Trans. 1* **1991**, 1329.

153. Billhardt, U.-M.; Stein, P.; Whitesides, G. M. *Bioorg. Chem.* **1989**, *17*, 1.

154. Rieke, E.; Barry, S.; Mosbach, K. *Eur. J. Biochem.* **1979**, *100*, 203.

155. Patel, S. S.; Conlon, H. D.; Walt, D. R. *J. Org. Chem.* **1986**, *51*, 2842; Patel, S. S.; Walt, D. R. *J. Biol. Chem.* **1987**, *262*, 7132.

156. Ouyang, T.; Walt, D. R.; Patel, S. S. *Bioorg. Chem.* **1990**, *18*, 131.

157 . Martin, D. P.; Drueckhammer, D. G. *J. Am. Chem. Soc.* **1992**, *114*, 7287.

158. Byrom, D. *TIBTECH* **1987**, *5*, 246 - 250; Suzuki, T.; Yamane, T.; Shimizu, S. *Appl. Microb. Biotech.* **1986**, *24*, 370.

159. a) Peoples,O. P.; Masamune, S.; Walsh, C. T.; Sinskey, A. J. *J. Biol. Chem.* **1987**, *262*, 97; Masamune, S.; Palmer, M. A. J.; Gamboni, R.; Thompson, S.; Davis, J. T.; Williams, S. F.; Peoples, O. P.; Sinskey, A. J.; Walsh, C. T. *J. Am. Chem. Soc.* **1989**, *111*, 1879. b) Seebach, D.; Zuger, M. F. *Tetrahedron Lett.* **1984**, *25*, 2747; Seebach, D.; Zuger, M. F. *Helv. Chim. Acta* **1982**, *65*, 495.

160. a) McDaniel, R.; Ebert-Khosla, S.; Hopwood, D. A.; Khosla, C.; *Science* **1993**, *262*, 1546. b) Shen, B.; Hutchinson, C. R. *Science* **1993**, *262*, 1535.

161. van Tamelen, E. E.; Leopold, E. C.; Marson, S. A.; Waespe, H. R. *J. Am. Chem. Soc.* **1982**, *104*, 6479.

162. J. Bujons, R. Guajardo, K. S. Kyler, *J. Am. Chem. Soc.* **1988**, *110*, 604; J. C. Medina, K.S. Kyler, *J. Am. Chem. Soc.* **1988**, *110*, 4818.

163. Medina, J. C.; Guajardo, R.; Kyler, K. S. *J. Am. Chem. Soc.* **1989**, *111*, 2310.

164. Grundy, S. M. *N. Eng. J. Med.* **1988**, *319*, 24 and references therein.

165   Koyama, T.; Matsubara, M.; Ogura, K. *J. Biochem. Jap.* **1985**, *98*, 457; Koyama, T.; Ogura, K.; Baker, F. C.; Jamieson, G.C.; Schooley, D. A. *J. Am. Chem. Soc.* **1987**, *109*, 2853.

166. Nagasawa, T.; Yamada, H. *App. Biochem. Biotech.* **1986**, *13*, 147.

167. Scheffold, R.; Rytz, G.; Walder, L.; Orlinski, R.; Chilmonczyk, Z. *Pure Appl. Chem.* **1983**, *55*, 1791.; Essig, S. Scheffold, R. *Chimia* **1992**, 4530.

Chapter 5:   Synthesis of Glycoside Bonds

1.   Background

Oligosaccharides and polysaccharides are found in nature as components of a broad range of molecular structures.[1-40] As components of cell surface glycoproteins and glycolipids, they play vital roles in cellular communication processes[1,3-5,8-12,18-34] and as points of attachment for antibodies and other proteins. They also serve as receptor sites for bacteria and viral particles.[22-23] A particular example of an oligosaccharide that functions in cellular communication is the sialyl-Lewis X substructure (Figure 1), which mediates the adhesion of neutrophils to endothelial cells, the initial event in many inflammatory responses.[39-41] The saccharide moieties of glycoproteins are also involved in modulating protein folding and in the sorting and trafficking of proteins to appropriate cellular sites.[1,35-36]

Nature employs two groups of enzymes in the biosynthesis of oligosaccharides: the enzymes of the Leloir pathway[42-45] and those of non-Leloir pathways. The Leloir pathway enzymes are responsible for the synthesis of most glycoproteins and other glycoconjugates in mammalian systems. Glycoproteins are typically classified as either N-linked or O-linked, depending on the nature of the attachment of the carbohydrate to the protein. The N-linked are characterized by a β-glycosidic linkage between a GlcNAc residue and the δ-amide nitrogen of an asparagine. The less common O-linked glycoproteins contain an α-glycosidic linkage between GalNAc (or less commonly mannose and xylose) and the hydroxyl group of a serine or threonine. The addition of oligosaccharide chains to glycoproteins occurs cotranslationally for both O-linked and N-linked types, and occurs in the endoplasmic reticulum and the Golgi apparatus.[43] The N-linked oligosaccharides all contain the same base structure of GlcNAc and mannose residues, the similarity of which stems from their origin. The biosynthesis of the N-linked type involves an initial synthesis of a dolichyl pyrophosphoryl oligosaccharide intermediate in the endoplasmic reticulum by the action of GlcNAc-transferases and mannosyltransferases. This structure is further glucosylated, presumably to signal for transfer of the oligosaccharide to the polypeptide. The entire oligosaccharide moiety is then transferred to an Asn residue of the growing peptide chain by the enzyme oligosaccharyltransferase.[43] The Asn is typically part of the amino acid sequence Asn-X-Ser(Thr), where X ≠ Pro or Asp.[14,43,46-48] Before transport into the Golgi apparatus, the glucose residues and some mannose residues are removed by the action of glucosidase I and II and a mannosidase, in a process called trimming, to reveal a core pentasaccharide (peptide-Asn-$(GlcNAc)_2$-$(Man)_3$). The resulting core structure is further processed by mannosidases and glycosyltransferases present in the Golgi apparatus to produce either the high-mannose type, the complex type, or the hybrid type oligosaccharides. Monosaccharides are then added sequentially to this core structure to provide the fully-elaborated oligosaccharide chain.

In contrast to the dolichyl pyrophosphate mediated synthesis of *N*-linked oligosaccharides, the glycosyltransferases necessary for the synthesis of *O*-linked oligosaccharides are located in the Golgi apparatus.[43] The biosynthetic route to the *O*-linked type is also different. Monosaccharide residues are added sequentially to the growing oligosaccharide chain.

All mammalian cells, with the exception of erythrocytes, contain the necessary elements for glycosylation. In certain secretory cells, however, the preponderance of transferases is greater.[49] The structures of some typical glycoproteins and glycolipids are illustrated in Figure 1.

The major classes of cell-surface glycolipids include the glycosphingolipids (GSLs) and glycoglycerolipids. Particularly significant are gangliosides,[50] or sialic acid-containing glycosphingolipids, which are especially abundant on neural cell surfaces.[51] These compounds play a role in the differentiation of cell types and in the regulation of cell growth. Additionally, sphingosine, the lipid component of GSLs, has been suggested to function as an intracellular second messenger.[52]

The glycosyltransferases of the Leloir pathway in mammalian systems utilize as glycosyl donors monosaccharides, which are activated as glycosyl esters of nucleoside mono- or diphosphates.[43] Non-Leloir transferases typically utilize glycosyl phosphates as activated donors. The Leloir glycosyltransferases utilize primarily eight nucleoside mono- or diphosphate sugars as monosaccharide donors for the synthesis of most oligosaccharides: UDP-Glc, UDP-GlcNAc, UDP-Gal, UDP-GalNAc, GDP-Man, GDP-Fuc, UDP-GlcUA, and CMP-NeuAc (Figure 2). Many other monosaccharides, such as the anionic or sulfated sugars of heparin and chondroitin sulfate, are also found in mammalian systems, but they usually are a result of modification of a particular sugar after it is incorporated into an oligosaccharide structure.[1,6] A very diverse array of monosaccharides (e.g. xylose, arabinose, KDO, deosysugars) and oligosaccharides is also present in microorganisms, plants, and invertebrates.[17,45,53-55] The enzymes responsible for their biosynthesis , however, have not been extensively exploited for synthesis, though they follow the same principles as do those in mammalian systems. Some sugar nucleotides used by enzymes of other pathways are also shown in Scheme 2.

Chemists have employed transferases from the Leloir and non-Leloir pathways for the synthesis of oligosaccharides and glycoconjugates.[56-57] Glycosidases have also been exploited for synthesis.[56-58] The function of glycosidases *in vivo* is to cleave glycosidic bonds, however under appropriate conditions they can be useful synthetic catalysts. Each group of enzymes has certain advantages and disadvantages for synthesis. Glycosyltransferases are highly specific in the formation of glycosides. The availability of many of the necessary transferases, however, is limited. Glycosidases have the advantage of wider availability and lower cost, but they are not as

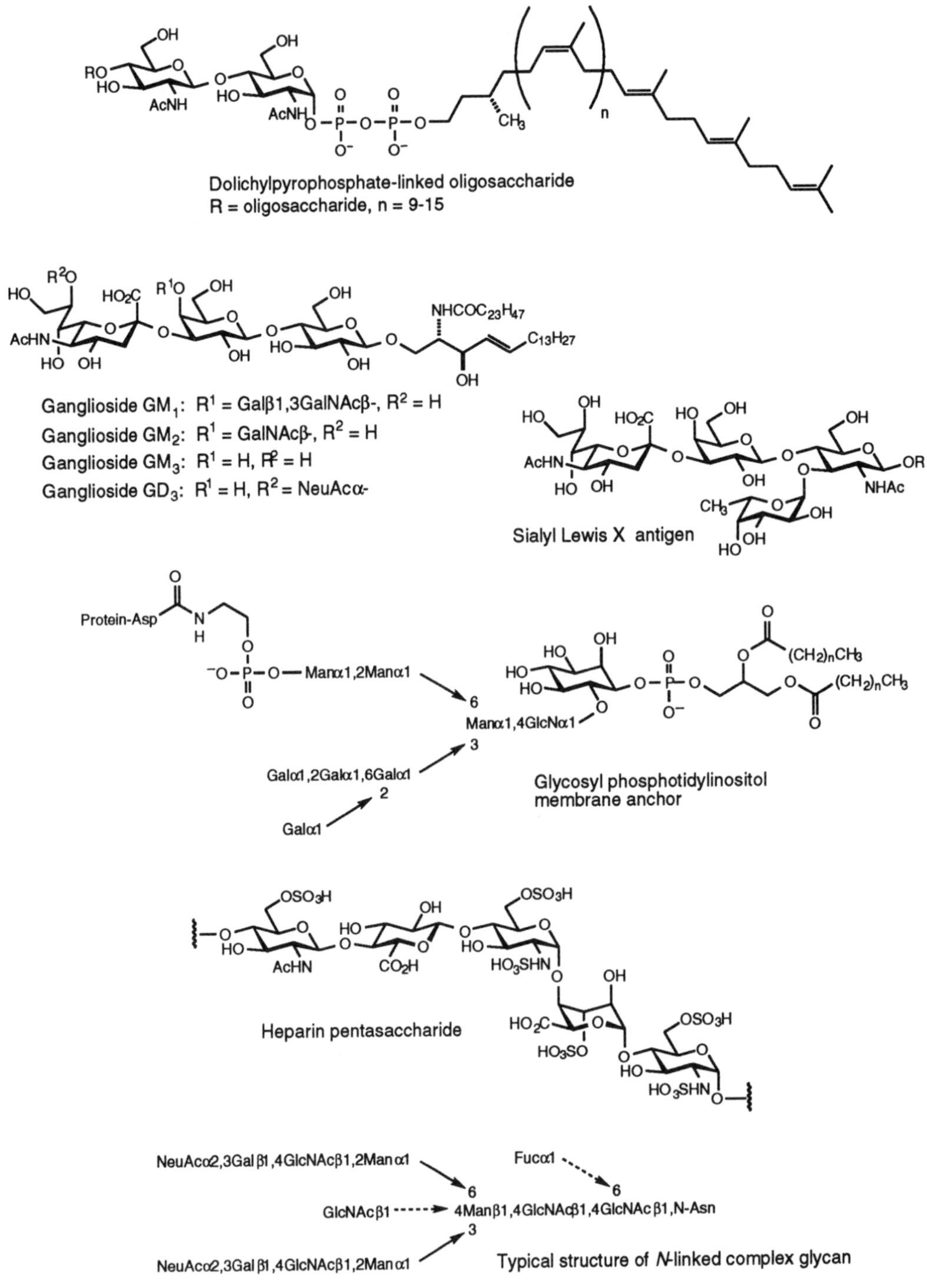

Figure 1. Structures of some biologically significant oligosaccharides and glycoconjugates.

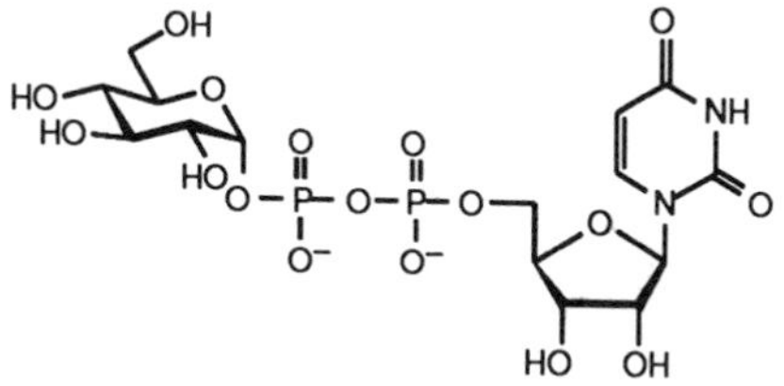
α-UDP-Glucose (UDP-Glc)

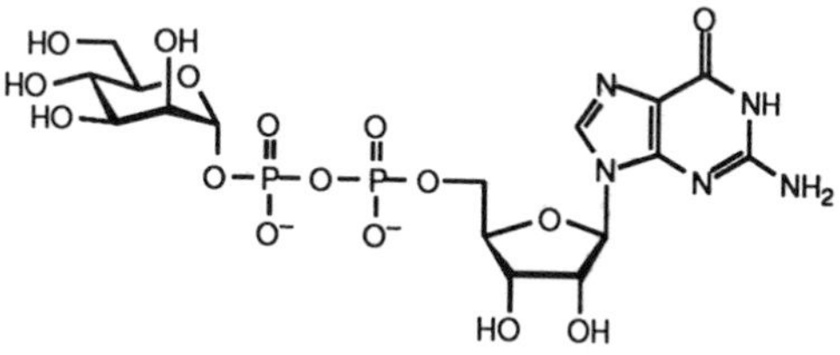
α-GDP-Mannose (GDP-Man)

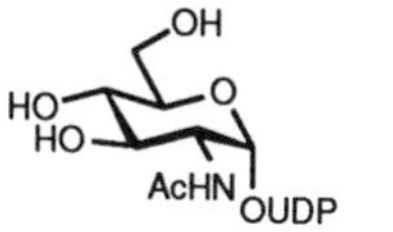
α-UDP-*N*-Acetylglucosamine (UDP-GlcNAc)

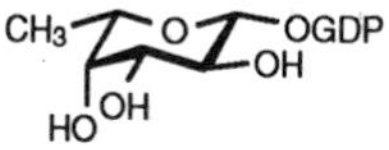
β-GDP-L-Fucose (GDP-Fuc)

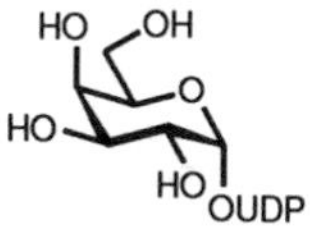
α-UDP-Galactose (UDP-Gal)

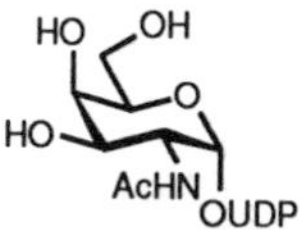
α-UDP-*N*-Acetylgalactosamine (UDP-GalNAc)

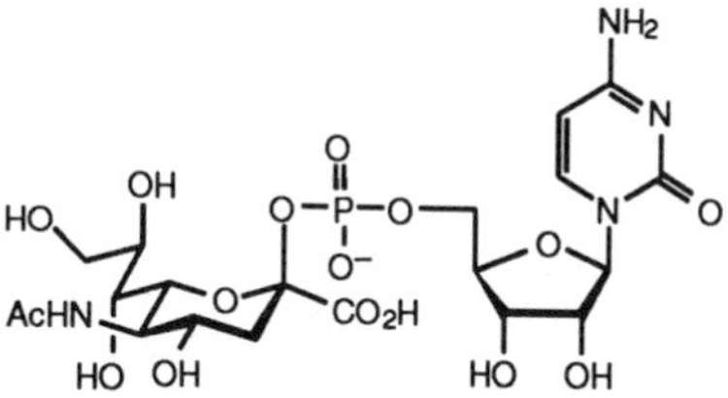
β-CMP-*N*-Acetylneuraminic acid (CMP-NeuAc)

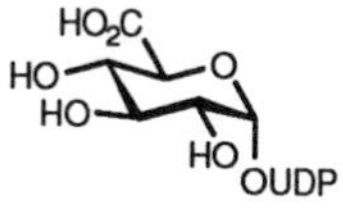
α-UDP-Glucuronic acid (UDP-GlcUA)

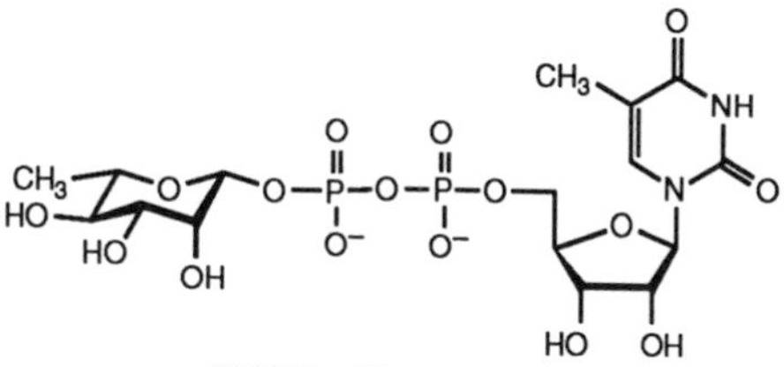
β-TDP-L-Rhamnose

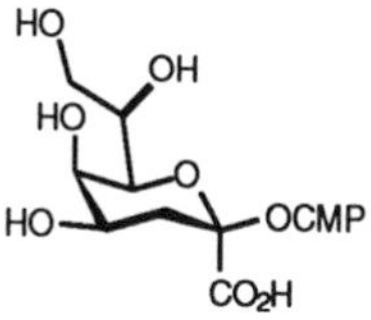
CMP-KDO

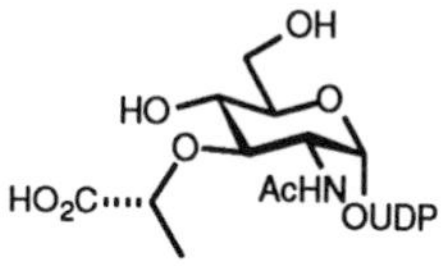
α-UDP-*N*-Acetylmuramic acid

Figure 2. Common structures of sugar nucleotides.

specific or high-yielding in synthetic reactions. Therefore the chemist must choose the enzyme that best suits his or her particular needs. Several other enzymatic methods have also been used to synthesize *N*-glycosides, such as nucleosides. Each of these strategies will be discussed in detail.

2.   Glycosyltransferases of the Leloir Pathway

Glycosyltransferases are highly regiospecific and stereospecific with respect to the formation of new glycosidic linkages. They also are usually substrate-specific. Minor chemical modifications, however, are tolerated on both the donor and acceptor components. The preparative use of glycosyltransferases has been somewhat limited in the past due to a lack of availability of these enzymes. Additionally, transferases are membrane-bound or membrane-associated enzymes, and some are difficult to handle and unstable. Recent isolation of many of these enzymes, as well as advances in genetic engineering and recombinant techniques, are rapidly alleviating these drawbacks.

Glycosyltransferases utilize glycosyl esters of nucleoside phosphates as activated monosaccharide donors.[43] Most of these sugar nucleoside phosphates are biosynthesized *in vivo* from the corresponding monosaccharides (Figure 3). The initial step is a kinase-mediated phosphorylation to produce a glycosyl phosphate. This glycosyl phosphate then reacts with a nucleoside triphosphate (NTP), catalyzed by a nucleoside diphosphosugar pyrophosphorylase, to afford an activated nucleoside diphosphosugar (eq. 1). Other sugar nucleoside phosphates, such as GDP-Fuc and UDP-GlcUA, are biosynthesized by further enzymatic modification of these existing key sugar nucleotide phosphates. Another exception is CMP-NeuAc, which is formed by the direct reaction of NeuAc with CTP (eq. 2). Some of the enzymes involved in the biosynthesis of sugar nucleotides also accept unnatural sugars as substrates. In general, however, the rates are quite slow, and limit the usefulness of this approach.

$$ \text{Sugar-1-P} \quad + \quad \text{NTP} \quad \rightarrow \quad \text{NDP-Sugar} \quad + \quad \text{PPi} \quad (1) $$

$$ \text{NeuAc} \quad + \quad \text{CTP} \quad \rightarrow \quad \text{CMPNeuAc} \quad + \quad \text{PPi} \quad (2) $$

2.1.   Synthesis of Sugar Nucleoside Phosphates

Chemical syntheses of some sugar nucleoside phosphates have been reported.[59] Most of these methods involve the reaction of an activated NMP[60-64] with a glycosyl phosphate to produce a sugar nucleoside diphosphate (Figure 4). Of the commonly used activated NMP derivatives, phosphoramidates such as phosphorimidazolidates[65-67] and phosphoromorpholidates[60-64] are considered the most effective. These activated NMPs may also be used to prepare NTPs by reaction with pyrophosphate (Figure 4).[67] A number of chemical

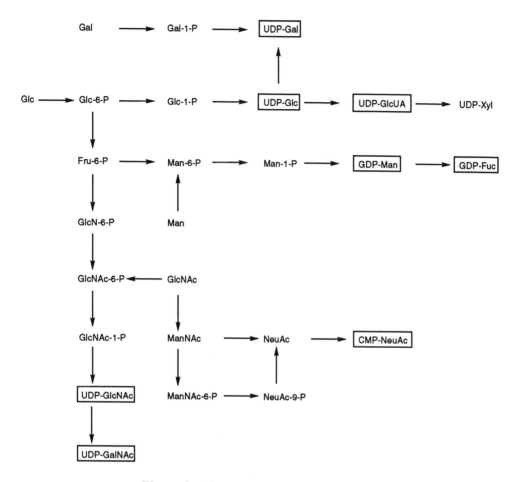

Figure 3. Biosynthesis of sugar nucleotides.

methods are available for the synthesis of glycosyl phosphates. Reactions of phosphates with activated glycosyl donors[68] or chemical phosphorylation of anomeric hydroxyl groups[63-65,69] have proven to be convenient. Additionally, routes via glycosyl phosphites are useful.[70] Enzymatic procedures are also available. Glycogen phosphorylase[71] and sucrose phosphorylase[72] were used to produce α-glucose-1-phosphate. Phosphoglucomutase can also be used to prepare glucose-1-phosphate from glucose-6-phosphate,[73] which is in turn synthesized from glucose by hexokinase.

2.1.1. Preparative-Scale Synthesis of Nucleoside Triphosphates

The appropriate nucleoside triphosphates are utilized as substrates for the biosynthesis of sugar nucleoside phosphates. Therefore biosynthesis-based enzymatic preparation of these donors for use in glycosylations requires a practical-scale synthesis of NTPs.

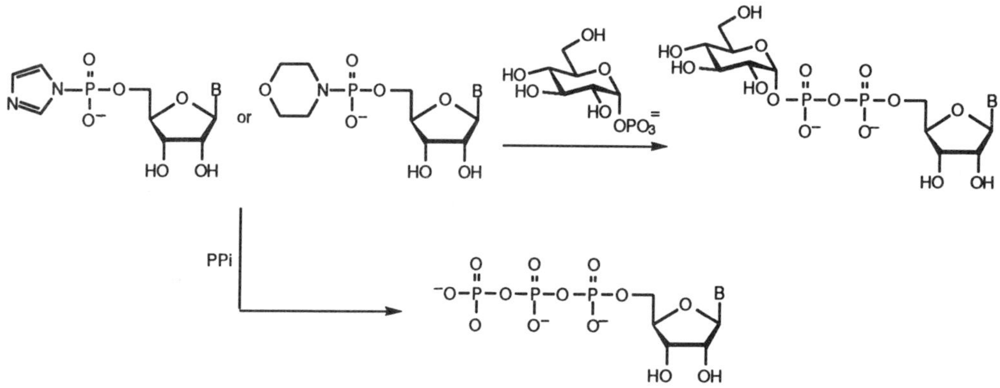

Figure 4. Typical chemical synthesis of NDP-sugar and NTP.

Most preparative-scale enzymatic syntheses of NTPs use commercially-available nucleoside monophosphates (NMPs) as starting materials. Alternatively, all of the NMPs can be obtained from yeast RNA digests at low cost,[74] or can be easily prepared chemically. In general, these methods involve the sequential use of two kinases to transform NMPs to NTPs, via the corresponding NDPs. Any of three kinases may be used to synthesize NTPs from the corresponding NDPs, each of which uses a different phosphoryl donor: pyruvate kinase (EC 2.7.1.40) uses phosphoenolpyruvate (PEP)[75-76] as a phosphoryl donor, acetate kinase (EC 2.7.2.1) uses acetyl phosphate, and nucleoside diphosphate kinase (EC 2.7.4.6) uses ATP. Pyruvate kinase is generally the enzyme of choice because it is less expensive than nucleoside diphosphate kinase,[56,67,77] and because PEP is more stable and provides a more thermodynamically favorable driving force for phosphorylation than does acetyl phosphate (Figure 5).

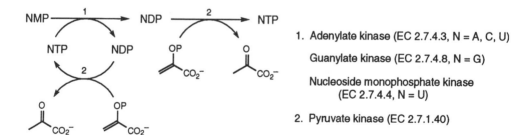

1. Adenylate kinase (EC 2.7.4.3, N = A, C, U)

   Guanylate kinase (EC 2.7.4.8, N = G)

   Nucleoside monophosphate kinase
   (EC 2.7.4.4, N = U)

2. Pyruvate kinase (EC 2.7.1.40)

Figure 5. Synthesis of NTPs.

The preparation of NDPs from NMPs is more complicated, and requires different enzymes for each NMP. Adenylate kinase (EC 2.7.4.3) phosphorylates AMP[67] and CMP,[78] and also slowly phosphorylates UMP. Guanylate kinase (EC 2.7.4.8) catalyzes the phosphorylation of GMP. Nucleoside monophosphate kinase (EC 2.7.4.4) uses ATP to phosphorylate AMP, CMP, GMP, and UMP; the enzyme is, however, relatively expensive and unstable.[67] Both CMP and UMP kinases exist but are not commercially available. For those kinases requiring ATP as a phosphorylating agent, ATP is usually used in a catalytic amount and recycled from ADP using pyruvate kinase/PEP or acetate kinase/acetylphosphate.[56,79] Phosphoenolpyruvate may be prepared chemically from pyruvate[75] or generated enzymatically from D-3-phosphoglyceric acid[76] (Figure 6).

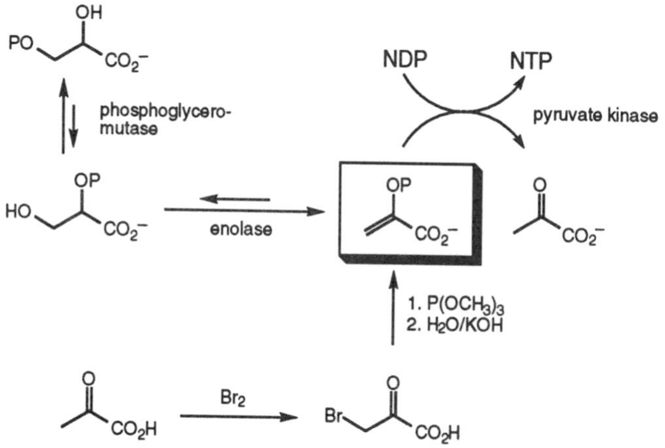

Figure 6. Synthesis of PEP.

Comparisons of chemical and enzymatic methods for the synthesis of NTPs[67] conclude that enzymatic methods provide the most convenient route to CTP and GTP. Chemical deamination of CTP is the best method for preparing UTP.[67] ATP is relatively inexpensive from commercial sources, although it has been synthesized enzymatically from AMP on 50 mmol scale. Mixtures of NTPs can be prepared from RNA by sequential reactions catalyzed by nuclease P1, polynucleotide phosphorylase, and pyruvate kinase.[81] This mixture can be selectively converted to a sugar nucleotide using a particular sugar nucleoside diphosphate pyrophosphorylase.[81]

2.1.2. UDP-Glucose (UDP-Glc) and UDP-Galactose (UDP-Gal)

UDP-Glucose has been prepared from UTP and glucose-1-phosphate under catalysis by UDP-glucose pyrophosphorylase (Figure 7).[67,73,82] UDP-Gal can be synthesized in an analogous fashion using UDP-Gal pyrophosphorylase.[45] Additionally, UDP-Gal can be generated from UDP-Glc by epimerization of C-4 with UDP-glucose epimerase[73] (Figure 8). Although the equilibrium for this reaction favors UDP-Glc, the reaction can be coupled to an *in situ* glycosylation with galactosyltransferase to shift the equilibrium. The latter process has been applied to a large-scale synthesis of *N*-acetyllactosamine.[73] UDP-Gal has been prepared on a gram scale from Gal-1-phosphate and UDP-Glc using UDP-Gal uridyltransferase,[83] and generated *in situ* for glycosylation (see below). UDP-Gal has also been prepared from UMP and Gal using dried cells of *Torulopsis candida*.[82] In this system, Gal-1-phosphate and UTP were generated *in situ* as substrates for UDP-Gal pyrophosphorylase. A recent chemical synthesis of UDP-2-deoxygalactose and its use in glycosylations has been recently reported.[84]

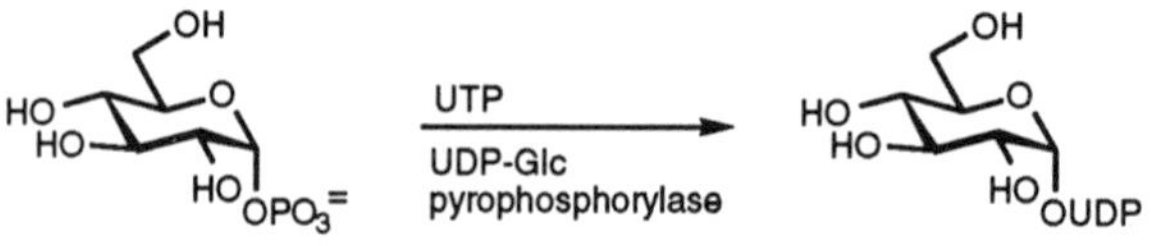

Figure 7. Synthesis of UDP-Glc.

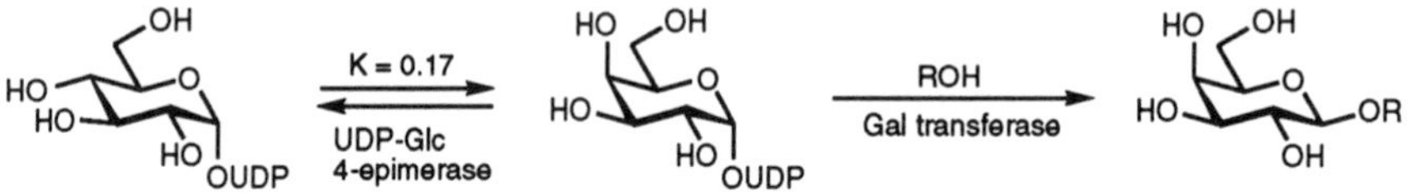

Figure 8. Synthesis of UDP-Gal and its use in glycosylations.

2.1.3.　　UDP-N-Acetylglucosamine (UDP-GlcNAc)

Two enzymatic methods have been developed for the synthesis of UDP-GlcNAc. The first involves a reaction between GlcNAc-1-phosphate and UTP, catalyzed by UDP-GlcNAc pyrophosphorylase.[82] UDP-GlcNAc pyrophosphorylase is currently not commercially available, although a whole-cell process using Baker's yeast can be employed.[82a-c] The enzyme from calf liver[82d] can also be used in the synthesis. The second procedure exploits UDP-Glc pyrophosphorylase to catalyze a condensation between UTP and glucosamine-1-phosphate (GlcN-1-P) to afford UDP-glucosamine[85] (Figure 9). The product UDP-GlcN can then be selectively *N*-acetylated to provide UDP-GlcNAc. GlcN-1-P was synthesized from GlcN by phosphorylation of the 6-position with hexokinase to give GlcN-6-P, followed by a phosphoglucomutase-catalyzed isomerization to provide GlcN-1-P. The second sequence can be performed in a hollow fiber reactor.

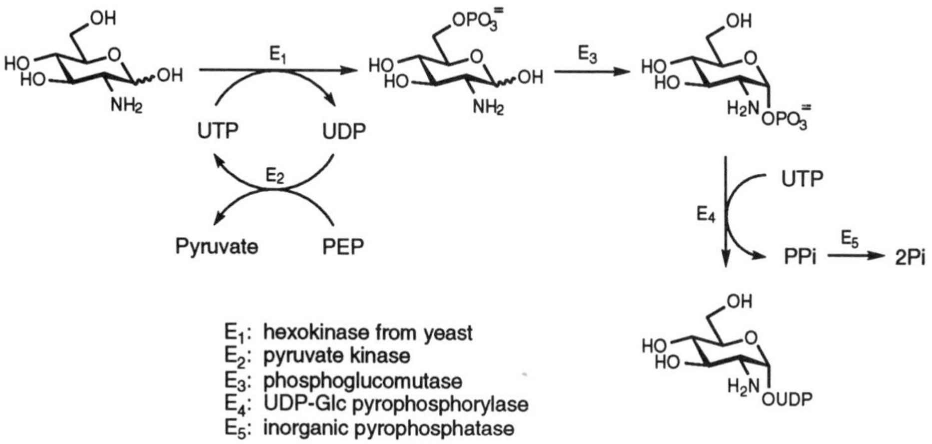

Figure 9. Synthesis of UDP-GlcNAc.

2.1.4. UDP-*N*-Acetylgalactosamine (UDP-GalNAc)

UDP-GalNAc can be prepared from GalNAc-1-P and UTP using UDP-GalNAc pyrophosphorylase, or from UDP-GlcNAc using an epimerase, although the necessary enzymes are not readily available. An alternative procedure has been reported which is based on a UMP exchange reaction between UDP-Glc and GalN-1-P, under catalysis by UDP-glucose:galactosylphosphate uridyltransferase (EC 2.7.7.12); this enzyme is commercially available (Figure 10).[45,83,86] Galactose-1-phosphate is the natural substrate for the enzyme, but 2-deoxygalactose-1-phosphate, 2-deoxyglucose-1-phosphate, and galactosamine-1-phosphate are also accepted (see below). The equilibrium constant for the exchange reaction is about 1, therefore phosphoglucomutase was added to remove the product Glc-1-P and shift the equilibrium toward UDP-galactosamine. The UDP-GalN thus produced was acetylated with acetic anhydride in a subsequent step to give UDP-GalNAc.

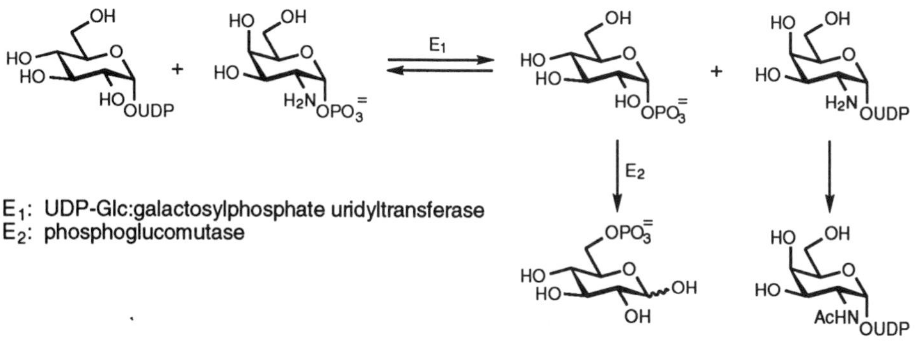

Figure 10. Synthesis of UDP-GalNAc.

A modification of the latter procedure has been adapted for a large-scale synthesis of UDP-GalNAc.[83] In this procedure, galactosamine-1-phosphate was prepared with galactokinase coupled with an ATP regeneration system, and used as a substrate for the uridyltransferase. UDP-Glc was regenerated *in situ* from UTP and the product Glc-1-P, under catalysis by UDP-Glc pyrophosphorylase. This coupling also shifts the equilibrium toward the formation of UDP-GalN. Finally, a chemical *N*-acetylation of the UDP-GalN thus produced with *N*-acetoxysuccinimide provides UDP-GalNAc. This procedure has also been adapted to an *in situ* synthesis of UDP-GalN and UDP-2-deoxy-Gal for glycosylation, with cofactor regeneration, to provide β-glycosides of GalN and 2-deoxy-Gal, respectively (see below).

2.1.5. GDP-Mannose (GDP-Man) and GDP-Fucose (GDP-Fuc)

GDP-mannose has been prepared from Glc and GMP using dried Baker's yeast cells.[82] The procedure involves the biocatalytic conversion of glucose to Man-1-P, and a subsequent conversion to GDP-Man using GDP-Man pyrophosphorylase. A cell-free extract from Baker's yeast has also been used to synthesize GDP-Man from mannose.[87] A direct synthesis from chemically-prepared Man-1-P and GTP, catalyzed by GDP-Man pyrophosphorylase (EC 2.7.7.13) is useful for large scale (Figure 11).[67]

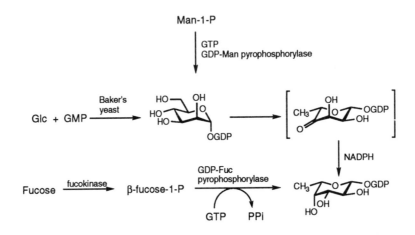

Figure 11. Synthesis of GDP-Fuc.

GDP-fucose is biosynthesized *in vivo* from GDP-Man by an NADPH-dependent oxidoreductase enzyme system. Such systems have also been utilized for *in vitro* syntheses of GDP-Fuc. For example, the conversion of GDP-Man to GDP-Fuc was accomplished using a crude enzyme preparation from *Agrobacterium radiobacter*.[88] NADPH was regenerated *in situ* from NADP using glucose-6-phosphate dehydrogenase and Glc-6-P.[89] Using a similar procedure, GDP-Fuc has been generated *in situ* for use in a glycosylation reaction with α–1,3-

fucosyltransferase (see below).[90a] Enzymes from a minor biosynthetic pathway that synthesize GDP-Fuc from L-fucose-1-phosphate[90a] or L-fucose[90b] have also been exploited for synthesis.[90a,b] Fucose was phosphorylated by fucokinase [EC 2.7.1.52] to produce Fuc-1-P, which subsequently underwent a GDP-fucose pyrophosphorylase-catalyzed reaction with GTP to provide GDP-Fuc. Several practical chemical syntheses of GDP-Fuc have also been reported.[91]

2.1.6. UDP-Glucuronic Acid (UDP-GlcUA)

UDP-Glucuronic acid is biosynthesized by oxidation of C-6 of UDP-Glc with UDP-Glc dehydrogenase, an NAD-dependent enzyme. Enzyme preparations from bovine liver have been employed for gram-scale syntheses of UDP-GlcUA (Figure 12).[67,92] The NAD cofactor was regenerated with lactate dehydrogenase in the presence of pyruvate. Additionally, extracts from guinea pig liver have been used to generate UDP-GlcUA *in situ* for use in enzymatic glycosylations with glucuronyl transferases.[93]

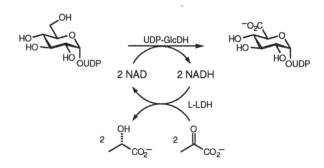

Figure 12. Synthesis of UDP-GlcUA.

2.1.7. CMP-*N*-Acetylneuraminic Acid (CMP-NeuAc)

CMP-*N*-acetylneuraminic acid has been prepared enzymatically on small scales (<0.5 mmol) from CTP and NeuAc, under catalysis of CMP-NeuAc synthetase (EC 2.7.7.43).[94] An improvement in this procedure which is suitable for multigram-scale synthesis[95] has been developed which utilized CTP, itself synthesized *in situ* from CMP using adenylate kinase and pyruvate kinase. Adenylate kinase catalyzes the equilibration of CTP and CMP to CDP, which is subsequently phosphorylated by pyruvate kinase to provide CTP. Another procedure has been employed in which the NeuAc used in the synthesis of CMP-NeuAc was prepared in a NeuAc aldolase-catalyzed reaction of pyruvate with *N*-acetylmannosamine, which was itself generated from *N*-acetylglucosamine by a base-catalyzed epimerization.[78] A one-pot synthesis of CMP-NeuAc which is based on the latter procedure involves the *in situ* synthesis of NeuAc from *N*-acetylmannosamine and pyruvate, catalyzed by sialic acid aldolase (Figure 13).[96] Chemical syntheses of CMP-NeuAc have also been reported.[97]

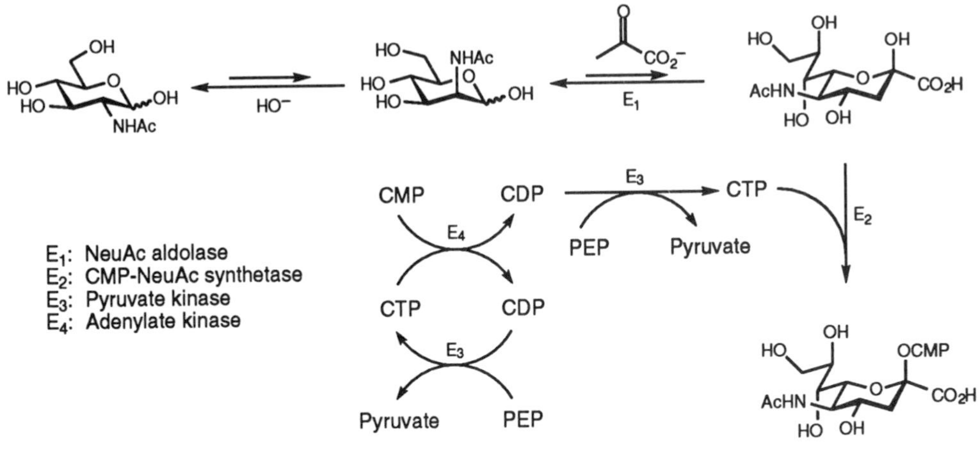

Figure 13. Synthesis of CMP-NeuAc.

E_1: NeuAc aldolase
E_2: CMP-NeuAc synthetase
E_3: Pyruvate kinase
E_4: Adenylate kinase

The gene encoding *E. coli* CMP-NeuAc synthetase has been cloned[98-99] and overexpressed in *E. coli* using Lambda-ZAP vector-LacZ promotor[100-101] or PKK223 vector-tac promotor.[102] The enzyme from calf brain has also been cloned and overexpressed. CMP-NeuAc synthetase was shown to accept several NeuAc derivatives as substrates. For example, 9-deoxy-, 7,9-dideoxy-, and 4,7,9-trideoxy-NeuAc are all converted to the corresponding CMP-NeuAc derivative.[103] On the other hand, neither the 4-oxo, 7-oxo, or 8-oxo derivatives of NeuAc, nor their respective dimethylacetals, are substrates for CMP-NeuAc synthetase.[104] The enzyme accepts a variety of modifications at the 9 position, and the hydroxyl group can be replaced with several different groups with little effect on the K_m value.[96,105-107]

2.2. Substrate Specificity and Synthetic Applications of Glycosyltransferases

For each sugar nucleotide glycosyl donor, many glycosyltransferases exist, each of which transfers the particular donor to different acceptors. These enzymes are generally considered to be specific for a given glycosyl donor and acceptor, as well as for the stereochemistry and the linkage position of the newly-formed glycoside bond. This specificity has led to the "one enzyme-one linkage" concept.[7,44,108] In other words, the specificity of the glycosyltransferases ensures fidelity in oligosaccharide sequences *in vivo* without the use of a template scheme. Though systematic investigations of the *in vitro* substrate specificity of most glycosyltransferases have not been carried out, some deviations from this picture of absolute specificity have been observed, both in the glycosyl donors and acceptors. Additionally, studies toward the design of inhibitors of glycoprotein biosynthesis[109] have also shown that the specificities of glycosyltransferases are not absolute.

2.2.1. Galactosyltransferase

Because of its availability, β1,4-galactosyltransferase (UDP-Gal:*N*-acetylglucosamine β1,4-galactosyltransferase, EC 2.4.1.22)[110-111] is one of the most extensively studied mammalian glycosyltransferases with regard to synthesis and substrate specificity. This enzyme catalyzes the transfer of galactose from UDP-Gal to the 4-position of β-linked GlcNAc residues to produce the Galβ1,4GlcNAc substructure. In the presence of lactalbumin, however, glucose is the preferred acceptor, resulting in the formation of lactose, Galβ1,4Glc. The enzyme has been employed in the *in vitro* synthesis of *N*-acetyllactosamine and glycosides thereof, as well as other galactosides (Table 1).

Table 1. Products of galactosyltransferase reactions.

UDP-Gal (or analogs) + GalTase	Scale[a]	Ref.
Galβ1,4Glc	C	110b
Galβ1,4GlcNAc	A	110b,73
Galβ1,4GlcNAc-Agarose	C	110b
Galβ1,4GlcNAc-hexylamine	C	110b
Galβ1,4GlcNAcβ1,4Gal	C	110b
Galβ1,4GlcNAcβ1,6Gal	C	112
Galβ1,4GlcNAcβ1,3Gal	C	112
Galβ1,4GlcβOCH$_2$C$_6$H$_4$(NO$_2$)-CONH-Polymer	D	113
Galβ1,4Glcβ1,4GlcβOCH$_2$C$_6$H$_4$(NO$_2$)-CONH-Polymer	D	113
Galβ1,4Glcβ1,4GlcβOCH$_2$NH-L-Phe-CONH-Polymer	D	113
Galβ1,4GlcNAcβ1,3(Galβ1,4GlcNAcβ1,6)Galβ1,4GlcβOMe	C	114
Galβ1,4GlcNAcβ1,6(GlcNAcβ1,3)Galβ1,4GlcβOMe	C	114
Galβ1,4(Fucα1,6)GlcNAcβO(CH$_2$)$_8$CO$_2$Me	D	112
Galβ1,4(NeuAc(OMe)α2,6)GlcNAcβO(CH$_2$)$_8$CO$_2$Me	D	112
Galβ1,4GlcNAcβR; R=N-Ac-Asn(OMe)	C	115
Galβ1,4GlcNAcβ1,4GlcNAc	C	115
Galβ1,4GlcNAcβ1,4GlcNAcβR; R=N-Ac-Asn(OMe)	C	115
Glcβ1,4GlcNAcβO(CH$_2$)$_8$CO$_2$Me	D	116
GalNAcβ1,4GlcNAcβO(CH$_2$)$_8$CO$_2$Me	D	116
GalNAcβ1,4GlcNAcβ1,2ManβO(CH$_2$)$_8$CO$_2$Me	D	116
GalNAcβ1,4GlcNAcβ1,2Manα1,6(GalNAcβ1,4GlcNAcβ1,2Manα1,3)ManβO(CH$_2$)$_8$CO$_2$Me	D	116
GlcNAcβ1,4GlcNAcβO(CH$_2$)$_8$CO$_2$Me	D	116
Galβ1,4GlcNAcβR; R=GlyGlyAsnGlyGly or N-Alloc-PheAsnSerThrIle	C	117
Galβ1,3Galβ1,4Glc	D	118
Galα1,3Galβ1,4GlcNAc	D	118

a: A, >1g; B, 0.1-1g; C, 10-100mg; D, <10mg.

Galactosyltransferase utilizes as acceptor substrates *N*-acetylglucosamine and glucose and β-glycosides thereof, 2-deoxyglucose, D-xylose, 5-thioglucose, *N*-acetylmuramic acid, and myo-inositol.[110] Modifications at the 3- or 6-position of the acceptor GlcNAc are also tolerated. For example, Fucα1,6GlcNAc and NeuAcα2,6GlcNAc are substrates.[112] Acceptor substrates which are derivatized at the 3-position include 3-*O*-methyl-GlcNAc,[112] 3-deoxy-GlcNAc, 3-*O*-allyl-GlcNAcβOBu, and 3-oxo-GlcNAc.[119] All glycosides of GlcNAc that are reported to be substrates for the galactosyltransferase have β-glycosidic linkages. Both α- and β-glycosides of glucose are acceptable; the presence of lactalbumin, however, is required for galactosyl transfer onto α-glycosides. Neither D-mannose, D-allose, D-galactose, D-ribose, nor D-xylose are substrates. Monosaccharides that have a negative charge, such as glucuronic acid and α-glucose-1-phosphate, are also not accepted as substrates. Figure 14 illustrates several disaccharides that can be synthesized with galactosyltransferase. The β1,4-galactosides of 5-thioglucose, deoxynojirimycin, and glucal have been prepared and characterized.[119] Particularly interesting examples are β,β-1,1-linked disaccharides, in which the anomeric hydroxyl of 3-acetamido-3-deoxyglucose or of 3-acetamido-3-deoxy-5-thio-D-xylose serves as the acceptor moiety.[120] The acetamido function apparently controls the position of glycosylation.

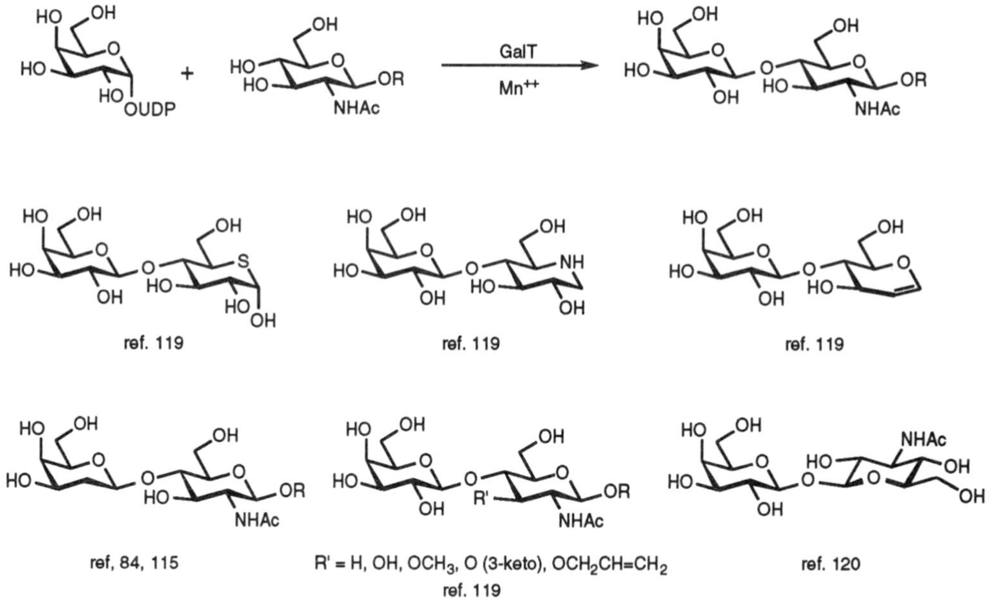

Figure 14. Some disaccharides synthesized using β1,4-galactosyltransferase.

β1,4-Galactosyltransferase has also been employed in solid-phase oligosaccharide synthesis, and has been used to galactosylate *gluco* or *cellobio* subunits of polymer-supported oligosaccharides and polysaccharides (Figure 15).[113] The resulting oligosaccharides can then be removed from the support by either a photochemical cleavage or a chymotrypsin-mediated hydrolysis. The types of polymer supports employed include polyacrylamide and a water-soluble poly(vinyl alcohol). *N*-Acetylglucosaminyl amino acids and peptides have also been used as substrates for galactosyltransferase to afford galactosylated glycopeptides (Figure 16).[115,117,121] The carbohydrate chain can then be further extended with other transferases, such as sialyltransferase.[115,117,121] Similarly, the enzymatic synthesis of a ceramide glycoside of Galβ1,4GlcNAc which was subsequently enzymatically sialylated provided a GM3 analog.[122]

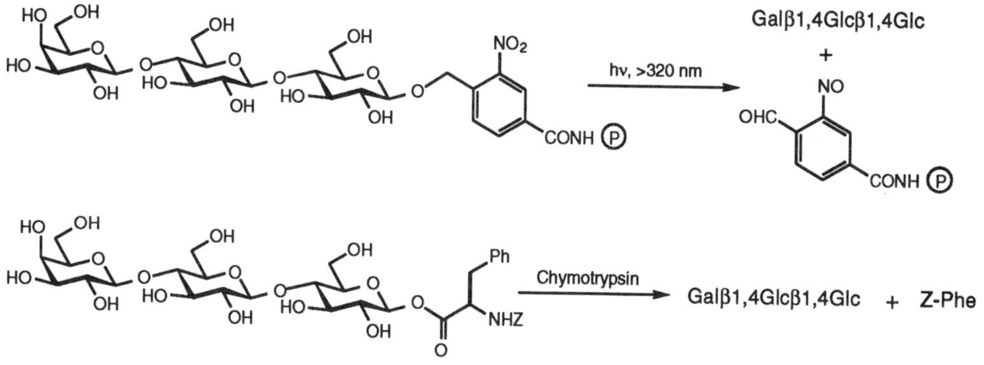

Figure 15. Polymer-supported enzymatic synthesis of oligosaccharides.

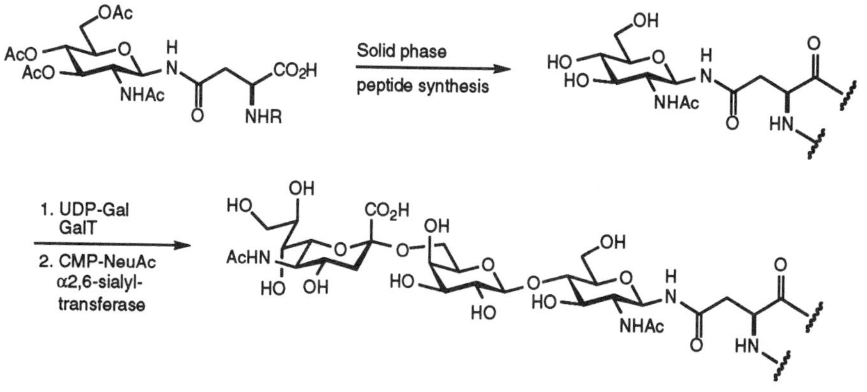

Figure 16. Synthesis of glycopeptides.

With regard to the donor substrate, the β-galactosyltransferase also transfers glucose, 4-deoxygalactose, arabinose, glucosamine, galactosamine, *N*-acetylgalactosamine, 2-deoxygalactose, and 2-deoxyglucose from their respective UDP-derivatives, providing an enzymatic route to oligosaccharides which terminate in β1,4-linked residues other than galactose (Table 2).[116,123,124] An example worthy of note is the transfer of 5-thiogalactose to an acceptor (entry 8).[123] Although the rate of the enzyme-catalyzed transfer of many of these unnatural donor substrates is quite slow, this method is useful for milligram-scale synthesis. The α-1,3-galactosyltransferase responsible for the biosynthesis of the B blood group antigen has been

Table 2. Relative rates of β1,4-galactosyltransferase catalyzed transfer of donor substrates.

Donor Substrate		Rel. rate	Ref.
UDP-Gal	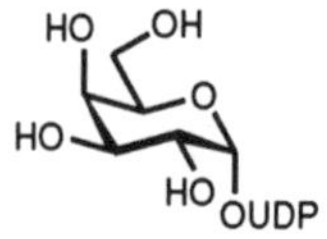	100	124a
UDP-Glc	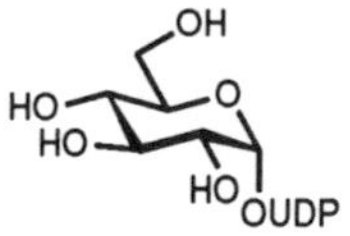	0.3	116,124a
UDP-4-deoxy-Glc	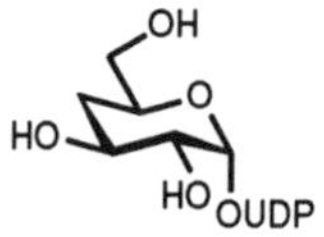	5.5	124a
UDP-Ara		4.0 R = H 1.3 R = CH$_3$ 0.2 R = CH$_2$F	124a 124b 124b
UDP-GalNAc	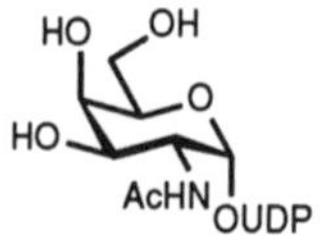	4.0	116
UDP-GlcNAc	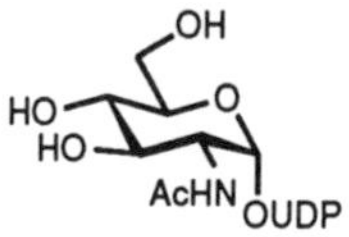	0.00	116
UDP-GlcN	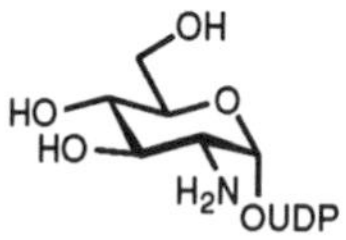	0.09	116
UDP-5-thio-Gal	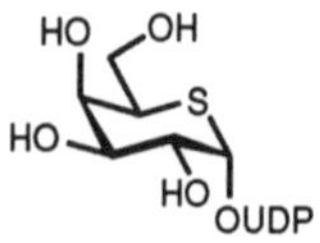	5.0	123
UDP-2-deoxy-Gal	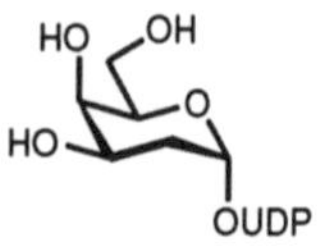	90	84[a]

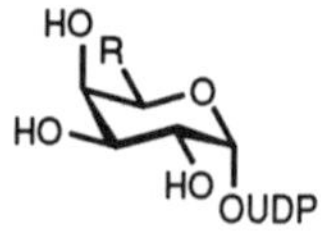

[a]Generated in situ from 2-deoxygalactose with galactokinase/uridyltransferase,[157] or from 2-deoxyglucose with hexokinase/mutase/UDP-Glc pyrophatase/epimerase[115]

studied regarding its acceptor specificity.[124c] The enzyme transfers Gal with an α linkage to OH-3 of the Gal residue in Fucα1,2GalβOR. It also accepts the 6-deoxy and 6-deoxy-6-fluoro analogs, but not derivatives that are modified at the 4-position. A similar result was observed for the α-1,3-GlcNAc transferase involved in the biosynthesis of the A blood group antigen.

2.2.2. Sialyltransferase

Several sialyltransferases, classified as either α2,6- and α2,3-sialyltransferases, have been used for oligosaccharide synthesis.[125-127] Sialyltransferases generally transfer *N*-acetylneuraminic acid to either the 3- or 6-position of terminal Gal or GalNAc residues. Table 3 lists some compounds thus prepared. Some sialyltransferases have been shown to accept CMP-NeuAc analogs that are derivatized at the 9-position of the sialic acid side chain,[92a,105-107,127-129] such as those in which the hydroxyl group at C-9 is replaced with an amino, fluoro, azido, acetamido, or benzamido group. Analogs of the acceptors Galβ1,4GlcNAc and Galβ1,3GalNAc,

Table 3. Products of sialyltransferase reactions.

CMP-NeuAc + α2,6–SialylTase	Scale[a]	Ref.
NeuAcα2,6GalβOMe	D	125
NeuAcα2,6Galβ1,4GlcβOMe	D	125
NeuAcα2,6Galβ1,4GlcNAc	C	125-6,131,78,9
NeuAcα2,6Galβ1,4GlcNAcβOMe	C	125
NeuAcα2,6Galβ1,4GlcNAcβ1,3Galβ1,4Glc	C	125
NeuAcα2,6Galβ1,4GlcNAcβ1–N–Asn	C	121
NeuAcα2,6Galβ1,4GlcNAcβ1,2ManαOMe	C	132
NeuAcα2,6Galβ1,4GlcNAcβ1,3(Galβ1,4GlcNAcβ1,6)Galβ1,4GlcβOMe	C	123
NeuAc(9-O-Ac)α2,6Galβ1,4GlcNAc	C	127
NeuAcα2,6Galβ1,4GlcNAcβR; R=OH, N$_3$, GlyGlyAsnGlyGly or N-Alloc-PheAsnSerThrIle	C	117
NeuAcα2,6Galβ1,4GlcNAcβ1,4(NeuAcα2,6Galβ1,4GlcNAcβ1,2/3)GalβO(CH$_2$)$_5$CO$_2$Me	D	133

CMP-NeuAc + α2,3–SialylTase		
NeuAcα2,3Galβ1,4GlcβOMe	D	125
NeuAcα2,3Galβ1,4GlcNAcβOMe	D	125
NeuAcα2,3Galβ1,3GlcNAcβOR; R=Me, Ph, (CH$_2$)$_5$CO$_2$Me	D	125
NeuAcα2,3Galβ1,3GlcNAcβ1,3Galβ1,4Glc	D	125
NeuAcα2,3Galβ1,3GlcNAcβ1,3GalβO(CH$_2$)$_8$CO$_2$Me	D	134
NeuAcα2,3Galβ1,3GlcNAcβ1,6GalβO(CH$_2$)$_8$CO$_2$Me	D	134
NeuAcα2,3Galβ1,3GalNAcβOR (R = Et,)	C	135
(R = H, (CH$_2$)$_5$CO2Me	D	126
NeuAcα2,3Galβ1,3(NeuAcα2,6)GalNAcβOPh	D	136
3-O-MeGalβ1,4Glcβ1,6(NeuAcα2,3Galβ1,4)GlcNAcβ1,3Galβ1,4Glcβ1,6-		
(NeuAcα2,3Galβ1,4)GlcNAcβOMe	D	137
NeuAcα2,3GlcβOCH$_2$CH(N$_3$)CH(OH)CH=CH(CH$_2$)$_{12}$CH$_3$	--	130b

a: A, >1g; B, 0.1-1g; C, 10-100mg; D, <10mg.

in which the acetamido function is replaced by an azide, phthalimide, carbamate, or pivaloyl functionality are also substrates for the enzymes.[130a] The newly isolated α-2,8-sialyltransferase catalyzes the biosynthesis of α-2,8-linked polysialic acids which are components of neural cell adhesion molecules.[138] Both α-2,8-linked[138b] and α-2,9-linked[138c] polysialic acid are also found on bacterial cell surfaces.

2.2.3. Fucosyltransferase

Fucosyltransferases are involved in the biosynthesis of many oligosaccharide structures such as blood-group substances and cell-surface and tumor-associated antigens. Fucosylation is one of the last modifications of oligosaccharides *in vivo*. Several fucosyltransferases have been isolated and used for *in vitro* synthesis (Table 4). For example, α1,3-fucosyltransferase has been used to L-fucosylate the 3-position of the GlcNAc of *N*-acetyllactosamine and of sialyl α2,3-*N*-acetyllactosamine to provide the Lewis X and sialyl Lewis X structural motifs, respectively.[90a,134] Several other acceptor substrates with modifications in the GlcNAc residue can also be fucosylated.[139] Galβ1,4Glc, Galβ1,4Glucal, and Galβ1,4(5-thioGlc) are all substrates. A similar enzyme, α1,3/4-fucosyltransferase, has also been used for synthesis. This enzyme fucosylates either the GlcNAc 3-position of Galβ1,4GlcNAc or the GlcNAc 4-position of Galβ1,3GlcNAc to afford Lewis X or Lewis A, respectively.[134,140] The corresponding sialylated substrates have also been employed as acceptors.[134]

Table 4. Products of fucosyltransferase reactions.

GDP-Fuc + α1,2 or α1,3/4FucTase	Scale[a]	Ref.
Fucα1,2GalβOR; R= CH$_2$CH$_3$, (CH$_2$)$_6$NH$_2$	C	141
Fucα1,2Galβ1,4GlcNAcβOR; R= H, (CH$_2$)$_6$NH$_2$	C	141
Fucα1,3(NeuAcα2,3Galβ1,4)GlcNAcβO(CH$_2$)$_5$CO$_2$Me	D,C	134,140
Fucα1,3(Galβ1,4)-5-thio-Glc	C	140
Fucα1,4(Galβ1,3)GlcNAc	C	140
Fucα1,3(NeuAcα2,3Galβ1,4)Glucal	D	123
Fucα1,4(NeuAcα2,3Galβ1,3)GlcNAcβ1,6GalβO(CH$_2$)$_5$CO$_2$Me	D	142
Fucα1,4(NeuAcα2,3Galβ1,3)GlcNAcβ1,3GalβO(CH$_2$)$_5$CO$_2$Me	D	142
Fucα1,4(Galβ1,3)GlcNAcβO(CH$_2$)$_8$CO$_2$Me	D	142
3-deoxy-Fucα1,4(Galβ1,3)GlcNAcβO(CH$_2$)$_8$CO$_2$Me	D	142
5-desmethyl-Fucα1,4(Galβ1,3)GlcNAcβO(CH$_2$)$_8$CO$_2$Me	D	142

a: A, >1g; B, 0.1-1g; C, 10-100mg; D, <10mg.

The Lewis A α1,4-fucosyltransferase has been shown to transfer unnatural fucose derivatives from their GDP esters. 3-Deoxyfucose and L-arabinose are transferred to Galβ1,4GlcNAcβO(CH$_2$)$_8$CO$_2$CH$_3$ at a rate of 2.3% and 5.9%, respectively, relative to L-fucose.[142] Furthermore, this enzyme will transfer a fucose residue that is substituted on C-6 by a very large sterically demanding structure. In particular, a synthetic blood group antigen can be attached, and the resulting "oligosaccharide" can be transferred to an acceptor from its GDP derivative by fucosyltransferase.[143] This approach has been used to alter the antigenic properties of cell-surface glycoproteins.

2.2.4. *N*-Acetylglucosaminyltransferase

In vivo, the *N*-acetylglucosaminyl transferases control the branching pattern of *N*-linked glycoproteins.[144-145] Each of the enzymes transfers a β-GlcNAc residue from the donor UDP-GlcNAc to a mannose or other acceptor. The GlcNAc transferases I-VI, which catalyze the addition of the GlcNAc residues to the core pentasaccharide of asparagine glycoproteins as outlined in Figure 17, have been identified and characterized.[144-146] These, as well as other GlcNAc transferases, have been exploited for purposes of oligosaccharide synthesis (Table 5).

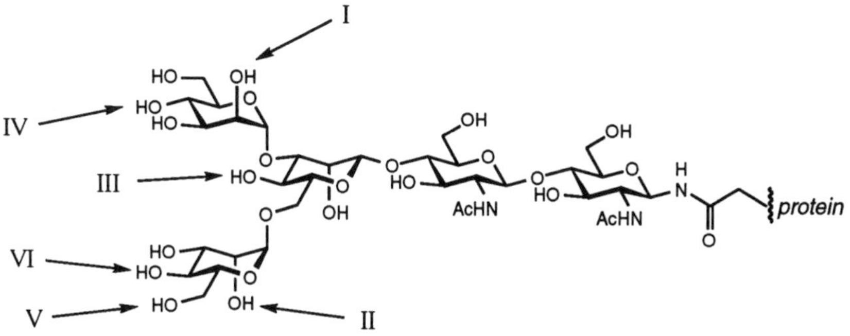

Figure 17. Specificity of GlcNAc transferases I-VI.

GlcNAc transferases have also been utilized to transfer non-natural residues onto oligosaccharides. In addition to transferring GlcNAc, *N*-acetylglucosaminyl transferase I from human milk catalyzes the transfer of 3-, 4-, or 6-deoxy-GlcNAc from its respective UDP derivative to Manα1,3(Manα1,6)ManβO(CH$_2$)$_8$CO$_2$CH$_3$.[147] The 4- and 6-deoxy-GlcNAc analogs can also be transferred by GlcNAc transferase II, however UDP-3-deoxy-GlcNAc is not a substrate for this enzyme.[147] In addition to the synthetic applications of GlcNAc transferases shown in Table 5, a GlcNAc transferase has been used to attach the terminal GlcNAc of GlcNAcβ1,4GlcNAcα dolichyl pyrophosphate, a substance employed in the study of oligosaccharyl transferase.[148] The core-2 GlcNAc transferase from mouse kidney acetone powder was used for the synthesis of a hexasaccharide containing sialyl Le[x].[151b] The enzyme catalyzes the transfer of β-GlcNAc onto the 6-OH of Galβ1,3GlcNAc.

2.2.5. Mannosyltransferase

Various mannosyltransferases have been shown to transfer mannose and 4-deoxymannose from their respective GDP adducts to acceptors.[152] α1,2-Mannosyltransferase was employed to transfer mannose to the 2-position of various derivatized α-mannosides and α-mannosyl peptides to produce the Manα1,2Man structural unit.[153] A recent report indicates that mannosyltransferases from pig liver accept GlcNAcβ1,4GlcNAc phytanyl pyrophosphate, an analog of the natural substrate in which the phytanyl moiety replaces dolichol.[154]

Table 5. Oligosaccharides synthesized by other glycosyltransferases.

UDP-GlcNAc (or analogs) + GlcNAcTase I	Scale[a]	Ref.
GlcNAcβ1,2Manα1,3(Manα1,6)ManβO(CH$_2$)$_8$CO$_2$Me	C	42,149a
3-deoxy, 4-deoxy or 6-deoxy-GlcNAcβ1,2Manα1,3(Manα1,6)ManβO(CH$_2$)$_8$CO$_2$Me	D	147
UDP-GlcNAc + GlcNAcTase II		
GlcNAcβ1,2Manα1,6(GlcNAcβ1,2Manα1,3)ManβO(CH$_2$)$_8$CO$_2$Me	D	149a
UDP-GlcNAc + GlcNAcTase V[b]		
GlcNAcβ1,2(GlcNAcβ1,6)Manα1,6GlcβO(CH$_2$)$_7$CH$_3$	D	149b[c]
UDP-GlcNAc + GlcNAcTase		
GlcNAcβ1,6(Galβ1,3)GlcNAc	D	150
UDP-Glc + GlcTase		
GlcβOR; R= CH$_2$CH$_3$, (CH$_2$)$_6$NH$_2$	C/D	151a

a: A, >1g; B, 0.1-1g; C, 10-100mg; D, <10mg.
b: Recognizes the gg conformation of the acceptor (ref. 149c).
c: None of the OH groups on the Glc residue are important.

2.2.6. Sucrose Synthetase

The fructose derivatives 1-azido-1-deoxy-, 1-deoxy-1-fluoro-, 6-deoxy-, 6-deoxy-6-fluoro-, and 4-deoxy-4-fluorofructose have been used as glycosyl acceptors in the sucrose synthetase-catalyzed synthesis of sucrose analogs (Figure 18).[155] 6-Deoxy- and 6-deoxy-6-fluorofructose were generated *in situ* from the corresponding glucose derivatives under catalysis by glucose isomerase.[155a] Because of the reversible nature of the reaction, the sucrose synthetase from rice was used for the preparation and regeneration of UDP-Glc in the synthesis of *N*-acetyllactosamine.[155b] The enzyme also accepts TDP, ADP, and GDP.[155b]

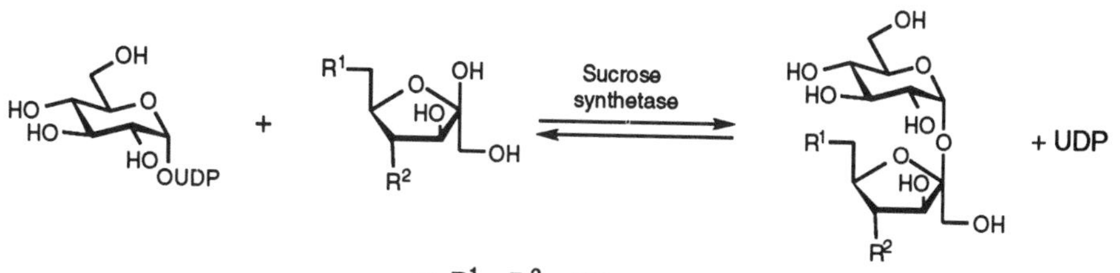

a: $R^1 = R^2 = OH$
b: $R^1 = F, R^2 = OH$
c: $R^1 = OH, R^2 = F$

Figure 18. Synthesis of sucrose analogs using sucrose synthetase.

2.2.7. Oligosaccharyltransferase

As mentioned earlier, oligosaccharyltransferase catalyzes the transfer of an oligosaccharide consisting of two glucose, nine mannose, and three glucose units from a dolichyl pyrophosphate intermediate to an Asn residue of a nascent peptide or protein.[43] The enzyme also transfers the minimal structure GlcNAcβ1,4GlcNAc from the corresponding dolichyl pyrophosphate donor or from a derivative in which the lipid component is truncated or simplified.[48d] The minimal peptide structure which will serve as an acceptor is the tripeptide Asn-X-Ser/Thr. Oligosaccharyltransferase has been utilized for the *in vitro* synthesis of several peptides containing glycosylated Asn residues. The glycopeptides Bz-Asn(oligosaccharyl)-Leu-Thr-NH$_2$,[156a] Bz-Asn(GlcNAc$_2$)-Leu-Thr-NH$_2$,[156b] and Ac-Asn(GlcNAc$_2$)-Leu-Thr-OCH$_3$/NHCH$_3$[47a] were synthesized as well as glycosylated cyclic peptides such as oxid(Cys-Als-Asn(GlcNAc$_2$)-Cys)-Thr-Ser-Ala[47a] have been prepared.

2.3. *In Situ* Cofactor Regeneration

Though analytical- and small-scale synthesis using glycosyltransferases is extremely powerful, the high cost of sugar nucleotides and the product inhibition caused by the released nucleoside mono- or diphosphates present major obstacles to large-scale synthesis. A simple solution to both of these problems is to use a scheme in which the sugar nucleotide is regenerated *in situ* from the released nucleoside diphosphate. The first example of the use of such a strategy is the galactosyltransferase-catalyzed synthesis of *N*-acetyllactosamine.[73] (Figure 19). A catalytic amount of UDP-Gal is initially used to glycosylate GlcNAc; UDP-Gal is regenerated from the product UDP and galactose using an enzyme-catalyzed reaction sequence which requires stoichiometric amounts of a phosphorylating agent. Several oligosaccharides have been prepared using routes based on this concept.[112] A second regeneration system for UDP-Gal, which is based on the use of galactose-1-phosphate uridyltransferase, has also been developed,[157] and has been used in the preparation of analogs such as 2'-deoxy-LacNAc and 2'-amino-2'-deoxy-

(a)

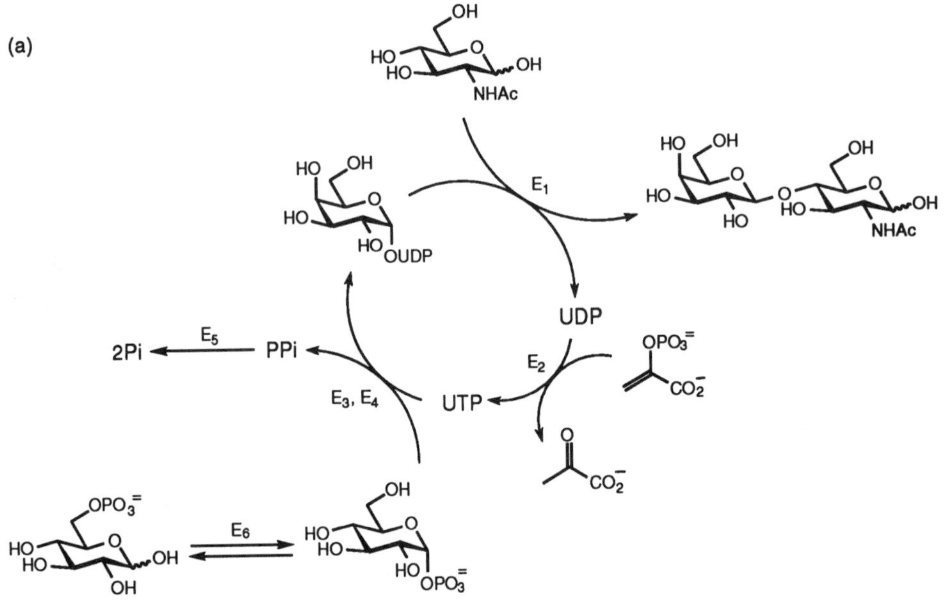

E$_1$: β1,4-galactosyltransferase; E$_2$: pyruvate kinase; E$_3$: UDP-Glc pyrophosphorylase
E$_4$: UDP-Glc epimerase; E$_5$: pyrophosphorylase; E$_6$: phosphoglucomutase

(b)

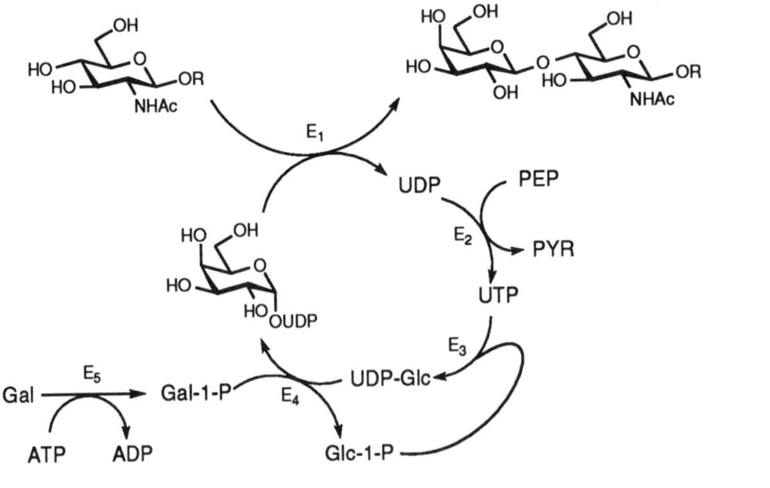

E$_1$: β1,4-galactosyltransferase; E$_2$: pyruvate kinase; E$_3$: UDP-Glc pyrophosphorylase
E$_4$: Galactose-1-phosphate uridyltransferase; E$_5$: galactokinase

Figure 19. Galactosyltransferase-catalyzed glycosylation with *in situ* regeneration of UDP-Gal.
(a) UDP-Glc epimerase-based method. (b) Gal-1-P uridyltransferase-based method.

LacNAc. UDP-Glc and UDP-Gal can also be regenerated *in situ* based on sucrose synthetase.[155b]

   In situ cofactor regeneration offers several advantages. First, a catalytic amount of nucleoside diphosphate and a stoichiometric amount of monosaccharide can be used as starting materials rather than a stoichiometric quantity of sugar nucleotide, thus tremendously reducing costs. Second, product inhibition by the released NDP is minimized due to its low concentration in solution. Third, isolation of the product is greatly facilitated.

   A regeneration system for CMP-NeuAc has also been developed, and is illustrated in Figure 20.[100,158] This system follows the same basic principles as that for the regeneration of UDP-Gal. The UDP-Gal and CMP-NeuAc regeneration schemes have been combined in a one-pot reaction and applied to the synthesis of sialyl Lewis X.[90a] The development of these regeneration systems, as well as the more recent development of regeneration schemes for UDP-GlcNAc[150], GDP-Man,[153] and GDP-Fuc,[90a] and UDP-GlcUA[93] should facilitate the more widespread use of glycosyltransferases for oligosaccharide synthesis.

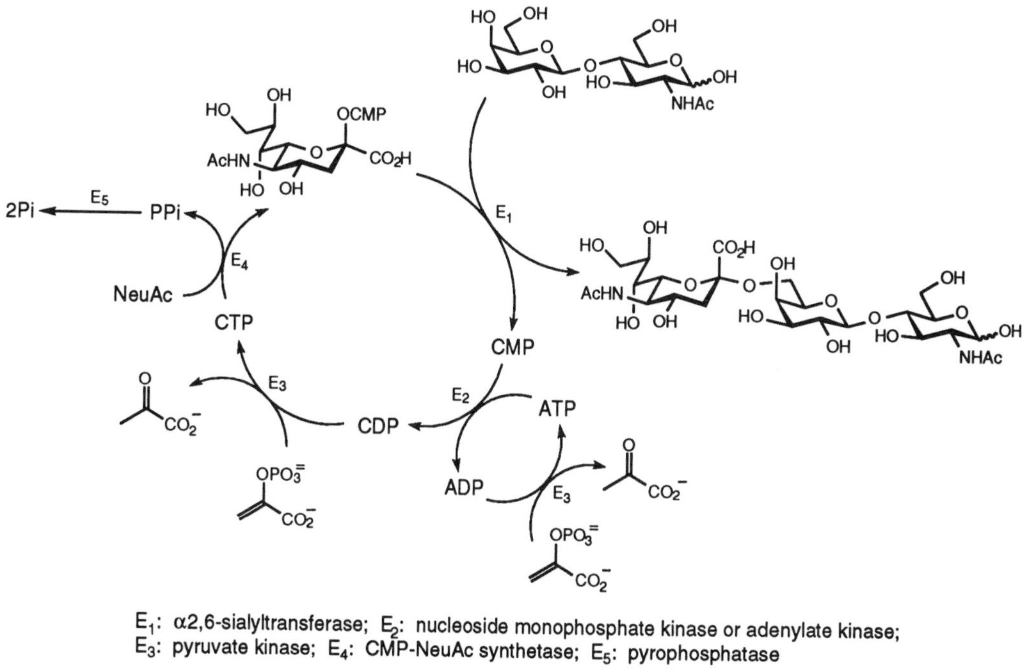

E_1: $\alpha 2,6$-sialyltransferase; E_2: nucleoside monophosphate kinase or adenylate kinase; E_3: pyruvate kinase; E_4: CMP-NeuAc synthetase; E_5: pyrophosphatase

Figure 20. Enzymatic sialylation with *in situ* regeneration of CMP-NeuAc.

2.4. Cloning and Expression of Glycosyltransferases

While many glycosyltransferases catalyze similar reactions and in many cases use the same donor substrate, there appears to be little sequence homology among the different transferases. There is, however, a significant cross species homology between the same enzymes. For instance, one finds an 86% identity in comparing the β1,4-galactosyltransferase from humans to the protein sequence from rat. The different glycosyltransferases do exhibit some similarity in that all the cDNA sequences determined to date encode regions consistent with a short N-terminal tail, a hydrophobic transmembrane sequence, a short stem sequence and a large C-terminal catalytic domain.[159] In addition to the membrane bound form of the glycosyltransferases, soluble forms of the enzymes have also been identified in various body fluids such as the blood, milk, and colostrum. Indeed, these fluids have been the sources for the purification of some of these enzymes.[160-162] A comparison of the cDNA sequences of these soluble enzymes with the N-terminal protein sequence of the glycosyltransferases which have been sequenced suggests that the stem region has been cleaved to release the large catalytic domain from the membrane. Presumably, this theme of signal sequence cleavage is consistent for all the glycosyltransferases (Figure 21).[163]

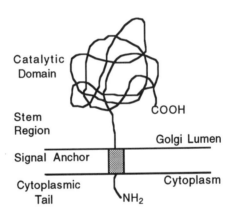

Figure 21. Schematic diagram for the structure of a typical glycosyltransferase.

The amount of a glycosyltransferase that can be isolated from a natural source is often limited by the low concentrations of these enzymes present in most tissues and body fluids. The purification of glycosyltransferases is further complicated by the relative instability of this group of enzymes.[44] For this reason, a great deal of interest has been directed toward the cloning of the glycosyltransferase genes into convenient expression systems. The general strategy involved in this procedure is outlined in Figure 22. The glycosyltransferase gene must first be identified and isolated from the mRNA pool via the cloning of the cDNA to make a cDNA library. This library is then screened to identify the glycosyltransferase gene of interest among the ~106 different sequences present in the library. Once identified, the gene is sequenced and a more complete cloning strategy is developed in order to incorporate the gene into an expression vector. This laborious path has successfully been practiced by several groups, many of whom are referenced in Table 6. The nuances to the general cloning scheme used by these groups are discussed below.

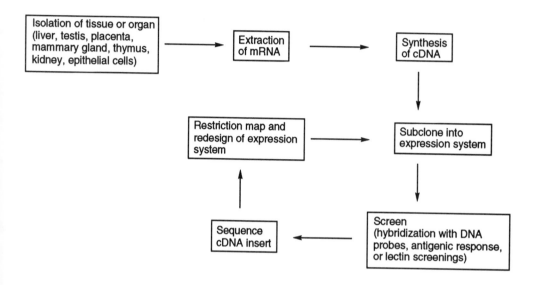

Figure 22. Generalized approach to cloning glycosyltransferases.

Among the organs that have been used for the isolation of mRNA are the liver,[164-165] placenta,[166] mammary gland,[167] testis,[168] and the thymus.[169] In addition, tissue cultures have been used in place of the organ.[170-171] From these sources, cDNA is made, and, for example, the double stranded cDNA is ligated in λ phage via a convenient linker and packed into bacteriophages. The bacteriophages are then plated onto a lawn of *E. coli*, and screened for the desired gene or gene product. The identification of the glycosyltransferase gene has most frequently been achieved by the hybridization of the gene to specific radiolabelled DNA probes.[164-167] Screening in this manner requires a previous knowledge of the gene sequence - information that in some cases may be obtained by extrapolation from a partial protein sequence or from the DNA sequence of the glycosyltransferase from a related source. Two other approaches have been used to screen glycosyltransferase cDNA libraries, both requiring successful transcription and translation of the gene product. In the cloning of the α2,6-sialyltransferase from rat liver, Weinstein et al. used polyclonal antibodies raised to the purified enzyme to screen the λ plaques.[172] The other approach used by Larsen et al. alleviated the need for a previous knowledge of the sequence.[170] This method made use of the specificity of a lectin that recognizes the surface-expressed glycoconjugate product of the α1,3-galactosyltransferase.

The transfected cells were then panned in dishes coated with the lectin. The adherent cells were isolated and re-panned for further purification. Each of these techniques makes use of libraries in which there are very few copies of the desired gene. A greater chance of success may be possible if the number of copies of the genes could be amplified. The introduction, in 1985, of an *in vitro* amplification method based on the polymerase chain reaction (PCR) fulfilled this need.[173-174] Of course PCR (and ECPCR),[174] like the hybridization screening, require a specific knowledge of the sequence.

Once identified, the genes are sequenced using standard procedures. An enlightened approach to the recloning of the gene into an expression vector is then used to develop an expression system. This recloning into expression systems has been performed on only a few of the glycosyltransferases. Toghrol et al. have inserted the mouse liver glucuronyltransferase gene into the yeast vector pEVP11 and expressed the enzyme in *Saccharomyces cerevisiae*.[164] The rat liver glucuronyltransferase, on the other hand, has been expressed in COS cells using the SV40 vector.[165] Expression in COS cells using SV40 was also applied to the cloning of bovine β1,4 galactosyltransferase.[166] A noteworthy approach toward the first expression of the glycosyltransferase in *E. coli* has been developed by Aoki et al. to obtain human β1,4 galactosyltransferase.[175] A unique RsrII restriction site in the galactosyltransferase gene allowed the dissection of the sequence at the location of signal peptidase cleavage. The cohesive terminus was digested with Klenow fragment, and the blunt-end ligated to pINIII-ompA$_2$[176] at a Klenow fragment treated EcoRI site. This construction generated the code for a fusion protein of the soluble form of galactosyltransferase with the ompA signal sequence. Transcription and translation of this sequence in *E. coli* produced an active enzyme that was released into the periplasmic space. Purification and *N*-terminal sequencing of the enzyme verified the expression of the soluble form of galactosyltransferase with an added three amino acid *N*-terminal tail. The kinetic parameters of this enzyme appear to be identical to the isolated native enzyme. This system, however, only produced a very small amount of enzyme (< 0.014 U/L). Though a 30-fold increase of productivity has been achieved with a modified *E. coli* host, the level of productivity is still too low to be of practical synthetic value.[119] More practical expression systems perhaps are those based on *baculo* virus,[90a] and yeast.[177]

To date, very few glycosyltransferases have been cloned, expressed, and produced in quantities sufficient for enzymatic synthesis.[187] Given the advantages of enzymatic synthesis of oligosaccharides over traditional schemes, research into the overexpression of glycosyltransferases will continue to flourish.

Table 6. Cloned glycosyltransferases of the glycoprotein and glycolipid pathways.

Enzyme	Source	Reference
UDP-Glucuronosyltransferase	murine liver	164
	rat liver	165
Mannosyltransferase	yeast	176, 153
α2,6 Sialyltransferase	rat liver	172
α2,3 Sialyltransferase	porcine submaxillary gland	178
β1,4 Galactosyltransferase	bovine placenta	166
	bovine mammary gland	167
	murine mammary gland	179
	bovine liver	180
	murine F9 cells	181
	bovine kidney epithelial cells	182
	murine testes	168
	human placenta	183
	human milk	177
	murine F9 cells	170
α1,2 Fucosyltransferase	human A431 cells	171
α1,3/1,4 Fucosyltransferase	human A431 cells	184a
α1,3 Fucosyltransferase IV-VI	human A431 cells	184b-d
polypeptide N-acetylgalactos-aminyltransferase	bovine colostrum	185
β1,6-*N*-acetylglucosaminyl transferase	rat kidney	186

3. Non-Leloir Glycosyltransferases: Transfer of Glycosyl Donors from Glycosyl Phosphates and Glycosides

3.1. Glycosyl Phosphorylases

Oligosaccharides can also be prepared using non-Leloir glycosyltransferases. Phosphorolysis of polysaccharides is catalyzed by a group of these enzymes called glucan phosphorylases. The latter reaction is reversible and can be used in the synthesis of oligo- or polysaccharides. Two particularly important examples are the syntheses of sucrose and trehalose, catalyzed by sucrose phosphorylase[188] and trehalose phosphorylase,[188] respectively (Figure 23). Examples of other enzymes of this class are those involved in synthesis of dextrans and levans.[189]

Non-Leloir transferases have also been used to synthesize a variety of polysaccharides. The synthesis of modified polysaccharides may provide materials with more desirable physical and biological properties than their natural counterparts. Approaches to controlling the characteristics of polymers include the control of genes encoding the enzymes responsible for their production, regulation of the activity of these enzymes, or the influence of their *in vivo* synthesis.[190] Potato phosphorylase has been used *in vitro* to prepare maltose oligomer.[71] The substrate α-glucose 1-phosphate can be generated *in situ* from sucrose, catalyzed by sucrose phosphorylase.[72] Potato phosphorylase has also been used in the synthesis of a family of linear, star- and comb-shaped polymers.[191]

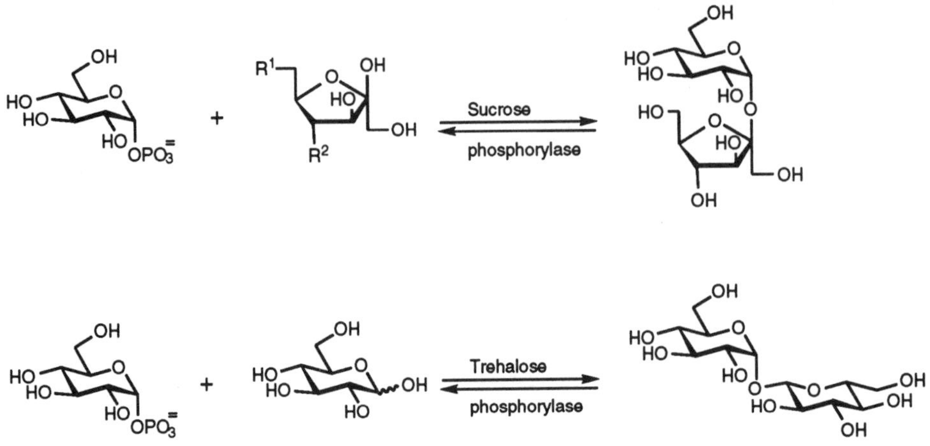

Figure 23. Phosphorylase-catalyzed synthesis of sucrose and trehalose.

Potato phosphorylase (EC 2.4.1.1), for example, has been used in the presence of primers to synthesize polysaccharides.[71] Improvement of this system has recently been accomplished with the use of a coupled enzyme system where glucose-1-phosphate is generated *in situ* from sucrose and inorganic phosphate catalyzed by sucrose phosphorylase.[72] The inorganic phosphate liberated by potato phosphorylase is used by sucrose phosphorylase to drive

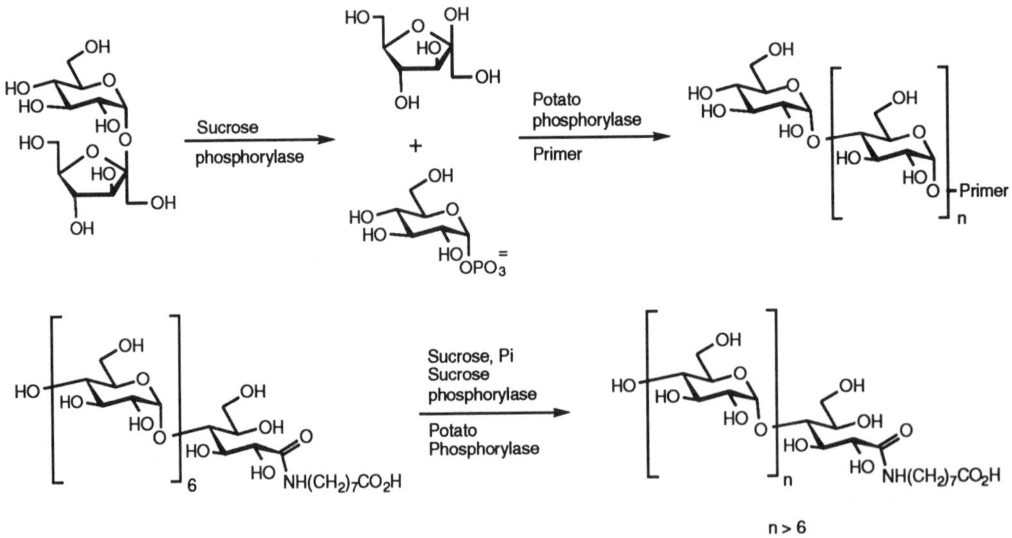

Figure 24. Phosphorylase-catalyzed polysaccharide synthesis.

the formation of polymer, thereby increasing the yield. This coupled-enzyme system also allows for regulation of the molecular weight of the polysaccharide product by control of the concentration of the primer. Unnatural primers bearing functional groups can also be used to prepare tailor-made polysaccharides for further manipulation, e.g. attachment to proteins or other compounds (Figure 24).

3.2. Cyclodextrin Glycosyltransferases

Cyclodextrin α1,4-glucosyltransferase (EC 2.4.1.19) from *Bacillus macerans* catalyzes the cyclization of oligomaltose to form α-, β- and δ-cyclodextrin, and the transfer of sugars from cyclodextrin to an acceptor to form oligosaccharides.[192-193] This enzyme is also able to transform α-glucosyl fluoride into a mixture of α- and β-cyclodextrins and malto-oligomers in almost equal amounts (Figure 25).[194] When immobilized on a silica gel support, which was functionalized with glutaraldehyde, the enzyme was very stable, and no loss of activity was observed after 4 weeks when stored at 4 °C. With an appropriate choice of acceptor, this type of enzymatic catalysis may provide a new route to cyclodextrin analogs and novel oligosaccharides.

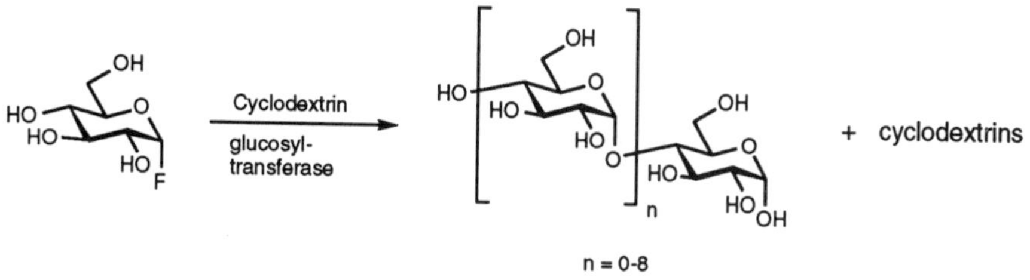

Figure 25. Synthesis of oligosaccharides and cyclodextrins from glycosyl fluorides.

Unnatural sugar acceptors that are structurally similar to glucose are also substrates for cyclodextrin-glucosyl transferase. Using deoxynojirimycin as acceptor and cyclodextrin as donor, an oligoglucosyl deoxynojirimycin was produced, which was subsequently hydrolyzed by glucoamylase to give 4-*O*-α-D-glucopyranosyl deoxynojirimycin in ~60% yield (Figure 26).[195] A number of *N*-substituted derivatives of deoxynojirimycin were also good substrates

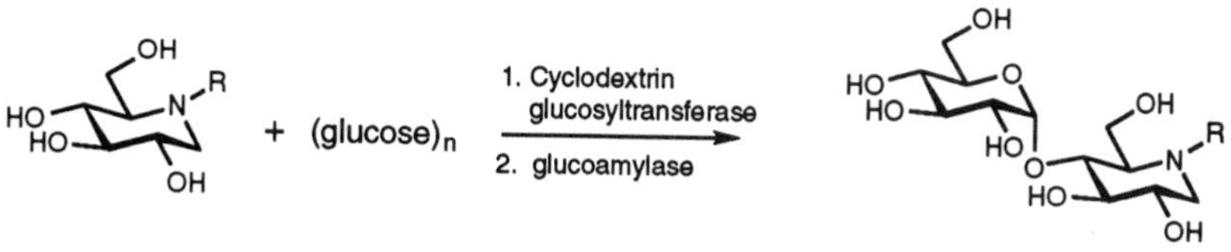

Figure 26. Synthesis of glycosyl *N*-alkyldeoxynojirimycin derivatives.

for the transferase, and the products can be degraded with glucoamylase to glucosylazasugars. One of these glucosylazasugars, 4-*O*-α-D-glucopyranosyl-*N*-methyldeoxynojirimycin, was reported to be a potent inhibitor of glucosidase.

In spite of the progress that has been made, several difficulties limit the use of cell-free enzymes for the synthesis of polysaccharides. The major problem is the complexity of many polysaccharide-synthesizing systems. Isolation, purification, and stabilization of the required enzymes are often difficult. Many enzymes lose activity when they are no longer membrane-associated, and their concentration in cells is very low. The problem is especially significant with enzymes isolated from eukaryotic sources. It is unlikely that cell-free enzymatic synthesis will provide better routes to most natural polysaccharides than do fermentation and isolation. The use of genetic engineering, both using classical genetics and recombinant DNA technology, is an approach being used to prepare modified carbohydrate polymers.[196]

3.3. Lipid A Synthetase

Lipid A is a component of the outer membrane lipopolysaccharides of Gram-negative bacteria, and is responsible for many of the pathological effects of endotoxin.[197a] A key step in the biosynthesis of lipid A is the glycosylation of the 6-position of a 2,3-diacylglucosamine-1-phosphate by the donor UDP-2,3-diacylglucosamine to produce a lipid A precursor. This transformation is catalyzed by the enzyme lipid A synthetase (Figure 27). Lipid A synthetase has been cloned and overexpressed,[197b] and used for the *in vitro* synthesis of the lipid A precursor and analogs thereof. Some examples include *C*-glycoside[198a] and phosphate[198b] analogs (Figure 27).

$$X = OPO_3^{2-} \quad \text{(Lipid A precursor)}$$
$$X = CH_2CO_2^-$$
$$X = PO_3^{2-}$$

Figure 27. Synthesis of the lipid A precursor and analogs thereof using lipid A synthetase.

4. Glycosidases

Glycosidases play an important role in catalyzing the hydrolysis of glycosidic linkages. Cleavage of glycosidic bonds by glycosidases most typically occurs with retention of configuration at the anomeric carbon (β-galactosidase and lysozyme) but can also proceed with inversion (trehalase and β-amylase).[199]

Enzymatic hydrolysis is thought to be mechanistically similar to the acid-catalyzed hydrolysis of glycosides. Both proceed *via* an oxonium ion intermediate or *via* a transition state having oxonium ion character.[199] A proximal carboxylate appears as a common structural motif among glycosidases and presumably acts to stabilize this intermediate or transition state. Whether the oxocarbenium ion exists as a stabilized ion pair or if it is trapped by the carboxylate to form a glycosyl ester has been the subject of a number of investigations and is not yet resolved. Recently, however, an α-glycosyl enzyme intermediate has been observed by ^{19}F NMR in the β-glucosidase-catalyzed hydrolysis of 2-deoxy-2-fluoro-D-glucosyl fluoride and shown to be a catalytically competent species.[200]

Glycosidase-catalyzed glycoside synthesis is quite analogous to protease catalyzed peptide synthesis. As with enzymatic peptide synthesis, glycosidases may be used under either equilibrium or kinetically controlled conditions (Figure 28).[58,201]

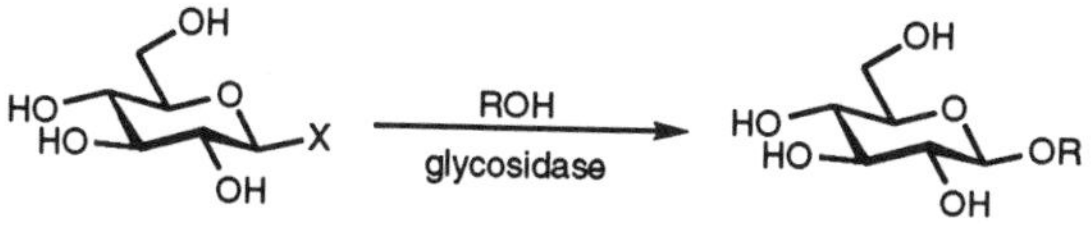

Equilibrium conditions: X = OH
Kinetic conditions: X = F, *o*-NO$_2$C$_6$H$_4$, OR'

Figure 28. Glycosidase-catalyzed glycosylation.

4.1. Equilibrium Controlled Synthesis

Perhaps the most obvious approach to glycosidase catalyzed synthesis of glycosidic linkages involves the direct reversal of the catabolic role of the enzyme. Indeed, examples of equilibrium controlled synthesis were reported by Bourquelot at the early part of this century.[201] Synthesis via this approach involves an endergonic process with the free energy change under ambient aqueous conditions favoring bond cleavage by approximately 4 kcal/mol. Reaction conditions must therefore be manipulated in order to drive the reaction to product glycoside.

In efforts to shift the equilibrium toward product, the addition of water-miscible organic cosolvents and the use of high substrate concentrations have been explored. Preliminary investigations of organic cosolvent addition have suggested enzyme inactivation and decreasing K$_m$ for the glycoside acceptor.[202] This approach awaits more detailed study. The utility of high substrate concentrations and elevated reaction temperatures to increase the reaction rate have been reported by two groups. Johansson et al.[203] reported the synthesis of mannose disaccharides with jack bean α-mannosidase while Ajisaka et al.[204] utilized almond β-glucosidase with glucose concentrations as high as 90% w/v to obtain 40% yield of a mixture of glucose disaccharides. It is important to note that even in 90% w/v glucose the water concentration is still quite high, 22 M, as compared to glucose at 5 M. Two groups have reported the use of carbon-Celite[205] and active carbon columns as molecular traps which selectively absorb the product as the reaction mixture is circulated through the column. Yields, however, are still only about 15%.

Though experimentally simple, the equilibrium approach provides poor yields (generally not exceeding 15%). Also, the low yield of the desired product and the formation of side products generally make purification difficult.

4.2.   Kinetically Controlled Synthesis

Kinetically controlled synthesis relies on the trapping of a reactive intermediate generated from an activated glycosyl donor with exogenous nucleophiles to form a new glycosidic bond.[58,202] Suitable glycosyl donors for this transglycosylation reaction include di- or oligosaccharides, aryl glycosides, and glycosyl fluorides. This approach depends on the more rapid trapping of the reactive intermediate by the glycosyl acceptor than by water. Indeed, enhanced rates of glycosidase catalyzed glycosyl cleavage have been observed in the presence of alcohols.[206] This enhancement could be due to more effective binding of the alcohol relative to water in the active site. Another proposed rationalization is that the mechanism involves a solvent separated ion pair toward which an alcohol is a better nucleophile than water.[206] While under the proper conditions, the glycoside formation may be favored kinetically, hydrolysis is favored thermodynamically. The reaction must therefore be carefully monitored, and arrested when the glycosyl donor is consumed in order to minimize glycoside hydrolysis.

In a comparative study of kinetically vs. thermodynamically controlled synthesis of Galβ1,6GalNAc, the kinetic approach afforded of 10-fold increase in product yield (20% vs. 2%).[207] Yields in kinetically controlled synthesis generally range from 20 to 40%. Although addition of organic solvent might be expected to increase product yields, as in the equilibrium controlled approach, this effect has not generally been observed. An increase in acceptor or donor concentration, however, seems to be quite effective for increasing the yield of the products. Polyethylene glycol-modified β-galactosidase, however, is soluble in organic solvents and seems to have transglycosidase activity.[208]

The kinetically controlled approach has primarily been applied to the retaining glycosidases. Using glycosyl fluorides as glycosyl donors, however, an inverting glycosidase has been used to afford products having the configuration at the anomeric position which is opposite to that of the donor.[209] For example, the α,α-linkage of α-D-glucopyranosyl-α-Dxylopyranoside has been prepared utilizing β-glucosylfluoride and α-trehalase.[210]

4.3. Selectivity

The primary goal of enzymatic glycoside formation or oligosaccharide synthesis is to achieve a selectivity that is difficult to accomplish by chemical methods. Regioselective reaction of one hydroxyl group of an unprotected sugar with the glycosyl donor has been observed, although the selectivity is not necessarily absolute or predictable. It is perhaps likely that kinetically controlled synthesis will be more successful than thermodynamically controlled synthesis in achieving selectivity in reasonably good yields. In general, the primary hydroxyl group of the acceptor reacts preferentially over secondary hydroxyl groups, resulting in a 1,6-glycosidic linkage. Some control of selectivity has been demonstrated by the selection of an appropriate donor/acceptor combination.[211] For example, the α-galactosidase catalyzed reaction of α-Gal-O-*p*-NO$_2$C$_6$H$_4$ with α-Gal-OMe (or allyl α-galactoside) and β-Gal-OMe form predominantly α-1,3 and α-1,6 linkages, respectively.[211a,b] The configuration of the anomeric center of the acceptor controls, to some extent, the position of glycosylation. A similar situation was observed in the β-*N*-acetylgalactosiminidase reactions.[211c] αGal-O-*p*-NO$_2$C$_6$H$_4$ acting both as donor and acceptor, forms preferentially the α-1,3 linkage whereas the *ortho*-nitrophenyl glycoside reacts in a similar fashion to form predominantly the α-1,2 linkage.[211] With β-galactosidase, the β1,3-linked disaccharides were formed predominantly when benzyl and allyl β-galactoside[58,211] or glucal and galactal[212] were used as acceptors.

One can also use glycosidases from different species to control the regioselectivity. For example, the β-galactosidase from testes catalyzes the formation of Galβ1,3GlcNAc or Galβ1,3GlcNAcβSEt[213a] from lactose and either GlcNAc or GlcNAcβSEt, respectively. The minor products produced in this preparation were then hydrolyzed by the *E. coli* β-galactosidase which preferentially hydrolyze β1,6-linked galactosyl residue. In a typical β-glucosidase reaction, incubation of *o*-nitrophenyl 6-deoxy-β-D-*xylo*-hex-5-enopyranoside with the enzyme gave the β-1,3-linked self condensed product in ~9 % yield.[213b] The overall yield of the β1,3-linked disaccharides was around 10-20%. Synthesis of polysaccharides based on kinetically controlled glycosidase reactions have been accomplished, as exemplified by the cellulase catalyzed polymerization of β-cellobiosyl fluoride to form cellulose, with degree of polymerization >22.[214]

Glycosyl transfer to non-sugar acceptors has also been demonstrated. These reactions are especially interesting with chiral, racemic or meso alcoholic acceptors, as one might expect some degree of diastereoselectivity due to the asymmetric microenvironment of an enzyme active site. Such selectivity has indeed been observed, with diastereoselectivities ranging from moderate to exceptional. Some selected examples of the use of glycosidases for synthesis are illustrated in Table 7.

Table 7. Oligosaccharides and other glycosides synthesized using glycosidases.

Substrate	Product	Scale	Ref.
α-Galactosidase			
Raffinose + $CH_2=CHCH_2OH$	$Gal\alpha OCH_2CH=CH_2$	A	211b
$Gal\alpha O\text{-}p\text{-}NO_2Ph + Gal\alpha OCH_2CH=CH_2$	$Gal\alpha1,3Gal\alpha OCH_2CH=CH_2$	B	211b
+ $Gal\alpha OMe$	$Gal\alpha1,3/6Gal\alpha OMe$	B	211a
+ $Gal\beta OMe$	$Gal\alpha1,3/6Gal\alpha OMe$	B	211a
+ $Gal\alpha O\text{-}p\text{-}PhNO_2$	$Gal\alpha1,2/3Gal\alpha O\text{-}p\text{-}PhNO_2$	C	211a
$Gal\alpha O\text{-}o\text{-}PhNO_2 + Gal\beta O\text{-}p\text{-}PhNO_2$	$Gal\alpha1,2/3Gal\alpha O\text{-}p\text{-}PhNO_2$	C	211a
β-Galactosidase			
$Glc\beta OPh + ROH$	$Glc\beta OR$	C	215
$ROH = CH_3(CH_2)_nOH\ (n = 3\text{-}7), [CH_3(CH_2)_3]_2CHOH$			
$(CH_3)_2CHOH, C_6H_{11}OH, HO(CH_2)_2OH$			
$(CH_3)_2CH(CH_2)_3OH, (CH_3)_2C(OH)CH(CH_3)OH$			
$PhCH_2CH_2OH$			
$Gal\beta1,4Glc + GlcNAc$ or $GlcNAc\beta SEt$	$Gal\beta1,3GlcNAc$ or $Gal\beta1,3GlcNAc\beta SEt$	B	213
$Gal\beta1,4Glc + GlcNAc$	$Gal\beta1,4GlcNAc$	A	217
+ GalNAc or GlcNAc	$Gal\beta1,6GalNAc$ or $Gal\beta1,3GlcNAc$	B	207,216
+ $CH_2=CHCH_2OH$	$Gal\beta OCH_2CH=CH_2$	A	211b
	$Gal\beta1,3/6Gal\beta OCH_2CH=CH_2$	B	
+ $PhCH_2OH$	$Gal\beta OCH_2Ph$	A	211b
	$Gal\beta1,3/6Gal\beta OCH_2Ph$	B	211b
+ $Me_3Si(CH_2)_2OH$	$Gal\beta O(CH_2)_2SiMe_3$	B	

$Gal\beta OPh$ +

B 218

1. $R^1 = R^4 = H, R^2 = OH, R^3 = CH_3$ (Gitoxigenin)
2. $R^1 = R^2 = R^4 = H, R^3 = CH_3$ (Digitoxigenin)
3. $R^1,R^2 = O, R_3 = CH_3, R_4 = H$ (16β, 17β-epoxy-17α-digitoxigenin)
4. $R^1 = R^2 = H, R^3 = CHO, R^4 = OH$ (Strophanthidin)

$Gal\beta O\text{-}o\text{-}NO_2Ph + Gal\alpha OMe$	$Gal\beta1,6Gal\alpha OMe$	C	211a
+ $Gal\beta OMe$	$Gal\beta1,6/3Gal\beta OMe$	C	211a
$Gal\beta OPh$ + ROH	$Gal\beta OR$	B	219
$(ROH = CH_3(CH_2)_nOH\ (n = 0,3), (CH_3)_3CCH_2OH$			
$Glc\beta OPh$ + $C_6H_{11}OH$	$Glc\beta OC_6H_{11}$	B	219
$Gal\beta1,4Glc$ + sucrose	$Gal\beta1,6Glc\alpha1,2Fru$	E	220

$Gal\beta1,4Glc$ or $Gal\beta OPh$ +

89% de   E   221

Substrate	Product	Scale	Ref.

Galβ1,4Glc or GalβOPh +

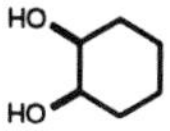

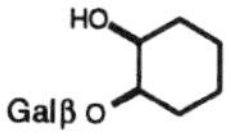

Galβ O— ... 90% de   E   221

GalβO ... 75% de   E   221

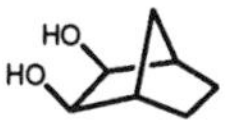

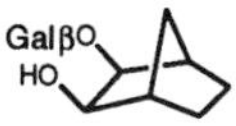

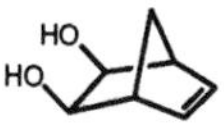

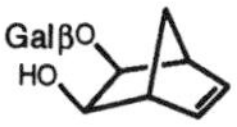

GalβO ... 75% de   E   221

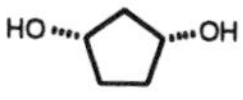

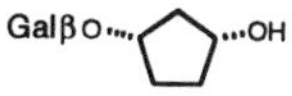

Galβ O...⟨ ⟩...OH   50% de   E   221

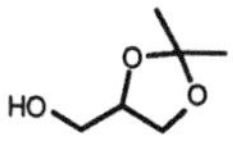

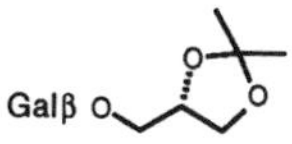

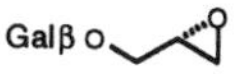

Galβ O... 20% de   A   222

Galβ O... 40% de   A   222

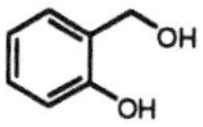

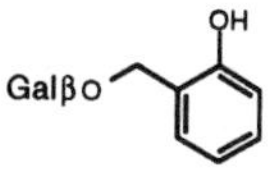

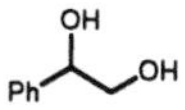

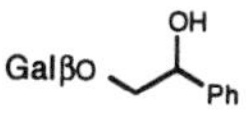

GalβOPh +   GalβO...   B   223

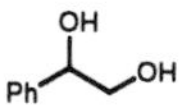

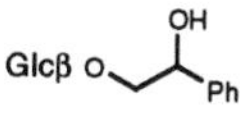

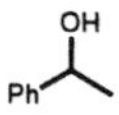

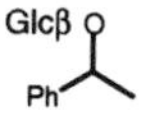

GalβO...   B   223

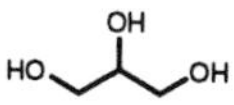

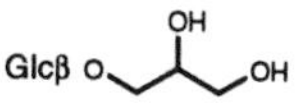

GlcβOPh +   Glcβ O...   B   223

Glcβ O...   B   223

Glcβ O...   B   224

GlcβO...   B   225

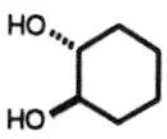

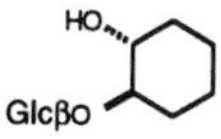

Galactal + ROH	2-deoxy-GalβOR	E	226
Galactal + Galactal	2-deoxy-Galβ1,3/6Galactal + 2-deoxy-Galβ1,3-2-deoxy-Galβ1,6- Galactal	C	227
Galβ0-*p*-NO₂Ph + glucal	Galβ1,3/6Glucal	B	212
Galβ0-*p*-NO₂Ph + 6-0-Ac-glucal	Galβ1,3-(6-0-Ac-Glucal)	B	212
Galβ0-*o*-NO₂Ph + Z-Ser-OR	GalβO-Z-Ser-OR	C	228
Galβ0-*o*-NO₂Ph + Ser	GalβO-Ser	-	229
Galβ1,4Glc + Z-Ser-OMe	GalβO-Z-Ser-OMe	B	230
Galβ0-*o*-NO₂Ph + HONCR₂	GalβONCR₂	C	231

Substrate	Product	Scale	Ref.

α-Mannosidase

ManαO-*p*-NO$_2$Ph + ManαOMe	Manα1,2/6ManαOMe	B	211a
ManαO-*p*-NO$_2$Ph + ManαO-*p*-NO$_2$Ph	Manα1,2/6ManαO-*p*-NO$_2$Ph	B	211a

α-Glucosidase

Glc + Fru	Glcα1,1Fru	D	232

Glcα1,4Glc

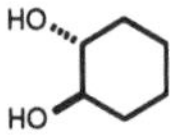

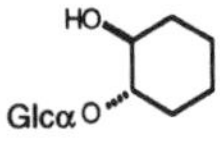

     100 % de   C   215,225

β-Glucosidase

Glc	Glcβ1,4/6Glc	C	233
Glcβ1,4Glc	Glcβ1,4Glcβ1,4Glc	C	233

β-*N*-Acetylhexosaminidase

GalNAcβO-*p*-NO$_2$Ph + GlcβOMe	GalNAcβ1,3/4GlcβOMe	C	234a
GalNAcβO-*p*-NO$_2$Ph + GlcαOMe	GalNAcβ1,4/6GlcαOMe	C	234a
GlcNAcβO-*p*-NO$_2$Ph + GlcβOMe	GlcNAcβ1,3/4GlcβOMe	C	234a
GlcNAcβO-*p*-NO$_2$Ph + GlcαOMe	GlcNAcβ1,4/6GlcαOMe	C	234a
GalNAcβO-*o*-NO$_2$Ph + GlcNAcαOMe	GalNAcβ1,4/6GlcNAcαOMe	C	234b
GalNAcβO-*o*-NO$_2$Ph + GlcNAcβOMe	GalNAcβ1,4GlcNAcβOMe	C	234b

α-Fucosidase

FucαO-*p*-NO$_2$Ph + GalαOMe	Fucα1,3GalαOMe	E	58
+ GalβOMe	Fucα1,3GalβOMe	E	58
FucαF + GalβOMe	Fucα1,2/6GalβOMe	C	235b

Neuraminidase

NeuAcα-*p*-NO$_2$Ph + GalβOCH$_3$	NeuAcα2,3/6GalβOCH$_3$	D	235a
NeuAcα-*p*-NO$_2$Ph + GalαOCH$_3$	NeuAcα2,3/6GalβOCH$_3$.	D	235a
NeuAcα-*p*-NO$_2$Ph + GlcβOCH$_3$	NeuAcα2,3/6GlcβOCH$_3$	D	235a
NeuAcα-*p*-NO$_2$Ph + Galβ1,4GlcβOCH$_3$	NeuAcα2,3/6Galβ1,4GlcβOCH$_3$	D	235a
NeuAcα-*p*-NO$_2$Ph + Galβ1,4GlcNAc	NeuAcα2,3/6Galβ1,4GlcNAc	D	235a

Invertase

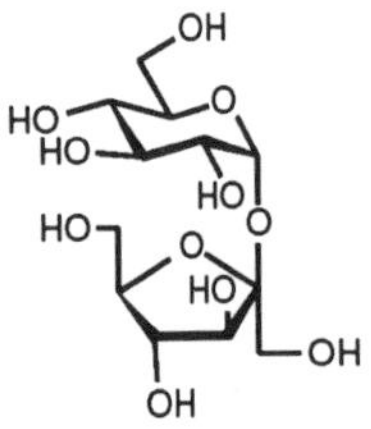

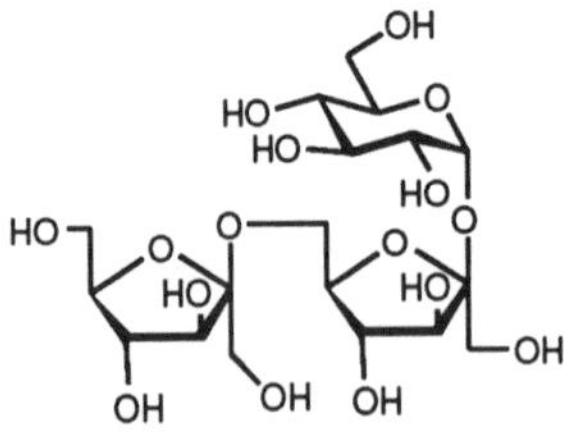

     B   236

+ ROH
R = CH$_3$(CH$_2$)$_n$O, n = 0-7   B   237,238

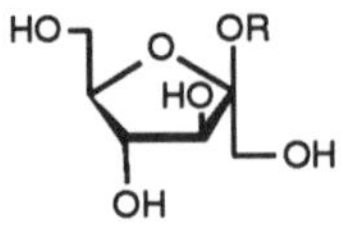

Substrate	Product	Scale	Ref.

Trehalase

| 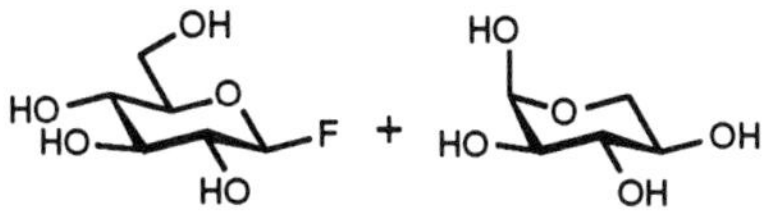 | 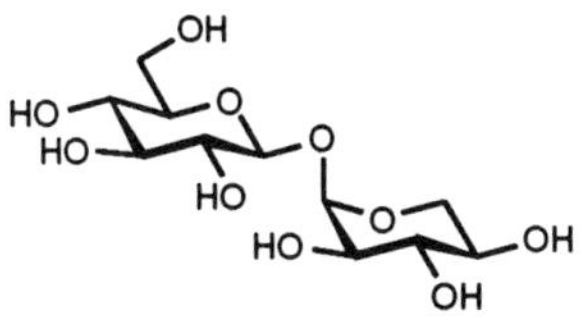 | E | 210 |

Cellulase

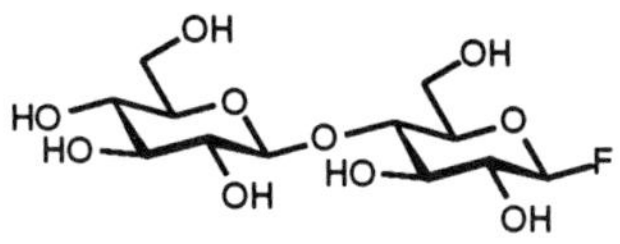

(Glcβ1,4Glc)$_n$, n ≥ 22   B   214

Scale: A, > 1g; B, 0.1-1 g; C, 10-100 mg; D < 10 mg; E, not reported.

Glycosidases can also be combined with glycosyltransferases in the same reaction, as illustrated in Figure 29.[239a] In this example, β-galactosidase from *Bacillus circulans* was used to synthesize Galβ1,4GlcNAc,[239b] which was in turn sialylated with α2,6-sialyltransferase. This provided a rapid synthesis of a trisaccharide from the three monosaccharide units and avoids the secondary hydrolysis of the galactosidase product.

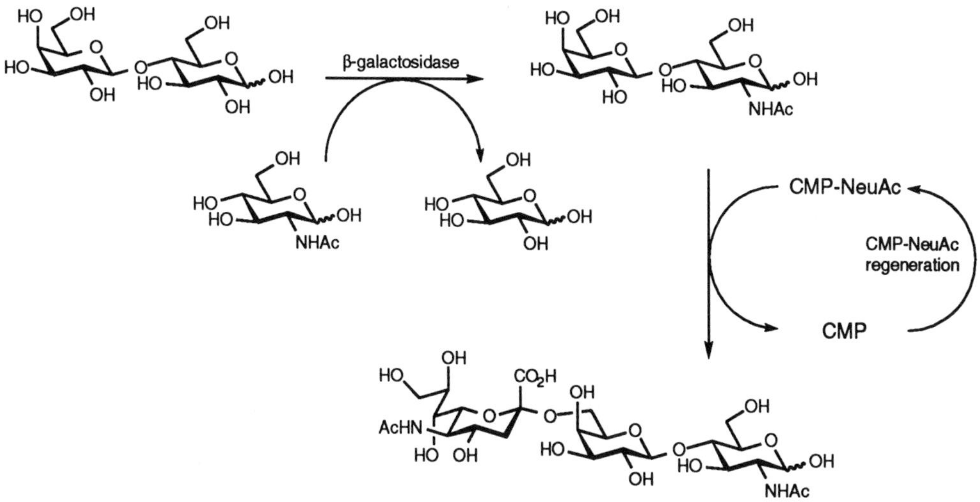

Figure 29. A galactosidase-catalyzed glycosylation coupled with a sialyltransferase reaction.

5.   Transglycosidases

Transglycosidases are related to glycosidases in that they cleave glycoside bonds; they differ, however, in that they usually transfer the glycosyl moiety to another acceptor with a minimal amount of hydrolysis. Transglycosidases have been found to be useful catalysts for glycosylation. For example, a β-fructofuranosidase from *Antherobacter* sp. K-1 has been used to transfer fructose from sucrose to the 6-position of the glucose residues of stevioside and rubusoside.[240] A sucrase from *Bacillus subtilis* catalyzes the reversible transfer of fructose from sucrose to the 6-hydroxyl of a fructose unit at the nonreducing end of a levan chain.[241] Several unnatural sucrose derivatives have been prepared by taking advantage of this process.[241c]

A transsialidase from *Trypanosoma cruzi* has been shown to transfer sialic acid reversibly to and from the 3-position of terminal β-Gal residues.[242a] Chains terminating in α-linked galactose are not substrates. A number of oligosaccharides containing the NeuAcα2,3Galβ-substructure have been synthesized using this transsialidase.[242b] Namely, sialic acid has been transferred to simple galactosides that are not substrates for sialyltransferases.[242d] Additionally, this enzyme has been shown to resialylate the terminal galactose units of the cell-surface glycoproteins and glycolipids of sialidase-treated erythrocytes.[242c] Thus, the *T. cruzi* transsialidase potentially provides a useful alternative to α2,3-sialyltransferase.

6.   Synthesis of *N*-Glycosides

Nucleosides and their derivatives are ubiquitous in nature, and are involved in a myriad of biochemical phenomena, most notably the storage and transfer of genetic information. Interest in this class of compounds has been stimulated by the efficacy of certain nucleosides as anti-parasitic[243] (e.g. 1-β-D-ribofuranosyl-1H-pyrazolo [e,4-d] pyrimidin-4-one, allopurinol riboside) and antiviral (e.g. 3'-azido-3'-deoxythymidine, azidovudine, AZT) agents.[244-245] Nucleosides have traditionally been prepared by various chemical methods.[246] These methods require protection and deprotection steps and glycosyl activation, and they dictate in a multi-step synthesis. Other problems include control of anomeric configuration, especially when preparing β-D-arabinofuranosyl- and 2'-deoxy-β-D-ribofuranosylnucleosides, and regiospecific C-N glycoside formation when there are several possible nucleophilic groups in the purine or pyrimidine base.

6.1.   Nucleoside Phosphorylase

Enzymatic preparations of both natural and unnatural nucleosides have been reported using nucleoside phosphorylases as catalysts.[247] These enzymes catalyze the reversible formation of a purine or pyrimidine nucleoside and inorganic phosphate from ribose-1-phosphate (R-1-P) and a purine or pyrimidine base, with the equilibrium lying well in favor of nucleoside formation (Figure 30). Nucleoside synthesis has relied on the transfer of the ribose moiety of a readily available nucleoside to a different purine or pyrimidine base or analogs through the intermediacy of R-1-P. This work has been done primarily with isolated enzymes[248] but whole cells have also been employed in a few cases.[249] The deleterious hydrolases present in whole cells could be largely neutralized by conducting the reactions at 60 °C, a temperature at which the nucleoside phosphorylases maintain >70% of their activity for 3-5 days.[249]

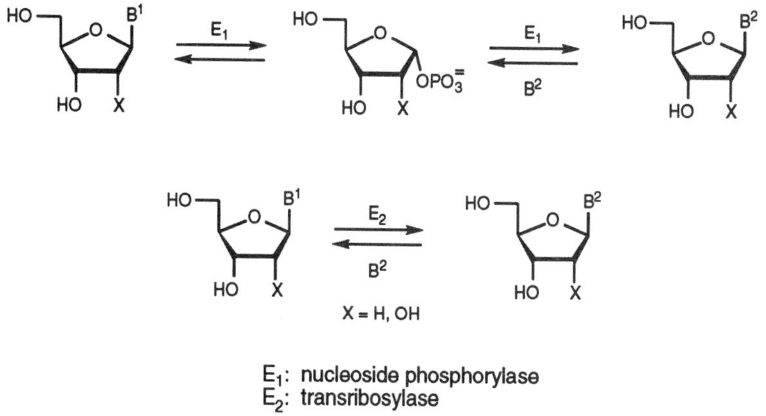

E$_1$: nucleoside phosphorylase
E$_2$: transribosylase

Figure 30. Enzyme-catalyzed nucleoside synthesis.

Two basic strategies have generally been employed. The first involves isolation of R-1-P, which is prepared in good yield from a nucleoside in the presence of a high concentration of phosphate.[250] The isolated R-1-P is then used as the glycosyl donor in an enzymatic coupling reaction with added purine or pyrimidine bases or analogs. The second strategy involves a one-pot exchange of one base for another in the presence of a catalytic amount of inorganic phosphate without isolation of R-1-P. The first strategy (isolation of R-1-P) is the most general, and any heterocycle that is a substrate for a nucleoside phosphorylase can generally be glycosylated by this method. The second strategy (*in situ* generation of R-1-P) is more limited. At best, it results in formation of an equilibrium mixture of the substrate and the product nucleosides, from which the product must be isolated. In less favorable cases, the natural purine or pyrimidine base released from the glycosyl donor may be a potent competitive inhibitor versus the purine or pyrimidine analog, for which the enzyme has lower affinity. For example, the competitive inhibition by hypoxanthine (K_m = 5.6 mM) was the cause for the lack of glycosylation of 1,2,4-triazole-3-carboxamide (TCA, the aglycon component of virazole, K_m = 167 mM) when inosine was used as the ribosyl donor and purine nucleoside phosphorylase (PNPase) as the catalyst.[251] It was, however, possible to synthesize virazole by isolating of R-1-P and subsequently using it as the ribosyl donor.[251] An alternative way to circumvent the inhibition problems is to employ a pyrimidine nucleoside as the glycosyl donor and a purine (or purine analog) as the acceptor, since the released pyrimidine base does not inhibit the purine nucleoside phosphorylase.[252] By this method, both pyrimidine nucleoside phosphorylase and purine nucleoside phosphorylase are required.

Recently, direct purine-to-purine exchange reactions have been conducted without isolation of R-1-P using activated purine derivatives as the ribosyl donors.[253] The activated purine derivatives were prepared by 7-*N*-methylation of inosine and guanosine to provide derivatives that are excellent substrates for phosphorolytic cleavage by PNPase. The cleaved 7-*N*-methylpurines do not show any measurable product inhibition and the equilibrium of the reaction greatly favors product. The effectiveness of this approach was demonstrated by the one-pot synthesis of virazole from 1,2,4-triazole-3-carboxamide and 7-*N*-methyl inosine.

The nucleoside phosphorylases have been found to accept a wide range of nucleoside analogs, with modifications in both the base and glycosyl components, as substrates. The synthesis of ribosides of unnatural purine and pyrimidine bases are summarized in Table 8, while Table 9 shows the synthesis of nucleosides containing modified glycosyl moieties with either natural or unnatural bases. Most of these reactions have been carried out in one step without isolation of the intermediate sugar phosphate, although involvement of the sugar phosphate intermediate has been demonstrated.

The use of unnatural bases has met with much success using both natural and unnatural glycosyl donors. A few limitations have, however, been observed. For example, in the purine nucleoside phosphorylase-catalyzed synthesis of certain imidazole [4,5-*c*] pyridine (3-deazapurine) nucleosides, the normally observed regiospecificity was lost, and a mixture of *N*-1 and *N*-3 glycosylated products was isolated. This problem was not encountered with purines that retain the nitrogen at the 3 position, as 2'-deoxyribosylation of unsubstituted purine gave only the *N*-9 glycosyl product.[252] It appears that either the *N*-3 of the purine base or an appropriate substituent at C-6 is necessary for proper orientation of the base for regiospecific glycosylation.

The synthesis of sugar-modified nucleosides has used glycosyl donors that are prepared by chemical modification of readily available nucleosides, such as uridine and cytidine. Good yields of arabino-[249] and 2'-amino-2'-deoxyribonucleosides[245] have also been obtained enzymatically, although the enzymatic synthesis of 3'-amino-2',3'-dideoxyribonucleosides has given only low yields.[256] The low yields in the latter case may be due to competition of the two pyrimidine heterocycles for enzyme binding or an overall decreased reaction rate due to alteration of the 3'-position, which may be important in substrate binding by the enzyme.[257]

6.2. *N*-Transribosylase

N-Transribosylases have been employed in the synthesis of nucleoside analogs (Scheme 30).[247,258] Two classes of transribosylases have been identified: type I enzymes catalyze the transfer of the sugar moiety between two purine bases, and type II catalyze the transfer between

Table 8. Nucleoside phosphorylase-catalyzed synthesis of nucleosides with various heterocycles as acceptors.

Donor	Acceptor		Method[1]	Yield (%)[2]	References
Uridine		X = MeS, Y = H	B	59-76	259, 260
		X = NH$_2$, Y = Cl			
Thymidine		X = Me$_2$N, Y = H	B	81	259, 260
7-*N*-Methyl Guanosine		X = NH$_2$, Y = H	B	100	253
Inosine		X = C$_5$H$_{11}$S, Y = H	A	59	259
Uridine		X = NH$_2$, BnNH	B	18-79	259, 252
Thymidine		X = Cl	B	18-71	259, 252
7-*N*-Methyl Guanosine		X = NH$_2$	B	53	253
Uridine		X = OH, PhCONH	B	23-63	259, 261
Inosine			A	47	251
7-*N*-Methyl Guanosine			B	44	253

[1]Method A: α-Glycosyl-1-phosphate generated and isolated prior to addition of acceptor heterocycle. Method B: *In situ* generation of α-glycosyl-1-phosphate.
[2]Yields are based on the initial amount of heterocycle acceptor.

Table 9. Nucleoside phosphorylase- and transribosylase-catalyzed synthesis of sugar-modified nucleosides.

Donor	Acceptor		Method[1]	Yield (%)[1]	References
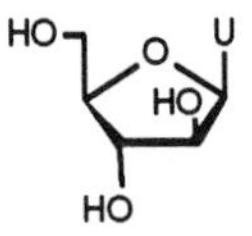 1-(β-D-arabinosyl)uracil			B	34-92	262
	X	Y			
	NH₂	H			
	OH	Cl			
	NH₂	NH₂			
	NH₂	CH₃			
	H	NH₂			
	OH	H, NH₂, CH₃			
	SH	NH₂			
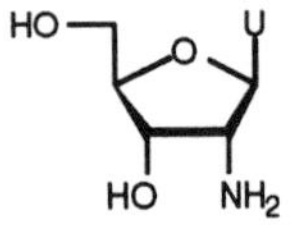 2'-amino-2'-deoxyuridine	NH₂	H	B	20-50	254-255, 263
	OH	H			
	OH	NH₂			
	OH	Cl			
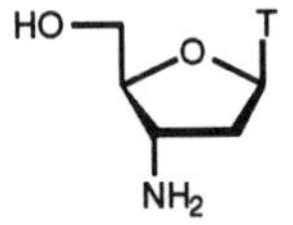 3'-amino-3'-deoxythymidine	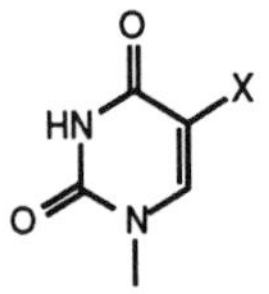 X = F, Cl, Br		B	7-29	256
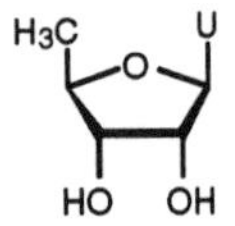 5'-deoxyuridine			B	12	252
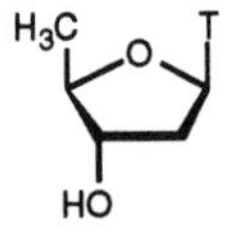 5'-deoxythymidine			B	17	252

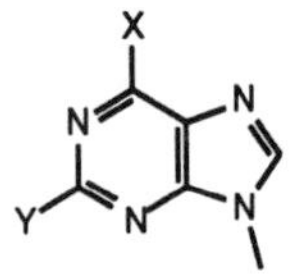

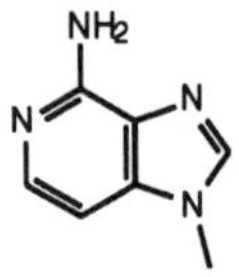

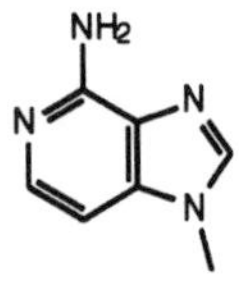

[1]See Table 8.

any two bases.[258a] Like nucleoside phosphorylases, transribosylases are stereospecific for the β anomer of the nucleoside product. Thymidine and 2'-deoxycytidine are the best glycosyl donors, and a reasonable amount of variation in the acceptor bases is tolerated. The transribosylase from *L. leichmanii* has been used to prepare 2-chloroadenosine, an antileukemic and immunosuppressive nucleoside.[258c]

6.3. NAD Hydrolase

The enzyme NAD glycohydrolase has been used in an exchange reaction for the preparation of NAD analogs.[264] The enzyme accepts nicotinamide analogs with modifications at the amide functionality as substrates (Figure 31). Depending on the structures of the nicotinamide analogs used, the reaction may be either reversible or irreversible. NADH and its 6-hydroxyl derivative are not substrates for the enzyme. When 4-amino, 4-methylamino, or 4-dimethylamino nicotinamide or nicotinate was used as substrate, the product NAD analog existed as a 1,4-dihydro-type tautomer.[265]

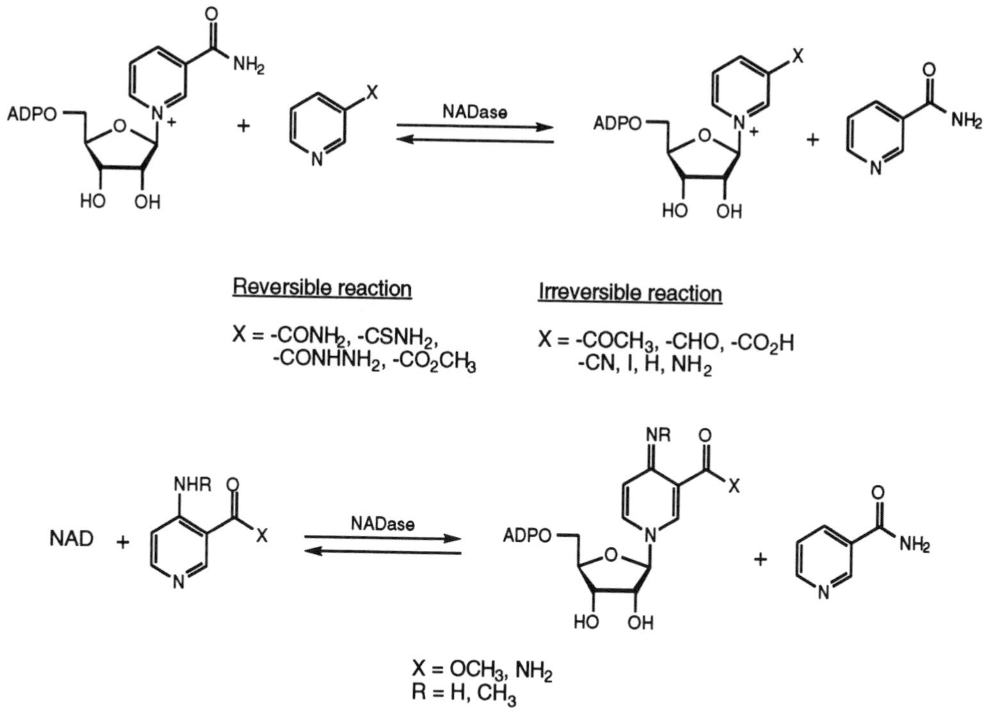

Figure 31. NADase-catalyzed synthesis of nucleosides.

7. Biological Applications of Synthetic Glycoconjugates

7.1. Glycosidase and Glycosyl Transferase Inhibitors

A range of analogues and derivatives of carbohydrates are proving to be of interest in studying the biosynthesis and modification of oligosaccharides: deoxynojirimycin, swainsonine, and castanospermine inhibit trimming of the *N*-linked oligosaccharides of glycoproteins;[266] tunicamycin and streptovirudin also inhibit protein glycosylation in the Leloir pathway;[267] acarbose inhibits amylase.[268] These inhibitor systems have attracted much attention for two reasons. First, they provide a way of exploring cell-surface oligosaccharide chemistry, a topic of central interest in differentiation and development, as well as other areas. Second, most are relatively easily understood as transition state analogues, and there is a good chance that the design of other, new sugar analogues to inhibit other glycosidases and glycosyltransferases[139,269] can be accomplished.

The syntheses of these types of structures are not straightforward using classical synthetic methods. Enzymatic methods have already been proven to be very useful in syntheses of deoxynojirimycin and related materials,[270] and will probably prove widely applicable to other preparations in this series.

7.2. Glycoprotein Remodeling and Synthesis

A number of the proteins of interest as human pharmaceuticals (tissue plasminogen activator, juvenile human growth hormone, CD4) are glycoproteins. There is substantial interest in developing methods that will permit modification of oligosaccharide structures on these glycoproteins by removing and adding sugar units ("remodeling") and in making new types of protein-oligosaccharide conjugates.[271-272] The motivation for these efforts is the hope that modification of the sugar components of naturally-occurring or unnatural glycoproteins might increase serum lifetime, increase solubility, decrease antigenicity, and promote uptake by target cells and tissues.

Enzymes are plausible catalysts for manipulating the oligosaccharide content and structure of glycoproteins. The delicacy and polyfunctional character of proteins, and the requirement for high selectivity in their modification, indicate that classical synthetic methods will be of limited use. The major problems in the widespread use of enzymes in glycoprotein remodeling and generation are that many of the glycosyl transferases that are plausible candidates for this area are not available, and the uncertainty in whether glycosyl transferases that probably act on unfolded or partially folded proteins *in vivo* will be active at the surface of a completely folded protein. Perhaps *in vitro* mutagenesis would allow the incorporation of glycosl amino acids into proteins. Additional sugars may be added, using glycosyltransferases, to prepare homogeneous glycoproteins. Coupling of glycopeptides in aqueous solution using engineered proteases is also available, and techniques for the extension of the sugar chain based on glycosyltransferases have also been developed.[273]

8. Future Opportunities

The pace of development of carbohydrate-derived pharmaceutical agents has, in general, been slower than that of more convenient classes of materials. The difficulties in the synthesis and analysis of carbohydrates have undoubtedly contributed to this slow pace, but at least three areas of biology and medicinal chemistry have redirected attention to carbohydrates. First, interfering with the assembly of bacterial cell walls[17,273] remains one of the most successful strategies for the development of antimicrobials. As bacterial resistance to penams and cephams becomes more widespread, there is increasing interest in interfering with the biosynthesis of the characteristic carbohydrate components of the cell wall, especially KDO, heptulose, lipid A and related materials. Interest in cell-wall constituents is also heightened by their relevance to vaccines and to leads toward non-protein immunomodulating compounds. Second, cell-surface carbohydrates are central to cell communication, cell adhesion, and differentiation and development, and may be relevant to abnormal states of differentiation, such as those characterizing some malignancies.[1-40] Third, the broad interest in diagnostics has finally begun to generate interest in carbohydrates as markers of human health. In addition, there are a number of other possible applications of carbohydrates, for example as dietary constituents, in antivirals, or as components of liposomes, which warrant attention. Enzymatic methods of synthesis, by rendering carbohydrates more accessible, will contribute to further research in all of these areas.

References

1. *Carbohydrate Chemistry* (Parts I and II); Royal Society of Chemistry Specialist Periodical Reports; Burlington House: London.

2. Kennedy, J. F.; White, C. A. *Bioactive Carbohydrates;* Ellis Harwood Ltd.: West Sussex, 1983.

3. Sharon, N. *Complex Carbohydrates*; Addison-Wesley: Reading, MA, 1975.

4. *The Glycoconjugates*; Horowitz, M. I.; Pigman, W., Eds.; Academic Press: New York, 1977-1978; Vol. I-II.

5. *The Glycoconjugates*; Horowitz, M. I., Ed.; Academic Press: New York, 1982; Vol. III-IV.

6. *The Polysaccharides*; Aspinall, G. O., Ed.; Academic Press: New York, 1982-1985; Vol. I-III.

7. Beyer, A. T.; Sadler, J. E.; Rearick, J. I.; Paulson, J. C.; Hill, R. L. *Adv. Enzymol.* **1981**, *52*, 24.

8. *Carbohydrate Recognition in Cellular Function*; Ciba Foundation Symposium 145; Wiley:

New York, 1989.

9.   *The Molecular Immunology of Complex Carbohydrates*; Wu, A. M., Ed.; Plenum Press: 1988.

10.   Paulson, J. C. *Trends Biochem. Sci.* **1989**, 272.

11.   (a) Feizi, T.; Larkin, M. *Glycobiology* **1990**, *1*, 17. (b) Feizi, T. *Trends Biochem. Sci.* **1991**, 84.

12.   Varki, A. *Glycobiology* **1993**, *3*, 97.

13.   Doering, T. L.; Masterson, W. J.; Hart, G. W.; Englund, P. T. *J. Biol. Chem.* **1990**, *265*, 611.

14.   Lennarz, W. J. *Biochemistry* **1987**, *26*, 7205.

15.   Ryan, C. A. *Biochemistry* **1988**, *27*, 8879.

16.   Thomas, J. R.; Dwek, R. A.; Rademacher, T. W. *Biochemistry* **1990**, *29*, 5413.

17.   *Bacterial Lipopolysaccharides*; Anderson, L.; Unger, F. M., Eds.; ACS Symposium Series 231; American Chemical Society: Washington, D.C., 1983.

18.   For a special issue on glycoconjugates, see *Carbohydr. Res.* **1987**, 164.

19.   Schauer, R. *Adv. Carbohydr. Chem. Biochem.* **1982**, *40*, 131.

20.   *The Lectins: Properties, Functions and Applications in Biology and Medicine*; Liener, I. E.; Sharon, N.; Goldstein, I. J., Eds.; Academic Press: New York, 1986.

21.   Feizi, T. *Nature* **1985**, *314*, 53.

22.   Paulson, J. C. In *The Receptors*; Cohn, P. M., Ed.; Academic Press: New York 1985, Vol. 2, p131.

23.   Sairam, M. R. In *The Receptors*; Cohn, P. M., Ed.; Academic Press: New York, 1985; Vol. 2, p307.

24.   Hakomori, S. *Ann. Rev. Biochem.* **1981**, *50*, 733

25.   Feizi, T.; Childs, R. A. *Biochem. J.* **1987**, *245*, 1.

26.   Rademacher, T. W.; Parekh, R. B.; Dwek, R. A. *Ann. Rev. Biochem.* **1988**, *57*, 785.

27.   Sariola, H.; Aufderheide, E.; Berhard, H.; Henke-Fahle, S.; Dippold, W.; Ekblom, P. *Cell* **1988**, *54*, 235.

28.   Nojiri, H.; Kitagawa, S.; Nakamura, M.; Kirito, K.; Enomoto, Y.; Saito, M. *J. Biol. Chem.* **1988**, *263*, 7443.

29.   Seyfried, T. N. *Developmental Biology* **1987**, *123*, 286.

30.   Usuki, S.; Lyu, S.-C.; Sweeley, C.C. *J. Biol. Chem.* **1988**, *263*, 6847.

31.   Hakomori, S. *Cancer Res.* **1985**, *45*, 2405.

32.   Kornfeld, R.; Kornfeld, S. In *The Biochemistry of Glycoproteins and Proteoglycans*; Lennarz, W. J. Ed.; Plenum: New York, 1980; p 1.

33.   Neufeld, E. F.; Ashwell, G. In *The Biochemistry of Glycoproteins and Proteoglycans*; Lennarz, W. J., Ed.; Plenum: New York 1980; p 266.

34. Roseman, S. In *Cell Membranes: Biochemistry, Cell Biology and Pathology*; Weissman, G.; Clairborne, R., Eds.; Hospital Practice: New York 1975; p 55.

35. Ashwell, G.; Harford, J. *Ann. Rev. Biochem.* **1982**, *61*, 531.

36. Berger, E. G.; Greber, U. F.; Mosbach, K. *FEBS Lett.* **1986**, *203*, 64.

37. (a) Parekh, R. B.; Dwek, R. A.; Edge, C. J.; Rademacher, T. W. *TIBTECH* **1989**, *7*, 117; (b) Sasaki, H.; Bathner, B.; Dell, A.; Fukuda, M. *J. Biol. Chem.* **1987**, *262*, 12059; (c) Tsuda, E.; Kawanishi, G.; Ueda, M.; Masuda, S.; Sasaki, R. *Eur. J. Biochem.* **1990**, *188*, 405.

38. *Enzymatic Bases of Detoxification*; Jakoby, W. B., Ed.; Academic Press: New York 1980; Vol. II.

39. (a) Phillips, M. L.; Nudelman, E.; Gaeta, F. C. A.; Perez, M.; Singhal, A. K.; Hakomori, S.-I.; Paulson, J. C. *Science* **1990**, *250*, 1130. (b) Walz, G.; Aruffo, A.; Kolanus, W.; Bevilacqua, M.; Seed, B. *Science* **1990**, *250*, 1130. (c) Lowe, J. B.; Stoolman, L. M.; Nair, R. P.; Larsen, R. D.; Berhend, T. L.; Marks, R. M. *Cell* **1990**, *63*, 475.

40. Springer, T. A.; Lasky, L.A. *Nature* **1991**, *349*, 196.

41. Lasky, L. A. *Science* **1992**, *258*, 964.

42. Leloir, L. F. *Science* **1971**, *172*, 1299.

43. Kornfeld, R.; Kornfeld, S. *Ann. Rev. Biochem.* **1985**, *54*, 631.

44. Sadler, J. E.; Beyer, T. A.; Oppenheimer, C. L.; Paulson, J. C; Prieels, J.-P.; Rearick, J. I.; Hill, R. L. *Method. Enzymol.* **1982**, *83*, 458.

45. (a) Ginsburg, V. *Adv. Enzymol.* **1964**, *26*, 35. (b) *Method. Enzymol.* Vol. 8 (1966), Vol. 138 (1987) and Vol. 179 (1989).

46. Hubbard, S. C. *Ann. Rev. Biochem.* **1981**, *50*, 555.

47. (a) Imperiali, B.; Shannon, K. L.; Rickert, K. W. *J. Am. Chem. Soc.* **1992**, *114*, 7942. (b) Imperiali, B.; Shannon, K. L.; Unno, M.; Rickert, K. W. *J. Am. Chem. Soc.* **1992**, *114*, 7944.

48. (a) Kaplan, H. A.; Welply, J. K.; Lennarz, W. J. *Biochim. Biophys. Acta* **1987**, *906*, 161. (b) Abbadi, A.; Mcharfi, M.; Aubry, A.; Prémilat, S.; Boussard, G.; Marrand, M. *J. Am. Chem. Soc.* **1991**, *113*, 2729. (c) Imperiali, B.; Shannon, K. L. *Biochemistry* **1991**, *30*, 4374. (d) Sharma, C. B.; Lehle, L.; Tanner, W. *Eur. J. Biochem.* **1981**, *116*, 101.

49. Roth, J. *Biochem. Biophys. Acta.* **1987**, *906*, 405.

50. (a) *Structure and Function of Gangliosides*; Svennerholm, L.; Mande., P.; Dreyfus, P.; Urbum, P.-F., Eds.; Plenum: New York, 1980. (b) *Ganglioside Structure, Function, and Biomedical Potential*; Ledeen, R. W.; Yu, R. K.; Rapport, M. M.; Suzuki, K.; Eds., Plenum: New York, 1984. (c) *New Trends in Ganglioside Research*; Ledeen, R. W.;

Hogan, E. L.; Tettamanti, G.; Yates, A. J.; Yu, R. K., Eds.; Liviana Press: Padova, 1988.

51.   (a) Svennerholm, L. In *Cellular and Pathological Aspects of Glycoconjugate Metabolism*; Dreyfus, H.; Massarelli, R.; Freysz, L.; Rebel, G., Eds.; INSERM: France, 1984; Vol. 126, pp21-44. (b) Ledeen, R. W.; Yu, R. K. *Methods Enzymol.* **1982**, *83*, 139. (c) van Echten, G.; Sandhoff, K. *J. Neurochem.* **1989**, *52*, 207.

52.   Merill, A. H., Jr. *J. Bioenerg. Biomembr.* **1991**, *23*, 83.

53.   Shematek, E. M.; Cabib, E. *J. Biol. Chem.* **1980**, *255*, 895.

54.   Grisebach, H. *Adv. Carbohydr. Chem. Biochem.* **1978**, *35*, 80.

55.   (a) Hash, J. H. *Method. Enzymol.* **1975**, Vol. 43. (b) Russell, R. N.; Liu, H.-W. *J. Am. Chem. Soc.* **1991**, *113*, 7777. (c) Vara, J. A.; Hutchinson, C. R. *J. Biol. Chem.* **1988**, *263*, 14992. (d) Oths, P. J.; Hayer, R. M.; Floss, H. G. *Carbohydr. Res.* **1990**, *198*, 91.

56.   (a) Toone, E. J.; Simon, E. S.; Bednarski, M. D.; Whitesides, G. M. *Tetrahedron* **1989**, *45*, 5365. (b) Hindsgaul, O. *Seminars in Cell Biology* **1991**, *2*, 319. (c) Stangier, P.; Thiem, J. In *Enzymes in Carbohydrate Synthesis*; Bednarsky, M. D.; Simon, E. S., Eds.; ACS Symposium Series 466; American Chemical Society: Washington, D. C., 1991, pp 63-78.

57.   (a) Drueckhammer, D. G.; Hennen, W. J.; Pederson, R. L.; Barbas, C. F., III; Gautheron, C. M.; Krach, T.; Wong, C.-H. *Synthesis* **1991**, 499. (b) Ichikawa, Y.; Look, G. C.; Wong, C.-H. *Anal. Biochem.* **1992**, *202*, 215.

58.   Nilsson, K. G. I. *TIBTECH* **1988**, *6*, 256.

59.   (a) Heidlas, J. E.; Williams, K. W.; Whitesides, G. M. *Acc. Chem. Res.* **1992**, *25*, 307. (b). Kochetkov, N. K.; Shibaev, V. N. *Adv. Carbohydr. Chem. Biochem.* **1973**, *28*, 307.

60.   Khorana, H. G. *Some Recent Developments in the Chemistry of Phosphate Esters of Biological Interests*; Wiley: New York, 1961.

61.   Michelson, A. M. *The Chemistry of Nucleosides and Nucleotides*; Academic Press: New York; 1963.

62.   Clark, V. M.; Hutchinson, D. W.; Kirby, A. J.; Warren, S. G. *Angew. Chem. Int. Ed. Engl.* **1984**, *76*, 704.

63.   Slotin, L. A. *Synthesis* **1977**, 737.

64.   Scheit, K. H. *Nucleotides Analogs, Synthesis and Biological Function*; Wiley: New York, 1980.

65.   Cramer, F.; Neunhoeffer, H. *Chem. Ber.* **1962**, *95*, 1664.

66.   Hoard, D. E.; Ott, D. G. *J. Am. Chem. Soc.* **1965**, *87*, 1785.

67.   Simon, E. S.; Grabowski, S.; Whitesides, G. M. *J. Org. Chem.* **1990**, *55*, 1834.

68.   (a) Schmidt, R. R. *Angew. Chem. Int. Ed. Engl.* **1986**, *25*, 212.   (b) Pale, P.; Whitesides, G. M. *J. Org. Chem.* **1991**, *56*, 4547.

69.   Gokhale, U. B.; Hindsgaul, O.; Palcic, M. M. *Can. J. Chem.* **1990**, *68*, 1063.

70.   Sim, M. M.; Kondo, H.; Wong, C.-H. *J. Am. Chem. Soc.* **1993**, *115*, 2260.

71.   (a) Pfannemuller, B. *Staerke* **1968**, *11*, 341.   (b) Praznik, W.; Ebermann, R. *Staerke* **1979**, *31*, 288.

72.   Waldmann, H.; Gygax, D.; Bednarski, M. D.; Shangraw, W. R.; Whitesides, G. M. *Carbohydr. Res.* **1986**, *157*, C4.

73.   Wong, C.-H.; Haynie, S. L.; Whitesides, G. M. *J. Org. Chem.* **1982**, *47*, 5416.

74.   Leucks, H. J.; Lewis, J. M.; Rios-Mercadillo, V. M.; Whitesides, G. M. *J. Am. Chem. Soc.* **1979**, *101*, 5829.

75.   Hirschbein, B. L.; Mazenod, F. P.; Whitesides, G. M. *J. Org. Chem.* **1982**, *47*, 3765.

76.   Simon, E. S.; Grabowski, S.; Whitesides, G. M. *J. Am. Chem. Soc.* **1989**, *111*, 8920.

77.   Kim, M.-J.; Whitesides, G. M. *App. Biochem. Biotech.* **1987**, *16*, 95.

78.   Simon, E. S.; Bednarski, M. D.; Whitesides, G. M. *J. Am. Chem. Soc.* **1988**, *110*, 7159.

79.   Chenault, H. K.; Simon, E. S.; Whitesides, G. M.   In *Biotechnology and Genetic Engineering Reviews*; Russell, G. E., Ed.; Intercept, Wimborne, Dorset: 1988; Vol. 6, Chapter 6.

81.   Wong, C.-H.; Haynie, S. L.; Whitesides, G. M. *J. Am. Chem. Soc.* **1983**, *105*, 115.

82.   (a) Kawaguchi, K.; Kawai, H.; Tochikura, T. *Method. Carbohydr. Chem.* **1980**, *8*, 261; (b) Tochikura, T.; Kawaguchi, K.; Kawai, H.; Mugibayashi, Y.; Ogata, K. *J. Fermentl. Technol.* **1968**, *46*, 970; (c) Tochikura, T.; Kawai, H.; Tobe, S.; Kawaguchi, K.; Osugi, M.; Ogata, K. *J. Fermentl. Technol.* **1968**, *46*, 957.   (d) Korf, U.; Thimm, J.; Thiem, J. *Synlett* **1991**, 313.

83.   Heidlas, J. E.; Lees, W. J.; Whitesides, G. M. *J. Org. Chem.* **1992**, *57*, 152.

84.   Srivastava, G.; Hindsgaul, O.; Palcic, M. M. *Carbohydr. Res.* **1993**, *245*, 137.

85.   (a) Ropp, P. A.; Cheng, P.-W. *Anal. Biochem.* **1990**, *187*, 104.   (b) Heidlas, J. E.; Lees, W. J.; Pale, P.; Whitesides, G. M. *J. Org. Chem.* **1992**, *57*, 146.

86.   Maley, F. *Method. Enzymol.* **1972**, *28*, 271.

87.   Grier, T. J.; Rasmussen, J.R. *Anal. Biochem.* **1982**, *127*, 100.

88.   Ginsberg, V. *J. Biol. Chem.* **1960**, *235*, 2196.

89.   Yamamoto, K.; Maruyama, T.; Kumagai, H.; Tochikura, T.; Seno, T.; Yamaguchi, H. *Agric. Biol. Chem.* **1984**, *48*, 823.

90.   (a) Ichikawa, Y.; Lin, Y.-C.; Dumas, D. P.; Shen, G.-J.; Garcia-Junceda, E.; Williams, M. A.; Bayer, R.; Ketcham, C.; Walker, L. E.; Paulson, J. C.; Wong, C.-H. *J. Am. Chem. Soc.* **1992**, *114*, 9283.   (b) Stiller, R.; Thiem, J. *Liebigs Ann. Chem.* **1992**,

467.

91.   (a) Nunez, H. A.; O'Connor, J. V.; Rosevear, P. R.; Barker, R. *Can. J. Chem.* **1981**, *59*, 2086. (b) Gokhale, V. B.; Hindsgaul, O.; Palcic, M. M. *Can. J. Chem.* **1990**, *68*, 1063. (c) Schmidt, R. R.; Wegmann, B.; Jung, K.-H. *Liebigs Ann. Chem.* **1991**, *191*, 121. (d) Veeneman, G. H.; Broxterman, H. J. G.; van der Marel, G. H.; van Boom, J. H. *Tetrahedron Lett.* **1991**, *32*, 6175. (e) Ichikawa, Y.; Sim, M. M.; Wong, C.-H. *J. Org. Chem.* **1992**, *57*, 2943. (f) Adelhorst, K.; Whitesides, G. M. *Carbohydr. Res.* **1993**, *242*, 69.

92.   Toone, E. J.; Simon, E. S.; Whitesides, G. M. *J. Org. Chem.* **1991**, *56*, 5603.

93.   Gygax, D.; Spies, P.; Winkler, T.; Pfarr, U. *Tetrahedron* **1991**, *28*, 5119.

94.   (a) Higa, H. H.; Paulson, J. C. *J. Biol. Chem.* **1985**, *260*, 8838. (b) Thiem, J.; Treder, W. *Angew. Chem. Int. Ed. Engl.* **1986**, *25*, 1096. (c) van den Eijnden, D. H.; van Dijk, W. *Hoppe-Seyler's Z. Physiol. Chem.* **1972**, *353*, 1817. (d) Haverkamp, J.; Beau, J. M.; Schauer, R. *Hoppe-Seyler's Z. Physiol. Chem.* **1972**, *360*, 159; (e) Kean, E. L. *J. Biol. Chem.* **1970**, *9*, 2391. (f) Auge, C.; Gautheron, C. *Tetrahedron Lett* **1988**, *29*, 789.

95.   Thiem, J.; Stangier, P. *Liebigs Ann. Chem.* **1990**, 1101.

96.   Liu, J. L.-C.; Shen, G.-J.; Ichikawa, Y.; Rutan, J. F.; Zapata, G.; Vann, W. F.; Wong, C.-H. *J. Am. Chem. Soc.* **1992**, *114*, 3901.

97.   (a) Martin, T. J.; Schmidt, R. R. *Tetrahedron Lett.* **1993**, *34*, 1765. (b) Makino, S.; Ueno, Y.; Ishikawa, M.; Hayakawa, Y.; Hata, T. *Tetrahedron Lett.* **1993**, *34*, 2775. (c) Kondo, H.; Ichikawa, Y.; Wong, C.-H. *J. Am. Chem. Soc.* **1992**, *114*, 8748.

98.   Warren, L.; Blacklow, R. *J. Biol. Chem.* **1962**, *237*, 3527. See also: Vann, W. F.; Silver, R.P.; Abeiyon, C.; Chang, K.; Aaronson, W.; Sutton, A.; Finn, C. W.; Lindner, W.; Kotsatos, M. *J. Biol. Chem.* **1987**, *262*, 17562.

99.   Zapata, G.; Vann, W. F.; Aaronson, W.; Lewis, M. S.; Moos, M. *J. Biol. Chem.* **1989**, *264*, 14769.

100.   Ichikawa, Y.; Shen, G.-J.; Wong, C.-H. *J. Am. Chem. Soc.* **1991**, *113*, 4698.

101.   Shen, G.-J.; Liu, J. L.-C.; Wong, C.-H. *Biocatalysis* **1992**, *6*, 31.

102.   Shames, S. L.; Simon, E. S.; Christopher, C. W.; Schmid, W.; Whitesides, G. M.; Yang, L.-L. *Glycobiology* **1991**, *1*, 187.

103.   Schreiner, E.; Christian, R.; Zbiral, E. *Liebigs Ann. Chem.* **1990**, 93.

104.   Hartmann, M.; Christian, R.; Zbiral, E. *Liebigs Ann. Chem.* **1990**, 83.

105.   Gross, H. J.; Buensch, A.; Paulson, J. C.; Brossmer, R. *Eur. J. Biochem.* **1987**, *168*, 595.

106.   Gross, H. J.; Brossner, R. *Eur. J. Biochem.* **1988**, *177*, 583.

107.   Schauer, R.; Wember, M.; do Amaral, C. F. *Hoppe-Seyler's Z. Physiol. Chem.* **1972**,

353, 883.

108. Hagopian, A.; Eylar, E. H. *Arch. Biochem. Biophys.* **1968**, *128*, 422.

109. Schwarz, R.-T.; Datema, R. *Adv. Carbohydr. Chem. Biochem.* **1982**, *40*, 287.

110. (a) Schanbacher, F. L.; Ebner, K. E. *J. Biol. Chem.* **1970**, *245*, 5057. (b) Berliner, L. J.; Davis, M. E.; Ebner, K. E.; Beyer, T. A.; Bell, J. E. *Mol. Cell. Biochem.* **1984**, *62*, 37. (c) Nunez, H. A.; Barker, R. *Biochemistry* **1980**, *19*, 489.

111. (a) Trayer, I. P.; Hill, R. L. *J. Biol. Chem.* **1971**, *246*, 6666. (b) Andrews, P. *FEBS Lett.* **1970**, *9*, 297. (c) Barker, R.; Olsen, K. W.; Shaper, J. H.; Hill, R. L. *J. Biol. Chem.* **1972**, *247*, 7135. (d) Rao, A. K.; Garver, F.; Mendicino, J. *Biochemistry* **1976**, *15*, 5001.

112. (a) Palcic, M. M.; Srivastava, O. P.; Hindsgaul, O. *Carbohydr. Res.* **1987**, *159*, 315. (b) Auge, C.; David, S.; Mathieu, C.; Gautheron, C. *Tetrahedron Lett.* **1984**, *25*, 1467.

113. (a) Zehavi, U.; Herchman, M. *Carbohydr. Res.* **1984**, *133*, 339. (b) Zehavi, U.; Sadeh, S.; Herchman, M. *Carbohydr. Res.* **1983**, *124*, 23.

114. Auge, C.; Mathieu, C.; Merienne, C. *Carbohydr. Res.* **1986**, *151*, 147.

115. Thiem, J.; Wiemann, T. *Angew. Chem. Int. Ed. Engl.* **1990**, *29*, 80.

116. Palcic, M. M.; Hindsgaul, O. *Glycobiology* **1991**, *1*, 205.

117. Unverzagh, C.; Kunz, H.; Paulson, J. C. *J. Am. Chem. Soc.* **1990**, *112*, 9308.

118. Joziasse, D. H.; Shaper, N. L.; Salyer, L. S.; van den Eijnden, D. H.; van der Spoel, A. C.; Shaper, J. H. *Eur. J. Biochem.* **1990**, *191*, 75.

119. Wong, C.-H.; Ichikawa, Y.; Krach, T.; Gautheron-Le Narvor, C.; Dumas, D. P.; Look, G. C. *J. Am. Chem. Soc.* **1991**, *113*, 8137.

120. (a) Nishida, Y.; Wiemann, T.; Thiem, J. *Tetrahedron Lett.* **1992**, *33*, 8043. (b) Nishida, Y.; Wiemann, T.; Sinwell, V.; Thiem, J. *J. Am. Chem. Soc.* **1993**, *115*, 2536. (c) Nishida, Y.; Wiemann, T.; Thiem, J. *Tetrahedron Lett.* **1993**, *34*, 2905.

121. Auge, C.; Gautheron, C.; Pora, H. *Carbohydr. Res.* **1989**, *193*, 288.

122. Guilbert, B.; Khan, T. H.; Flitsch, S. L. *J. Chem. Soc., Chem. Commun.* **1992**, 1526.

123. Yuasa, H.; Hindsgaul, O.; Palcic, M. M. *J. Am. Chem. Soc.* **1992**, *114*, 5891.

124. (a) Bferliner, L. J.; Robinson, R. D. *Biochemistry* **1982**, *21*, 6340. (b) Kodama, H.; Kajihara, Y.; Endo, T.; Hashimoto, H. *Tetrahedron Lett.* **1993**, *34*, 6419. (c) Lowary, T. L.; Hindsgaul, O. *Carbohydr. Res.* **1993**, *249*, 163.

125. Sabesan, S.; Paulson, J. C. *J. Am. Chem. Soc.* **1986**, *108*, 2068.

126. Thiem, J.; Treder, W. *Angew. Chem. Int. Ed. Engl.* **1986**, *25*, 1096.

127. Auge, C.; Gautheron, C. *Tetrahedron Lett.* **1988**, *29*, 789.

128. Condradt, H. S.; Bunsch, A.; Browwmer, R. *FEBS Lett.* **1984**, *170*, 295.

129. Petrie, C. R.; Sharma, M.; Simmons, O. D.; Korytnyk, W. *Carbohydr. Res.* **1989**, *186*, 326.

130. (a) Ito, Y.; Gaudino, J. J.; Paulson, J. C. *Pure Appl. Chem.* **1993**, *65*, 753. (b) Liu, K. K.-C.; Danishefsky, S. J. *J. Am. Chem. Soc.* **1993**, *115*, 4933.

131. David, S.; Auge, C. *Pure Appl. Chem.* **1987**, *59*, 1501.

132. Auge, C.; Fernandez-Fernandez, R.; Gautheron, C. *Carbohydr. Res.* **1990**, *200*, 257.

133. Sabesan, S.; Duus, J.; Domaille, P.; Kelm, S.; Paulson, J. C. *J. Am. Chem. Soc.* **1991**, *113*, 5865.

134. Palcic, M. M.; Venot, A. P.; Ratcliffe, R. M.; Hindsgaul, O. *Carbohydr. Res.* **1989**, *190*, 1.

135. Nilsson, K. G. I. *Carbohydr. Res.* **1989**, *188*, 9.

136. de Heij, H. T.; Kloosterman, M.; Koppen, P. L.; van Boom, J. H.; van den Eijnden, D. H. *J. Carbohydr. Chem.* **1988**, *7*, 209.

137. Pozsgay, V.; Gaudino, J. J.; Paulson, J. C.; Jennings, H. J. *Bioorg. Med. Chem. Lett.* **1991**, *1*, 391.

138. (a) McCoy, R. D.; Vimr, E. R.; Troy, F. A. *J. Biol. Chem.* **1985**, *260*, 12695. (b) Vimr, E. R.; Bergstrom, R.; Steenbergen, S. M.; Boulnios, G.; Roberts, I. *J. Bacteriol.* **1992**, *174*, 5127; Troy, F. A.; McCloskey, M. *J. Biol. Chem.* **1979**, *254*, 7377. (c) Vann, W. F.; Liu, T.-Y.; Robbins, J. B. *J. Bacteriol.* **1978**, *133*, 1300. (d) Review: Finne, J. *Trends Biochem. Sci.* **1985**, 129.

139. Wong, C.-H.; Dumas, D. P.; Ichikawa, Y.; Koseki, K.; Danishefsky, S. J.; Weston, B. W.; Lowe, J. B. *J. Am. Chem. Soc.* **1992**, *114*, 7321.

140. Dumas, D. P.; Ichikawa, Y.; Wong, C.-H.; Lowe, J. B.; Nair, R. P. *Bioorg. Med. Chem. Lett.* **1991**, *1*, 425.

141. Rosevear, P. R.; Nunez, H. A.; Barker, R. *Biochemistry* **1982**, *21*, 1421.

142. Gokhale, U. B.; Hindsgaul, O.; Palcic, M. M. *Can. J. Chem.* **1990**, *68*, 1063.

143. Srivastava, G.; Kaur, K. J.; Hindsgaul, O.; Palcic, M. M. *J. Biol. Chem.* **1992**, *267*, 22356.

144. Schachter, H. *Biochem. Cell. Biol.* **1986**, *64*, 163.

145. Brackhausen, I.; Hull, E.; Hindsgaul, O.; Schachter, H.; Shah, R. N.; Michnick, S. W.; Carver, J. P. *J. Biol. Chem.* **1989**, *264*, 11211.

146. Brockhausen, I.; Carver, J.; Schacter, H. *Biochem. Cell. Biol.* **1988**, *66*, 1134.

147. Strivastava, G.; Alton, G.; Hindsgaul, O. *Carbohydr. Res.* **1990**, *207*, 259.

148. Imperiali, B.; Zimmerman, J. W. *Tetrahedron Lett* **1990**, *31*, 6485.

149. (a) Kaur, K. J.; Alton, G.; Hindsgaul, O. *Carbohydr. Res.* **1990**, *210*, 145. (b) Linker, T.; Crawley, S. C.; Hindsgaul, O. *Carbohydr. Res.* **1993**, *245*, 323. (c) Lindh, I.; Hindsgaul, O. *J. Am. Chem. Soc.* **1991**, *113*, 216.

150. Look, G. C.; Ichikawa, Y.; Shen, G.-J.; Cheng, G.-J.; Wong, C.-H. *J. Org. Chem.* **1993**, *58*, 4326.

151. (a) Weisesmann, S.; Denzel, K.; Schilling, G.; Gross, G. G. *Bioorg. Chem.* **1988**, *16*, 29. (b) Oehrlein, R.; Hindsgaul, O.; Palcic, M. M. *Carbohydr. Res.* **1993**, *244*, 149.

152. McDowell, W.; Grier, T. J.; Rasmussen, J. R.; Schwarz, R. T. *Biochem. J.* **1987**, *248*, 523.

153. Wang, P.; Shen, G.-J.; Wang, Y.-F.; Ichikawa, Y.; Wong, C.-H. *J. Org. Chem.* **1993**, *58*, 3985.

154. (a) Flitsch, S. L.; Taylor, J. P.; Turner, N. J. *J. Chem. Soc., Chem. Commun.* **1991**, 380, 382. (b) Flitsch, S. L.; Pinches, H. L.; Taylor, J. P.; Turner, N. J. *J. Chem. Soc., Perkin Trans. 1* **1992**, 2087.

155. (a) Card, P.J.; Hitz, W.D. *J. Am. Chem. Soc.* **1984**, *106*, 5348; Card, P. J.; Hitz, W. D.; Ripp, K. G. *J. Am. Chem. Soc.* **1986**, *108*, 158. (b) Elling, L.; Grothus, M.; Kula, M.-R. *Glycobiology* **1993**, *3*, 349.

156. (a) Clark, R. S.; Banerjee, S.; Coward, J. K. *J. Org. Chem.* **1990**, *55*, 6275. (b) Lee, J.; Coward, J. K. *J. Org. Chem.* **1992**, *57*, 4126.

157. Wong, C.-H.; Wang, R.; Ichikawa, Y. *J. Org. Chem.* **1992**, *57*, 4343.

158. Ichikawa, Y.; Liu, J. J.-C.; Shen, G.-J.; Wong, C.-H. *J. Am. Chem. Soc.* **1991**, *113*, 6300.

159. Paulson, J. C.; Weinstein, J.; Ujita, E. L.; Riggs, K. J.; Lai, P.-H. *Biochem. Soc. Trans.* **1987**, *15*, 618.

160. Brew, K.; Castellino, F. J.; Vanaman, J. C.; Hill, R. L. *J. Biol. Chem.* **1970**, *245*, 4570.

161. Bartholomew, B. A.; Jourdian, G. W.; Roseman, S. *J. Biol. Chem.* **1973**, *248*, 5751.

162. Nagai, M.; Dave, V.; Kaplan, B. E.; Yoshida, A. *J. Biol. Chem.* **1978**, *253*, 377.

163. Paulson, J. C.; Colley, K. J. *J. Biol. Chem.* **1989**, *264*, 17615.

164. Toghrol, F.; Kimura, T.; Owens, I. S. *Biochemistry* **1990**, *29*, 2349.

165. MacKenzie, P. I. *J. Biol. Chem.* **1986**, *261*, 6119.

166. Masibay, A. S.; Qasba, P. K. *Proc. Natl. Acad. Sci.* **1989**, *86*, 5733.

167. Narimatsu, H.; Sinha, S.; Brew, K.; Okayama, H.; Qasba, P.K. *Proc. Natl. Acad. Sci.* **1986**, *83*, 4720.

168. Shaper, N. L.; Wright, W. W.; Shaper, J. H. *Proc. Natl. Acad. Sci.* **1990**, *87*, 791.

169. Joziassae, D. H.; Shaper, J. H., van den Eijnden, D. H.; Van Tunen, A. J.; Shaper, N. L. *J. Biol. Chem.* **1989**, *264*, 14290.

170. Larsen, R. D.; Rajan, V. P.; Ruff, M. M.; Kukowska-Latallo, J.; Cummings, R. D.; Lowe, J. B. *Proc. Natl. Acad. Sci.* **1989**, *86*, 8227.

171. Larsen, R. D.; Ernst, L. K.; Nair, R. P.; Lowe, J. B. *Proc. Natl. Acad. Sci. USA* **1990**, *87*, 6674.

172. Weinstein, J.; Lee, E. U.; McEntee, K.; Lai, P.-H.; Paulson, J. C. *J. Biol. Chem.*

1987, *262*, 17735.

173. (a) Saiki, R. K.; Scharf, S.; Faloona, F.; Mullis, K. B.; Horn, G. T.; Erlich, H. A.; Arnheim, N. *Science* **1985**, *230*, 1350. (b) Arnheim, N.; Levenson, C. H. *Chem. Eng. News* **1990**, Oct. 1, p 36.

174. MacFerrin, K. D.; Terranova, M. P.; Schreiber, S. L.; Verdine, G. L. *Proc. Natl. Acad. Sci.* **1990**, *87*, 1937.

175. Aoki, D.; Appert, H. E.; Johnson, D.; Wong, S. S.; Fukuda, M. N. *EMBO J.* **1990**, *9*, 3171.

176. Ghrayab, J.; Kimura, H.; Takahara, M.; Hsiung, H.; Masui, Y.; Inouye, M. *EMBO J.* **1984**, *3*, 2437.

177. Yeast: Krezdorn, C. H.; Watsele, G.; Kleene, R. B.; Ivanov, S. X.; Berger, E. G. *Eur. J. Biochem.* **1993**, *212*, 113. *E. coli:* Nakazawa, K.; Furukawa, K.; Narimatsu, H.; Kobata, A. *J. Biochem.* **1993**, *113*, 747.

178. (a) Gillespie, W.; Kelms, S.; Paulson, J. C. *J. Biol. Chem.* **1992**, *267*, 21004. (b) Wen, D. X.; Livingston, B. D.; Medzihradszky, K. F.; Burlingame A. L.; Paulson, J. C. *J. Biol. Chem.* **1992**, *267*, 21011.

179. Shaper, N. L.; Hollis, G. F.; Douglas, J. G.; Kirsch, I. R.; Shaper, J. H. *J. Biol. Chem.* **1988**, *263*, 10420.

180. D'Agostaro, G.; Bendiak, B.; Tropak, M. *Eur. J. Biochem.* **1989**, *183*, 211.

181. Nakazawa, K.; Ando, T.; Kimura, T.; Narimatsu, H. *J. Biochem.* **1988**, *104*, 165.

182. Shaper, N. L.; Shaper, J. H.; Meuth, J. L.; Fox, J. L.; Chang, H.; Kirsch, I. R.; Hollis, G. F. *Proc. Natl. Acad. Sci.* **1986**, *83*, 1573.

183. Masri, K. A.; Appert, H. E.; Fukuda, M. N. *Biochem. Biophys. Res. Commun.* **1988**, *157*, 657.

184. (a) FucTase III: Kukowska-Latallo, J. F.; Larsen, R. D.; Nair, R. P.; Lowe, J. B. *Genes and Development* **1990**, *4*, 1288. (b) FucTase IV: Kumar, R.; Potvin, B.; Muller, W. A.; Stanley, P. *J. Biol. Chem.* **1991**, *266*, 21777. (c) FucTase V: Weston, B. W.; Nair, R. P.; Larsen, R. D.; Lowe, J. B. *J. Biol. Chem.* **1992**, *267*, 4152. (d) FucTase VI: Weston, B. W.; Smith, P. L.; Kelly, R. J.; Lowe, J. B. *J. Biol. Chem.* **1992**, *267*, 24575.

185. Homa, F. L.; Hollander, T.; Sehman, D. J.; Thomsen, D. R.; Elhammer, A. P. *J. Biol. Chem.* **1993**, *268*, 12609.

186. Shoreibah, M.; Perng, G.-S.; Adler, B.; Weinstein, J.; Basu, R.; Cupples, R.; Wen, D.; Browne, J. K.; Buckhaults, P.; Fregien, N.; Pierce, M. *J. Biol. Chem.* **1993**, *268*, 15381.

187. For a review, see: Lowe, J. B. *Seminars in Cell Biology* **1991**, *2*, 289.

188. Haynie, S. L.; Whitesides, G. M. *Appl. Biochem. Biotech.* **1990**, *23*, 205.

189. Dedonder, R. *Method. Enzymol.* **1966**, *8*, 500.

190. (a) Botstein, D.; Davis, R. W. In *Molecular Biology of the Yeast Saccharomyces; Metabolism and Gene Expression*; Strathern, J. N.; Jones, E. W.; Broach, J. R., Eds.; Cold Spring Harbor Laboratory: 1981; p 607. (b) Carlson, D. M. *Pure Appl. Chem.* **1987**, *59*, 1489.

191. Ziegast, G.; Pfannemuller, B. *Carbohydr. Res.* **1987**, *160*, 185.

192. (a) French, D. *Adv. Carbohydr. Chem. Biochem.* **1957**, *12*, 189. (b) Saenger, W. *Angew. Chem. Int. Ed. Engl.* **1980**, *19*, 344.

193. (a) Bender, H. *Carbohydr. Res.* **1980**, *78*, 133. (b) Wallenfels, K.; Foldi, B.; Niermann, H.; Bender, H.; Linder, D. *Carbohydr. Res.* **1978**, *61*, 359.

194. Treder, W.; Thiem, J.; Schlingmann, M. *Tetrahedron Lett.* **1986**, *27*, 5605.

195. (a) Ezure, Y. *Agric. Biol. Chem.* **1985**, *49*, 2159. (b) Ezure, Y.; Maruo, S.; Ojima, N.; Konno, K.; Yamashita, H.; Miyazaki, K.; Seto, T.; Yamada, N.; Sugiyama, M. *Agr. Biol. Chem.* **1989**, *53*, 61.

196. (a) Jamas, S.; Rha, C. K.; Sinskey, A. J. *Biotechnol. Bioeng.* **1986**, *28*, 769. (b) Brunt, J. V. *Bio/Technology* **1986**, *4*, 780.

197. (a) Raetz, C. R. H. *Annu. Rev. Biochem.* **1990**, *59*, 129. (b) Crowell, D. N.; Anderson, M. S.; Raetz, C. R. H. *J. Bacteriol.* **1986**, *168*, 152.

198. (a) Vypel, H.; Scholz, D.; Macher, I.; Schindlmaier, K.; Schütze, E. *J. Med. Chem.* **1991**, *34*, 2759. (b) Scholz, D.; Bednarik, K.; Ehn, G.; Neruda, W.; Janzek, E.; Loibner, H.; Briner, K.; Vasella, A. *J. Med. Chem.* **1992**, *35*, 2070.

199. Sinnott, M. L. *Chem. Rev.* **1990**, *90*, 1171.

200. (a) Withers, S. G.; Street, I. P. *J. Am. Chem. Soc.* **1988**, *110*, 8551. (b) Withers, S. G.; Warren, R. A.; Street, I. P.; Rupitz, K.; Kempton, J. B.; Aebersol, R. *J. Am. Chem. Soc.* **1990**, *112*, 5887.

201. (a) Bouquelot, E. *Ann. Chim.* **1913**, *29*, 145. (b) Bouquelot, E. *J. Pharm. Chem.* **1914**, *10*, 361.

202. Nilsson, K. G. I. In *Biocatalysis in Organic Media*; Laane, C.; Tramper, J.; Lilly, M.D., Eds.; Elsevier: 1987; p 369.

203. Johansson, E.; Hedbys, L.; Lorsson, P. O.; Mosbach, K.; Gunnarsson, A.; Suensson, S. *Biotechnol. Lett.* **1986**, *8*, 421.

204. Ajisaka, K.; Nishida, H.; Fujimoto, H. *Biotechnol. Lett.* **1987**, *4*, 243.

205. (a) Wallenfels, K. *Bull. Soc. Chim. Belg.* **1960**, *42*, 1715. (b) Ajisaka, K.; Nishida, H.; Fujimoto, H. *Biotechnol. Lett.* **1987**, *9*, 387.

206. Umezurike, G. M. *Biochem. J.* **1987**, *241*, 455.

207. Hedbys, L.; Larsson, P.; Mosbach, K.; Svensson, S. *Biochem. Biophys. Res. Commun.* **1984**, *123*, 8.

208. Beecher, J. E.; Andrews, A. T.; Vulfson, E.N. *Enzyme Microb. Technol.* **1990**, *12*, 955.

209. (a) Hehre, E. S.; Sawai, T.; Brewer, C. F.; Nakano, M.; Kanda, T. *Biochemistry* **1982**, *21*, 3090. (b) Kasumi, T.; Tsumuraya, Y.; Brewer, C. F.; Kersters-Hilderson, H.; Claeyssens, M.; Hehre, E. S. *Biochemistry* **1987**, *26*, 3010.

210. Kasumi, T.; Brewer, C. F.; Reese, E. T.; Hehre, E. S. *Carbohydr. Res.* **1986**, *146*, 39.

211. (a) Nilsson, K.G.I. *Carbohydr. Res.* **1987**, *167*, 95. (b) Nilsson, K.G.I. *Carbohydr. Res.* **1988**, *180*, 53. (c) Crout, D. H. G.; MacManus, D. A.; Ricca, J.-M.; Singh, S.; Crithley, P.; Gibson, W. T. *Pure Appl. Chem.* **1992**, *64*, 1079.

212. Look, G. C.; Wong, C.-H. *Tetrahedron Lett.* **1992**, *33*, 4253.

213. (a) Hedbys, L.; Johansson, E.; Mosbach, K.; Larsson, P. O.; Gunnarsson, A.; Svensson, S.; Lonn, H. *Glycoconjugate J.* **1989**, *6*, 161. (b) Lehmann, J.; Schroter, E. *Carbohydr. Res.* **1979**, *71*, 65.

214. Kobayashi, S.; Koshiwa, K.; Kawasaki, T.; Shoda, S.-I. *J. Am. Chem. Soc.* **1991**, *113*, 3079.

215. Mitsuo, N.; Takeichi, H.; Satoh, T. *Chem. Pharm. Bull.* **1984**, *32*, 1183.

216. Larsson, P. O.; Hebdys, L.; Svensson, S.; Mosbach, K. *Methods Enzymol.* **1987**, *136*, 230.

217. Sakai, K.; Katsumi, R.; Ohi, H.; Usui, T.; Ishido, Y. *J. Carbohydr. Chem.* **1992**, *11*, 553.

218. Ooi, Y.; Hashimoto, T.; Mitsuo, N.; Satoh, T. *Tetrahedron Lett.* **1984**, *25*, 2241.

219. Ooi, Y.; Hashimoto, T.; Mitsuo, N.; Satoh, T. *Chem. Pharm. Bull.* **1985**, *33*, 1808.

220. Suyama, K.; Adachi, S.; Toba, T.; Sohma, T.; Hwang, C. J.; Itoh, T. *Agric. Biol. Chem.* **1986**, *50*, 2069.

221. Gais, H. J.; Zeissler, A.; Maidonis, P. *Tetrahedron Lett.* **1988**, *29*, 5743.

222. Bjorkling, F.; Godtfredsen, S. E. *Tetrahedron* **1988**, *44*, 2957.

223. Itano, K.; Yamasaki, K.; Kihara, C.; Tanaka, O. *Carbohydr. Res.* **1980**, *87*, 27.

224. Abe, J. I.; Mizowaki, N.; Hizukuri, S.; Koizumi, K.; Utamura, T. *Carbohydr. Res.* **1986**, *154*, 81.

225. Ooi, Y.; Mitsuo, N.; Satoh, T. *Chem. Pharm. Bull.* **1985**, *33*, 5547.

226. Petit, J.-M.; Paquet, F.; Beau, J.-M. *Tetrahedron Lett.* **1991**, *32*, 6125.

227. Bay, S.; Cantacuzene, D. *Bioorg. Med. Chem. Lett.* **1992**, *2*, 423.

228. Holla, E. W.; Schudok, M.; Weber, A.; Zulauf, M. *J. Carbohydr. Chem.* **1992**, *11*, 659.

229. Sauerbrei, B.; Thiem, J. *Tetrahedron Lett.* **1992**, *33*, 201.

230. Cantacuzene, D.; Attal, S. *Carbohydr. Res.* **1991**, *211*, 327.

231. Pozo, M.; Gotor, V. *J. Chem. Soc., Perkin Trans. 1* **1993**, 1001.

232. Fujimoto, H.; Ajisaka, K. *Biotechnol. Lett.* **1988**, *10*, 107.

233. Tanaka, T.; Oi, S. *Agric. Biol. Chem.* **1985**, *49*, 1267.

234. (a) Crout, D. H. G.; Howarth, O. W.; Singh, S.; Swoboda, B. E. P.; Critchley, P.; Gibson, W. T. *J. Chem. Soc., Chem. Commun.* **1991**, 1550. (b) Crout, D. H. G.; Singh, S.; Swoboda, B. E. P.; Critchley, P.; Gibson, W. T. *J. Chem. Soc., Chem. Commun.* **1992**, 704.

235. (a) Thiem, J.; Sauerbrei, B. *Angew. Chem. Int. Ed. Engl.* **1991**, *30*, 1503. (b) Svensson, S. C. T.; Thiem, J. *Carbohydr. Res.* **1990**, *200*, 391.

236. Straathof, A. J. J. J.; Kieboom, A. P. G.; van Bekkum, H. *Carbohydr. Res.* **1986**, *146*, 81.

237. Andersen, B.; Thiesen, N.; Broe, P. E. *Acta Chem. Scand.* **1969**, *23*, 2367.

238. Straathof, A. J. J. J.; Vrijenhoef, J. P.; Sprangers, E. P. A. T.; van Bekkum, H.; Kieboom, A. P. G. *J. Carbohydr. Chem.* **1988**, *7*, 223.

239. (a) Herrmann, G. F.; Ichikawa, Y.; Wandrey, C.; Gaeta, F. C. A.; Paulson, J. C.; Wong, C.-H. *Tetrahedron Lett.* **1993**, *34*, 3091. (b) Herrmann, G. F.; Kragl, U.; Wandrey, C. *Angew. Chem. Int. Ed. Engl.* **1993**, *32*, 1342.

240. Ishikawa, H.; Kitahata, S.; Ohtani, K.; Ikuhara, C.; Tanaka, O. *Agric. Biol. Chem.* **1990**, *54*, 3137.

241. (a) Dedonder, R. *Methods Enzymol.* **1964**, *8*, 500. (b) Kunst, F.; Pascal, M.; Lepesant, J.-A.; Walle, J.; Dedonder, R. *Eur. J. Biochem.* **1974**, *42*, 611. (c) Rathbone, E. B.; Hacking, A. J.; Cheetham, P. S. J. U. S. Patent 4,617,269; 1986.

242. (a) Schenkman, S.; Man-Shiow, J.; Hart, G. W.; Nussenzweig, V. *Cell* **1991**, *65*, 1117. (b) Vanderkerckhove, F.; Schenkman, S.; de Carvalho, L. P.; Tomlinson, S.; Kiso, M.; Yoshida, M.; Hasegawa, A.; Nussenzweig, V. *Glycobiology* **1992**, *2*, 541; Ito, Y.; Paulson, J. C. *J. Am. Chem. Soc.* **1993**, *115*, 7862. (c) Tomlinson, S.; de Carvalho, L. P.; Vanderkerckhove, F.; Nussenzweig, V. *Glycobiology* **1991**, *2*, 549. (d) Ito, Y.; Paulson, J. C. *J. Am. Chem. Soc.* **1993**, *115*, 7862.

243. Hupe, D. J. *Ann. Rep. Med. Chem.* **1986**, *21*, Chapter 23.

244. Mansuri, M. M.; Martin, J. C. *Ann. Rep. Med. Chem.* **1987**, *22*, Chapter 15.

245. Mansuri, M. M.; Martin, J. C. *Ann. Rep. Med. Chem.* **1988**, *23*, Chapter 17.

246. (a) *Nucleic Acid Chemistry, Part 3*, Townsend, L. B.; Tipson, R. S., Eds.; Wiley: New York, 1986. (b) *Nucleoside Analogs; Chemistry, Biology and Medicinal Applications*; Walker, R. T.; Declerez, E.; Eckstein, F., Eds.; Plenum: New York; 1979.

247. Hutchinson, D.W. *TIBTECH* **1990**, *8*, 348.

248. Krenitsky, T. A.; Koszalka, G. W.; Tuttle, J. V.; Rideout, J. L.; Elion, G. B. *Carbohydr. Res.* **1981**, *97*, 139.

249. (a) Utagawa, T.; Morisawa, H.; Yoshinaga, F.; Yamazaki, A.; Mitsugi, K.; Hirose, Y. *Agric. Biol. Chem.* **1985**, *49*, 1053. (b) Utagawa, T.; Morisawa, H.; Yamanaka, S.; Yamazaki, A.; Yoshinaga, F.; Hirose, Y. *Agric. Biol. Chem.* **1985**, *49*, 2167.

250. (a) Tener, G. M.; Khorana, H. G. *J. Chem. Soc.* **1957**, *79*, 437. (b) Inoue, Y.; Ling, F.; Kimura, A. *Agric. Biol. Chem.* **1991**, *55*, 629.

251. Utagawa, T.; Morisasa, H.; Yamanaka, S.; Yamazaki, A.; Yoshinaga, F.; Hirose, Y. *Agric. Biol. Chem.* **1986**, *50*, 121.

252. Krenitsky, T. A.; Rideout, J. L.; Chao, E. Y.; Koszalka, G. W.; Gurney, F.; Crouch, R. C.; Cohn, N. K.; Wolberg, G.; Vinegar, R. *J. Med. Chem.* **1986**, *29*, 138.

253. Hennen, W. J.; Wong, C.-H. *J. Org. Chem.* **1989**, *54*, 4692.

254. Utagawa, T.; Morisawa, H.; Yamanaka, S.; Yamazaki, A.; Yoshinaga, F.; Hirose, Y. *Agric. Biol. Chem.* **1985**, *49*, 2711.

255. Morisawa, H.; Utagawa, T.; Yamanaka, S.; Yamazaki, A. *Chem. Pharm. Bull.* **1981**, *29*, 3191.

256. Krenitsky, T. A.; Freeman, G. A.; Shaver, S. R.; Beacham, L. M.; Hurlbert, S.; Cohn, N. K.; Elwell, L. P.; Selway, J. W. T. *J. Med. Chem.* **1983**, *26*, 891.

257. Stoecker, J. D.; Ealick, S. E.; Bugg, C. E.; Parks, R. E., Jr. *Proc. Fed. Am. Soc. Exp. Biol.* **1986**, *45*, 2773.

258. (a) Holguin, J.; Cardinaud, R. *Eur. J. Biochem.* **1975**, *54*, 505. (b) Holguin, J.; Cardinaud, R. *Eur. J. Biochem.* **1975**, *54*, 575. (c) Carson, D. A.; Wasson, D. B.; Beutler, E. *Proc. Natl. Acad. Sci. USA* **1984**, *81*, 2232. (d) Carson, D.A.; Wasson, D. B. *Biochem. Biophys. Res. Commun.* **1988**, *155*, 829. (e) Betbeder, D.; Hutchinson, D. W.; Richards, A. O. L. *Nucleic Acids Res.* **1989**, *17*, 4217.

259. Krenitsky, T. A.; Rideout, J. L.; Koszalka, G. W.; Inmon, R. B.; Chao, E. Y.; Elion, G. B. *J. Med. Chem.* **1982**, *25*, 32.

260. Krenitsky, T. A.; Koszalka, G. W.; Tuttle, J. B. *Biochemistry* **1981**, *20*, 3615.

261. Rideout, J. L.; Krenitsky, T. A.; Koszalka, G. W.; Cohn, N. K.; Chao, E. Y.; Elion, G. B.; Latter, V. S.; Williams, R. B. *J. Med. Chem.* **1982**, *25*, 1040.

262. (a) Utagawa, T.; Morisawa, H.; Yoshinaga, F.; Yamazaki, A.; Mitsugi, K.; Hirose, Y. *Agric. Biol. Chem.* **1985**, *49*, 1053. (b) Morisawa, H.; Utagawa, T.; Miyoshi, T.; Yoshinaga, F.; Yamazaki, A.; Mitsugi, K. *Tetrahedron Lett.* **1980**, *21*, 479. (c) Utagawa, T.; Morisawa, H.; Miyoshi, T.; Yoshinaga, F.; Yamazaki, A.; Mitsugi, K. *FEBS Lett.* **1980**, *109*, 261.

263. Utagawa, T.; Morisawa, H.; Nakamatsu, A. *FEBS Lett.* **1980**, *119*, 101.

264. Schuber, F. *Bioorg. Chem.* **1979**, *8*, 83.

265. Tono-oka, F. *Bull. Chem. Soc. Jpn.* **1982**, *55*, 1531.

266. Winchester, B.; Fleet, G. W. J. *Glycobiology* **1992**, *2*, 199.

267. (a) Elbein, A. D. *Ann. Rev. Biochem.* **1987**, *56*, 497. (b) Schwarz, R. T.; Datema, R. *Trends Biotechnol.* **1984**, 932.

268. Muller, L. In *Biotechnology*, Rehm, H.-J.; Reed, G., Eds.; VCH: Verlagsgesellschaft Weinheim, 1985, Vol. 4, Chapter 18.

269. Wang, Y.-F.; Dumas, D. P.; Wong, C.-H. *Tetrahedron Lett.* **1993**, *34*, 403.

270. (a) Ziegler, T.; Straub, A.; Effenberger, F. *Angew. Chem. Int. Ed. Engl.* **1988**, *29*, 716. (b) Pederson, R. L.; Kim, M. J.; Wong, C.-H. *Tetrahedron Lett.* **1988**, *29*, 4645. (c) von der Osten, C. H.; Sinskey, A. J.; Barbas, C. F.; Pederson, R. L.; Wang, Y.-F.; Wong, C.-H. *J. Am. Chem. Soc.* **1989**, *111*, 3924. (d) Kajimoto, T.; Liu, K. K.-C.; Pederson, R. L.; Zhong, Z.; Ichikawa, Y.; Porco, J. A., Jr.; Wong, C.-H. *J. Am. Chem. Soc.* **1992**, *113*, 6187. (e) Look, G. C.; Fotsch, C. H.; Wong, C.-H. *Acc. Chem. Res.* **1993**, *26*, 182.

271. (a) Conradt, H. S.; Egge, H.; Peter-Katalinic, J.; Reiser, W.; Siklosi, T.; Schaper, K. *J. Biol. Chem.* **1987**, *262*, 14600. (b) Little, S. P.; Bang, N. U.; Harms, C. S.; Marks, C. A.; Mattler, L. E. *Biochemistry* **1984**, *23*, 6191.

272. Livingston, B. D.; Robertis, E. M. D.; Paulson, J. C. *Glycobiology* **1990**, *1*, 39.

273. Wong, C.-H.; Schuster, M.; Wang, P.; Sears, P. *J. Am. Chem. Soc.* **1993**, *115*, 5893.

274. Walsh, C. T. *J. Biol. Chem.* **1989**, *264*, 2393.

Chapter 6. Addition, Elimination and other Group Transfer Reactions (Phosphoryl-, Methyl-, Sulfo- and Amino-Transfer Reactions)

1. Addition of Water to Alkenes: Fumarase

Addition of water to an olefin can create new chiral centers. An example is the stereospecific hydration of fumarate to L-malate,[1] catalyzed by fumarase (EC 4.2.1.2). This reaction has been used in a microbial whole cell process for the commercial synthesis of L-malic acid. The enzyme from pig heart has been examined for synthetic utility, and found to have narrow substrate specificity. No other nucleophiles can replace water for the addition reaction, and only a few fumarate analogs are acceptable.[2] Chlorofumarate and difluorofumarate have been used as substrates and converted to L-*threo*-chloromalic acid and 2,3-difluoromalate respectively.[2] The latter substance spontaneously eliminates HF and forms 3-fluorooxalate, which can be further converted to L-*threo*-fluoromalate in a process catalyzed by malate dehydrogenase. L-*threo*-Chloromalic acid was used in the synthesis of 2-deoxyribose and *trans*-D-*erythro*-sphingosine[2] (Figure 1).

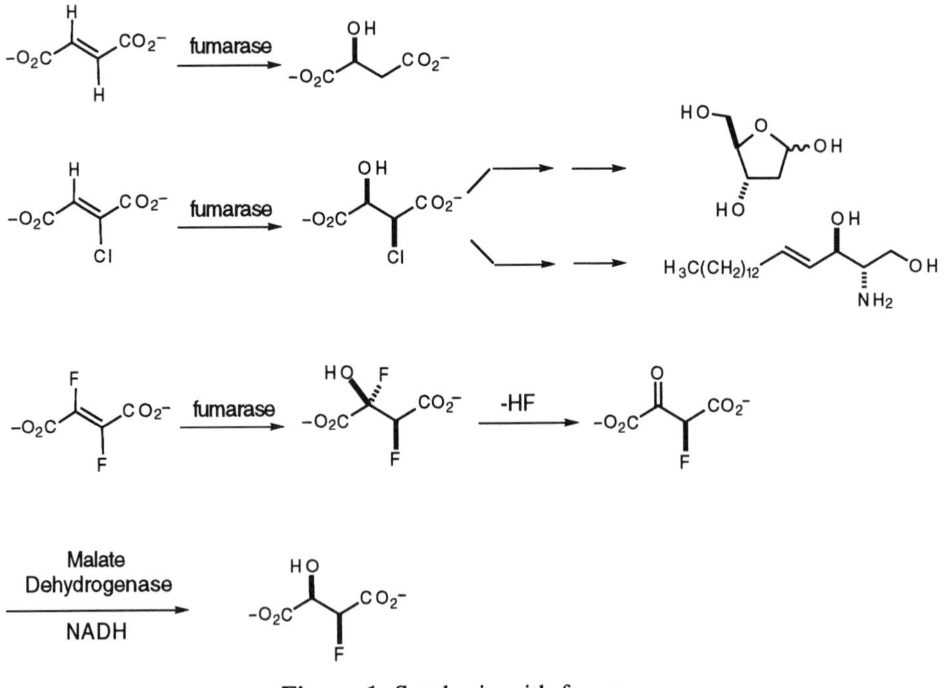

Figure 1. Synthesis with fumarase.

312

2. Addition of Ammonia to Double Bonds: Ammonia Lyases

The enzymes that catalyze the reversible addition of ammonia to alkenes are named ammonia lyases.[3] They include aspartate ammonia lyase (aspartase, EC 4.3.1.1), 3-methylaspartate ammonia lyase (EC 4.3.1.2), and other amino acid ammonia lyases such as L-histidine, L-tyrosine, and L-phenylalanine ammonia lyases. L-Aspartase from *E. coli* catalyzes the addition of ammonia to the C-2 *si*-face of fumarate to form L-aspartate. The enzyme is specific for its amino acid substrate.[4] Addition of ammonia to mesaconic acid is catalyzed by 3-methylaspartate ammonia lyase[5] (EC 4.3.1.2). Replacement of the methyl group with H, Cl, or Br is also acceptable, and gives the corresponding 3-substituted aspartic acid (Figure 2).

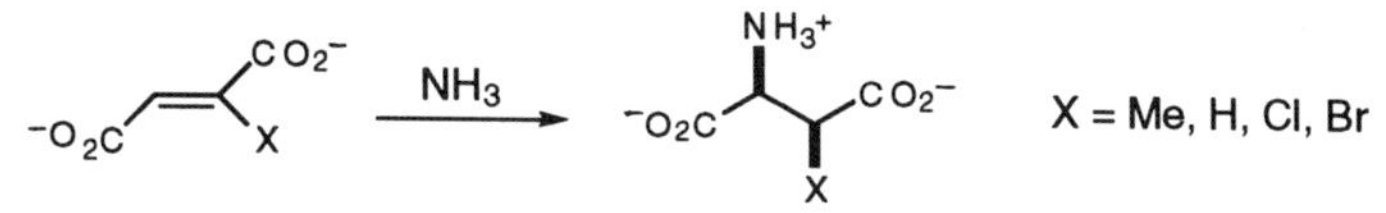

Figure 2. Ammonium lyase reaction.

3. Transamination: Aminotransferases

Aminotransferases are pyridoxal 5'-phosphate dependent enzymes that catalyze the reversible transfer of the amino group of an amino acid donor to a ketoacid acceptor. The reaction comprises two half-reactions: the first involves transfer of the amino group of an L-amino acid donor to pyridoxal 5'-phosphate to give a 2-ketoacid product, which is released from the enzyme and an enzyme-bound pyridoxamine 5'-phosphate; the second is the binding of another 2-keto acid and the transfer of the amino group from pyridoxamine 5'-phosphate to the 2-ketoacid to produce an L-amino acid and regenerate the cofactor[6] (Figure 3). Of the many known

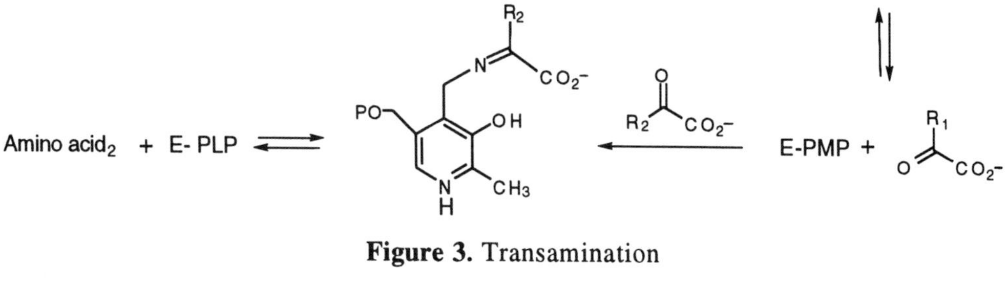

Figure 3. Transamination

transaminases, aspartate aminotransferase (EC 2.6.1.1) from *E. coli* is the most useful, because the oxaloacetate generated from aspartate undergoes spontaneous decarboxylation to pyruvate, and this makes the reaction irreversible. The enzyme has been cloned and overexpressed in *E. coli*[7,8] and used in the synthesis of a number of L-α-amino acids.[9,10]

Several new mutant enzymes derived from aspartate aminotransferase with interesting substrate specificity have been prepared *via* site-directed mutagenesis.[11] A highly active L-aspartate-phenylpyruvate transaminase from *Pseudomonas putida*,[12] and a thermostable aspartate aminotransferase from a *Bacillus sp.*[13] have been isolated. A D-amino acid aminotransferase from *Bacillus sp.* has also recently been isolated that catalyzes transamination between various D-amino acids and α-keto acids.[14] The aspartate aminotransferase from pig heart and bacteria has been studied with regard to its enantioselectivity for the amination of 4-hydroxy-4-methyl, and 4-ethyl-2-ketoglutaric acids. The enzyme prefers the 4R enantiomer to the 4S isomer in a ~8:2 ratio. Synthesis of the 4-substituted glutamic acids has been accomplished[15] (Figure 4). Other

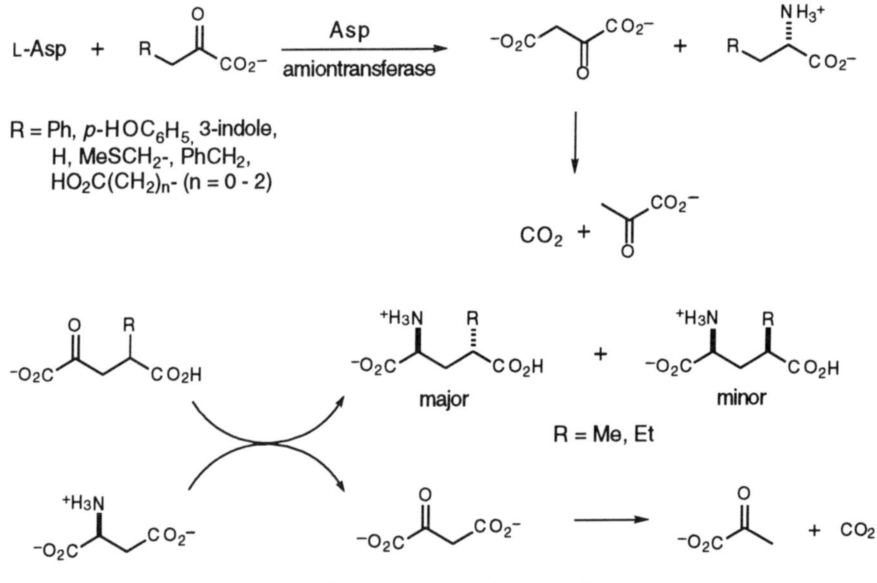

Figure 4. Aspartate aminotransferase

transaminases used in synthesis include L-lysine: 2-oxoglutarate 6-aminotransferase[16] (EC 2.6.1.36) and 4-aminobutyrate: 2-ketoglutarate transaminase[17] (EC 2.6.1.19); the latter was used in the synthesis of L-phosphinothricin (L-homoalanine-4-yl(methyl)phosphinic acid) (Figure 5). In an enzymatic approach to the synthesis of L-phenylalanine, acetamidocinnamate amidohydrolase (EC 3.5.1.-)[18] was used to catalyze the hydrolysis of acetamidocinnamic acid to phenylpyruvic acid, which was in turn converted to L-phenylalanine *via* enzymatic

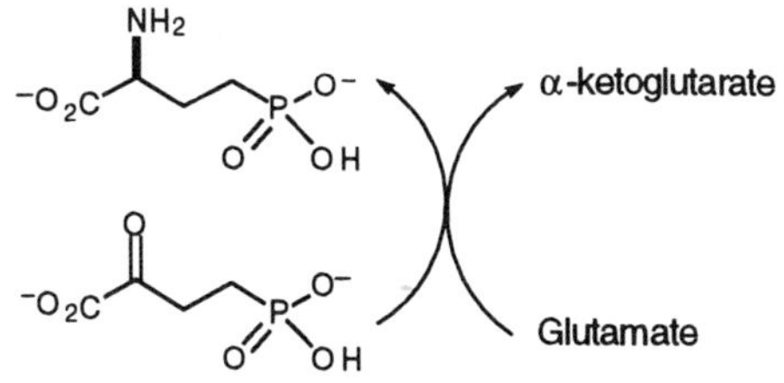

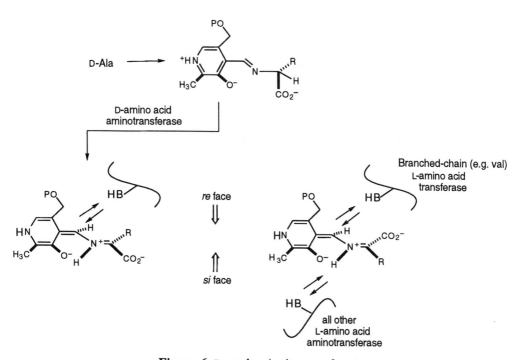

Figure 5.

transamination. The D-Amino acid aminotransferase from *Bacillus sp.*[19] and the branched-chain L-amino acid aminotransferase from *E. coli*[20] show a significant sequence homology. Both enzymes catalyze the stereospecific catalysis of pro-*R* C-4' hydrogen transfer through the coenzyme-substrate Schiff base intermediate, while all other aminotransferases catalyze the pro-*S* hydrogen transfer[21] (Figure 6).

Figure 6. D- and L-Aminotransferase

4. Addition and Elimination of the Carboxyl Group

These types of reactions are also synthetically interesting. A decarboxylase from *Alcaligenes bronchisepticus* was used in the asymmetric decarboxylation of a number of aryl methyl malonic acids to (*R*)-aryl propionic acids in very high ee[22,23] (Figure 7). Pyruvate decarboxylase is thiamine diphosphate dependent[24] and is the main enzyme involved in acyloin condensation (Figure 8). Other related enzymes include gluconate 6-phosphate dehydrogenase,[25] isocitrate dehydrogenase,[26] and ribulose-1,5-diphosphate carboxylase.[27]

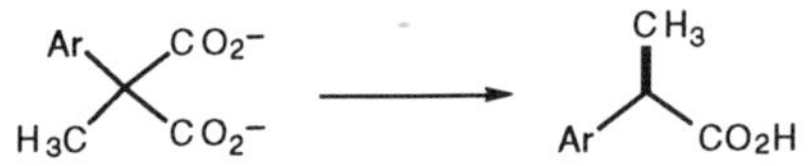

Figure 7. Asymmetric decarboxylase

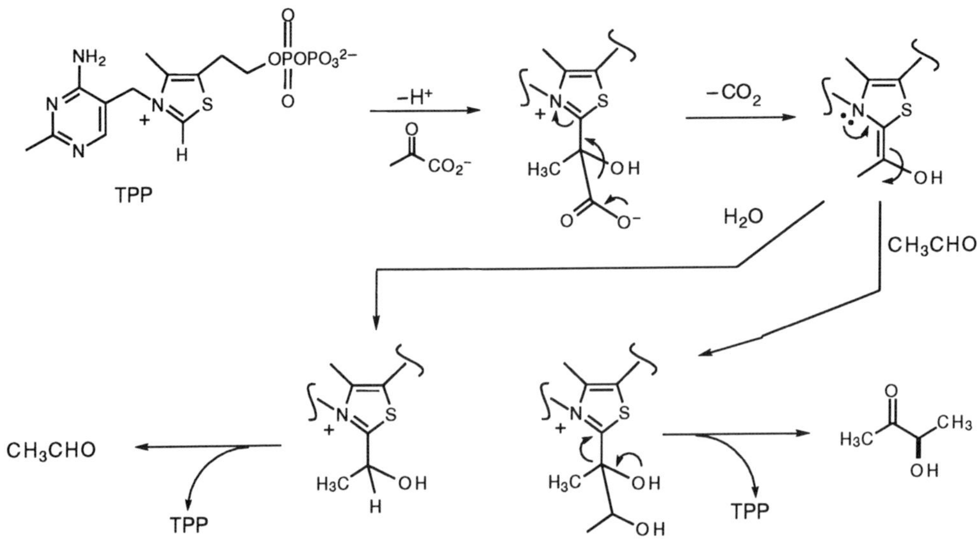

Figure 8. Thiamine diphosphate dependent decarboxylase reaction.

5. Nucleoside Triphosphate Requiring Enzymatic Reactions

Many synthetically important reactions are catalyzed by enzymes requiring nucleoside triphosphate cofactors, including adenosine-5'-triphosphate. These "high energy" compounds are sources of nucleotides, nucleoside phosphates and phosphates in bioorganic reactions; phosphate esters are commonly found in biosynthetic sequences as groups activating oxygen as a leaving group. Recent developments have made possible the use of such enzymes in organic syntheses and many reactions have been described in Chapter 5. More details are provided in this section.

While ATP is most often involved as phosphorylating agent, other nucleoside triphosphates (NTP), including GTP, CTP and UTP transfer the nucleoside phosphate moiety.

In vivo, complex biosynthetic processes regulate the concentration of cofactors available for reaction. As it is necessary to maintain one molecule of cofactor per turnover, the concentrations of cofactors are controlled with high efficiency. In enzyme-catalyzed organic synthesis, the high cost of cofactors has inhibited their use as stoichiometric reagents. In order to achieve large scale production with these enzymes it has been necessary to develop methodologies for generating cofactors economically, and for regenerating them *in situ*. Most research efforts on NTP regeneration have been focused on ATP, as its role in enzymatic reactions is widespread.[28]

Three basic strategies are conceivable for NTP regeneration: chemical, whole-cell and enzymatic. Chemical methods lack specificity and are often incompatible with enzymatic process. Biological systems using whole cells or organelles have been successful.[29] These systems usually require separation of the desired products from the byproducts of the system, and are inconvenient for complex syntheses. Enzyme-catalyzed systems have been the most successful; they offer both higher specificities and higher volumetric productivity than the whole cell systems, and are generally more convenient. It is, however, obvious that effort and expense is requried to obtain the purified enzyme. As the interest is increased for such enzymes, developments in isolation and preparation techniques should facilitate their use as synthetic catalysts.

5.1 Enzymatic Systems for ATP Regeneration

The enzyme systems for ATP regeneration (Figure 9) are well developed and efficient. Regeneration of ATP is based on the selective transfer of phosphate from a high energy donor to ADP by use of a kinase enzyme. Several factors must be considered in choosing a suitable system. The phosphate donor should be readily available or accessible and should be relatively stable under the conditions of enzymatic reactions; the transfer of phosphate should be regioselective and have a favorable equilibrium constant. The enzymes that are involved should have sufficient activity and stability to achieve practical turnover numbers throughout the reaction, and should be inexpensive.

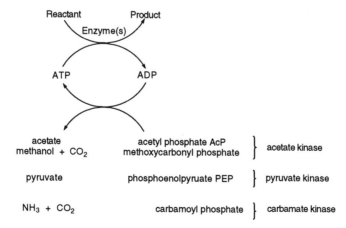

Figure 9. Regeneration of triphosphates via enzymatic synthesis.

5.1.1 Phosphoenol Pyruvate/Pyruvate Kinase (PEP/PK)[30]

This system, using PEP as the phosphate donor in a pyruvate kinase (PK; EC 2.7.1.40) catalyzed reaction, is the best enzymatic system for ATP regeneration. Pyruvate kinase from rabbit muscle is commercially available and inexpensive. The enzyme has a high specific activity (~500 U/mg protein) and is stable upon immobilization.

Phosphoenol pyruvate has a high phosphate donor capability with a $\Delta G^{o'}$ hydrolysis = -12.8 Kcal/mol (see Table 1). The stability of PEP from in solution is excellent. Although the synthesis of PEP[30] is slightly more difficult and expensive than those of other phosphate donors, the higher stability and donor capability offset this minor disadvantage. A more serious disadvantage of the PEP/PK system is the problem of product inhibition. Pyruvate is a competitive inhibitor of the PK reaction. In order to minimize this problem, the reaction must be run in dilute solutions, high concentrations of PEP must be used, or pyruvate must be removed from the system. Pyruvate is also reactive toward NAD(P) in some circumstances, and may interfere with some nicotinamide cofactor-requiring reactions.

Table 1

Property	Phosphoryl Donor		
	PEP	AcP	MCP
Synthesis	+	+++	++
$\Delta G^{o'}$ hyd (Kcal/mol)[a]	-12.8	-10.1	-12.4
Half-life for hydrolysis (h)[c]	ca. 10^3	21	0.3
Product Inhibition[d]	$CH_3C(O)CO_2^-$	$CH_3CO_2^-$	
(K$_i$, mM/Type)	10,C	400 NC	

a) Jenks, W. P. In *Handbook of Biochemistry*, 2nd ed.; CRC: Cleveland, 1970; pp 1-89.
b) @ pH = 9.5
c) pH = 7.5, 25 °C
d) C = competitive inhibitor; NC = non-competitive inhibitor

5.1.2 Acetyl phosphate/acetate kinase (AcP/AcK)

Acetyl phosphate is also very useful as the ultimate source of phosphate in ATP regeneration.[31] With this system, acetate kinase (EC 2.7.2.1) is used as the catalyst for conversion of ADP to ATP. The enzyme from *E. coli* is commercially available, although slightly more expensive than pyruvate kinase. (In both methods of regenerating ATP, however, the enzymes may be recovered and reused. Therefore, the costs of the catalysts contribute only a small portion of the overall cost of the systems.) Acetate kinase is oxygen sensitive due to free cysteine thiol groups. Treatment with methyl methanethiolsulfonate and immobilization on polymers both decrease autooxidation sensitivity and increase stability.[32] Another commercially available acetate kinase is a thermostable enzyme from *Bacillus Stearothermophilus*; this enzyme may be a better choice.[33] Unlike the *E. coli* enzyme it is thiol free and considerably more stable. It is also more expensive than the *E. coli* enzyme.

As far as synthesis of phosphoryl donors is concerned, acetyl phosphate is ideal. The material is easily prepared in multimolar quantities by reaction of acetic anhydride with phosphoric acid. This ease of synthesis makes acetyl phosphate/acetate kinase the most economical system for large scale production in most cases. A major drawback to the use of the AcP/AcK system is the instability of acetyl phosphate. Acetyl phosphate hydrolyzes rapidly in solution ($t_{1/2}$ for hydrolysis is 21 hours at pH 7, 25°C). The phosphoryl donor strength of acetyl phosphate is good but not as strong as that of PEP ($\Delta G^{o'}$ for AcP = -10.1 Kcal/mol). Product inhibition is not a serious problem for AcP/AcK.

The ease of synthesis of acetyl phosphate makes it the economical choice for large-scale work. However, if high phosphoryl donor strength or a long reaction time is required, then PEP is a more suitable choice.

Methoxy-carbonyl phosphate (MCP) is also a substrate for acetate kinase.[34] It is easily prepared in large quantities and has a phosphoryl donor strength comparable to PEP (Table 1). Unfortunately, MCP hydrolyzes spontaneously more rapidly than AcP. An advantage of MCP is that the product of the phosphate donation, methyl carbonate, decarboxylates spontaneously and rapidly to methanol and carbon dioxide. This decarboxylation makes the reaction irreversible, simplifies reaction work-up, and eliminates product inhibition.

The regenerative schemes for ATP based on AcK and PK are generally the most applicable. Other systems have been demonstrated, but are less convenient than these two. A system using carbamyl phosphate as a donor catalyzed by carbamyl kinase, for example, has been used.[35] Though synthesis of CP is not a problem, its rapid hydrolysis in solution yields an ammonium ion that may cause difficulties with enzymatic activity and complicate work-up. Another potentially useful method utilizes creatine phosphate as a donor with creatine kinase as a catalyst; this system has been demonstrated in the enzymatic synthesis of fructose 1,6-diphosphate.[36]

5.1.3 Regeneration of ATP from AMP

For enzymatic reactions requiring ATP in which AMP (rather than ADP) is the ultimate product, there is also a need for regeneration. AMP is not a substrate for acetate or pyruvate kinase, and a different method is required. Adenylate kinase (EC 2.7.4.3) catalyzes the equilibration of AMP, ADP and ATP. This enzyme can, therefore, be used in conjunction with acetate or pyruvate kinase to generate ATP from AMP (see Figure 10). Adenylate kinase has also been found to accept other nucleoside monophosphates including CMP.[37a] Its use in regeneration of other nucleoside triphosphates is discussed in the next section. Note that the system for

regeneration of CTP from CMP requires the use of a catalytic amount of ATP and AMP in order
to initiate the cycle.

A useful demonstration of the enzymatic synthesis of ATP starts from adenosine and
acetyl phosphate as the ultimate source of phosphate.[37b] The system uses a catalytic amount of
ATP, AMP, and ADP to generate ATP on >100 mmol scale. The conversion is catalyzed by 3
enzymes: adenosine kinase (EC 2.7.1.20 from brewers yeast[38]) catalyzes the phosphorylation of
adenosine to the monophosphate. As just mentioned, the combination of adenylate kinase and
acetate, or pyruvate kinase generate ATP from the mono-phosphate.[37c] Adenosine kinase is,
however, not commercially available; this fact inhibits its use in practical organic synthesis.

5.1.4 Regeneration of Other Nucleoside Triphosphates

As mentioned earlier, nucleoside triphosphates other than ATP are usually involved in the
transfer of a nucleoside phosphate moiety, and create inorganic phosphate or pyrophosphate as

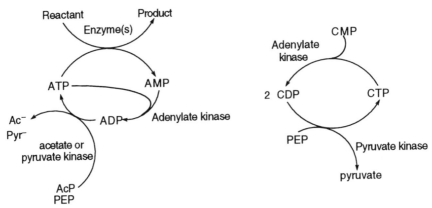

Figure 10. Regeneration of ATP from AMP.   **Figure 11.** Regeneration of CTP.

a product. Useful regenerative syntheses of nucleoside triphosphates such as CTP and UTP have
been accomplished using the relatively readily available monophosphates as the starting
reagent.[37] Regeneration of CTP (Figure 11) has been demonstrated[39] and used in a
multienzymatic synthesis of *N*-acetyl neuraminic acid derivatives (see Chapter 5). Similarly,
acetate and pyruvate kinase have been used in the *in situ* regeneration of UTP from UDP for use
in enzyme catalyzed synthesis of carbohydrate derivatives.[40]Regeneration of nucleoside
triphosphates from the corresponding diphosphates is not a problem because both acetate and
pyruvate kinase have broad substrate specificities. Preparation of nucleoside triphosphates from
unphosphorylated nucleosides is possible using nucleoside kinases. Unfortunately, these
enzymes are not yet commercially available. Large-scale synthesis of acetyl phosphate[41-43] has
been developed, and the best is that from acetic anhydride.[42]

5.2 Synthesis of Phosphorylated Carbohydrates

Enzymatic syntheses based on nucleoside triphosphate dependent kinases have provided a number of useful phosphorylated saccharides. The utility of such enzymatic systems is increased by the fact that a number of these enzymes will accept unnatural substrates. This breadth in specificity allows for the synthesis of a number of interesting compounds. For example, glycerol kinase[44] (EC 2.7.1.30, ATP: glycerol 3-phosphotransferase) catalyzes the synthesis of *sn*-glycerol 3-phosphate from ATP and glycerol. It was also used in the preparation of dihydroxy acetone phosphate, a useful phosphate in aldolase-catalyzed reactions.[45] The preparation of *sn*-glycerol-3-phosphate and several chiral phosphate derivatives of glycerol—potential precursors to interesting phospholipid derivatives—has been demonstrated.[44] Figure 12 details the analogs of glycerol that function as substrates for the kinase reaction. The phosphorylation produced chiral phosphate derivatives from racemic alcohols (90% ee) in good to excellent chemical yields. As with most examples discussed in this section, the ATP dependent reaction was accomplished using *in situ* regeneration of the ATP.

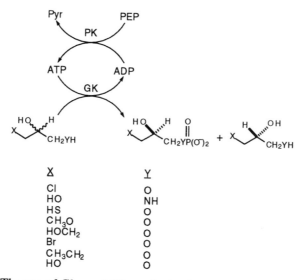

X	Y
Cl	O
HO	NH
HS	O
CH₃O	O
HOCH₂	O
Br	O
CH₃CH₂	O
HO	O

Figure 12. The use of Glycerol Kinase in the Synthesis of Organic Phosphates.

Although hexokinase is involved biosynthetically in the production of glucose-6-phosphate, its breadth of specificity has allowed its use in the synthesis of other phosphate sugars. Its use as a catalyst in the production of arabinose-5-phosphate from arabinose and ATP, provided the key step in the synthesis of 3-deoxy-D-manno-2-octulosonate-8-phosphate (KDO-8-P)[46] (Figure 13). Hexokinase has also been used in the synthesis of several uncommon sugar

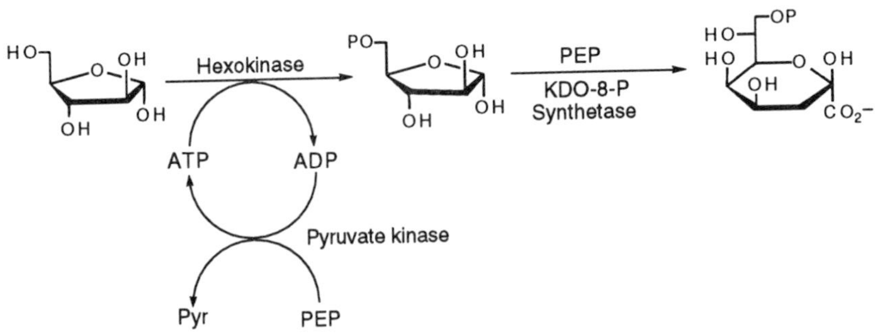

Figure 13.

phosphates including fluoro-derivatives of sugar phosphates (such as 2-deoxy-2-fluorogluco, 3-deoxy-3-fluoroallo, 3-deoxy-3-fluorogluco, 4-deoxy-4-fluorogluco, and 4-deoxy-4-fluorogalacto derivatives), and 5-thioglucose and aza sugar phosphates.[47a] The mechanism and specificity of hexokinases have been reviewed.[47b,c]

The enzyme ribokinase from *Lactobacillus plantarum*[48] (RK: EC 2.7.1.17) has been used to synthesize ribose-5-phosphate, a key intermediate in the synthesis of 5-phosphoribosyl-α-1-pyrophosphate and ribulose-1,5-bisphosphate.[49]

5.3   Other Organic Phosphates

Many other kinases are potentially useful for synthesis. Arginine kinase (from lobster tail muscle: EC 2.7.3.3) catalyzes the synthesis of arginine phosphate from the corresponding amino acid and ATP, and has been used in synthesis on a two mole scale.[50] Arginine phosphate is a high energy phosphate source among invertebrates.

Creatine phosphate, another naturally occurring phosphoryl donor *in vivo*, has been synthesized on a synthetically useful scale in a creatine kinase catalyzed reaction based on acetyl phosphate as the ultimate phosphate donor for ATP regeneration.[51] The instability of creatine phosphate, however, and acetyl phosphate and the equilibria of the reactions make this procedure less than ideal; a better procedure could involve PEP.

Ribose 5-phosphate is a key intermediate in the synthesis of 5'-phosphoribosyl-α-1-pyrophosphate (PRPP), which is a precursor to purine, pyrimidine and pyridine nucleotides in biosynthesis[49] (Figure 14).

Ribose-5-phosphate has also been used in the synthesis of ribulose-1,5-biphosphate (RuBP a key intermediate in the fixation of carbon dioxide in plant metabolism). Figure 15 summarizes two enzymatic schemes for the production of RuBP.[49,52] Although the route from AMP *via* ribose-5-phosphate is more direct and convenient, the route from glucose-6-phosphate is also an acceptable route, and which is more convenient depends on the local availability of starting materials for each route.

ATP
Pyr
PRPP
synthetase
AK
PK
AMP
PEP

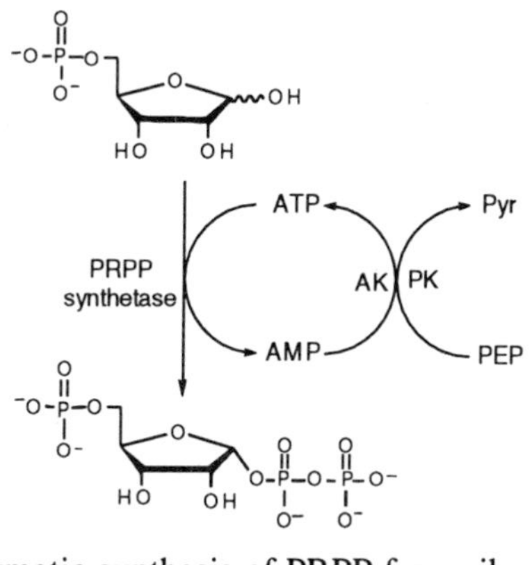

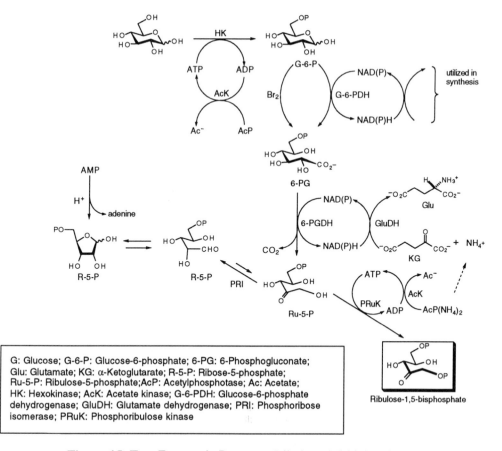

Figure 14. Enzymatic synthesis of PRPP from ribose-5-phosphate*.
*Ribose-5-phosphate can be obtained by any of three methods:
1) enzymatic phosphorylation of ribose by ribokinase
2) hydrolysis of purine nucleotides from RNA digest
3) acidic hydrolysis of AMP

HK
ATP
ADP
AcK
Ac⁻
AcP
G-6-P
Br2
G-6-PDH
NAD(P)
NAD(P)H
utilized in
synthesis
6-PG
AMP
H⁺
adenine
R-5-P
R-5-P
PRI
6-PGDH
GluDH
Glu
KG
CO2
Ru-5-P
ATP
Ac⁻
AcK
PRuK
ADP
AcP(NH4)2
Ribulose-1,5-bisphosphate
G: Glucose; G-6-P: Glucose-6-phosphate; 6-PG: 6-Phosphogluconate;
Glu: Glutamate; KG: α-Ketoglutarate; R-5-P: Ribose-5-phosphate;
Ru-5-P: Ribulose-5-phosphate;AcP: Acetylphosphotase; Ac: Acetate;
HK: Hexokinase; AcK: Acetate kinase; G-6-PDH: Glucose-6-phosphate
dehydrogenase; GluDH: Glutamate dehydrogenase; PRI: Phosphoribose
isomerase; PRuK: Phosphoribulose kinase

Figure 15. Two Enzymatic Routes to Ribulose-1,5-biphosphate.

Combined chemical and enzymatic techniques have been used to produce NAD and NADP.[53] Although it is not as efficient as fermentative procedures, enzymatic synthesis does provide a route to NAD analogs for research purposes. Ribose 5-phosphate is converted to nicotinamide mononucleotide (NMN) in two chemical steps. This intermediate is coupled with AMP by the ATP dependent NAD pyrophosphorylase (EC 2.7.7.1). The NAD can be converted into NADP by the action of NAD kinase[54] (Figure 16).

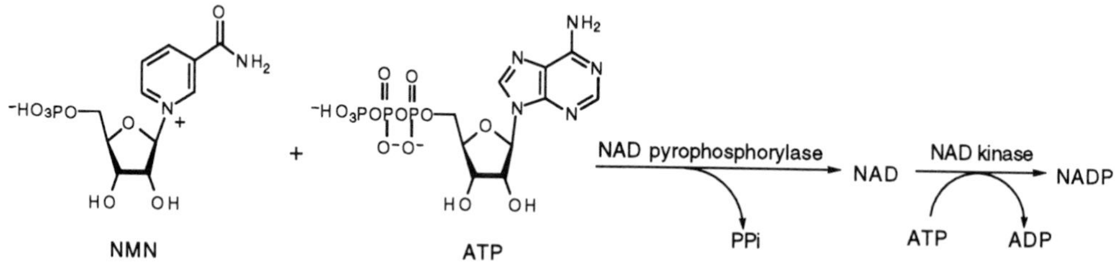

Figure 16. Enzymatic Synthesis of NAD^+ and $NADP^+$.

Aspartokinase from *E. coli* (EC 2.7.2.4) catalyzes the phosphorylation of the β-CO_2 group of aspartate to form β-aspartylphosphate. Studies of substrate specificity indicate that the α-amino group is essential for substrate recognition. The 1-carboxyl group is not required for substrate recognition: both the 1-amide and 1-esters are competent alternative substrates. In addition, β-derivatized structural analogs such as β-hydroxamate, the β-amide or β-esters are phosphorylated at the carboxyl group through a reversal of regioselectivity[55] (Figure 17).

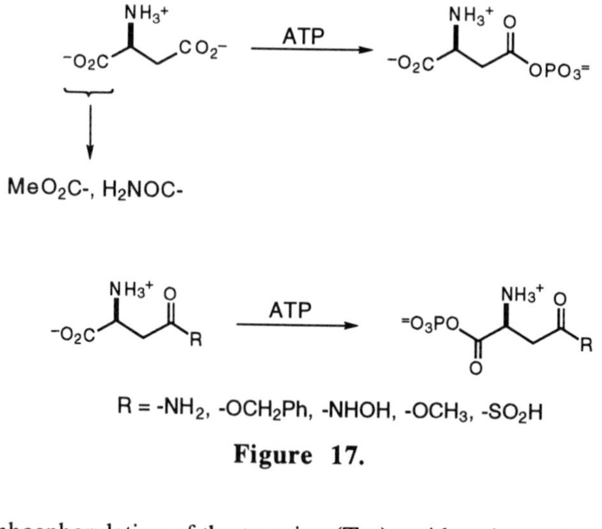

R = $-NH_2$, $-OCH_2Ph$, $-NHOH$, $-OCH_3$, $-SO_2H$

Figure 17.

Specific *O*-phosphorylation of the tyrosine (Tyr) residues in peptides was accomplished by two enzymatic reactions:[56] the first involves the enzymatic adenylation of the Tyr residue with

ATP catalyzed by *E. coli* glutamine synthetase adenyltransferase, and the second involves the enzymatic degradation of the adenylated intermediate catalyzed by micrococcal nuclease to produce the corresponding phosphotyrosine containing peptides. Peptides with sequence related to the phosphorylation site of glutamine synthetase are substrate.[56]

The glycolytic pathway from glucose to lactic acid has been used in regeneration of ATP for enzymatic phosphorylation.[57] The regeneration system is energetically very favorable and can be coupled with unfavorable phosphorylation reactions to drive the overall reaction. ATP dependent enzymes have also been used in synthesis of coenzyme A analogs.[58] The limitations of substrate specificity in these processes can be partially circumvented using appropriate substrate analogs. For example, enzymatic synthesis of an easily functionalized thioester analog of CoA, from a thioester derivative of pantetheine phosphate, provides a new route to different CoA analogs via aminolysis of the thioester (Figure 18).[58]

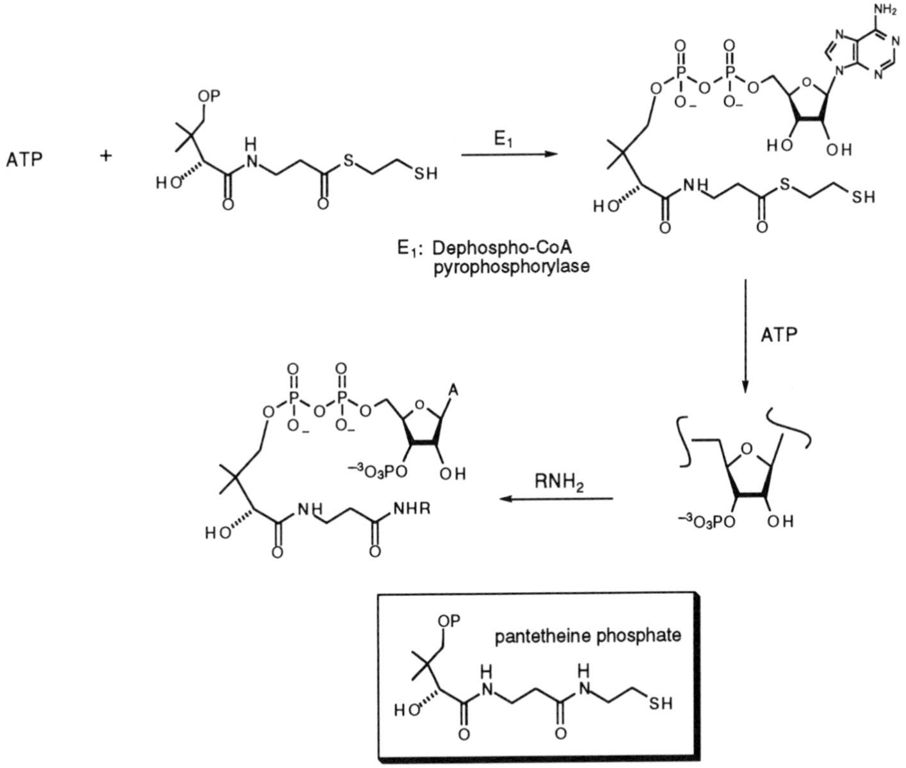

E₁: Dephospho-CoA pyrophosphorylase

pantetheine phosphate

Figure 18.

6. Preparation of Derivatives of ATP Chiral at the α-, β, or γ-Phosphorus

The α- and β-phosphorus groups of nucleoside triphosphates are prochiral and the γ-phosphorus group is pro-prochiral. To understand the stereochemical course of nucleoside triphosphate-dependent phosphoryltransfer reactions, the phosphorus group involved in the reaction must be chiral; to introduce the chirality required for mechanistic studies, one (or two) of the oxygen atom(s) is often replaced with S or isotopic oxygen (^{17}O or ^{18}O). The synthesis of nucleoside triphosphates, especially ATP, chiral at the phosphorus group has been well developed[59] and reviewed.[60] Of particular interest are the enzymatic syntheses of phosphorothioate analogs of ATP including ATPαS, ATPβS, and ATPγS (Figure 19). ATPαS with *S*-configuration at the α-position (Sp-ATPαS) can be prepared from AMPαS *via* adenylate kinase and pyruvate kinase reactions,[61] and Rp-ATPαS can be prepared from ADPαS in a reaction catalyzed by creatine kinase.[62] Pyruvate kinase also catalyzes the synthesis of Sp-ATPβS from ADPβS. The minor byproduct Rp-ATPβS can be decomposed selectively with hexokinase in the presence of glucose. Rp-ATPβS can be prepared from ADPβS in a reaction catalyzed by acetate kinase; the minor product Sp-ATPβS that is formed can be selectively decomposed with myosin. Both Rp- and Sp-AMP containing S and oxygen-18 can be prepared from diadenosine pyrophosphate using a process catalyzed by nucleotide pyrophosphatase.[63]

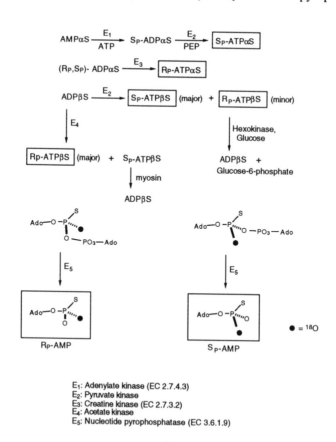

E$_1$: Adenylate kinase (EC 2.7.4.3)
E$_2$: Pyruvate kinase
E$_3$: Creatine kinase (EC 2.7.3.2)
E$_4$: Acetate kinase
E$_5$: Nucleotide pyrophosphatase (EC 3.6.1.9)

Figure 19. Synthesis of ATPαS and ATPβS.

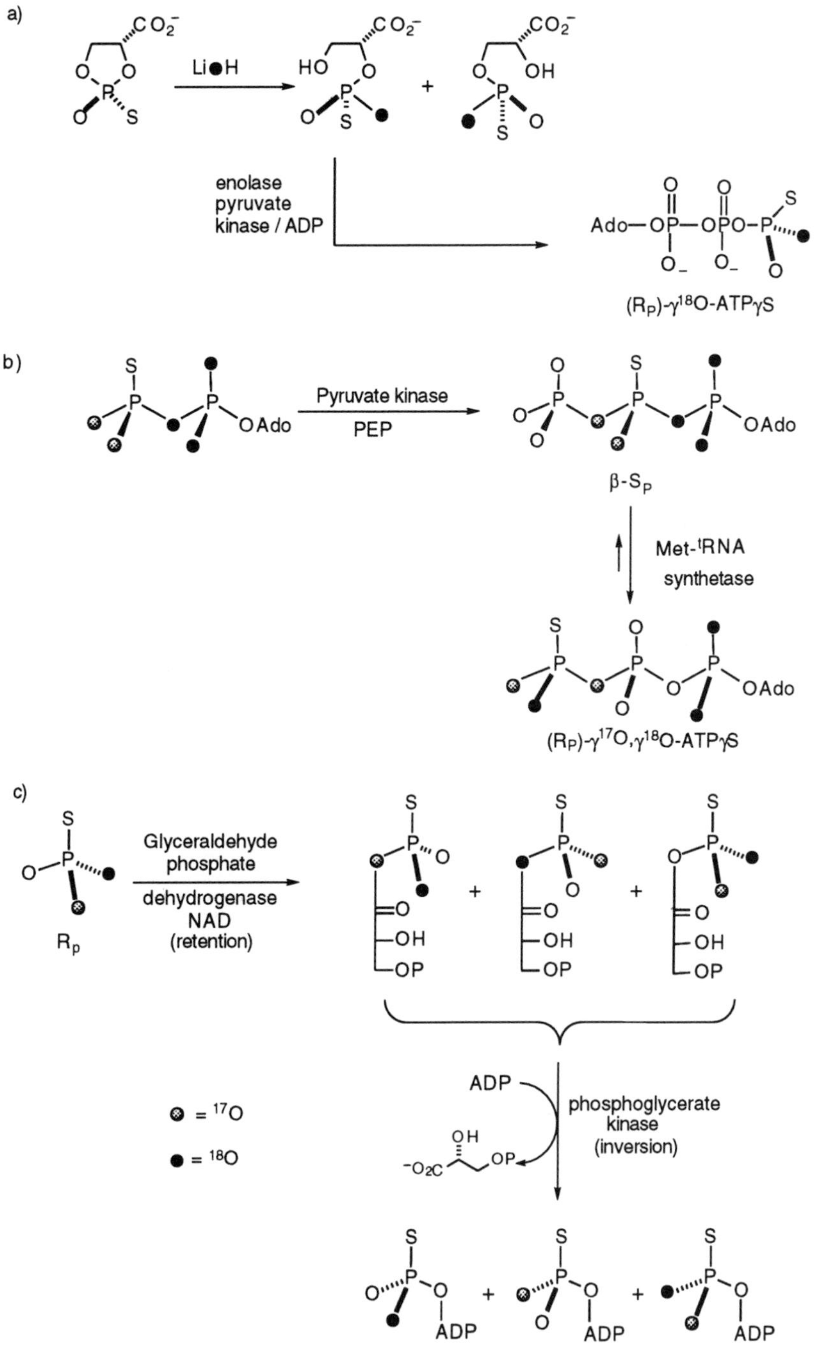

Webb, M. R. *Method Enzymol* **1982**, *87*, 301

Figure 20. Synthesis of ATPγS.

Several forms of ATP chiral at γ-phosphorus have been prepared (Figure 20). The first synthesis of [γ-^{18}O]ATPγS was reported in 1978[64] and the absolute stereochemistry was subsequently determined.[65] Pyruvate kinase coupled with Met-^{t}RNA synthetase was used for the preparation of (Rp)-γ^{17}O, γ^{18}O-ATPγS,[66] and glyceraldehyde 3-phosphate dehydrogenase coupled with phosphoglycerate kinase was used for the synthesis of ATPγS containing ^{17}O and/or ^{18}O.[67] A 50-mmol scale synthesis of ATPγS was accomplished enzymatically startingfrom dihydroxyacetone and inorganic thiophosphate[68] (Figure 21). [α-^{32}P]-ATP was

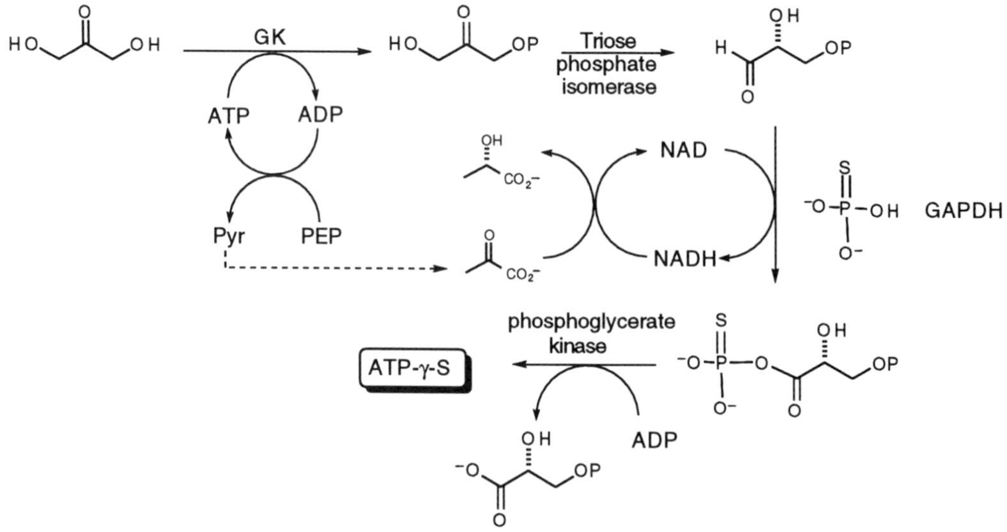

Figure 21. Preparation of ATP-γ-S (50 mmol).

also prepared from [γ-^{32}P]-ATP *via* adenosine kinase, adenylate kinase and creatine kinase-catalyzed reactions.[69] Polynucleotides containing ^{32}P was prepared from nucleoside triphosphates [α-^{32}P] via ATPase and polynucleotide phosphorylase reactions.[70]

7.   Phosphorothioate-Containing DNA and RNA (DNA-S and RNA-S)

Both polymers can be prepared from the monomers (Sp-NTPαS) enzymatically (Figure 22).[71] These polymers containing Rp-phosphorothioate are resistant to nuclease and are

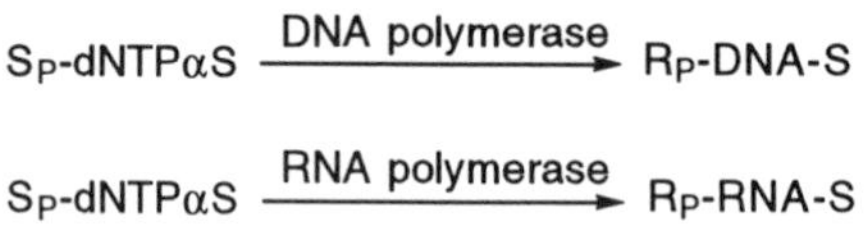

Sp-dNTPαS $\xrightarrow{\text{DNA polymerase}}$ Rp-DNA-S

Sp-dNTPαS $\xrightarrow{\text{RNA polymerase}}$ Rp-RNA-S

Figure 22.

recognized by kinases, ligases and restriction enzymes. Exonuclease III cannot hydrolyze the 3'-end of DNA with a phosphorothioate group. Nuclease S1 and nuclease P differentiate the two phosphorothioate internucleotidic linkages. DNase I, DNase II, staphylococcal nuclease, and spleen phosphodiesterase cannot accept either isomer of the phosphorothioate internucleotidic linkage. EcoRI digests the Rp-configuration but at a rate 1/15th that of the unmodified oligonucleotide. The Sp-isomer is not cleaved.

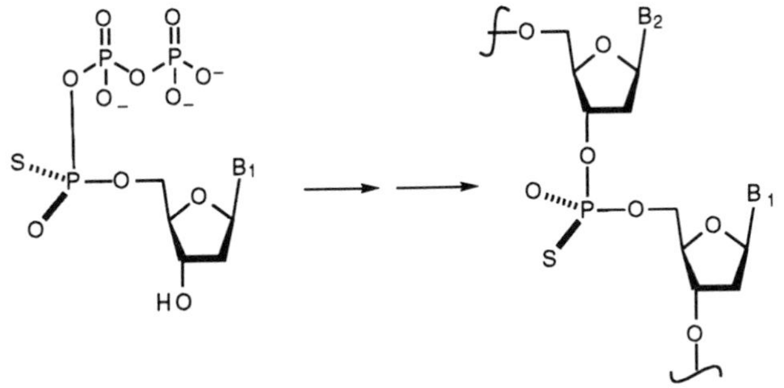

Sp-dNTPαS   Rp-DNA-S

Figure 23.

8. DNA and RNA Oligomers

Recombinant DNA technology depends on a number of enzymes that make it possible to introduce the desired genes into an organism. Synthesis of recombinant DNA involves chemical and/or enzymatic synthesis of primers, polymerase catalyzed reactions (PCR) to extend these primers, restriction enzymes and DNA ligases reactions.[72] These transformations cannot be accomplished by classical organic synthetic techniques. Although the enzymes used in recombinant DNA technology are very expensive, large-scale work is unnecessary because DNA can be amplified in cells. For the preparation of small oligonucleotides in large quantities for use in anti-sense and genetic engineering technology, enzymatic synthesis may ultimately prove useful, especially for the synthesis of small RNA oligomers. The formation of phosphodiester links often requires ATP (e.g. DNA and RNA ligases) and regeneration of ATP will probably be required. The enzyme T4 RNA ligase, for example, catalyzes the synthesis of single-stranded oligonucleotides of different lengths from a 3'-terminal hydroxyl acceptor and a 5'-terminal phosphate donor through the formation of a 3'→5' phosphodiester bond, with hydrolysis of ATP to AMP and pyrophosphate.[73] The 5'-phosphate donor can be mono- or polynucleotides, and the 3'-terminal hydroxyl acceptor can be a trimer or oligomer. The enzyme also accepts single-stranded DNA as a substrate. This enzyme has been used to couple short, chemically synthesized oligonucleotides to longer oligomers, to introduce radioactive probes, and to modify RNA[74] (Figure 24). Similarly, T4 phage DNA ligase is an ATP-requiring enzyme that catalyzes

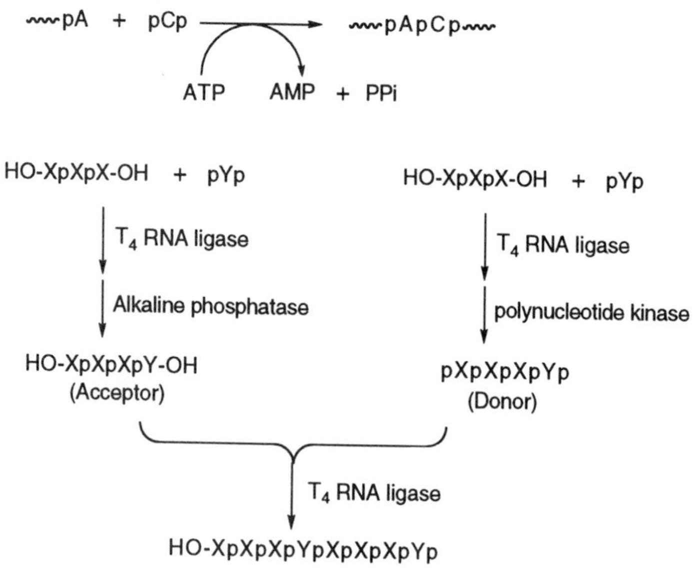

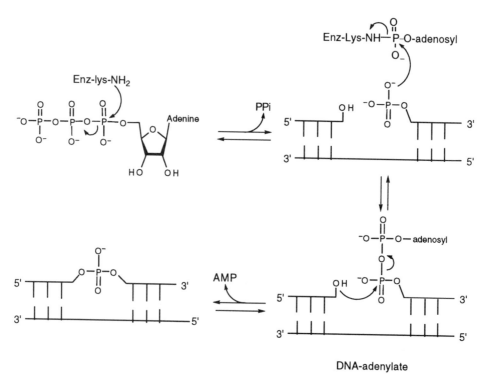

Figure 24. Synthesis of Oligomers with T4 RNA Ligase.

Figure 25. Mechanism of DNA Ligase.

the formation of phosphodiester bonds between nucleotide chains.[75] *In vivo* it acts to repair and replicate DNA strands. T_4 DNA ligase accepts a large range of deoxyribonucleotides, including single strand nucleotide chains (sticky ends) or fully base paired double strands (blunt ends). Figure 25[76] sketches the mechanism of DNA ligase.

T_4 DNA ligase may also be used in the synthesis of certain oligoribonucleotides. Since the chemical methods for the synthesis of RNA are not as well developed as those for DNA, RNA ligase could play an important role in RNA engineering. Two other examples of polynucleotide synthesis deserving mention are the polynucleotide phosphorylase-catalyzed synthesis of polynucleotides from nucleoside diphosphates[77a] and ribonuclease-catalyzed synthesis of trinucleotide codons.[77b] These two enzymes have a broad substrate specificity and are thus particularly useful for the synthesis of oligomers (Figure 26).

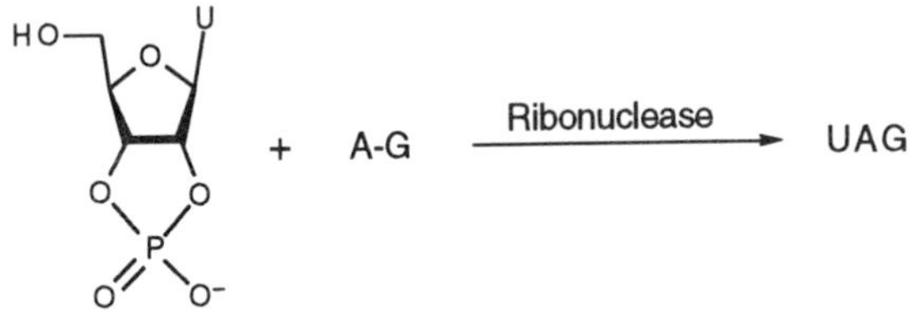

Figure 26. Ribonuclease-catalyzed Synthesis of Trinucleotides.

9. Incorporation of Modified or Unnatural Bases into DNA or RNA

Incorporation of modified or unnatural nucleosides into DNA or RNA provides new opportunities for the study of DNA and RNA recognition, for cross-linking, and for expanding the genetic code. A series of modified and fluorescent nucleosides and their triphosphates have been prepared and exploited as biological probes or as alternative substrates for enzymes that utilize nucleoside phosphates.[78] Etheno-bridged ATP, for example, was found to be a good substrate for several ATP-dependent kinases and RNA polymerase.[78c] N^7-Methyl-2'-deoxyguanosine has been incorporated into DNA *via* DNA polymerase and DNA ligase.[79] A dGTP analog containing an aziridine group was incorporated into DNA by *E. coli* DNA polymerase and avian myeloma virus reverse transcriptase in the place of dGTP and the aziridine group was rapidly opened by the N_4 of the complementary cytosine group to give cross-linked DNA.[80] The DNA polymerase from *E. coli* also accepted O^4-methylthymidine to replace thymidine.[81] The 2'-deoxy derivative of N^4-(6-aminohexyl)cytidine,[82] 8-azido-dATP[83] and several UTP and 2'-deoxy UTP derivatives with a substituent at 5-position[84] have been used as substrates for the polymerase reactions (Figure 27). Site-specific enzymatic incorporation of an unnatural base has also been reported. For example, a deoxy-5-methylisocytidine in the DNA template directs the T_7 RNA polymerase-catalyzed incorporation of N^6-(6-aminohexyl)isoguanosine into the transcribed RNA product to form a new complementary base pair.[85] isoG can also be specifically incorporated into RNA using isoG-containing DNA templates and T_7 RNA polymerase.[86]

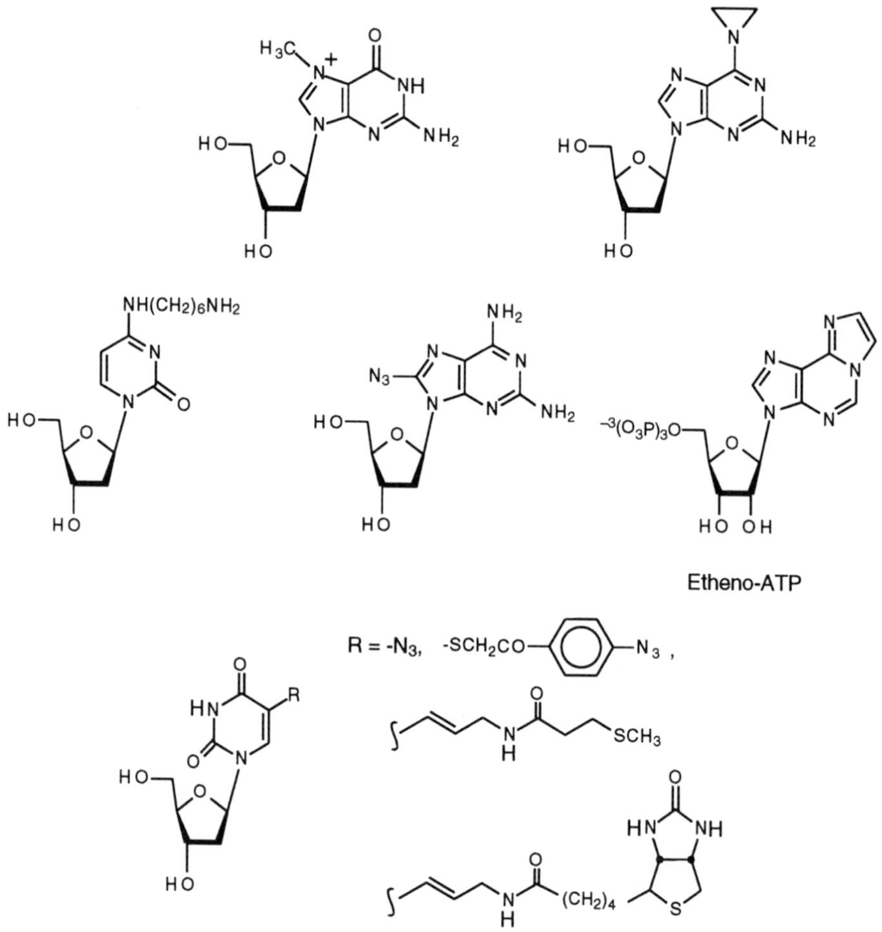

Figure 27. Modified Nucleosides Incorporated into DNA or RNA.

Other base pairs have also been enzymatically incorporated into DNA or RNA[87] (Figure 28). In the presence of non-standard tRNA containing the anticodon CUisodG and charged with iodotyrosine, there was a high degree (~90%) of read-through of the isoCAG codon, compared to ~35% for the iodotyrosine-tRNACUA- and the UAG nonsense codon system;[88] this observation suggests a new way of incorporating unnatural amino acids into proteins. Other reactions useful in DNA or RNA engineering include the replacement of the tRNA anticodon *via* chemical depurination at low pH followed by ribonuclease A digestion and ligation with another oligonucleotide using T_4 RNA ligase and T_4 polynucleotide kinase,[89] and synthesis of a knot from single strand[90] or double-strand DNA.[91] In the RNA engineering, the codon encoding the amino acid of interest was replaced with a nonsense codon, TAG, by oligonucleotide directed

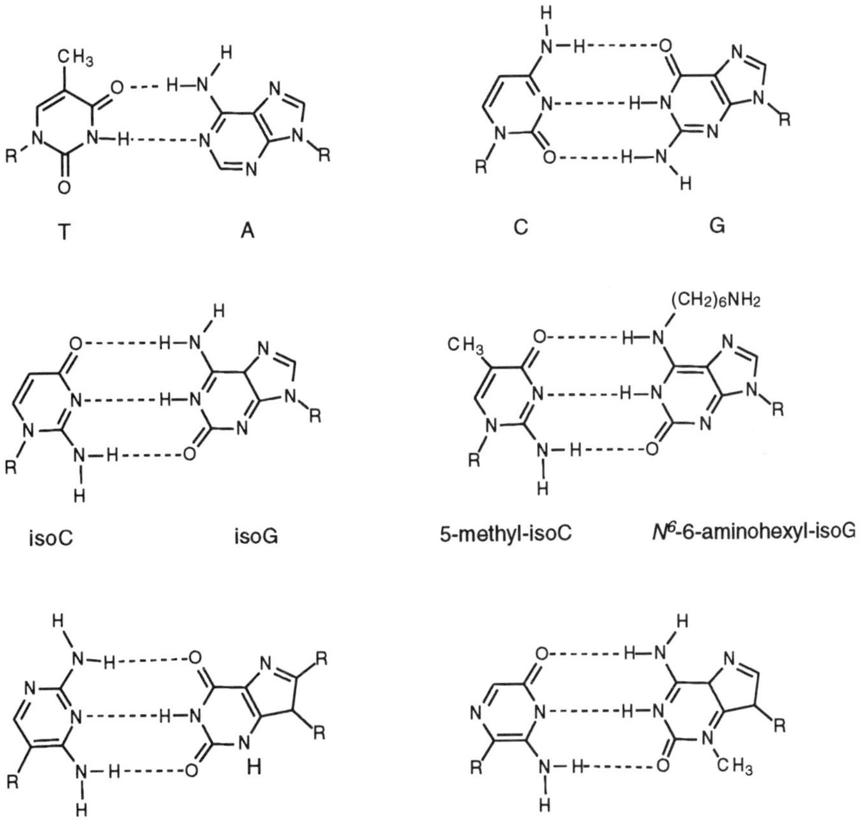

Figure 28. Existing and New Base Pairs Incorporated into DNA Enzymatically.

mutagenesis, then a suppressor tRNA directed against this codon was chemically aminoacylated *in vitro* with an unnatural amino acid to allow incorporation of the unnatural amino acid into the protein[92] (Figure 29).

10. Dehalogenation

Haloalkane dehalogenase from *Xanthobacter autotrophicus* contains Asp124, His289, and Asp260 as a catalytic triad. The crystal structure and the mechanism of its action has been elucidated.[93] Asp124 first attacks the carbon containing the halogen group to form a covalent intermediate. His289 is then involved as a general base catalyst to hydrolyze the intermediate and form the alcohol, a process similar to the deacylation mechanism of serine protease-catalyzed

a) Overall strategy

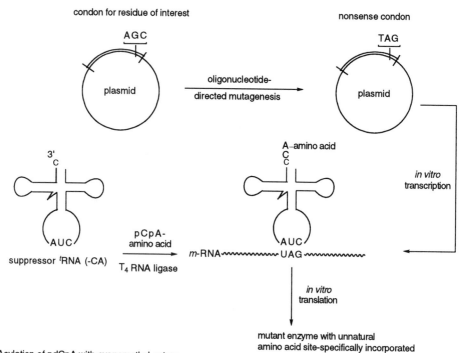

b) Acylation of pdCpA with cyanomethyl esters

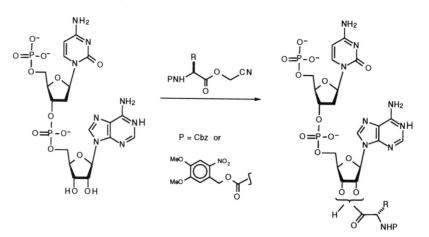

Figure 29. Strategy for biosynthetic incorporation of amino acids into proteins.

ester hydrolysis (Figure 30). Application of the dehalogenase from *Alcaligenes sp.* and *Pseudomonas sp.* to the synthesis of (*S*)- and (*R*)-glycidol from (*RS*)-3-chloro-1,2-propanediol has been reported[94] (Figure 31).

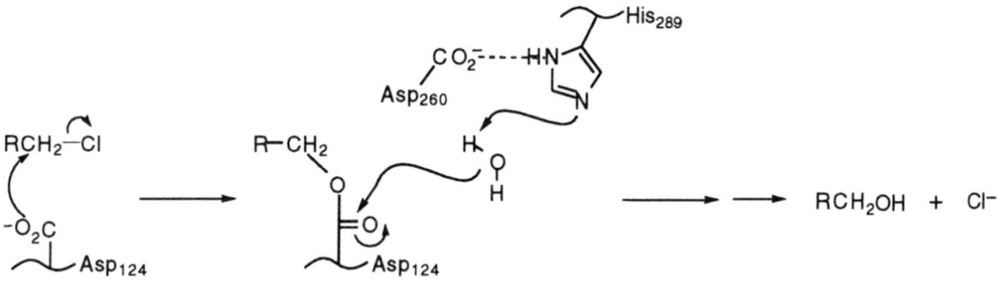

Figure 30. Mechanism for Haloalkane Dehalogenase.

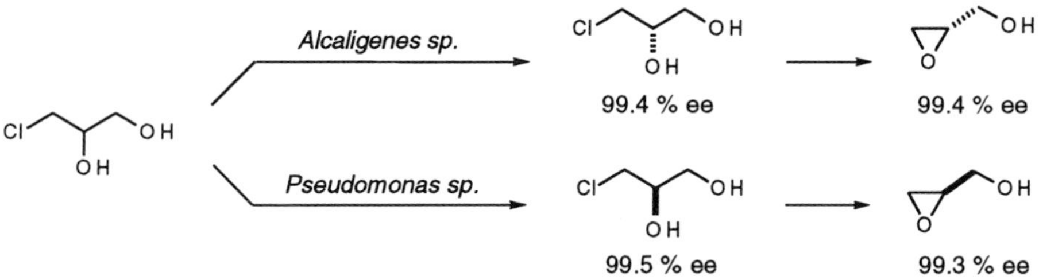

Figure 31.

11. Synthesis of Chiral Methyl Groups

Acetate and pyruvate with chiral methyl groups have been used to study the stereochemistry of biochemical methyl transfer reactions. The syntheses of chiral acetate were originally reported by Cornforth,[95,96] Arigoni[97] and their coworkers. New enzymatic or chemoenzymatic procedures more easily carried out in biochemical laboratories have recently been developed.[98-101] Figure 32 illustrates two representative procedures.[98,101]

12. S-Adenosylmethionine and Transmethylation

(-)-*S*-Adenosyl-L-methionine (AdoMet, SAM (Figure 33)) is a cofactor required in many enzyme reactions, particularly as an electrophilic methyl source in transmethylation reactions. Enzymatically active SAM has the (*S*)-configuration at the chiral amino acid center and at the sulfonium center. Racemization of the sulphonium center produces an inactive cofactor, making the stereo control in synthetic SAM reactions crucial to their utility. SAM is labile to both alkaline and acid hydrolysis, so a rapid turnover is a requirement for economical use of SAM in organic

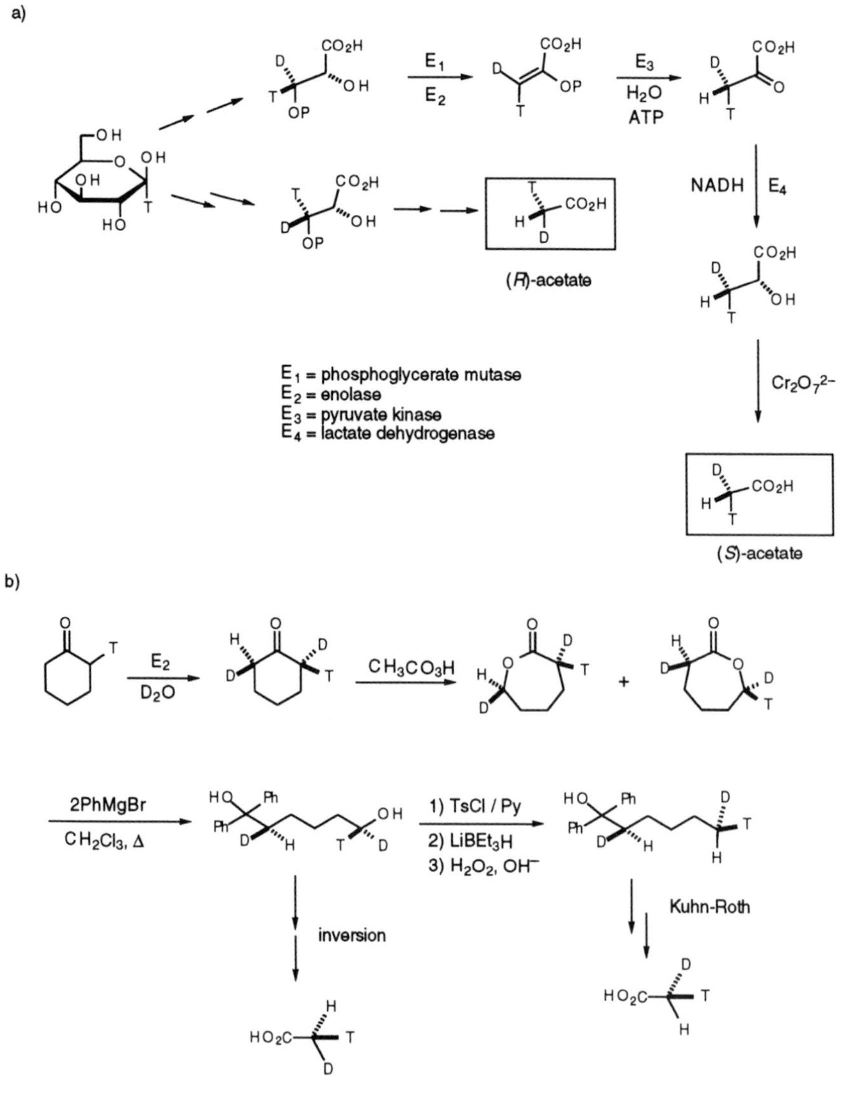

a)

E$_1$ = phosphoglycerate mutase
E$_2$ = enolase
E$_3$ = pyruvate kinase
E$_4$ = lactate dehydrogenase

(*R*)-acetate

(*S*)-acetate

b)

Figure 32. Synthesis of (*R*) and (*S*)-acetate.

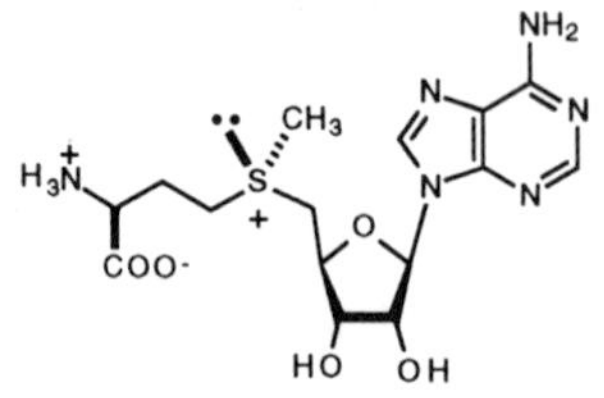

S-adenosyl-L-methionine

Figure 33.

synthesis. Chemical methods involving methylation of *S*-adenosyl homocysteine (SAH) with methyl iodide produce a slight majority of the inactive (+)-isomer.

SAM is synthesized in Nature from L-methione and ATP, in a reaction catalyzed by *S*-adenosyl methionine synthetase. Efforts to use this enzyme *in vitro* have had only marginal success; SAM is a strong inhibitor of the *S*-adenosylmethionine synthetase, and the reaction must, therefore, be carried out in dilute solutions.[102] The best methods for the production of SAM are fermentative techniques using microbial cells.[103]

The product of transmethylation, *S*-adenosyl homocysteine is not remethylated enzymatically, but is degradated by SAH hydrolase to L-homocysteine and adenosine. L-homocysteine may be catabolized to methionine and the SAM. Figure 34 shows two possible

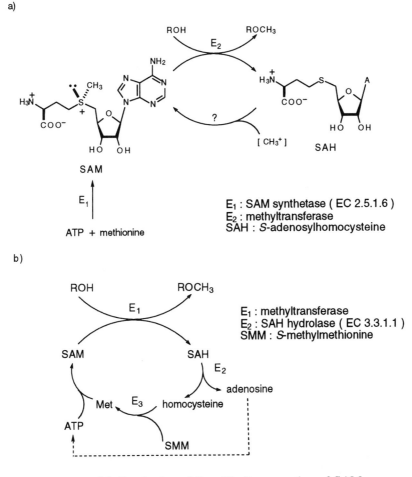

Figure 34. Synthesis and Possible Regeneration of SAM
(*S*-Adenosyl-L-methionine).

methods for the regeneration of *S*-adenosyl methionine. One would allow the catalytic use of 5-adenosyl homocysteine and rely on the use of 5-methylmethionine and a method of stereoselective methylation. Perhaps the methyl transferase enzyme homocysteine methyl transferase would catalyze the reaction.[28a] The other suggestion is considerably more complex and would require the use of several enzymes; these enzymes are not available for practical use at this time. Perhaps fermentation using genetically engineered microorganisms with appropriately enhanced enzyme activities would be feasible. In summary, at this point, no practical methods have been reported for regeneration of SAM.

5-Aminoleulinic acid dehydratase catalyzes the formation of porphobilinogen, in a pathway leading to vitamin B_{12} through a multienzyme-catalyzed reaction that included the sequential methylation catalyzed by *S*-adenosylmethionine dependent methyltransferase[104,105] (Figure 35).

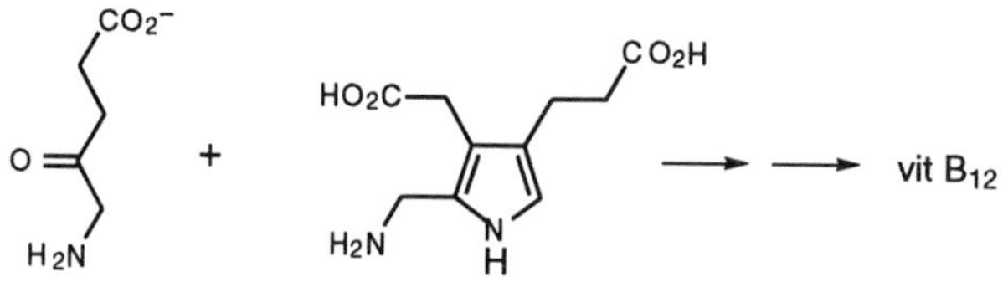

Figure 35.

13. Sulfate Activation and Transfer Reactions

Regioselective sulfation of polysaccharides is an important biochemical process, especially in the modification of heparin and chondroitin.[106] The reaction has not been utilized in preparative organic synthesis, because the enzymes involved in the activation of sulfate and its regeneration are not readily available.[28a] To be used in biochemical sulfation, sulfate must be activated to 3'-phosphoadenosine 5'-phosphosulfate (PAPS). Two enzymes generate PAPS from ATP and sulfate: ATP-sulphurylase (EC 2.7.7.4) catalyzes the reaction of ATP and sulfate to form adenosine 5'-phosphosulfate (APS) and APS-kinase (EC 2.7.1.25) catalyzes the phosphorylation of APS with ATP to form PAPS (Figure 36). Coupling of the ATP-sulphurylase reaction with pyrophosphatase drives the reaction toward APS formation. Both ATP-sulphurylase and APS-kinase[106,107] have been isolated from a variety of sources and studied. The sulphotransferases use PAPS in sulfation; 5'-phosphoadenosine 5'-phosphate (PAP) is generated as a byproduct. Because of the high cost of PAPS, regeneration of PAPS is necessary for multi-gram synthesis. A possible regeneration of PAPS from PAP is via AMP as illustrated in Figure 37. This regeneration system has not been demonstrated, however. For small-scale sulfation reactions, PAPS can be used stoichiometrically, and a number of chemical and enzymatic methods have been reported for the synthesis of PAPS.[106,108] A useful procedure is based on the chemical synthesis of adenosine 2',3'-cyclic phosphate 5'-phosphosulphate followed by T2-RNase catalyzed hydrolysis of the cyclic phosphate to form PAPS.[109] As more new sulfotransferases[110] are discovered, the enzymatic sulfation may become another interesting synthetic process.

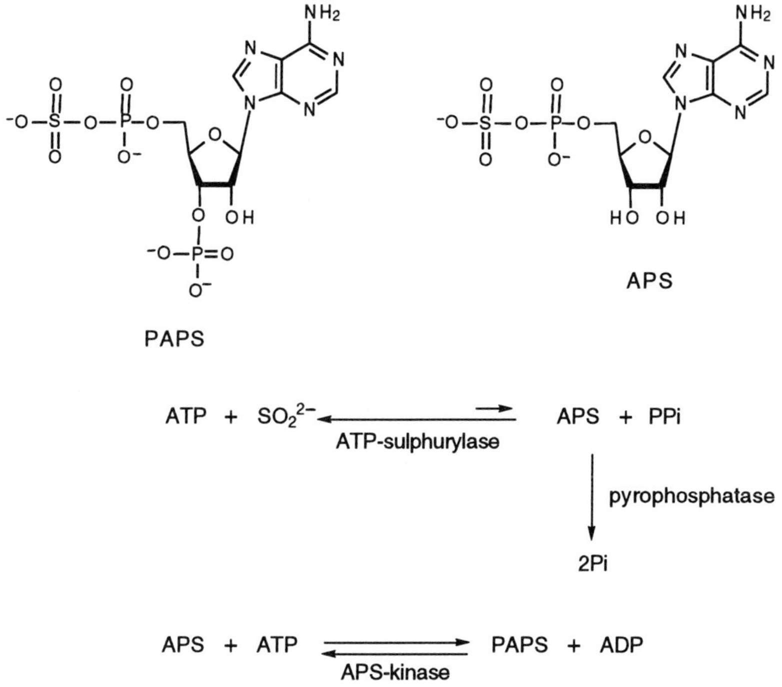

Figure 36.

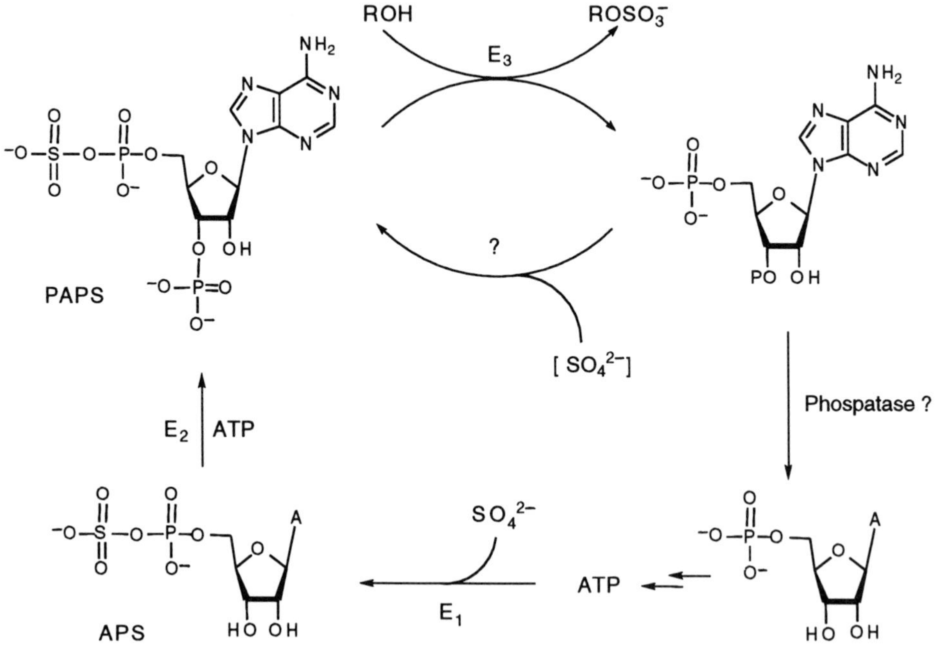

E$_1$: ATP-sulphurylase (EC 2. 7. 7. 4)
E$_2$: APS kinase (EC 2. 7. 1. 25)
E$_3$: Sulfotransferase

Figure 37. Synthesis and Possible Regeneration of PAPS
(3'-Phosphoadenosine-5'-phosphosulfate).

References

1. Chibata, I.; Tosa, T.; Takata, I. *Trends in Biotechnol.* **1983**, *1*, 9.
2. Findeis, M.A.; Whitesides, G.M. *J. Org. Chem.* **1987**, *52*, 2838.
3. Hanson, K.R.; Havir, E.A. *Enzymes*, 3rd ed., Boyer, P.D. Ed.; Academic Press: New York, 1972, Vol. 7, pp.75-166.
4. Falzone, C.J.; Karsten, W.E.; Conley, J.D.; Viola, R.E. *Biochemistry* **1988**, *27*, 9089.
5. Akhtar, M.; Cohen, M.A.; Gani, D. *Tetrahedron Lett.* **1987**, *28*, 2413.
6. Braunstein, A.E. *The Enzymes IX* (P.D. Bayer, ed.) Part B, p. 379, Academic Press: New York 1973.
7. Malcolm, B.A.; Kirsch, J.F. *Biochem. Biophys. Res. Commun.* **1985**, *132*, 915.
8. Fotheringham, I.G.; Dacey, S.A.; Taylor, P.P.; Smith, T.J.; Hunter, M.G.; Finlay, M.E.; Primrose, S.B; Parker, D.M.; Edwards, R.M. *Biochem. J.* **1986**, *234*, 593.
9. Rozzell, D. *Method Enzymol.* **1987**, *136*, 479.
10. Baldwin, J.E.; Dyer, R.L.; Ng, S.C.; Pratt, A.J.; Russell, M.A. *Tetrahedron Lett.* **1987**, *28*, 3745.
11. White, P.W.; Kirsch, J.F. *J. Am. Chem. Soc.* **1992**, *114*, 3567.
12. Ziehr, H.; Kula, M.-R. *J. Biotech.* **1985**, *3*, 19.
13. Sung, M.-H.; Tanizawa, K.; Tanaka, H.; Kuramitsu, S.; Kagamiyama, H.; Hirotsu, K.; Okamoto, A.; Higuchi, T.; Soda, K. *J. Biol. Chem.* **1991**, *266*, 2567.
14. Tanizawa, K.; Masu, Y.; Asano, S.; Tanaka, H.; Soda, K. *J. Biol. Chem.* **1989**, *264*, 2445.
15. Echalier, F.; Constant, O.; Bolte, J. *J. Org. Chem.* **1993**, *58*, 2747.
16. Yagi, T.; Misono, H.; Tanizawa, K.; Yoshimura, T.; Soda, K. *J. Biochem.* **1991**, *109*, 61.
17. Bartsch, K.; Dichmann, r.; Schmitt, P.; Uhlmann, E.; Schulz, A. *Appl. Environ. Microbiol.* **1990**, *56*, 7.
18. Nakamichi, K.; Nabe, K.; Tosa, T. *J. Biotechnol.* **1986**, *4*, 293.
19. Tanizawa, K.; Masu, Y.; Asano, S.; Tanaka, H.; Soda, K. *J. Biol. Chem.* **1989**, *264*, 2445.
20. Kuramitsu, S.; Ogawa, T.; Ogawa, H.; Kagamiyama, H. *J. Biochem.* **1985**, *97*, 993.
21. Yoshimura, T.; Nishimura, K.; Ito, J.; Esaki, N.; Kagamiyama, H.; Manning, J.M.; Soda, K. *J. Am. Chem. Soc.* **1993**, *115*, 3897.
22. Miyamoto, K.; Ohta, H. *J. Am. Chem. Soc.* **1990**, *112*, 4077.
23. Miyamoto, K. *Biocatalysis* **1991**, *5*, 49.

24.   Zehender, H.; Trescher, D.; Ullrich, J. *Eur. J. Biochem.* **1987**, *167*, 149; Alvarez, F.J.; Ermer, J.; Hubner, G.; Schellenberger, A.; Schowen, R.L. *J. Am. Chem. Soc.* **1991**, *113*, 8402.

25.   Wong, C.-H.; McCurry, S.D.; Whitesides, G.M. *J. Am. Chem. Soc.* **1980**, *102*, 7938.

26.   Wong, C.-H.; Daniels, L.; Orme-Johnson, W.H.; Whitesides, G.M. *J. Am. Chem. Soc.* **1981**, *103*, 6227.

27.   Lorimer, G.H.; Chen, Y.-R.; Hartman, F.C. *Biochemistry* **1993**, *32*, 9018.

28.   a) Chenault, H.K.; Simon, E.S.; Whitesides, G.M. *Biotech. Genetic Eng. Rev.* **1988**, *6*, 221. b) Chenault, H.K.; Whitesides, G.M. *Applied Biochem. Biotech.* **1987**, *14*, 147. c) Crans, D.C.; Kazlauskas, R.J.; Hirschein, R.L.; Abril, O.; Whitesides, G.M. *Methods Enzymol.* **1987**, *136*, 263. d) Whitesides, G.M.; Wong, C.-H. *Angew Chem. Int. Ed. Engl.* **1985**, *24*, 617.

29.   For applications of biological systems, NADP: Murata, K.; Tani, K.; Kato, J.; Chibata, I. *Enzyme Microbiol. Technol.* **1981**, *3*, 233. Glutathione: Murata, K.; Tani, K.; Kato, J.; Chibata, I. *Biochimie* **1980**, *62*, 347. CDP-choline: Kimura, A.; Tatsutomi, Y.; Matsuno, R.; Tanaka, A.; Fukuda, H. *J. Appl. Microbiol. Biotech.* **1980**, *11*, 78 and Ado, et al. *J. of Solid-Phase Biochem.* **1979**, *4*, 42.

30.   Hirschbein, B.L.; Mazenod, F.P.; Whitesides, G.M. *J. Org. Chem.* **1982**, *47*, 3765.

31.   Pollack, A.; Baughn, R.L.; Whitesides, G.M. *J. Am. Chem. Soc.* **1977**, *99*, 2366, and Rios-Mercadillos, V.M.; Whitesides, G.M. *J. Am. Chem. Soc.* **1979**, *101*, 5828.

32.   Whitesides, G.M.; Lamotte, A. L.; Adalsteinsson, O.; Baddur, R.F.; Chmvrny, A.C.; Cotton, C.K.; Pollack, A. *J. Molec. Catal.* **1979**, *6*, 177.

33.   Kondo, H.; Tomioka, H.; Nakajima, K.; Imahori, K. *J. Appl. Biochem.* **1984**, *6*, 29; Nakajima, H.; Nagata, K.; Kondo, H.; Imahori, K. ibid, 1984 6, 19; Kim, M.-J.; Whitesides, G.M. *Appl. Biochem. Biotech.* **1987**, *16*, 95.

34.   Whitesides, G.M.; Kazlauskas, R.J. *J. Org. Chem.* **1985**, *50*, 1069.

35.   Marshall, D.L. *Biotech. Bioeng.* **1973**, *15*, 447.; Marshall, D.L. in *Enzyme Engineering*, Pye, E.K.; Wignard, L.B. Eds., Vol. 2, 223-28, Plenum Press, NY 1973.

36.   Sakata, I.; Kitano, H.; Ise, N. *J. Appl. Biochem.* **1981**, *3*, 518.

37.   a) Simon, E.S.; Bednarski, M.D.; Whitesides, G.M. *Tetrahedron Lett.* **1988**, *29*, 1123. b) Baughn, R.L.; Adalsteinsson, O.; Whitesides, G.M. *J. Am. Chem. Soc.* **1978**, *100*, 304. c) Kim, M.-J.; Whitesides, G.M. *Appl. Biochem. Biotech.* **1987**, *16*, 95; Haynie, S.L.; Whitesides, G.M. *Appl. Biochem. Biotech.* **1990**, *23*, 205.

38.   Leibach, T.K. et al. *Z. Physiol. Chem.* **1971**, *352*, 228.

39.   See Chapter 5.

40.   Wong, C.-H.; Haynie, S.L.; Whitesides, G.M. *J. Org. Chem.* **1982**, *47*, 5416 and Chapter 5.

41. Clark, V.M.; Kirby, A.J. *Biochem. Prep.* **1966**, *11*, 101.

42. Crans, D.C.; Whitesides, G.M. *J. Org. Chem.* **1983**, *48*, 3130.

43. Lewis, J.M.; Haynie, S.L.; Whitesides, G.M. *J. Org. Chem.* **1979**, *44*, 864.

44. Crans, D.C.; Whitesides, G.M. *J. Am. Chem. Soc.* **1985**, *107*, 7019. For the substrate specificity of GK: Crans, D.C.; Whitesides, G.M. *J. Am. Chem. Soc.* **1985**, *107*, 7008.

45. Wong, C.-H.; Whitesides, G.M. *J. Org. Chem.* **1983**, *48*, 3199. Wong, C.-H.; Mazenod, F.P.; Whitesides, G.M. *J. Org. Chem.* **1983**, *48*, 3493.

46. Bednarski, M.D.; Crans, D.C.; DiCosimo, R.; Simon, E.S.; Stein, P.D.; Whitesides, G.M.; Schneider, M.J. *Tetrahedron Lett.* **1988**, *29*, 427.

47. a) Dreuckhammer, D.G.; Wong, C.-H. *J. Org. Chem.* **1985**, *50*, 5912. b) For a review: Villafranca, J. J.; Raushel, F. M. *Adv. Catalysis* **1979**, *28*, 323. c) Chen, M.; Whistler, R. L. *Arch. Biochem. Biophys.* **1975**, *169*, 392.

48. Schimmel, S.D.; Hoffee, P.; Horecker, B.L. *Arch. Biochem. Biophys.* **1974**, *164*, 560.

49. Gross, A.; Abril, O.; Lewis, J.M.; Geresh, S.; Whitesides, G.M. *J. Am. Chem. Soc.* **1983**, *105*, 7428.

50. Bolte, J.; Whitesides, G.M. *Bioorg. Chem.* **1984**, *12*, 170.

51. Sih, Y.-S.; Whitesides, G.M. *J. Org. Chem.* **1977**, *42*, 4165.

52. Wong, C.-H.; Pollack, A.; McCurry, S.D.; Sue, J.M.; Knowles, J.R.; Whitesides, G.M. *Methods Enzymol.* **1982**, *89*, 108.

53. Walt, D.R., Findeis, M.A.; Rios-Mercadillo, V.M.; Auge, J.; Whitesides, G.M. *J. Am. Chem. Soc.* **1984**, *106*, 234.

54. A review of MAD kinase. McGuinness, E.T.; Butler, R.J. *Int. J. Biochem.* **1985**, *17*, 1.

55. Angeles, T.S.; Hunsley, J.R.; Viola, R.E. *Biochemistry* **1992**, *31*, 799.

56. Gibson, B.W.; Hines, W.; Yu, Z.; Kenyon, G.L.; McNemer, L.; Villafranca, J. J. *J. Am. Chem. Soc.* **1990**, *112*, 8523.

57. Wei, L.L.; Goux, W.J. *Bioorg. Chem.* **1992**, *20*, 62.

58. Martin, D.P.; Druckhammer, D.G. *J. Am. Chem. Soc.* **1992**, *114*, 72878.

59. Frey, P.A.; Richard, J.P.; Ho, H.-T.; Brody, R.S.; Sammon, R.D.; Sheu, K.-F. *Method Enzymol.* **1982**, *87*, 213; Buchwald, S.L.; Hansen, D.E.; Hassett, A.; Knowles, J.R. *Method Enzymol.* **1982**, *87*, 279; Tsai, M.-D. *Method Enzymol.* **1982**, *87*, 235; Eckstein, F.; Romaniuk, P.J.; Connolly, B.A. *Method Enzymol.* **1982**, *87*, 197.

60. Eckstein, F. *Angew Chem. Int. Ed. Engl.* **1983**, *22*, 423; Eckstein, F. *Ann. Rev. Biochem.* **1985**, *54*, 367; Frey, P.A.; Sammons, R.D. *Science* **1985**, *228*, 541.

61. Shen, K.F.R.; Frey, P.A. *J. Biol. Chem.* **1977**, *252*, 4445; Jaffe, E.K.; Cohn, M. *Biochemistry* **1978**, *17*, 652.

62. Yee, D.; Armstrong, V.W.; Eckstein, F. *Biochemistry* **1979**, *18*, 4116.

63. Richard, J.P.; Frey, P.A. *J. Am. Chem. Soc.* **1982**, *104*, 3476.

64. Orr, G.A.; Simon, J.; Jones, S.R.; Chiu, G.J.; Knowles, J.R. *Proc. Natl. Acad. Sci.* **1978**, *75*, 2230.

65. Richard, J.P.; Ho, H.T.; Frey, P.A. *J. Am. Chem. Soc.* **1978**, *100*, 7756.

66. Bethell, R.C.; Lowe, G. *J. Chem. Soc., Chem. Commun.* **1986**, 1341.

67. Webb, M.R. *Method Enzymol.* **1982**, *87*, 301.

68. Abril, O.; Crans, D.C.; Whitesides, G.M. *J. Org. Chem.* **1984**, *49*, 1360.

69. Martin, B.R.; Voorheis, H.P. *Biochem. J.* **1977**, *161*, 555.

70. Kang, C.; Cantor, C.R. *Anal. Biochem.* **1985**, *144*, 291.

71. Eckstein, F.; Gish, G. *TIBS* **1989**, 97 and references cited therein.

72. Wu, R.; Grossman, L.; Moldave, K. *Method Enzymol.* **1979**, 65; **1980**, 68; **1983**, 100; **1983**, 101; Gassen, H.G.; Lang, A. (eds.): *Chemical and Enzymatic Synthesis of Gene Fragments*, Verlag Chemie: Weinheim, 1982.

73. Cranston, J.; Silber, R.; Malathi, V.G.; Hurwitz, J. *J. Biol. Chem.* **1974**, *249*, 7447; England, T.E.; Uhlenbeck, O.C. *Biochemistry* **1978**, *17*, 2069.

74. Middleton, T.; Herlihy, W.C.; Schimmel, P.R.; Munro, H.N. *Anal. Biochem.* **1985**, *144*, 110; Neilson, T.; Gregoire, R.J.; Fraser, A.R.; Kofoid, E.C.; Ganoza, M.C. *Eur. J. Biochem.* **1979**, *99*, 429; De Haseth, P.L.; Uhlenbeck, O.C. *Biochemistry* **1980**, *19*, 6138; Barrio, J. R.; Barrio, M. D.; Leonard, N. J.; England, T. E.; Uhlenbeck, O. C. *Biochemistry* **1978**, *17*, 2077.

75. Narang, S. A. *Tetrahedron* **1983**, *39*, 3.

76. Lehman, I. R. *Science* **1974**, *186*, 790; Willis, A. E.; Lindahl, T. *Nature* **1987**, *325*, 355; Chan, J. V. H.; Becker, F. F.; German, J.; Ray, J. H. *Nature* **1987**, *325*, 357.

77. a) Hoffman, C.-H.; Harris, E.; Chodroff, S.; Micholson, S.; Rothrock, J. W.; Peterson, E.; Reuter, W. *Biochem. Biophys. Res. Commun.* **1970**, *41*, 710; Ochoa, S.; Mii, S. *J. Biol. Chem.* **1961**, *236*, 3303. b) Gassen, H. G.; Nolte, R. *Biochem. Biophys. Res. Commun.* **1971**, *44*, 1410.

78. a) Leonard, N. J. *CRC Critical Review in Biochemistry* **1984**, *15*, 125. b) Leonard, N. J. *Acc. Chem. Res.* **1982**, 15, 128. c) Secrist, J. A.; Barrio, J. R.; Leonard, N. J. *Science* **1972**, *175*, 646.

79. Czaz-Nikpay, K.; Verdine, G.L. *J. Am. Chem. Soc.* **1992**, *114*, 6562.

80. Cowart, M.; Benkovic, S.J. *Biochemistry* **1991**, *30*, 788.

81. Preston, B.D.; Singer, B.; Loeb, L.A. *Proc. Natl. Acad. Sci.* **1986**, *83*, 8501.

82. Gillam, I.C.; Tener, G.M. *Anal. Biochem.* **1986**, *157*, 199.

83. Meffert, R.; Rathgeber, G.; Schafer, H.-J.; Dose, K. *Nucleic Acids Res.* **1990**, *18*, 6633.

84. Evans, R.K.; Haley, B.E. *Biochemistry* **1987**, *26*, 269; Hanna, M.M.; Dissinger, S.; Williams, B.D.; Colston, J.E. *Biochemistry* **1989**, *28*, 5814; Iverson, B.L.; Dervan, P.B. *J. Am. Chem. Soc.* **1987**, *109*, 1241; Langer, P.R.; Waldrop, A.A.; Ward, D.C. *Proc. Natl. Acad. Sci.* **1981**, *78*, 6633.

85. Tor, Y.; Dervan, P.B. *J. Am. Chem. Soc.* **1993**, *115*, 4461.

86. Switzer, C.; Moroney, S.e.; Benner, S.A. *J. Am. Chem. Soc.* **1989**, *111*, 8322.

87. Piccirilli, J.A.; Krauch, T.; Moroney, S.E.; Benner, S.A. *Nature* **1990**, *343*, 33.

88. Bain, J.D.; Switzer, C.; Chamberlin, A.R.; Benner, S.A. *Nature* **1992**, *356*, 537.

89. Bruce, A.G.; Uhlenbeck, O.C. *Biochemistry* **1982**, *21*, 855.

90. Mueller, J.E.; Du, S.M.; Seeman, N.C. *J. Am. Chem. Soc.* **1991**, *113*, 6306.

91. Wasserman, S.A.; Cozzarelli, N.R. *Science* **1986**, *232*, 951.

92. Robertson, S.A.; Ellman, J.A.; Schultz, P.G. *J. Am. Chem. Soc.* **1991**, *113*, 2722.

93. Verschueren, K.H.G.; Seljee, F.; Rozeboom, H.J.; Kalk, K.H.; Digkstra, B.W. *Nature* **1993**, *363*, 693.

94. Suzuki, T.; Kasai, N. *Bioorg. Med. Chem. Lett.* **1991**, *1*, 343.

95. Cornforth, J.W.; Redmond, J.W.; Eggerer, H.; Buckel, W.; Gutschow, C. *Nature* **1969**, *221*, 1212.

96. Cornforth, J.W.; Redmond, J.W.; Eggerer, H.; Buckel, W.; Gutshow, C. *Eur. J. Biochem.* **1970**, *14*, 1.

97. Luthy, J.; Retey, J.; Arigoni, D. *Nature* **1969**, *221*, 1213.

98. Floss, H.G. *Method Enzymol.* **1982**, *87*, 126.

99. Rose, I.A. *J. Biol. Chem.* **1970**, *245*, 6052.

100. Woodard, R.W.; Mascaro, L., Jr.; Horhammer, R.; Eisenstein, S.; Floss, H.G. *J. Am. Chem. Soc.* **1980**, *102*, 6314.

101. Rozzell, J.D., Jr.; Benner, S.A. *J. Org. Chem.* **1983**, *48*, 1190.

102. Matos, J.R.; Raushel, F.M.; Wong, C.-H. *Biotech. Appl. Biochem.* **1987**, *9*, 39; Gross, A.; Geresh, S.; Whitesides, G.M. *Appl. Biochem. Biotech.* **1983**, *8*, 415.

103. Shiozaki, S.; Shimizu, S.; Yamada, H. J. Biotech. 1986, 4, 345; Shiozaki, S.; Shimizu, S.; Yamada, H. *Trends Biotech.* **1984**, *2*, 137.

104. Warren, M.J.; Scott, A.I. *Trends Biochem. Sci.* **1990**, *15*, 486.

105. Warren, M.J.; Roessner, C.A.; Ozaki, S.-I.; Stolowich, N.J.; Santander, P.J.; Scott, A.I. *Biochemistry* **1992**, *31*, 603.

106. Schiff, J.A.; Saidha, T. *Method Enzymol.* **1987**, *143*, 329; Chenault, H.K.; Simon, E.S.; Whitesides, G.M. *Biotech. Genetic Engineering Reviews* **1988**, *6*, 221.

107. Renosto, F.; Seubert, P.A.; Segel, I.H. *J. Biol. Chem.* **1984**, *259*, 2113 and references cited therein.

108. De Meio, R.H. *Sulfate Activation and Transfer in Metabolic Pathways* (D.M. Greenberg, Ed.) Academic Press, New York, 1975, Vol. 7, pp. 287; Sekura, R.D. *Method Enzymol.* **1981**, *77*, 413.

109. Horowitz, J.P.; Neenan, J.P.; Misra, R.S.; Rozhin, J.; Huo, A.; Phillips, K.D. *Biochim. Biophys. Acta* **1977**, *480*, 376.

110. Hashimoto, Y.; Orellana, A.; Gil, G.; Hirschberg, C.B. *J. Biol. Chem.* **1992**, *267*, 15744.

Subject Index